Methods in Enzymology

Volume 423

TWO-COMPONENT SIGNALING SYSTEMS,
PART B

METHODS IN ENZYMOLOGY

EDITORS-IN-CHIEF

John N. Abelson   Melvin I. Simon

DIVISION OF BIOLOGY
CALIFORNIA INSTITUTE OF TECHNOLOGY
PASADENA, CALIFORNIA

FOUNDING EDITORS

Sidney P. Colowick and Nathan O. Kaplan

Methods in Enzymology

Volume 423

Two-Component Signaling Systems, Part B

EDITED BY

Melvin I. Simon

CALIFORNIA INSTITUTE OF TECHNOLOGY
DIVISION OF BIOLOGY
PASADENA, CALIFORNIA

Brian R. Crane

DEPARTMENT OF CHEMISTRY AND CHEMICAL BIOLOGY
CORNELL UNIVERSITY
ITHACA, NEW YORK

Alexandrine Crane

DEPARTMENT OF CHEMISTRY AND CHEMICAL BIOLOGY
CORNELL UNIVERSITY
ITHACA, NEW YORK

AMSTERDAM • BOSTON • HEIDELBERG • LONDON
NEW YORK • OXFORD • PARIS • SAN DIEGO
SAN FRANCISCO • SINGAPORE • SYDNEY • TOKYO
Academic Press is an imprint of Elsevier

ELSEVIER

Table of Contents

Section I. Structural Approaches

v

Section II. Reconstitution of Heterogeneous Systems

Section III. Intracellular Methods and Assays

Section IV. Genome-Wide Analyses of Two-Component Systems

Contributors to Volume 423

Article numbers are in parentheses following the names of contributors.
Affiliations listed are current.

PETER AMES (21), *Department of Biology, University of Utah, Salt Lake City, Utah*

JUDITH P. ARMITAGE (18), *Microbiology Unit, Department of Chemistry, University of Oxford, Oxford, United Kingdom*

ABDALIN E. ASINAS (12), *Department of Chemistry, University of Massachusetts, Amherst, Massachusetts*

RANDAL B. BASS (1, 2), *Analytical Sciences, Amgen, Inc., Seattle, Washington*

DAGMAR BEIER (25), *Theodor Boveri Institut fur Biowissenschaflen, Lehrstuhl fur Mikrobiologie, Universität Würzburg, Wurzburg, Germany*

HOWARD C. BERG (17), *Department of Molecular and Cellular Biology, Harvard University, Cambridge, Massachusetts*

JAYA BHATNAGAR (4), *Department of Chemistry and Chemical Biology, Cornell University, Ithaca, New York*

EMANUELE G. BIONDI (26), *Department of Biology, Massachusetts Institute of Technology, Cambridge, Massachusetts*

THOMAS BOLDOG (14), *Boehringer Ingelheim Austria GmbH, Vienna, Austria*

PETER P. BORBAT (3), *Department of Chemistry and Chemical Biology, Cornell University, Ithaca, New York*

YVONNE BRAUN (10), *School of Engineering and Science, International University Bremen, Bremen, Germany*

KERRY K. BRINKMAN (5), *Department of Biological Sciences, Bowling Green State University, Bowling Green, Ohio*

SCOTT L. BUTLER (2), *Virology Division, Pfizer, Inc., La Jolla Laboratories, San Diego, California*

STEPHEN A. CHERVITZ (2), *CIS Enterprise Data Group, Affymetrix, Inc., Emeryville, California*

BRIAN R. CRANE (4), *Department of Chemistry and Chemical Biology, Cornell University, Ithaca, New York*

JOSEPH J. FALKE (1, 2), *Department of Chemistry and Biochemistry, Molecular Biophysics Program, University of Colorado, Boulder, Colorado*

JACK H. FREED (3, 4), *Department of Chemistry and Chemical Biology, Cornell University, Ithaca, New York*

ALEXANDRA K. GARDINO (6), *Department of Biochemistry, Brandeis University, Waltham, Massachusetts*

SUSAN L. GLOOR (1, 2), *Department of Chemistry and Biochemistry, Molecular Biophysics Program, University of Colorado, Boulder, Colorado*

SUSAN GOTTESMAN (16), *Laboratory of Molecular Biology, National Cancer Institute, Bethesda, Maryland*

MARK GOULIAN (22), *Department of Biology and Department of Physics, University of Pennsylvania, Philadelphia, Pennsylvania*

GERALD L. HAZELBAUER (13, 14), *Department of Biochemistry, University of Missouri-Columbia, Columbia, Missouri*

FREDERICK M. HUGHSON (11), *Department of Molecular Biology, Princeton University, Princeton, New Jersey*

MASAYORI INOUYE (7, 8), *Department of Biochemistry, Robert Wood Johnson Medical School, Piscataway, New Jersey*

YINDUO JI (24), *Department of Veterinary and Biomedical Sciences, College of Veterinary Medicine, University of Minnesota, St. Paul, Minnesota*

BIJU JOSEPH (25), *Institut fur Hygiene und Mikrobiologie, Universität Würzburg, Wurzburg, Germany*

KIMBERLY L. KELLER (5), *Department of Biological Sciences, Bowling Green State University, Bowling Green, Ohio*

DOROTHEE KERN (6), *Department of Biochemistry, Brandeis University, Waltham, Massachusetts*

WING-CHEUNG LAI (13), *Department of Biochemistry, University of Missouri-Columbia, Columbia, Missouri*

TILMAN LAMPARTER (9), *Freie Universität Berlin, Pflanzenphysiologie, Berlin, Germany*

RAY A. LARSEN (5), *Department of Biological Sciences, Bowling Green State University, Bowling Green, Ohio*

MICHAEL T. LAUB (26), *Department of Biology, Massachusetts Institute of Technology, Cambridge, Massachusetts*

MINGSHAN LI (14), *Department of Biochemistry, University of Missouri-Columbia, Columbia, Missouri*

XUDONG LIANG (24), *Department of Veterinary and Biomedical Sciences, College of Veterinary Medicine, University of Minnesota, St. Paul, Minnesota*

NADIM MAJDALANI (16), *Laboratory of Molecular Biology, National Cancer Institute, Bethesda, Maryland*

AARON S. MILLER (1), *Department of Chemistry and Biochemistry, Molecular Biophysics Program, University of Colorado, Boulder, Colorado*

TIM MIYASHIRO (22), *Department of Physics, University of Pennsylvania, Philadelphia, Pennsylvania*

DAVID J. MONTEFUSCO (12), *Department of Chemistry, University of Massachusetts, Amherst, Massachusetts*

JENS PREBEN MORTH (23), *Department of Molecular Biology, Aarhus University, Denmark*

TRAVIS J. MUFF (15), *Department of Biochemistry, University of Illinois, Urbana, Illinois*

MATTHEW B. NEIDITCH (11), *Department of Molecular Biology, Princeton University, Princeton, New Jersey*

STEFFI NOACK (9), *Freie Universität Berlin, Pflanzenphysiologie, Berlin, Germany*

ELZBIETA NOWAK (23), *National Cancer Institute, Argonne National Laboratory, Argonne, Illinois*

GEORGE W. ORDAL (15), *Department of Biochemistry, University of Illinois, Urbana, Illinois*

JOHN S. PARKINSON (19, 20, 21), *Department of Biology, University of Utah, Salt Lake City, Utah*

SANGITA PHADTARE (7, 8), *Department of Biochemistry, Robert Wood Johnson Medical School, Piscataway, New Jersey*

STEVEN L. PORTER (18), *Microbiology Unit, Department of Chemistry, University of Oxford, Oxford, United Kingdom*

ANDREW R. S. ROSS (27), *Plant Biotechnology Institute, National Research Council of Canada, Saskatoon, Saskatchewan, Canada*

ALEXANDER SCHENK (10), *School of Engineering and Science, International University Bremen, Bremen, Germany*

THOMAS S. SHIMIZU (17), *Department of Molecular and Cellular Biology, Harvard University, Cambridge, Massachusetts*

JEFFREY M. SKERKER (26), *Department of Biology, Massachusetts Institute of Technology, Cambridge, Massachusetts*

ANGELA V. SMIRNOVA (10), *School of Engineering and Science, International University Bremen, Bremen, Germany*

VICTOR SOURJIK (17), *ZMBH (Center for Molecular Biology Heidelberg), University of Heidelberg, Heidelberg, Germany*

CLAUDIA A. STUDDERT (19), *Instituto de Investigaciones Biológicas, Universidad Nacional de Mar del Plata, Mar del Plata, Argentina*

PAUL A. TUCKER (23), *Hamburg Outstation, European Molecular Biology Laboratory (EMBL), Hamberg, Germany*

MATTHIAS S. ULLRICH (10), *School of Engineering and Science, International University Bremen, Bremen, Germany*

ADY VAKNIN (17), *Department of Molecular and Cellular Biology, Harvard University, Cambridge, Massachusetts*

GEORGE H. WADHAMS (18), *Microbiology Unit, Department of Chemistry, University of Oxford, Oxford, United Kingdom*

HELGE WEINGART (10), *School of Engineering and Science, International University Bremen, Bremen, Germany*

ROBERT M. WEIS (12), *Department of Chemistry, University of Massachusetts, Amherst, Massachusetts*

TAKESHI YOSHIDA (7, 8), *Department of Biochemistry, Robert Wood Johnson Medical School, Piscataway, New Jersey*

CHUANXIN YU (24), *Department of Veterinary and Biomedical Sciences, College of Veterinary Medicine, University of Minnesota, St. Paul, Minnesota*

METHODS IN ENZYMOLOGY

VOLUME 210. Numerical Computer Methods
Edited by LUDWIG BRAND AND MICHAEL L. JOHNSON

VOLUME 211. DNA Structures (Part A: Synthesis and Physical Analysis of DNA)
Edited by DAVID M. J. LILLEY AND JAMES E. DAHLBERG

VOLUME 212. DNA Structures (Part B: Chemical and Electrophoretic Analysis of DNA)
Edited by DAVID M. J. LILLEY AND JAMES E. DAHLBERG

VOLUME 213. Carotenoids (Part A: Chemistry, Separation, Quantitation, and Antioxidation)
Edited by LESTER PACKER

VOLUME 214. Carotenoids (Part B: Metabolism, Genetics, and Biosynthesis)
Edited by LESTER PACKER

VOLUME 215. Platelets: Receptors, Adhesion, Secretion (Part B)
Edited by JACEK J. HAWIGER

VOLUME 216. Recombinant DNA (Part G)
Edited by RAY WU

VOLUME 217. Recombinant DNA (Part H)
Edited by RAY WU

VOLUME 218. Recombinant DNA (Part I)
Edited by RAY WU

VOLUME 219. Reconstitution of Intracellular Transport
Edited by JAMES E. ROTHMAN

VOLUME 220. Membrane Fusion Techniques (Part A)
Edited by NEJAT DÜZGÜNEŞ

VOLUME 221. Membrane Fusion Techniques (Part B)
Edited by NEJAT DÜZGÜNEŞ

VOLUME 222. Proteolytic Enzymes in Coagulation, Fibrinolysis, and Complement Activation (Part A: Mammalian Blood Coagulation Factors and Inhibitors)
Edited by LASZLO LORAND AND KENNETH G. MANN

VOLUME 223. Proteolytic Enzymes in Coagulation, Fibrinolysis, and Complement Activation (Part B: Complement Activation, Fibrinolysis, and Nonmammalian Blood Coagulation Factors)
Edited by LASZLO LORAND AND KENNETH G. MANN

VOLUME 224. Molecular Evolution: Producing the Biochemical Data
Edited by ELIZABETH ANNE ZIMMER, THOMAS J. WHITE, REBECCA L. CANN, AND ALLAN C. WILSON

VOLUME 225. Guide to Techniques in Mouse Development
Edited by PAUL M. WASSARMAN AND MELVIN L. DEPAMPHILIS

Section I

Structural Approaches

[1] The PICM Chemical Scanning Method for Identifying Domain–Domain and Protein–Protein Interfaces: Applications to the Core Signaling Complex of *E. coli* Chemotaxis

By RANDAL B. BASS, AARON S. MILLER, SUSAN L. GLOOR, and JOSEPH J. FALKE

Abstract

    The number of known protein structures is growing exponentially (Berman *et al.*, 2000), but the structural mapping of essential domain–domain and protein–protein interaction surfaces has advanced more slowly. It is particularly difficult to analyze the interaction surfaces of membrane proteins on a structural level, both because membrane proteins are less accessible to high-resolution structural analysis and because the membrane environment is often required for native complex formation. The Protein-Interactions-by-Cysteine-Modification (PICM) method is a generalizable, *in vitro* chemical scanning approach that can be applied to many protein complexes, in both membrane-bound and soluble systems. The method begins by engineering Cys residues on the surface of a protein of known structure, then a bulky probe is coupled to each Cys residue. Next, the effects of both Cys substitution and bulky probe attachment are measured on the assembly and the activity of the target complex. Bulky probe coupling at an essential docking site disrupts complex assembly and/or activity, while coupling outside the site typically has little or no effect. PICM has been successfully applied to the core signaling complex of the *E. coli* and *S. typhimurium* chemotaxis pathway, where it has mapped out essential docking surfaces on transmembrane chemoreceptor (Tar) and histidine kinase (CheA) components (Bass and Falke, 1998; Mehan *et al.*, 2003; Miller *et al.*, 2006). The approach shares similarities with other important scanning methods like alanine and tryptophan scanning (Cunningham and Wells, 1989; Sharp *et al.*, 1995a), but has two unique features: (1) functional effects are determined for both small volume (Cys) and large volume (bulky probe) side chain substitutions in the same experiment, and (2) nonperturbing positions are identified at which Cys residues and bulky probes can be introduced for subsequent biochemical and biophysical studies, without significant effects on complex assembly or activity.

METHODS IN ENZYMOLOGY, VOL. 423   0076-6879/07 $35.00
          DOI: 10.1016/S0076-6879(07)23001-0

Introduction

Domain–domain and protein–protein interactions are essential to the functions of many, if not most, proteins. Such molecular contacts are especially crucial in signaling pathways, including the two-component signaling pathways of prokaryotic organisms. Signal transduction through a cellular circuit typically requires both intramolecular and intermolecular contacts. Domain–domain interactions within a single pathway component are often essential for the transmission of internal conformational signals, while protein–protein interactions between pathway components are needed for transmission of information throughout the cellular circuit. Thus, a molecular understanding of signal transduction requires methods capable of mapping and analyzing domain–domain and protein–protein contacts. More generally, such mapping methods can provide useful information about a wide array of cellular processes involving interactions between domains or different proteins, ranging from interdomain allostery within enzymes, to cooperative interactions between the subunits of homo-oligomers, to the assembly and regulation of multiprotein complexes.

Comparison of the PICM Method with Other Scanning Approaches

Several scanning methods have proven useful in analyzing the location, function, and physical–chemical parameters of protein interaction surfaces, including alanine scanning, tryptophan scanning, and the Protein-Interactions-by-Cysteine-Modifications (PICM) method (Bass and Falke, 1998; Cunningham and Wells, 1989; Mehan et al., 2003; Miller et al., 2006; Sharp et al., 1995a). These methods are most useful when they are applied to proteins of known structure, so that engineered mutations can be targeted exclusively to the protein surface where effects on the native fold of the modified protein are minimal, while effects on the docking interaction are maximal. The methods can all be applied effectively to soluble proteins, but, unlike many other structural methods, are also useful in the analysis of membrane proteins even in their native bilayer environments.

Alanine scanning substitutes Ala at selected surface positions, then measures the effects of each Ala substitution on the affinity of the docking interaction or on the activity of the docked complex (Cunningham and Wells, 1989; Wells, 1996). With the sole exception of Gly positions, substitution of Ala for a docking site residue truncates a larger side chain while having minimal effects on backbone flexibility. To a first approximation, then, the resulting effect of Ala substitution on docking affinity or activity reveals the contribution of an individual native side chain to the docking interaction. Alanine scanning is often used to analyze the physical chemistry of docking when the structure of the assembled complex is already known, but it can also map out the location of an unknown docking site on the surface of an isolated domain or protein as long as its docking partner is available for affinity and/or activity studies.

Tryptophan scanning substitutes Trp at selected surface positions, then determines the effect of each Trp substitution on complex assembly or activity (Sharp *et al.*, 1995a,b). Since Trp is the largest natural side chain, this substitution always increases side chain volume, thereby maximizing the probability of a dramatic effect on docking affinity. It follows that Trp substitutions within a docking site will generally yield measurable perturbations, making Trp scanning an efficient method of mapping out unknown docking sites on protein surfaces. For protein surfaces buried within a membrane bilayer, Trp scanning is particularly useful because the membrane environment often prevents the covalent coupling of extrinsic bulky probes, thereby largely eliminating the use of extrinsic probes in mapping the docking site.

The PICM method is complementary to the alanine and tryptophan scanning approaches, combining some of their strengths and offering additional advantages, particularly in systems where further biochemical and biophysical studies of the purified components are planned. PICM makes use of the unique chemical and physical properties of the Cys side chain (Bass and Falke, 1999; Falke *et al.*, 1986), which the method introduces at water-exposed surface positions scattered throughout the region where an unmapped docking site could reside. Subsequently, these engineered Cys residues are covalently modified with a bulky probe, and the effects of both the Cys substitution and bulky probe modification are determined on complex activity and assembly (Bass and Falke, 1998; Mehan *et al.*, 2003; Miller *et al.*, 2006). The engineered Cys side chain is smaller than all others except Gly, Ala, and Ser. Thus, Cys substitution typically replaces a native residue with a smaller side chain much as alanine scanning does. The bulky probe chosen for subsequent coupling to the Cys side chain is significantly larger than tryptophan, and thus replaces all native residues with a larger side chain that yields even better disruption of docking interactions than does tryptophan scanning. It follows that the PICM approach simultaneously determines the effects of smaller and larger side chain substitutions at most positions. Moreover, the PICM method yields an optimized labeling library of non-perturbing positions at which Cys substitution and bulky probe incorporation have little or no effect on docking interactions and activity. Such a library is of great utility in further biochemical and biophysical studies requiring sulfhydryl chemistry or the attachment of large spectroscopic probes or crosslinkers.

In practice, the PICM approach is limited primarily to the analysis of water-exposed docking sites on purified proteins or on the extracellular domains of proteins in living cells. Because the PICM method requires coupling of a bulky probe to a Cys residue, typically via an alkylation or disulfide exchange reaction involving the Cys sulfanion, the approach is easily applied to aqueous docking sites accessible to the probe. By contrast, the PICM method is less useful for (a) lipid-exposed docking sites where the coupling reaction proceeds slowly because the low dielectric environment raises the sulfhydryl pKa, and (b) cytoplasmically exposed docking sites in living cells where the plasma membrane barrier and high cytoplasmic glutathione concentration typically interfere with probe coupling. In such cases, alanine and tryptophan scanning methods are generally preferred.

PICM Studies of the Core Signaling Complex of Bacterial Chemotaxis

The PICM method was originally developed in studies of the core signaling complex of bacterial chemotaxis (Bass and Falke, 1998; Mehan *et al.*, 2003; Miller *et al.*, 2006). This complex transduces attractant binding into a transmembrane signal that regulates the activity of a cytoplasmic histidine kinase (Baker *et al.*, 2006; Bourret and Stock, 2002; Falke *et al.*, 1997; Parkinson *et al.*, 2005). The oligomeric receptor serves as the framework for the formation of the core complex, by providing docking surfaces on its cytoplasmic domain where the soluble cytoplasmic components dock to form a stable assembly. The core complex components required for reconstitution of receptor-regulated kinase activity are the transmembrane receptor, the histidine kinase CheA, and the coupling protein CheW. Once the complex is formed, the apo state of the receptor stimulates CheA auto-phosphorylation, while the attractant-occupied receptor inhibits CheA. Thus, the standard measure of core complex activity, termed the "reconstituted core complex kinase assay," begins by reconstituting the core complex under conditions where the CheA autophosphorylation reaction is the rate-limiting step in signal transduction.

The first PICM studies, using the initial version of the method now termed PICM-α, identified docking sites on the surface of the *Salmonella typhimurium* transmembrane aspartate receptor by introducing Cys residues at positions scattered uniformly over the protein surface (Bass and Falke, 1998; Bass *et al.*, 1999; Mehan *et al.*, 2003). These studies employed the standard reconstituted core complex kinase assay to determine the functional effects of each engineered surface Cys, both in its unmodified and bulky probe-labeled states. Essential docking surfaces were identified as regions where bulky probe attachment or, in fewer cases, the Cys substitution itself, blocked receptor-mediated kinase activation, indicating that the modification either prevented core complex assembly or prevented the flow of information from the receptor to the docked CheA kinase.

The surface positions where modifications caused large losses of receptor-mediated kinase activation were found to be clustered in two specific regions. Several perturbing positions were located in the vicinity of the adaptation sites, which are known to modulate kinase activity. The great majority of perturbing positions, however, were found within a large docking surface that contains the contact sites essential for receptor oligomerization as well as the CheA and the identified docking surfaces remain consistent with all current evidence.

In 2006, a PICM study employed an enhanced version of the method, termed PICM-β, to directly determine the location of four docking sites on the surface of the *S. typhimurium* CheA histidine kinase (Miller *et al.*, 2006). This PICM-β analysis utilized both enzyme assays and direct binding measurements to determine the effects of engineered surface cysteines and bulky probe modifications on the assembly and activity of the multi-protein complex. The use of direct protein–protein binding measurements greatly simplified the interpretation of the results, enabling direct identification of positions in or near essential docking sites. The resulting combination of activity assays and direct protein–protein binding measurements revealed the locations of four distinct docking sites on the surface of CheA: (1) the docking site on the substrate domain that associates with the catalytic domain during autophosphorylation, (2) the docking site on the catalytic domain that associates with the substrate domain during autophosphorylation, (3) the docking site for CheW on the regulatory domain, and (4) the putative docking site for the transmembrane chemoreceptor—a large site spanning portions of the CheA regulatory, catalytic, and dimerization domains. An independent X-ray crystal structure solved for the complex between a *Thermatoga thermophillus* CheA fragment and its CheW partner has confirmed that PICM-β correctly identified the location of the CheW docking site on the surface of the CheA (Park *et al.*, 2006), providing strong support for the accuracy of the PICM-β approach (Miller *et al.*, 2006). Thus, while the PICM-β approach cannot map out docking sites to atomic resolution, it can accurately map the general location of docking sites on the surfaces of proteins in their native environment, whether it be aqueous solution, a lipid bilayer, or a supermolecular complex.

Generalizing the PICM Method to Map Docking Sites in Other Systems

The present chapter focuses on the generalizable steps of the PICM procedure that can be used to map out domain–domain and protein–protein docking sites in a wide variety of systems. The initial steps in applying PICM to a specific protein are to generate a fully functional,

Cysless version of the protein and to select sites for surface Cys introduction. Next, high-throughput methods are used to introduce the engineered Cys residues by site-directed mutagenesis and to purify the resulting single-Cys containing proteins. Each single-Cys protein is then labeled with a bulky probe, and the level of probe incorporation is quantitated. The effects of Cys substitution and bulky probe attachment on docking affinity and activity are measured in functional assays. Finally, the PICM data are used to map out the location of the essential docking site(s). The remaining sections discuss the generalization of these steps to PICM studies of new protein systems, using published studies of the aspartate receptor and CheA kinase of bacterial chemotaxis as examples.

Incorporation of an Affinity Tag and Creation of a Cysless Protein

Overview

Ideally, the protein selected for PICM analysis should contain an affinity tag to facilitate rapid purification of modified proteins. Moreover, the protein should lack intrinsic Cys residues, or at least lack Cys residues accessible to the bulky probe. Alternatively, intrinsic Cys residues can be replaced to create a Cysless background for subsequent incorporation of single Cys residues. Generally, a Cysless construct that retains full activity can be engineered. The advantage of such a Cysless construct is that each Cys introduced at a selected surface position becomes a unique site for covalent attachment of the sulfhydryl-specific bulky probe, with no complications due to probe incorporation at alternate Cys positions. In previous PICM studies, the transmembrane aspartate receptor needed no affinity tag since isolation of *E. coli* membranes containing the overexpressed receptor yielded sufficient purity for PICM-α analysis (Bass and Falke, 1998; Bass *et al.*, 1999; Mehan *et al.*, 2003). Moreover, this receptor lacks intrinsic Cys residues, so the native protein was itself a perfect Cysless background in which to incorporate surface Cys residues (see following text). By contrast, the native CheA protein was prepared for PICM-β analysis both by fusing it to an N-terminal 6-His affinity tag for ease of isolation and by replacing three intrinsic Cys residues to generate a functional Cysless construct (Miller *et al.*, 2006). These modifications of CheA were carried out using standard subcloning and site-directed mutagenesis procedures; thus, the present discussion focuses on the design aspects most useful in applications to other systems.

Incorporation of a 6-His Affinity Tag

The simplest affinity tag to use is the 6-His tag, which can be attached at the N- or C-terminus of the target protein (Bornhorst and Falke, 2000), although an array of alternative, equally effective affinity tags are available

(Corbin and Falke, 2004). Standard subcloning methods are used to insert the target protein gene into a vector containing the 6-His tag and an appropriate linker. To create the N-terminal 6-His derivative of CheA, for example, the CheA gene was subcloned into the pET28 vector (Novagen) providing both the 6-His sequence and a long, flexible linker inserted between the tag and the native CheA sequence, yielding the N-terminal sequence **MGSSHHHHHHSSGLVPRGSHMASGGGGGGGGVSMD** (where the unbolded, nonnative residues derived from the vector were inserted into the bolded native sequence between Met 1 and Ser 2). The 21 residue linker includes a 7-residue poly-glycine region, ensuring that the tag is flexible and able to reach out into solution, away from the protein surface. This design maximizes the interaction between the tag and the Ni-NTA affinity matrix during purification, while minimizing potentially perturbing interactions between the highly charged tag and the surface of the protein. In a new system, it is best to create and compare N-terminal and C-terminal affinity tags, since the location of the tag may affect function. For example, while CheA possessing the above N-terminal 6-His tag is fully functional (Miller *et al.*, 2006), an analogous construct placing the tag at the C-terminus has little or no measurable activity (Bornhorst and Falke, unpublished). To generate affinity tag fusions, an array of vectors containing 6-His and other affinity tags at the N- and C-terminal ends of polylinkers are commercially available, and standard subcloning methods are used to insert target protein genes into these vectors (pET vectors, Novagen).

Creation of a Functional Cysless Protein

Following introduction of the affinity tag, if the target protein contains intrinsic Cys residues, a systematic approach is used to generate a functional Cysless construct (Frillingos *et al.*, 1998; Miller *et al.*, 2006). At each of the intrinsic Cys positions, standard site-directed mutagenesis is used to convert each Cys to both Ala and Ser. The functional effects of these single-Ala and single-Ser substitutions are determined by isolating each point mutant and subjecting them to activity studies. For CheA, which contains three intrinsic Cys residues, the resulting three Ala and three Ser point mutants were tested in the reconstituted core complex kinase assay. The three point mutations yielding the smallest effects on receptor-mediated CheA kinase activation were combined to give a Cysless triple mutant, CheA C120S/C218A/C432A (Miller *et al.*, 2006). In other systems, the same approach is generally useful, using standard methods to mutate the intrinsic Cys residues. Commercially available site-directed mutagenesis kits are available and provide efficient mutagenesis protocols (QuikChange kits, Stratagene).

Testing the Function of the Affinity-Tagged, Cysless Protein

It is important to compare the function of the final construct to the native protein, to ensure that the addition of the affinity tag and the replacement of the intrinsic Cys residues do not significantly alter the docking interactions and activities being studied. To test the function of the 6-His-tagged, Cysless CheA construct, this modified protein was compared to native CheA in both *in vitro* and *in vivo* functional assays (Miller *et al.*, 2006) (and Bornhorst and Falke, unpublished). First, the purified proteins were compared in the reconstituted core complex kinase assay wherein the modified and wild type proteins were found to be indistinguishable, within experimental error, in their abilities to be activated by the receptor apo state and inhibited by attractant binding. Next, the abilities of the two proteins to restore normal pathway function in an *E. coli* strain lacking CheA were compared in the swarm plate assay for cellular chemotaxis. Both proteins were found to restore chemotaxis *in vivo*, although the swarm rate of the 6-His-tagged, Cysless CheA construct was 0.7-fold that of the native construct. It follows that the 6-His-tagged, Cysless CheA protein is fully functional in the reconstituted core signaling complex *in vitro*, and its activity is only slightly perturbed *in vivo*. Since the 6-His-tagged, Cysless CheA protein is fully functional in the reconstituted core complex with CheA, CheW, and CheY, this modified construct is suitable for PICM studies of its interactions with these proteins (Miller *et al.*, 2006). In other systems, appropriate assays should be used to test the effects of the affinity tag and the removal of intrinsic Cys residues on target protein structure and function. If the replacement of a buried intrinsic Cys yields significant perturbations, or if two intrinsic Cys residues form an essential disulfide bond, such Cys residues can often be left in place since they typically react poorly with the bulky probe and thus do not interfere with the PICM method.

Choice of Positions for Cys Incorporation and Creation of a Mutant Library

Overview

The most rigorous type of PICM analysis scans the entire surface of the target protein for essential interaction sites. If additional information is available that limits the interaction site of interest to a specific domain or surface region, then the extent of the PICM analysis can be reduced. Either way, the approach is the same. Using the known structure of the target protein, sites are selected on the protein surface for Cys incorporation. High throughput methods are then employed to generate the single Cys mutants, express them, and purify them.

Selection of Engineered Cys Positions: General Considerations

When examining the candidate positions for Cys incorporation, it is important to select positions for which the side chains are fully exposed on the protein surface rather than partially buried in the protein interior. The goal is to avoid effects on protein folding and stability while focusing on surface residues that could be directly involved in a docking interaction. The positions selected should yield a distribution of engineered Cys residues scattered over the surface of the protein as uniformly as possible, with a spacing between Cys positions of approximately 10 Å or less to ensure that a typical docking site would contain several Cys positions. Conserved surface positions hypothesized to be involved in docking should not be avoided but rather should be included as targets for Cys substitution, since exclusion of these conserved positions could cause the PICM approach to miss an important docking site, while the design of the PICM method will ensure that side chains essential for docking will be directly identified. To be safe, in a multidomain protein for which the structure was determined by crystallographic methods, engineered Cys residues should be targeted even to the surfaces of domains that appear to be buried at domain–domain contacts, since such contacts can be artifacts of crystal constraints or of nonspecific associations between hydrophobic docking sites on different domains. (Domains are often coupled by flexible linkers, such that their locations and contacts can be easily perturbed in a crystal structure (Zhang *et al.*, 1995)). The final PICM-α analysis of the aspartate receptor selected 52 positions for Cys incorporation scattered over two distinct domains (Mehan *et al.*, 2003), while the PICM-β analysis of CheA utilized 70 positions scattered over five domains (Miller *et al.*, 2006).

High Throughput Cys Mutagenesis Protocol

1. Once a set of positions has been chosen for PICM analysis, it is useful to design a high-throughput strategy for site-directed mutagenesis. Typically, the most practical strategy is to run multiple mutagenesis reactions in parallel. The authors have found that 15 is a convenient number of reactions to run simultaneously using standard commercial mutagenesis protocols. Currently, we use the QuikChange II XL Site-Directed Mutagenesis Kit from Stratagene, which provides reagents for 30 mutagenesis reactions. After all the necessary reagents are obtained, an experienced researcher can generate at least two sets of 15 mutants per week.

2. Mutagenic primers are designed closely following the mutagenesis kit recommendations and ordered from an appropriate commercial DNA synthesis facility. We order primers in 25 nmol quantities, desalted but with no further purification, from Integrated DNA Technologies. Each primer

arrives from the manufacturer as a lyophilized film, and the manufacturer specifies the total mass of lyophilized DNA. An appropriate volume of sterile, deionized distilled H_2O is added to bring each primer up to a concentration of 1.25 $\mu g/\mu l$, which serves as a 10× stock for the subsequent mutagenesis reaction.

3. Template plasmid DNA is generated in *E. coli* strain DH5α, then is isolated using a standard commercial miniprep protocol. Currently, we use the Qiagen QIAprep Spin Miniprep Kit, which for our plasmids typically yield 50 μl of 300 ng/μl DNA. The miniprep DNA concentration is measured by its absorbance at 260 nm (1 A_{260} = 50 $\mu g/ml$), then a portion of the miniprep is diluted in H_2O to give a template stock of 10 ng/μl.

4. After obtaining both template and primers, parallel PCR mutagenesis reactions and transformations are carried out using the standard procedure recommended by the mutagenesis kit. Our PCR thermocycler is an Eppendorf MasterCycler, and we use the thermocycler settings specified by the mutagenesis kit for the length of the plasmid.

5. Following mutagenesis and transformation, mutations are confirmed by DNA sequencing. At least two colonies from each mutagenesis transformation are randomly selected for minipreps of plasmid DNA, carried out using the minilysate kit protocol, then quantitated as in step 3. We submit at least 250 ng DNA (typically, 1 to 2 μl of the miniprep) to a university facility for automated DNA sequencing. Sequencing primers are chosen to ensure that the region containing the intended mutation is adequately covered. Particularly when using PCR mutagenesis methods, it is important also to confirm that no unintended mutations have occurred elsewhere in the gene. Thus, one mutant isolate is chosen for full gene sequencing; alternatively, the sequenced region containing the desired mutation can be subcloned back into the starting plasmid.

Selection of a Cys-Specific Probe for Chemical Modification

Overview and General Considerations

A bulky probe suitable for PICM analysis is chosen based on probe size, charge, ease of incorporation, specificity for the Cys sulfhydryl, and reversibility. In general, the two most useful classes of Cys-specific coupling chemistries are (a) the maleimide alkylation reaction, which combines high sulfhydryl specificity and rapid reaction rate, and (b) the methanethiosulfonate disulfide exchange reaction, which combines high sulfhydryl specificity and reversibility.

In typical PICM applications where reversibility is not required, a suitable bulky probe is 5-fluorescein maleimide (5FM), which is significantly bulkier than any natural side chain (Fig. 1), reacts rapidly, specifically, and irreversibly with exposed Cys sulfhydryls, and yields bright, easily quantitated fluorescence on standard ultraviolet (UV) lightboxes. However, 5FM is highly sensitive to bleaching so care should be taken at all stages not to expose this probe, or proteins labeled with this probe, to UV light before the final quantitation step. Both the original PICM studies of the aspartate receptor and CheA kinase of bacterial chemotaxis used 5FM as the primary probe (Mehan *et al.*, 2003; Miller *et al.*, 2006).

For detailed PICM studies, it can be useful to investigate whether it is the size of the probe, or rather its charge, that dominates its effects on the docking interaction. For example, a set of three maleimide-based probes that can be utilized in such a study would include (a) 5FM, which is large and possesses two negative charges at neutral pH, (b) 5-tetramethylrhodamine maleimide (5TRM), which is approximately the same dimensions as 5FM but is a zwitterion with one positive and one negative charge, and

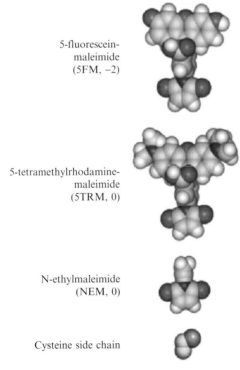

5-fluorescein-
maleimide
(5FM, −2)

5-tetramethylrhodamine-
maleimide
(5TRM, 0)

N-ethylmaleimide
(NEM, 0)

Cysteine side chain

Fig. 1. Space-filling structures of representative cysteine-specific probes and the cysteine side chain. All images to the same scale. For probes, the text indicates their full name, abbreviation, and net charge.

(c) N-ethylmaleimide (NEM), which is neutral and considerably smaller than both 5FM and 5TRM, being closer in size to a typical side chain (Fig. 1). Our most detailed PICM analysis of the aspartate receptor utilized these three probes to identify positions where probe size, charge, or both altered the interaction between the receptor and CheA kinase (Mehan *et al.*, 2003). We purchase 5FM and 5TRM from Invitrogen/Molecular Probes and NEM from Sigma.

In some cases, a reversible, disulfide-linked probe is preferable, for example, when the same sample will be used in activity studies of the labeled and unlabeled states. This approach can be utilized when all of the activity studies planned are insensitive to the presence of reducing agent, so that the labeled protein can be divided into two aliquots—one treated with reducing agent to reverse the labeling and the other untreated to retain the label. A wide selection of methanethiosulfonate (MTS) probes, varying in size, shape, and charge, is available from Toronto Research Chemicals. The MTS disulfide exchange reaction is significantly slower than the maleimide alkylation reaction, which must be considered when designing the labeling protocol (Frazier *et al.*, 2002; Malmberg *et al.*, 2003).

Probe Labeling and Purification of the Single Cys Mutants

Overview

Assuming that an affinity tag has been incorporated into the protein, it is straightforward to design a parallel purification procedure that will simultaneously label multiple single Cys mutants with a selected probe, then will purify both the labeled and unlabeled proteins. Here, we focus on the maleimide and MTS coupling chemistries for chemical labeling, and the 6-His affinity tag for purification (Bornhorst and Falke, 2000). The protocol used to express, label, and purify a given cloned, 6-His tagged protein should be optimized for the specific protein of interest, but the following procedure provides a generally useful starting point. Except where noted otherwise in this and other protocols, all chemicals, biochemicals, and reagents were obtained from Sigma.

6-His Affinity Tag Purification Protocol

1. The expression plasmid for the 6-His tagged protein is transformed into an appropriate *E. coli* line, then induction conditions are optimized for the given plasmid and protein. Using the optimized conditions, expression cultures are grown and induced in 500 ml of the appropriate media. If necessary for proper aeration, this volume is divided among multiple

growth flasks. We express and purify 6 to 12 mutant proteins in parallel, limited primarily by the number of bottle positions (typically, 6) in the centrifuge rotor, and the number of centrifuges available for simultaneous cell pelleting spins.

2. Cells are pelleted from 500 ml of media by spinning for 15 min at $6000 \times g$ in a 500 ml centrifuge bottle at $4°$. The pellet is resuspended by shaking on ice in 10 ml of buffer A (300 mM NaCl, 10% (v/v) glycerol, 50 mM NaH$_2$PO$_4$, pH to 8.0 with NaOH) to which are added fresh reducing agent and protease inhibitors (10 mM β-mercaptoethanol or βME, 1 mM phenylmethylsufonylfluoride or PMSF, which hydrolyzes rapidly in aqueous solutions and thus is added from a 200 mM stock in absolute EtOH, 1 μg/ml aprotinin, and 1 μg/ml leupeptin). If storage is necessary before beginning purification, the resuspended cells are best frozen by pouring them into liquid nitrogen, straining out the frozen cells, and immediately placing in a $-80°$ freezer.

3. When isolating, labeling, and purifying single Cys mutants from frozen cells, we typically prepare 3 to 6 mutants in parallel, limited again by the number of centrifuge rotor positions. Frozen cells are thawed by shaking on ice for approximately 30 min following addition of 10 ml ice cold buffer A, the latter containing βME and fresh protease inhibitors. Lysozyme is then added to 3 mg/ml. The mixture is incubated on ice for 30 min to allow the enzyme to hydrolyze the bacterial cell wall, with a single hand mixing at 15 min. Brij-58 detergent (polyoxyethylene-20-cetyl-ether) is added as a 7% solution (w/v) diluted 70-fold into the mixture to yield 0.1% final, then the lysate is spun for 15 min at $28,300 \times g$ in a 35 ml centrifuge tube at $4°$. The resulting supernatant (approximately 20 ml) is carefully removed and placed on ice. Note that cell lysis exposes the target protein to high concentrations of active proteases, even in the presence of the protease inhibitors, so subsequent steps should be carried out as rapidly as possible.

4. The supernatant is split into two aliquots (approximately 10 ml each), one for labeling with bulky probe and the other for unlabeled protein. Each aliquot is loaded into a BioRad EconoPac column containing 2.5 ml of Qiagen Ni-NTA beads pre-equilibrated in binding/wash buffer (15 mM imidazole in buffer A). Following loading, the column is sealed and placed on a rotator at $22°$ for 15 min to maximize mixing and binding of the 6-His-tagged protein to the beads. The column is then moved to the cold room ($4°$) and washed thrice with 15 ml aliquots of cold binding/wash buffer to remove the contaminating native proteins lacking the His tag, and also to remove the reducing agent and protease inhibitors. Following the third wash, the beads (2.5 ml) are moved out of the cold room and resuspended in 2.5 ml of room temperature binding/wash buffer together with an appropriate volume of 20 mM probe in solvent (dimethylformamide is suitable

for many probes), yielding a final probe concentration of 200 μM. To the other aliquot is added 2.5 ml of binding/wash buffer and the same volume of pure solvent. The columns are again sealed and placed on a rotator at 22° for 15 min (maleimide probes) or 1 hr (methanethiosulfonate probes) to allow the labeling reaction to proceed to completion. Then, for maleimide reactions, a 25 mM reduced glutathione stock is added to 250 μM final concentration and mixed to rapidly quench the remaining free probe (this step is omitted for methanethiosulfonate probes, to avoid reversing the labeling reaction). The columns containing the labeled and unlabeled proteins are then moved back to the cold room, the buffer is drained, and each column is washed twice more with 15 ml of cold binding/wash buffer.

5. Proteins are eluted from the beads by 3 to 5 washes with 2.5 ml of cold elution buffer (500 mM imidazole in buffer A) in the cold room. The washes from a given column are combined on ice, EDTA is added to 1 mM, and, for unlabeled and maleimide-labeled proteins, dithiothreitol (DTT) is also added to 10 mM (the last addition is omitted for methanethiosulfonate probes to avoid reduction of the disulfide linkage). Subsequently, the protein is concentrated by spinning 3220×g at 4° in a 15 ml Amicon Ultra Centrifuge Filter Device (with an appropriate membrane MW cutoff). Spinning is carried out in 30 min increments until the desired final volume of 0.5 ml is achieved (typically, 30 to 120 min total).

6. The concentrated protein is dialyzed overnight at 4° in a Slide-A-Lyzer Extra Strength Dialysis Cassette (0.5 ml capacity, appropriate membrane MW cutoff) against one liter of final buffer (50 mM Tris, pH to 7.5 with HCl, 10% glycerol (v/v), 1 mM EDTA, and, for unlabeled proteins or proteins labeled with reduction-proof probes, 10 mM DTT. Subsequently, the protein is dialyzed twice for 4 hours against 1 liter of final buffer and, for unlabled proteins or proteins labeled with reduction-proof probes, 10 μM DTT. Finally, the dialyzed protein is spun in an ultracentrifuge at 4° for 10 min at 500,000×g to remove any precipitates, and then is aliquoted and frozen in liquid nitrogen.

7. The purified protein is quantitated via a standard protein assay.

Quantitation of Probe Coupling

Overview

Following labeling and purification of the Cys mutants, it is important to measure the purity of the final proteins and their extent of labeling (Mehan *et al.*, 2003; Miller *et al.*, 2006). In particular, for each mutant, accurate PICM analysis requires virtually complete modification (preferably over 90%) of the engineered Cys residue. The protocol for measuring the extent of labeling carries out another labeling step with a fluorescent probe under denaturing, high-temperature conditions to ensure that this second labeling reaction goes to completion. Subsequently, the protein is resolved on SDS-PAGE and the labeling efficiency is determined by quantitative fluorescence imaging. This procedure also checks the purity of the final protein.

Procedure to Quantitate the Extent of Probe Coupling

1. Three reactions are prepared for each mutant protein: one for the unlabeled protein and two for the labeled protein, respectively. Typically, each of these reactions is run in triplicate to generate statistics. For each reaction, the protein stock is diluted into final buffer to a concentration of $5 \mu M$. To the unlabeled and one of the labeled samples, 20 mM 5FM in DMF is added to a final concentration of 125 μM, and to the other labeled sample, the same quantity of pure DMF is added. Each sample is diluted 1:1 with 2× Laemmli sample buffer, mixed quickly and heated at 95° for 1 min to denature the protein and drive the labeling reaction to completion. Typically, we analyze up to 8 mutants in parallel, which requires 8 SDS-PAGE gels.

2. The reactions for a given mutant are run side-by-side on a Laemmli SDS-PAGE gel of the appropriate acrylamide composition to yield a relative migration of 0.5, thereby maximizing resolution. Before staining with Coomassie or another protein stain, which typically quenches fluorescence, the fluorescent bands are visualized by placing the gel on a standard UV light box and imaged by digital photography, ensuring that none of the bands is saturated. Imaging software is used to quantitate the relative fluorescence of the bands. Following quantitation of fluorescence, the gel can be stained with Coomassie or other reagent and used to determine the purities and relative concentrations of the different protein samples by quantitation of the band absorption intensities. Care should be taken that the absorption spectrum of the bulky probe does not overlap that of the protein stain at the wavelength(s) used for imaging the stained bands.

3. To determine the extent of labeling, the relative fluorescence intensities of the three related bands are compared. These three bands are denoted LL (labeled protein that was subsequently subjected to a 5FM labeling reaction under denaturing conditions); UL (unlabeled protein that was subsequently subjected to a 5FM labeling reaction under denaturing conditions); and LU (labeled protein subjected to the control second labeling reaction without added label). The extent of labeling, which ranges between 0 for no labeling to 1 for complete labeling, can then be calculated from these fluorescence intensities. When the first and second labeling

reactions both use the same fluorophore, the extent of labeling can be calculated as LU/LL or, alternatively, as LU/UL. Generally, the LU/LL calculation is preferred since both of its fluorescence values are determined for the same protein sample treated in two different ways. In the case where the first label is nonfluorescent and the second label is a fluorophore, the extent of labeling is calculated as 1 − (LL/UL). All these calculations make use of the fact that the labeling reaction with a maleimide probe goes rapidly to completion under denaturing conditions at elevated temperatures.

Measuring Functional Effects of Cys Substitution and Bulky Probe Coupling

Overview and General Considerations

Once the library of single Cys mutants has been created and modified with the bulky probe, the effects of both the Cys substitution and bulky probe attachment on protein activity are determined using appropriate activity assays (Mehan et al., 2003; Miller et al., 2006). Often, the Cys substitution itself is relatively nonperturbing, since the Cys side chain is relatively small and sterically accommodating, while the sulfhydryl group can adapt to different electrostatic environments by varying between an apolar, protonated state and an anionic, depronated state (Falke et al., 1988, 1986). Thus, even when located within a surface docking site, Cys substitutions are often tolerated unless they replace an essential docking residue. By contrast, the bulky probe, which is significantly larger than even the largest native side chain, is designed to sterically prevent docking when introduced at a docking site. Thus, the PICM approach typically yields two categories of surface Cys positions: those outside of functionally important contact sites where neither the cyteine substution nor bulky probe is perturbing, and those within essential contact sites where the Cys may or may not be perturbing but the bulky probe is always perturbing. To resolve these two classes of positions, and thereby map out docking surfaces, it is essential to carry out suitable activity assays.

The activity assays selected to measure the functional effects of Cys substitution and bulky probe attachment are specialized for the system of interest. Ideally, the chosen assays would include a *stability or activity assay* for the modified proteins in their isolated state, to identify modifications that perturb the intrinsic folding or activity of the target protein itself. Modifications that perturb folding or stability are unsuitable for PICM analysis; thus, it is important to identify modified proteins exhibiting such perturbations so they can be excluded. Modifications that do not alter intrinsic folding or stability, but still block function, typically identify an essential domain–domain contact surface and thus provide useful information that can be gathered with the appropriate measurement. The chosen assays would also include a *complex formation assay* that quantitates the assembly and, if possible, the affinity of the complex formed by the target protein and its docking partner(s). This key measurement determines the effect of surface modifications on protein–protein interactions and is thus critical for mapping out docking sites. Finally, the chosen assays could include a *complex activity assay* to determine the effect of surface modifications on the activity of the multiprotein complex. This assay can double-check the results of the complex formation assay, since complexes that do not assemble should have no activity. More importantly, this assay can identify modifications that allow complex assembly but block complex function, usually by perturbing an essential regulatory surface or interaction.

The PICM-β analysis of CheA docking sites illustrates the selection of appropriate assays for a given system (Miller *et al.*, 2006). Since isolated CheA exhibits autokinase activity, a rapid and convenient ^{32}P autophosphorylation assay was chosen as an initial screen for native folding and activity of the isolated Cys mutants, both in their unmodified and bulky probe-modified states (Miller *et al.*, 2006). This assay revealed that most modified CheA proteins retained autophosphorylation activity, indicating that they were properly folded, functional kinases. The inactive proteins were eliminated from further PICM analysis, but the locations of their modifications provided useful information. All the perturbing modifications were found on two surfaces surrounding the ATP binding site on the CheA catalytic domain and the His 48 phosphorylation site on the CheA substrate domain, respectively, suggesting that contacts between these surfaces are essential for phosphotransfer from ATP to His 48 during the autophosphorylation reaction. To compare the intermolecular protein–protein interactions of the modified CheA proteins, two types of binding assays were chosen: a CheW binding assay utilizing fluorescence anisotropy to quantitate the docking of CheW to free CheA (Boukhvalova *et al.*, 2002), and a core complex formation assay that used centrifugation to detect the incorporation of modified CheA proteins into membrane-bound complexes with the transmembrane receptor and CheW (Levit *et al.*, 2002). Finally, to measure the ability of the modified CheA kinases to be activated by receptor in the reconstituted, membrane-bound core complex, a receptor-regulated CheA kinase assay monitoring the formation of [^{32}P]-phospho-CheY was utilized (Borkovich *et al.*, 1989; Chervitz *et al.*, 1995; Ninfa *et al.*, 1991). Together, these assays enabled resolution of four docking sites involved in (1) the contacts between the CheA catalytic and substrate domains during autophosphorylation, (2) the contact between CheA and CheW formed during core complex assembly, and (3) the contact between CheA and the receptor formed during core complex assembly.

Once a suitable set of assays has been selected, it is generally possible to design a streamlined experimental flow diagram that minimizes the number of measurements required. A streamlining procedure was implemented during the PICM-β analysis of CheA, yielding a significant reduction in the number of assays carried out on the 140 different modified proteins (Miller et al., 2006). To develop such a streamlined procedure, first, all of the modified proteins are tested in the assay for intrinsic folding or function, enabling elimination of perturbed proteins from further study. Second, the remaining proteins are tested for activity in the reconstituted multiprotein complex. Those exhibiting full activity need not be examined further, since they can assemble normally with the other components to form the active complex. Third, the modified proteins that exhibit perturbed activities in the reconstituted complex are subjected to binding assays, in order to ascertain which perturbing modifications directly inhibit a protein docking reaction. Overall, this approach provides the maximal unique information from the minimal number of measurements. Alternatively, when it is feasible to carry out all of the assays on each modified protein, this approach is preferred since it generates the maximum level of redundant, overdetermined data that can be used to check for self-consistency.

Interpretation of Results—Mapping Out Docking Sites

Overview and General Considerations

Once the PICM data has been obtained, the resulting information is used to map out one or more docking sites on the surface of the target protein. The analysis assumes that each docking site is located on a defined face or region of the protein surface, so that residues located within the same docking site will be clustered together. The key to developing an accurate PICM map of each docking site is the accurate identification of those positions at which Cys substitution and/or bulky probe modification generates a significant perturbation in a given assay. In general, a perturbation threshold must be defined for each assay to enable classification of one subset of modifications as nonperturbing (their activities lie above the threshold) and a second subset of modifications as perturbing (their activities lie below the threshold). The distribution of perturbing modifications on the protein surface will then map out the extent of the docking surface of interest. Since setting the perturbation threshold is a subjective procedure, a logical, systematic approach should be used to define this threshold.

It is useful to discuss separately the procedures used to identify essential intramolecular domain–domain contacts and those used to map out essential protein–protein docking sites.

Essential intramolecular, domain–domain contacts can exist within an isolated protein, and successful formation of these contacts can be required for native folding, stability, or activity. In such a case, there will be two distinct contact surfaces located within the protein, one on each of the contacting domains. These surfaces can be identified by focusing on the PICM data obtained for the isolated protein, which define the effects of Cys substitution and bulky probe modification on the stability or activity of the isolated protein in the absence of other components. To use PICM data to map out the two surfaces, the threshold that defines the perturbing modifications is first set quite high, so that the positions operationally defined as perturbing are widely distributed on the target protein surface. Since the domain–domain contact is assumed to be localized to two limited surfaces, the threshold is lowered until two distinct clusters of perturbing positions are apparent. These two clusters identify the contact sites. Typically, the observed effects of perturbing modifications within these sites are similar, since they inhibit the same domain–domain contact. Positions within the docking sites at which the Cys substitution itself is perturbing are strong candidates for essential docking residues. To double check the mapping process, additional PICM sites can be engineered and analyzed within the newly identified contact regions, if desired. The use of PICM to map out the contact surfaces involved in a functionally essential, intramolecular domain–domain interaction was illustrated by the 2006 PICM analysis of CheA kinase, which successfully identified the contact surfaces on the catalytic and substrate domains that must associate during the autophosphorylation reaction (Miller *et al.*, 2006).

The most important application of PICM is to identify docking sites involved in intermolecular, protein–protein contacts. Essential protein–protein contact surfaces are identified using the subset of PICM perturbations that have little or no effect on the stability or folding of the isolated protein. Again, the key to accurate identification of the docking site is the setting of the perturbation threshold for each binding and activity assay to the appropriate level. Since a protein–protein docking site is assumed to be localized to a single surface region or face, the threshold is lowered until a single cluster of perturbing positions appears, thereby mapping the docking site. Typically, perturbing modifications at different positions within the docking site have similar effects on the folding and activity assays, since they all inhibit the same protein–protein interaction. Positions within the docking site where Cys substitution itself perturbs the docking interaction are strong candidates for essential docking residues. Overall, this conservative method of analyzing PICM data tends to underestimate the total area of the docking site, since the actual docking site could possess a larger area than that detected. Once the general location of a docking site has been determined, additional PICM sites can be engineered and analyzed to more accurately determine the boundaries of the site if desired.

The use of PICM to map out protein–protein interaction surfaces has been illustrated by applications to the aspartate receptor, which have identified a large docking surface involved in the docking of CheA and CheW to the receptor, and in the assembly of receptor oligomers from isolated dimers (Mehan *et al.*, 2003). However, the fact that only activity assays were used in these applications, rather than both pairwise binding and activity assays, prevented the assignment of specific regions within the large docking surface to specific contacts with CheA, CheW, and other receptor dimers. The combined use of both binding and activity assays has been illustrated by an analysis of docking sites on CheA, which identified the specific docking sites for CheW and the receptor (Miller *et al.*, 2006). The CheW docking site defined by PICM analysis on the surface of CheA has been confirmed by an independent X-ray crystal structure of a complex formed between a thermophilic CheA fragment and its CheW partner (Park *et al.*, 2006), illustrating the ability of PICM to correctly map out a docking surface.

Finally, the PICM analysis typically identifies a large number of surface positions at which Cys substitution and bulky probe coupling have little or no effect on target protein stability, activity, and docking interactions. These nonperturbing Cys positions are useful in further biochemical and biophysical studies employing sulfhydryl chemistry, spectroscopic probes, or crosslinking chemistries to analyze the structure and mechanism of the assembled protein complex.

Acknowledgments

This work was supported by NIH R01 grant GM040731 to JJF.

References

Baker, M. D., Wolanin, P. M., and Stock, J. B. (2006). Signal transduction in bacterial chemotaxis. *Bioessays* **28,** 9–22.

Bass, R. B., Coleman, M. D., and Falke, J. J. (1999). Signaling domain of the aspartate receptor is a helical hairpin with a localized kinase docking surface: Cysteine and disulfide scanning studies. *Biochemistry* **38,** 9317–9327.

Bass, R. B., and Falke, J. J. (1998). Detection of a conserved alpha-helix in the kinase-docking region of the aspartate receptor by cysteine and disulfide scanning. *J. Biol. Chem.* **273,** 25006–25014.

Bass, R. B., and Falke, J. J. (1999). The aspartate receptor cytoplasmic domain: *In situ* chemical analysis of structure, mechanism, and dynamics. *Struct. Fold Des.* **7**, 829–840.

Berman, H. M., Westbrook, J., Feng, Z., Gilliland, G., Bhat, T. N., Weissig, H., Shindyalov, I. N., and Bourne, P. E. (2000). The Protein Data Bank. *Nucleic Acids Res.* **28**, 235–242.

Borkovich, K. A., Kaplan, N., Hess, J. F., and Simon, M. I. (1989). Transmembrane signal transduction in bacterial chemotaxis involves ligand-dependent activation of phosphate group transfer. *Proc. Natl. Acad. Sci. USA* **86**, 1208–1212.

Bornhorst, J. A., and Falke, J. J. (2000). Purification of proteins using polyhistidine affinity tags. *Methods Enzymol.* **326**, 245–254.

Boukhvalova, M. S., Dahlquist, F. W., and Stewart, R. C. (2002). CheW binding interactions with CheA and Tar. Importance for chemotaxis signaling in *Escherichia coli. J. Biol. Chem.* **277**, 22251–22259.

Bourret, R. B., and Stock, A. M. (2002). Molecular information processing: Lessons from bacterial chemotaxis. *J. Biol. Chem.* **277**, 9625–9628.

Chervitz, S. A., Lin, C. M., and Falke, J. J. (1995). Transmembrane signaling by the aspartate receptor: Engineered disulfides reveal static regions of the subunit interface. *Biochemistry* **34**, 9722–9733.

Corbin, J. A., and Falke, J. J. (2004). Affinity tags for protein immobilization and purification. *Encyclo. Biol. Chem.* **1**, 57–63.

Cunningham, B. C., and Wells, J. A. (1989). High-resolution epitope mapping of hGH-receptor interactions by alanine-scanning mutagenesis. *Science* **244**, 1081–1085.

Falke, J. J., Bass, R. B., Butler, S. L., Chervitz, S. A., and Danielson, M. A. (1997). The two-component signaling pathway of bacterial chemotaxis: A molecular view of signal transduction by receptors, kinases, and adaptation enzymes. *Annu. Rev. Cell Dev. Biol.* **13**, 457–512.

Falke, J. J., Dernburg, A. F., Sternberg, D. A., Zalkin, N., Milligan, D. L., and Koshland, D. E., Jr. (1988). Structure of a bacterial sensory receptor. A site-directed sulfhydryl study. *J. Biol. Chem.* **263**, 14850–14858.

Falke, J. J., Sternberg, D. E., and Koshland, D. E., Jr. (1986). Site-directed sulfhydryl chemistry and spectroscopy: Applications in the aspartate receptor system. *Biophys. J.* **49**, 20a.

Frazier, A. A., Wisner, M. A., Malmberg, N. J., Victor, K. G., Fanucci, G. E., Nalefski, E. A., Falke, J. J., and Cafiso, D. S. (2002). Membrane orientation and position of the C2 domain from cPLA2 by site-directed spin labeling. *Biochemistry* **41**, 6282–6292.

Frillingos, S., Sahin-Toth, M., Wu, J., and Kaback, H. R. (1998). Cys-scanning mutagenesis: A novel approach to structure function relationships in polytopic membrane proteins. *FASEB J.* **12**, 1281–1299.

Levit, M. N., Grebe, T. W., and Stock, J. B. (2002). Organization of the receptor-kinase signaling array that regulates *Escherichia coli* chemotaxis. *J. Biol. Chem.* **277**, 36748–36754.

Malmberg, N. J., Van Buskirk, D. R., and Falke, J. J. (2003). Membrane-docking loops of the cPLA2 C2 domain: Detailed structural analysis of the protein–membrane interface via site-directed spin-labeling. *Biochemistry* **42**, 13227–13240.

Mehan, R. S., White, N. C., and Falke, J. J. (2003). Mapping out regions on the surface of the aspartate receptor that are essential for kinase activation. *Biochemistry* **42**, 2952–2959.

Miller, A. S., Kohout, S. C., Gilman, K. A., and Falke, J. J. (2006). CheA kinase of bacterial chemotaxis: Chemical mapping of four essential docking sites. *Biochemistry* **45**, 8699–8711.

Ninfa, E. G., Stock, A., Mowbray, S., and Stock, J. (1991). Reconstitution of the bacterial chemotaxis signal transduction system from purified components. *J. Biol. Chem.* **266**, 9764–9770.

Park, S. Y., Borbat, P. P., Gonzalez-Bonet, G., Bhatnagar, J., Pollard, A. M., Freed, J. H., Bilwes, A. M., and Crane, B. R. (2006). Reconstruction of the chemotaxis receptor-kinase assembly. *Nat. Struct. Mol. Biol.* **13,** 400–407.

Parkinson, J. S., Ames, P., and Studdert, C. A. (2005). Collaborative signaling by bacterial chemoreceptors. *Curr. Opin. Microbiol.* **8,** 116–121.

Sharp, L. L., Zhou, J., and Blair, D. F. (1995a). Features of MotA proton channel structure revealed by tryptophan-scanning mutagenesis. *Proc. Natl. Acad. Sci. USA* **92,** 7946–7950.

Sharp, L. L., Zhou, J., and Blair, D. F. (1995b). Tryptophan-scanning mutagenesis of MotB, an integral membrane protein essential for flagellar rotation in *Escherichia coli. Biochemistry* **34,** 9166–9171.

Wells, J. A. (1996). Binding in the growth hormone receptor complex. *Proc. Natl. Acad. Sci. USA* **93,** 1–6.

Zhang, X. J., Wozniak, J. A., and Matthews, B. W. (1995). Protein flexibility and adaptability seen in 25 crystal forms of T4 lysozyme. *J. Mol. Biol.* **250,** 527–552.

[2]   Use of Site-Directed Cysteine and Disulfide
Chemistry to Probe Protein Structure and Dynamics:
Applications to Soluble and Transmembrane
Receptors of Bacterial Chemotaxis

By Randal B. Bass, Scott L. Butler, Stephen A. Chervitz,
Susan L. Gloor, and Joseph J. Falke

Abstract

Site-directed cysteine and disulfide chemistry is broadly useful in the analysis of protein structure and dynamics, and applications of this chemistry to the bacterial chemotaxis pathway have illustrated the kinds of information that can be generated. Notably, in many cases, cysteine and disulfide chemistry can be carried out in the native environment of the protein whether it be aqueous solution, a lipid bilayer, or a multiprotein complex. Moreover, the approach can tackle three types of problems crucial to a molecular understanding of a given protein: (1) it can map out $2°$ structure, $3°$ structure, and $4°$ structure; (2) it can analyze conformational changes and the structural basis of regulation by covalently trapping specific conformational or signaling states; and (3) it can uncover the spatial and temporal aspects of thermal fluctuations by detecting backbone and domain dynamics. The approach can provide structural information for many proteins inaccessible to high-resolution methods. Even when a high-resolution structure is available, the approach provides complementary information about regulatory mechanisms and thermal dynamics in the native environment. Finally, the approach can be applied to an entire protein, or to a specific domain or subdomain within the full-length protein, thereby facilitating a divide-and-conquer strategy in large systems or multiprotein complexes.

Rigorous application of the approach to a given protein, domain, or subdomain requires careful experimental design that adequately resolves the structural and dynamical information provided by the method. A full structural and dynamical analysis begins by scanning engineered cysteines throughout the region of interest. To determine $2°$ structure, the solvent exposure of each cysteine is determined by measuring its chemical reactivity, and the periodicity of exposure is analyzed. To probe $3°$ structure, $4°$ structure, and conformational regulation, pairs of cysteines are identified that rapidly form disulfide bonds and that retain function when induced to form a disulfide bond in the folded protein or complex. Finally, to map out thermal fluctuations in a protein of known structure, disulfide formation

METHODS IN ENZYMOLOGY, VOL. 423   0076-6879/07 $35.00
                 DOI: 10.1016/S0076-6879(07)23002-2

rates are measured between distal pairs of nonperturbing surface cysteines. This chapter details these methods and illustrates applications to two proteins from the bacterial chemotaxis pathway: the periplasmic galactose binding protein and the transmembrane aspartate receptor.

Introduction

The structures of many important proteins, as well as key aspects of protein regulation and dynamics, are not yet accessible to high-resolution methods. Site-directed cysteine and disulfide chemistry (Falke and Koshland, 1987; Falke et al., 1986, 1988) provides an approach that can, in many cases, address these challenging questions by analyzing the structure and dynamics of a given protein in its native environment. The most prevalent application of site-directed cysteine and disulfide chemistry has been in structural studies of membrane protein systems and large multiprotein complexes (Akabas et al., 1992; Bass and Falke, 1998, 1999; Chervitz and Falke, 1995, 1996; Danielson et al., 1997; Falke and Koshland, 1987; Falke et al., 1986, 1988; Hughson and Hazelbauer, 1996; Hughson et al., 1997; Lee et al., 1994; Miller et al., 2006; Pakula and Simon, 1992; Sahin-Toth and Kaback, 1993; Todd et al., 1989; van Iwaarden et al., 1991; Wu et al., 1996). The structures of such large, insoluble, complicated systems are often the most difficult to analyze by high-resolution structural methods, making the structural constraints provided by the present chemical approach quite valuable. More broadly, the ability of site-directed cysteine and disulfide chemistry to analyze proteins in their native solution, membrane, or supermolecular environment enables rigorous investigation of the relationship between structural changes and regulation or function. Even in proteins of known high-resolution structure, site-directed cysteine and disulfide chemistry can provide important information about thermal fluctuations, yielding insights into the timescales and trajectories of long-range backbone and domain motions that are currently inaccessible to other methods (Bass and Falke, 1999; Butler and Falke, 1996; Careaga and Falke, 1992; Careaga et al., 1995).

Site-Directed Cysteine and Disulfide Chemistry: History

Site-directed cysteine and disulfide chemistry was originally developed as a generalizable approach to protein structure and dynamics in the bacterial chemotaxis pathway (for example, Bass and Falke, 1998, 1999; Butler and Falke, 1996; Careaga and Falke, 1992; Careaga et al., 1995; Chervitz and Falke, 1995, 1996; Danielson et al., 1997; Falke and Koshland, 1987; Falke et al., 1986, 1988; Hughson and Hazelbauer, 1996; Hughson et al., 1997;

Lee *et al.*, 1994; Miller *et al.*, 2006; Pakula and Simon, 1992), but the method is broadly applicable. As a result, novel extensions of the approach have been developed and implemented in a wide array of other systems (for example, Akabas *et al.*, 1992; Sahin-Toth and Kaback, 1993; Todd *et al.*, 1989; van Iwaarden *et al.*, 1991; Wu *et al.*, 1996). The present chapter focuses on the use of site-directed cysteine and disulfide chemistry to analyze 2° structure, 3° structure, 4° structure, the coupling between structure and regulation, and the nature of thermal backbone dynamics. To illustrate applications of these methods to specific proteins, published studies of components from the bacterial chemotaxis pathway are discussed. However, the central goal of this chapter is to facilitate the application of site-directed cysteine and disulfide methods to new protein systems. Thus, the chapter emphasizes the general design and implementation of these methods and key factors that should be considered when interpreting results.

The two components of the *Escherichia coli* and *Salmonella typhimurium* chemotaxis pathway that are used to illustrate the methods of the present chapter include a soluble binding protein and a transmembrane receptor (reviewed in Baker *et al.*, 2006; Falke *et al.*, 1987). The periplasmic galactose binding protein is a 36 kDa, monomeric soluble receptor that specifically binds the chemoattractants D-galactose and D-glucose in a cleft between its two domains connected by a flexible hinge (Vyas *et al.*, 1988). The transmembrane aspartate receptor is a 120 kDa homodimeric protein that assembles into larger oligomers (Ames *et al.*, 2002; Falke and Hazelbauer, 2001; Falke and Kim, 2000; Kim *et al.*, 1999; Milburn *et al.*, 1991; Studdert and Parkinson, 2004). This receptor binds the chemoattractant aspartate at a site on its periplasmic domain and sends a transmembrane signal through membrane-spanning helices to its cytoplasmic domain, where the signal regulates the bound histidine kinase CheA.

Site-directed cysteine and disulfide chemistry has revealed important structural, functional, and dynamical features in both these chemotaxis proteins. The approach was first used to analyze helix–helix packing in the periplasmic and transmembrane domains of the aspartate receptor and other chemoreceptors (Chervitz and Falke, 1995, 1996; Chervitz *et al.*, 1995; Falke and Koshland, 1987; Falke *et al.*, 1988; Hughson and Hazelbauer, 1996; Hughson *et al.*, 1997; Lee *et al.*, 1994; Pakula and Simon, 1992; Winston *et al.*, 2005), and a nonperturbing disulfide bond discovered in these studies was used to stabilize a dimeric construct of the isolated periplasmic domain for crystallization and X-ray structure determination (Falke and Koshland, 1987; Milburn *et al.*, 1991). The chemical studies were then extended to the cytoplasmic domain where they revealed two regions of helical secondary structure in the HAMP subdomain and an extended 4-helix bundle forming the majority of the cytoplasmic domain

(Bass and Falke, 1998, 1999; Bass *et al.*, 1999; Butler and Falke, 1998; Danielson *et al.*, 1997; Winston *et al.*, 2005). Overall, the 2°, 3°, and 4° structural constraints determined by site-directed cysteine and disulfide chemistry in the full-length, membrane-bound receptor have been confirmed by high-resolution X-ray structures of the isolated receptor periplasmic and cytoplasmic domains, and by an NMR structure of a distantly related HAMP subdomain (Hulko *et al.*, 2006). Together, these chemically defined constraints have greatly aided the development of the current working atomic model for the full-length receptor dimer (Kim *et al.*, 1999; Milburn *et al.*, 1991), although the current model of HAMP subdomain structure (Hulko *et al.*, 2006) has yet to be incorporated.

The discovery of engineered disulfide bonds that lock the receptor in its kinase-activating and -inactivating signaling states, termed "lock-on" and "lock-off" disulfides, respectively, were crucial to the development of current mechanistic models describing signal transmission through the transmembrane and cytoplasmic regions of the receptor (Bass and Falke, 1999; Chervitz and Falke, 1995, 1996; Starrett and Falke, 2005). These studies were carried in full-length, membrane-bound receptor in its working complex with CheA kinase. Independent work carried out in several laboratories has strongly supported the mechanistic conclusions of these site-directed cysteine and disulfide chemistry studies (Draheim *et al.*, 2005, 2006; Falke and Hazelbauer, 2001; Hughson and Hazelbauer, 1996; Lai *et al.*, 2006; Miller and Falke, 2004).

Finally, studies of backbone and domain motions in the known structures of the galactose binding protein and the aspartate receptor have shed light on the thermal backbone and domain motions that occur within these proteins in their native aqueous and membrane environments, respectively (Bass and Falke, 1999; Butler and Falke, 1996; Careaga and Falke, 1992; Careaga *et al.*, 1995). The findings reveal dramatic, long-range motions that occur on biologically meaningful timescales but are difficult to detect by other methods, suggesting that proteins may be significantly more dynamic than is generally believed.

Site-Directed Cysteine and Disulfide Chemistry: Applications and Limitations

The present chapter presents detailed procedures for three types of cysteine and disulfide chemistry: chemical reactivity scanning, disulfide mapping, and disulfide trapping. *Chemical reactivity scanning* measures the reactivities of adjacent residues with an aqueous probe, revealing local patterns of solvent exposure. The resulting patterns can identify 2° structure elements in an unknown structure or conformational changes in a known structure. *Disulfide mapping* measures disulfide bond formation rates, as well as the effects of disulfide formation on protein function, to map out contacts between pairs of positions in the 3° and 4° structure. The resulting information can provide proximity constraints that map out folding and packing in an unknown structure. In addition, the effects of engineered disulfides on function can resolve the structures of different signaling or activity states, and define the conformational changes that connect different states. *Disulfide trapping* measures disulfide formation rates between pairs of positions separated by different distances in a protein of known structure. The resulting data reveal the amplitudes, trajectories, and frequencies of long-range backbone and domain motions within the protein of interest.

When undertaking a study involving site-directed cysteine and disulfide chemistry, it is important to consider the caveats and limitations of this chemical approach. In particular, the method provides information about both structure and dynamics, and it is important to separate these two classes of information when designing and interpreting experiments (Bass and Falke, 1999; Careaga and Falke, 1992). For example, analysis of 2° structure via cysteine scanning and chemical reactivity analysis assumes that the element selected for study possesses distinct buried and exposed faces that will give rise to a periodic pattern of solvent exposure. While this limitation is compatible with 2° elements on a protein surface, it prevents application of the method to completely buried elements in the protein interior. Many studies of 3° or 4° structure in various laboratories have attempted to detect packing contacts by measuring rates or extents of disulfide bond formation between pairs of engineered cysteines. Often, such rates are correlated with spatial proximities, but the fundamental determinants of disulfide formation rates are sulfhydryl–sulfhydryl collision frequencies and reactivities rather than simple proximities. Thus, it is important to design approaches that can resolve structural from dynamical contributions. In short, each of the caveats and limitations of site-directed cysteine and disulfide chemistry can be overcome by suitable experimental design and controls, as discussed below in the relevant sections.

Incorporation of an Affinity Tag and Creation of a Cysless Protein

Ideally, the protein selected for analysis by site-directed cysteine and disulfide chemistry should contain an affinity tag to facilitate rapid purification and should lack intrinsic cysteine residues. Many native proteins lack cysteine (Cys), while those possessing Cys residues can often be modified by a systematic approach to generate a fully functional cysteine-less (Cysless) derivative. The advantage of such a Cysless construct is that each Cys introduced at a selected surface position becomes a unique site for sulfhydryl and disulfide chemistry. In previous studies of bacterial chemotaxis components, the galactose binding protein and the transmembrane aspartate receptor required no affinity tag since both could be easily isolated to sufficient purity for sulfhydryl and disulfide chemistry; and, moreover, both of these proteins lack intrinsic Cys residues (Bass and Falke, 1999; Butler and Falke, 1996; Careaga and Falke, 1992; Careaga *et al.*, 1995; Chervitz and Falke, 1995, 1996; Chervitz *et al.*, 1995; Danielson *et al.*, 1997; Falke and Koshland, 1987; Falke *et al.*, 1988; Pakula and Simon, 1992; Winston *et al.*, 2005). By contrast, the histidine kinase CheA was more difficult to purify and thus was fused to an N-terminal 6-His affinity tag for ease of isolation (Miller *et al.*, 2006). In addition, native CheA possesses three intrinsic Cys residues that were substituted with an optimized combination of alanine and serine residues to generate a fully functional Cysless construct (Miller *et al.*, 2006). Chapter 1 in this volume (Bass *et al.*, 2007) outlines general strategies that can be employed to prepare a new protein for site-directed cysteine and disulfide chemistry, including the incorporation of a 6-His affinity tag, the creation of an optimized Cysless construct, and the importance of careful tests to exclude engineered proteins that exhibit perturbed structure, stability, or activity.

Choice of Positions for Cys Incorporation and Creation of a Mutant Library

One advantage of cysteine and disulfide chemistry is that it can be applied in a highly targeted fashion to a protein subdomain or domain if desired, or, at the other extreme, broadly applied to an entire protein. A typical study focuses on a specific region within the intact protein. Such adaptability facilitates divide-and-conquer strategies. To maximize the probability of a definitive conclusion within the target region, it is advisable to generate as many Cys substitutions as possible since the degree of certainty increases with the number of independent measurements. Moreover, whenever possible, experimental studies of the resulting Cys mutants are carried out on the intact protein in its native solution, membrane, or multiprotein complex environment.

Once a protein subdomain or domain has been chosen for analysis, single Cys substitutions are introduced at all positions within the selected region using high throughput methods to generate, express, and purify each Cys mutant. These methods are described in Chapter 1 of this volume (Bass *et al.*, 2007). Note that to avoid oxidation of the engineered Cys sulfhydryl groups during purification and storage, Cys mutant purification methods include reducing agent and minimize exposure to transition metals that catalyze oxidation, particularly Cu(II) and Fe(III). During the final stages of purification, the concentration of reducing agent is decreased to a low level (typically, 10 μM dithiothreitol, DTT) to protect the engineered sulfhydryls while minimizing interference with subsequent cysteine and disulfide reactions. In addition, a metal chelator is included (typically, 1 mM ethylene diamine tetraacetic acid, EDTA) for protection against transition metal-catalyzed oxidation, and mutants are flash-frozen in liquid nitrogen and stored at $-80°$ to further minimize oxidative damage.

The resulting library of single-Cys mutants is carefully tested for perturbations introduced by individual Cys substitutions. Often, the Cys substitution itself is relatively nonperturbing, since the Cys side chain is relatively small and sterically accommodating, while the sulfydryl group can adapt to different electrostatic environments by varying between an apolar, protonated state and an anionic, depronated state (Careaga and Falke, 1992; Falke and Koshland, 1987; Falke et al., 1986, 1988). Even when located at a buried position, Cys substitutions are often tolerated unless they replace a residue essential for folding stability. In a large library of Cys mutants, however, several mutants will typically be found to exhibit significant perturbations of stability or function. These perturbed mutants are best identified by appropriate stability and/or functional assays developed specifically for the target protein. Mutants found to be perturbed should then be eliminated from further studies.

Analysis of 2° Structure by Chemical Reactivity Scanning

Overview of Chemical Reactivity Scanning

In a typical protein, most secondary structure elements possess one face exposed to protein exterior, and another face exposed to the protein interior. When the solvent-exposed face of such an element is in contact with aqueous solution, its secondary structure can be elucidated by measuring the chemical reactivity of single Cys residues scanned throughout its length (Bass and Falke, 1998; Bass et al., 1999; Butler and Falke, 1998; Danielson et al., 1997; Winston et al., 2005). Those positions exposed to aqueous solution will react rapidly with a bulky, charged, water-soluble alkylating reagent, while those positions exposed to the buried positions will react significantly more slowly. The periodicity of exposed and buried positions is characteristic of the type of secondary structure, ranging from 3.5 to 3.6 residues per cycle for α-helices to 2 residues per cycle for β-strands. To detect this periodicity, reaction conditions are chosen that enable initial rates of alkylation by the aqueous probe to be accurately measured in the region of interest. Several parameters of the reaction conditions can be adjusted as needed for the local environment to ensure that the initial rates can be determined on a convenient time scale. The method possesses three limitations that should be considered. First, the target secondary structure element must possess distinct aqueous and buried faces to ensure that the periodicity of chemical reactivity can be interpreted. Second, the method cannot easily be applied to transmembrane domains exposed to the hydrocarbon core of the bilayer, since alkylation reactions proceed too slowly to measure in such an environment. Third, the method requires sufficient quantities of protein that the alkylated product can be directly detected and quantitated. The present approach utilizes the sulfhydryl-specific aqueous probe 5-iodoacetamido-fluorescein (5-IAF) as the alkylating agent, which yields an easily quantitated fluorescent band on an SDS-PAGE gel for quantities of labeled protein exceeding 5 pmol.

The accuracy of chemical reactivity scanning in mapping out secondary structure elements has been tested and confirmed in the full-length, membrane-bound aspartate receptor (Bass and Falke, 1998; Bass et al., 1999; Butler and Falke, 1998; Danielson et al., 1997; Winston et al., 2005). First, using the known high-resolution structure of the isolated periplasmic domain (Milburn et al., 1991), positive control studies were carried out focusing on helix $\alpha 4$ of the periplasmic domain in the intact receptor. Chemical reactivity scans of this region revealed a periodic pattern of reactivity that closely matched the oscillating pattern of solvent exposure observed for helix $\alpha 4$ in the crystal structure, thereby directly establishing the validity of the approach (Bass and Falke, 1998; Butler and Falke, 1998; Danielson et al., 1997). Furthermore, the approach was applied to the unknown structure of the receptor cytoplasmic domain, again in the intact, membrane-bound receptor. Chemical reactivity scans revealed four local regions of α-helical periodicity, two of them separated by a region proposed to be a hairpin turn (Bass and Falke, 1998; Bass et al., 1999; Danielson et al., 1997; Winston et al., 2005). The putative helical regions, with their distinct exposed and buried faces, were later confirmed to be amphiphilic and α-helical in crystal structures of the isolated cytoplasmic domain (Falke and Kim, 2000; Kim et al., 1999; Park et al., 2006). Similarly, this structure confirmed that the putative turn was indeed a hairpin. Another two regions of α-helical periodicity were observed in the cytoplasmic HAMP region, and one region lacking a defined secondary structure was detected (Butler and Falke, 1998). These three assignments were confirmed in a subsequent NMR structure of an isolated HAMP domain from a thermophilic bacterium (Hulko et al., 2006), although the relevance of this structure to full-length, membrane-bound chemoreceptors has not yet been experimentally tested. Overall, however, the evidence to date indicates that the chemical reactivity scan accurately identifies water-exposed secondary structure elements on the surfaces of full-length proteins in their native environments. In principle, chemical reactivity scanning could also be employed to map out conformation changes via measurements of changing solvent exposure, but such changes have not yet been detected in the aspartate receptor, presumably because signal transduction triggers a subtle structural change too small to be easily detected by this method (Bass and Falke, 1998; Bass *et al.*, 1999; Butler and Falke, 1998; Danielson *et al.*, 1997; Falke and Hazelbauer, 2001; Winston *et al.*, 2005).

Chemical Reactivity Scanning: General Considerations

The following protocol for cysteine reactivity scanning was developed for the aqueous domains of the full-length, membrane-bound aspartate receptor (Bass and Falke, 1998; Bass *et al.*, 1999; Butler and Falke, 1998; Danielson *et al.*, 1997; Winston *et al.*, 2005) but is generalizable to other protein systems. For typical protein surface elements exposed to the bulk aqueous phase, the indicated 5-IAF reaction conditions yield a rate of alkylation easily measured by quenching the reaction at specific timepoints after probe addition, then quantitating the extent of cysteine labeling with fluorescein. However, if the protein surface targeted for study is highly charged, near a membrane surface, or poorly accessible to bulk solvent, the alkylation reaction could be too fast or slow to conveniently measure. In such a case, it is simple to adjust the reaction conditions to bring the alkylation time scale into the convenient range by increasing the pH (which increases the fraction of cysteine sulfhydryl in the reactive sulfanion state), temperature, and/or 5-IAF concentration to speed the reaction, or by decreasing one of these variables to slow the reaction.

Chemical Reactivity Scanning: Detailed Procedure

1. Each of the single-Cys engineered proteins is diluted to a final concentration of 5 μM in labeling reaction buffer (10 mM NaH$_2$PO$_4$ pH to 6.5 with NaOH, 50 mM NaCl, 50 mM KCl, and 1 mM EDTA) on ice, yielding a total volume of 25 μl per sample in a microfuge tube. Care should be taken that protein sulfhydryls are protected from oxidation during purification of the protein stock, and that the concentrations of residual reducing agents such as DTT are negligible (less than 25 μM total sulfhydryl) relative to the 5-IAF concentration during the following alkylation reaction. Before proceeding with the 5-IAF labeling reaction, each sample is carefully mixed to ensure a homogenous solution and spun for 10 sec in a microfuge to bring the entire volume to the bottom of the tube.

2. The samples are moved to a 25° temperature bath, allowed to equilibrate for 1 min, then the alkylation reaction is initiated by adding 1.25 μl of
a 5-IAF stock (5 mM 5-IAF in N,N'-dimethylformamide, DMF), yielding a
final 5-IAF concentration of 250 μM, with careful mixing before quickly
returning to the 25° bath.

3. At timepoints of 2.5, 5, and 10 min, aliquots of 5 μl are removed and
quenched by addition to a microfuge tube containing 1 μl of quench solution
(500 mM stock of β-mercaptoethanol in water) with careful mixing.

4. Immediately after the 10 min aliquot is quenched, another 5 μl
aliquot is removed and transferred to a microfuge tube containing 1 μl of
denaturation solution (4% SDS in water) with careful mixing. This endpoint
sample is heated to 95° for 2 min to ensure complete unfolding and reaction
of all free Cys residues with 5-IAF. Then, the endpoint sample is removed
from the heat and 1 μl quench solution is added with careful mixing.

5. The four quenched aliquots are each pelleted in a microfuge to
deposit the sample at the bottom of the tube. Then, they are carefully
mixed with an equal volume of 2× Laemmli nonreducing sample buffer
and heated to 95° for 2 min. Longer heating times and higher temperatures
are avoided to minimize heat-triggered membrane protein aggregation.

6. The samples are loaded and run side-by-side on a nonreducing
Laemmli SDS-PAGE gel. To ensure that equal quantities of protein are
compared, 10 μl of each timepoint sample and 12 μl of the endpoint sample
are loaded. The gel is formulated so that the target protein exhibits a
relative migration (R_f) of approximately 0.5, thereby maximizing resolution
from other proteins and from free fluor at the dyefront. Before staining with
Coomassie or another protein stain, which typically quenches fluorescence,
the fluorescent bands are visualized by placing the gel on a standard
ultraviolet (UV) light box and imaged by digital photography, ensuring
that none of the bands are saturated. Imaging software is used to quantitate
the relative fluorescence of the bands. Following quantitation of fluorescence, the gel can be stained with Coomassie and used to determine the
purities and relative concentrations of the different protein samples by
digital photography quantitation of the resulting Coomassie absorbance.

7. To determine the extent of 5-IAF labeling at each timepoint, the
fluorescence intensity of each sample for a given timepoint is divided by
the intensity of the endpoint sample. If the Coomassie analysis reveals
significant discrepancies in the amount of protein in different lanes, the
Coomassie intensities can be used to normalize the fluorescence intensities
of each lane to the same protein level prior to division by the endpoint
fluorescence intensity.

8. Once the initial 5-IAF reaction time courses have been determined
for a subset of positions, it may be necessary to adjust the reaction rate. It is

useful to optimize the reaction rate so that virtually all the positions examined yield between 5 and 75% labeling at either the 2.5 or 5 min timepoint. This optimization ensures that each position is labeled to a sufficient extent to be quantitated, while the reaction has not proceeded to completion and thus remains in or near its linear range. To decrease or increase the speed of the reaction, one can decrease or increase, respectively, either the pH or temperature of the reaction.

9. After optimization, the labeling reaction is carried out in triplicate for each position to enable comparison of means and standard deviations for different Cys positions. Multiple positions can be tested in a set of simultaneous reactions, with the exact number depending on the timepoint chosen for sampling the reaction. In each set of reactions, two control reactions are carried out for a selected, well-characterized Cys position that serves as an internal standard when comparing different sets. For each reaction in a set, a given single-Cys protein is diluted to 5 μM final concentration, as in step 1, except that the total volume is now halved to 12.5 μl and the newly optimized reaction buffer is utilized. After mixing, spinning, and equilibration to the optimized temperature, 5-IAF is added to a final concentration of 250 μM as in step 2, and the reaction is allowed to proceed. At the optimized timepoint, two 5 μl aliquots are removed from each reaction. One is quenched as in step 3, the other is denatured and reacted to completion before quenching as in step 4. Then, the quenched samples are analyzed by SDS-PAGE and digital imaging as described in steps 5 through 7, yielding a quantitative extent of 5-IAF labeling for each reaction. By running multiple sets of reactions consecutively, as many as 20 engineered Cys positions can be analyzed per day. Care should be taken that the control Cys position included in all sets yields similar extents of 5-IAF labeling in different sets run on the same or different days.

Chemical Reactivity Scanning: Interpretation of Results

To analyze the chemical reactivity data and identify 2° structure elements, the mean chemical reactivities and standard deviations of consecutive engineered Cys positions are plotted as a function of residue number. Secondary structure elements with distinct water-exposed and buried faces are recognized via characteristic periodicities imposed by their oscillating solvent exposures. β-strands, for example, exhibit a periodicity of 2 residues in their pattern of chemical reactivities, since consecutive positions alternately face aqueous solution and the protein interior and thereby yield high and low relative reactivities, respectively. By contrast, α-helices exhibit a significantly longer periodicity of 3.5 (coiled-coil helix) or 3.6 (standard helix) residues and are thus easily distinguished from strands. Regions that lack a defined periodicity may represent linkers between structural elements. Published studies of the transmembrane aspartate receptor discussed previously further illustrate the use of chemical reactivity data to identify 2° structure elements, turns, and linkers (Bass and Falke, 1998; Bass et al., 1999; Butler and Falke, 1998; Danielson et al., 1997; Winston et al., 2005) in the full-length receptor embedded in its native membrane.

Disulfide Mapping of Spatial Proximity and Conformational Changes

Overview of Disulfide Mapping

In principle, disulfide bond formation can map out contacts between pairs of positions in the folded protein structure. For example, disulfide mapping approaches first shed light on the helix–helix packing within the periplasmic, transmembrane, and cytoplasmic domains of the full-length, membrane-bound aspartate receptor (Bass and Falke, 1996; Bass et al., 1999; Butler and Falke, 1998; Chervitz and Falke, 1995; Chervitz et al., 1995; Danielson et al., 1997; Falke and Koshland, 1987; Falke et al., 1988; Hughson et al., 1997; Lee et al., 1994; Pakula and Simon, 1992; Winston et al., 2005). The well-defined distance and angular constraints of the disulfide bond ensure that each Cys pair that forms a disulfide bond lies in close proximity and also satisfies strict angular constraints during the disulfide formation reaction (Careaga and Falke, 1992). Thus, the ability of two Cys positions to rapidly form a disulfide bond has been widely assumed to be a signature of proximity within the equilibrium structure. While this is often the case, it is important to note that disulfide bonds are sometimes formed between Cys residues that are distal in the equilibrium structure but able to collide during structural fluctuations away from this structure (Bass et al., 1999; Butler and Falke, 1996; Careaga and Falke, 1992). (Such thermal fluctuations can be analyzed by the disulfide trapping method described later.) Moreover, the environmental dependence of Cys chemical reactivities can also modulate disulfide formation rates; for example, a highly reactive pair could form a disulfide more rapidly than a more proximal pair with a higher collision rate. (Such unusual reactivities can generally be detected by comparison of 5-IAF reactivities measured for Cys residues on the same face of a structural element, as described in the preceding section on chemical reactivity measurements.) To rigorously detect proximities in the active equilibrium structure, it is useful to measure both (1) the rate of disulfide formation and (2) the effect of disulfide formation on protein activity. The information provided by the latter functional analysis of engineered disulfide bonds can also, in many cases, elucidate conformational changes among different signaling or functional states.

The disulfide mapping approach described herein can be applied to most proteins, except those that contain large numbers of intrinsic Cys residues that would complicate the analysis. Useful disulfide mapping strategies are illustrated by published and ongoing studies of the aspartate receptor and other bacterial chemoreceptors (Bass and Falke, 1998; Bass et al., 1999; Butler and Falke, 1998; Chervitz and Falke, 1995; Chervitz et al., 1995; Danielson et al., 1997; Hughson et al., 1997; Lee et al., 1994; Pakula and Simon, 1992; Winston et al., 2005). Disulfide mapping studies first eluci-dated the packing of the four membrane-spanning helices in the transmem-brane region of the receptors (Pakula and Simon, 1992), and the resulting packing model has been strongly supported by subsequent chemical, spec-troscopic, and modeling studies in several chemoreceptors (Chervitz and Falke, 1995; Chervitz et al., 1995; Kim et al., 1999; Lee et al., 1994; Ottemann et al., 1999), although a high-resolution structure is not yet available for the transmembrane region. Other studies used disulfide mapping to determine the helix–helix packing geometry within the cytoplasmic domain of the full-length, membrane-bound receptor, yielding an extended 4-helix bundle model for this domain (Bass and Falke, 1998; Bass et al., 1999; Butler and Falke, 1998; Danielson et al., 1997; Hughson et al., 1997; Winston et al., 2005), which has been confirmed by high-resolution X-ray crystal structures of cytoplasmic domain fragments (Kim et al., 1999; Park et al., 2006). Together, these and other studies of bacterial chemoreceptors have demon-strated the accuracy of disulfide mapping in the determination of various packing interactions within full-length proteins in their native environment, including both transmembrane and aqueous domains.

Disulfide Mapping: General Considerations

The following disulfide mapping protocol was developed for the aspar-tate receptor bacterial chemotaxis (Bass and Falke, 1998; Bass et al., 1999; Butler and Falke, 1998; Chervitz and Falke, 1995; Chervitz et al., 1995; Danielson et al., 1997; Hughson et al., 1997; Lee et al., 1994; Pakula and Simon, 1992; Winston et al., 2005), but is generalizable to other membrane and soluble proteins. For proteins containing intrinsic Cys residues, the first step is generally to remove them by substitution with Ala or Ser, especially in the domain targeted for disulfide mapping, using the procedure described in Chapter 1 of this volume (Bass et al., 2007). Intrinsic Cys residues buried in other domains will generally not interfere and need not be removed.

One key to successful implementation of the disulfide mapping appro-ach is the choice of suitable Cys pairs for proximity testing. The pairs chosen are defined by the nature of the target protein system. The simplest application of disulfide mapping is the analysis of a symmetric packing interface between two identical subunits, since the introduction of a single Cys in the primary structure of the subunit generates a pair of symmetric Cys residues at the interface, as illustrated in early disulfide mapping studies of the aspartate receptor (Bass and Falke, 1998; Bass *et al.*, 1999; Butler and Falke, 1998; Chervitz and Falke, 1995; Chervitz *et al.*, 1995; Danielson *et al.*, 1997; Hughson *et al.*, 1997; Lee *et al.*, 1994; Pakula and Simon, 1992; Winston *et al.*, 2005). In this special case, a simple, model-independent scanning approach is sufficient to map out contacts between the two subunits. To carry out such a dimer analysis, single Cys residues are introduced at consecutive positions in the region being analyzed, using the site-directed mutagenesis procedures described in Chapter 1 of this volume (Bass *et al.*, 2007). Each single-Cys mutant is expressed and purified, yielding a homo-dimer possessing two Cys residues at symmetric positions. The resulting library of Cys mutants is tested for perturbations due to individual Cys substitutions using stability and/or activity assays developed specifically for the target protein, and perturbed mutants are discarded. The surviving pool of active, stable mutants is subjected to the disulfide mapping procedure.

More generally, however, a model-independent scanning analysis of 3° or 4° structure within a protein or complex is not practical, since all possible combinations between different pair of positions would need to be tested for proximity, yielding a problem of enormous combinatorial complexity. Instead, a model-dependent approach is much more efficient in such applications, utilizing disulfide mapping to test specific predictions of the model by examining carefully chosen pairs of engineered Cys residues for proximity. To choose appropriate pairs of positions for proximity testing, the model structure is examined to find candidate adjacent positions. If an atomic model is available, each selected pair should possess a β-carbon to β-carbon distance in the range 4 to 8 Å (Careaga and Falke, 1992a,b). Depending on the level of confidence regarding the model, it is often advisable to test other registers of adjacent structural elements by examining additional pairs that would be closer in an alternate register. Finally, it is important to include negative controls, including pairs of Cys positions predicted by the model to be too distant for disulfide bond formation, with a β-carbon to β-carbon distance exceeding 12 Å (although this distance should be larger if the protein is unusually dynamic) (Careaga and Falke, 1992). Before disulfide mapping studies are carried out, the effects of individual Cys pairs on protein stability and/or function are determined by appropriate assays, and perturbed proteins are discarded.

Two different types of chemistry can be used to carry out disulfide mapping reactions, depending on the specific protein system. For transmembrane domains, in which pairs of Cys residues are buried deep in the hydrocarbon core of a phospholipid bilayer, the use of molecular iodine (I_2) as the oxidation agent often yields the most efficient disulfide production (Chervitz and Falke, 1995; Chervitz et al., 1995; Pakula and Simon, 1992). In this chemistry, the apolar I_2 molecule is added to initiate the reaction, whereupon it partitions selectively into the apolar membrane core and efficiently oxidizes pairs of Cys residues to form disulfide bonds. For aqueous proteins or domains, the use of ambient molecular oxygen (O_2) as the oxidation agent and Cu(II) (1,10-phenanthroline)$_3$ as a redox catalyst is generally preferred (Bass and Falke, 1998; Bass et al., 1999; Butler and Falke, 1998; Chervitz and Falke, 1995; Chervitz et al., 1995; Danielson et al., 1997; Falke and Koshland, 1987; Falke et al., 1988; Hughson et al., 1997; Lee et al., 1994; Winston et al., 2005). Here, the catalyst is added to initiate the reaction, whereupon it catalyzes the production of diffusible reactive intermediates that can penetrate protein interiors and drive disulfide bond formation (Careaga and Falke, 1992). Since the catalyst is charged it does not partition well into membranes, and although it can catalyze disulfide formation near the aqueous surface of a bilayer, it is generally not effective deeper in the membrane.

In both the I_2 and O_2/Cu(II) (1,10-phenanthroline)$_3$ chemistries, the detailed reaction mechanism appears to proceed through a series of free radical and anionic intermediates, generating a disulfide bond as one product and I^- or HO^- anions as the other product, respectively (Careaga and Falke, 1992). Additional sulfhydryl oxidation reactions that produce sulfinic and sulfenic acids compete with disulfide formation; thus, only the most proximal Cys pairs with the highest disulfide formation rates yield final extents of disulfide product approaching 100% (Careaga and Falke, 1992). Since the goal is to compare the initial rates of the disulfide formation reactions between different Cys pairs, it may be necessary to adjust the rate of the oxidation reaction to yield a time course that is more convenient for initial rate measurements. One advantage of the O_2/Cu(II) (1,10-phenanthroline)$_3$ chemistry is the ease with which it can be tuned to generate different rates in a variety of protein systems and environments. Disulfide mapping using O_2/Cu(II) (1,10-phenanthroline)$_3$ chemistry has even been successfully applied in living cells (Hughson et al., 1997), although this application is beyond the scope of the current chapter.

The final step in a full disulfide mapping analysis is carried out after oxidation reactions have revealed a set of Cys pairs that rapidly form disulfide bonds, and thus potentially identify proximities in the protein structure. To rigorously identify Cys pairs that are adjacent in the native, protein fold, rather than formed by rapid fluctuations away from the equilibrium structure, the effects of these disulfides on protein activity are tested. Cys pairs that rapidly form disulfide bonds retaining protein activity provide strong evidence for proximity between specific positions in the equilibrium structure. Moreover, analysis of functional effects can reveal mechanistically important disulfide bonds that covalently, but reversibly, trap the protein in specific signaling or other functional states, such as the function-retaining, lock-on, and lock-off disulfide bonds identified in disulfide mapping studies the aspartate receptor (Bass and Falke, 1998; Bass *et al.*, 1999; Butler and Falke, 1998; Chervitz and Falke, 1995; Chervitz *et al.*, 1995; Danielson *et al.*, 1997; Winston *et al.*, 2005). Such disulfides place strong structural constraints on the mechanisms of conformational changes, including transitions between different functional states such as on–off switching.

Disulfide Mapping: Detailed Procedure

1. Each di-Cys protein of interest is diluted to a final concentration of 2 μM in disulfide reaction buffer (buffered with either 20 mM NaH$_2$PO4, pH to 7.0 with NaOH or 20 mM Tris base, pH to 7.5 with HCl, and also containing 10% (v/v) glycerol, 100 mM NaCl, and 5 mM EDTA), yielding a total volume of 22.5 μl per sample in a microfuge tube. Published studies of the aspartate receptor and the galactose binding protein have successfully utilized the phosphate buffer at pH 7.0, while current studies of CheA use Tris buffer at pH 7.5 to increase the basicity and thus the disulfide formation rate (Gloor and Falke, unpublished). Ideally, both buffer systems should be tested on a subset of mutants to see whether one yields more convenient disulfide formation time courses or better data. In all cases, care should be taken that protein sulfhydryls are protected from oxidation during purification of the protein stock, and that the concentrations of residual reducing agents, such as DTT, are negligible (less than 5 μM total sulfhydryl) during the following disulfide reaction. Note that protein concentrations higher than low micromolar may lead to undesired intermolecular collisions and disulfide formation, particularly for small soluble proteins that diffuse rapidly. Before proceeding with the disulfide reaction, each sample is carefully mixed to ensure a homogenous solution and spun for 10 sec in a microfuge to bring the entire volume to the bottom of the tube.

2. The samples are moved to a 25° temperature bath, allowed to equilibrate for 1 min, then the disulfide reaction is initiated by adding 2.5 μl of the appropriate initiator stock with careful mixing before quickly returning to the 25° bath. The initiator stock varies with the type of oxidation chemistry chosen for the target system. For structural studies of aqueous proteins or domains, including regions near a membrane–water interface, the O$_2$/Cu(II) (1,10-phenanthroline)$_3$ chemistry is preferred because its disulfide

formation rate can be easily adjusted as needed for different systems. In this case, ambient O_2 (approx. 200 μM) within the sample is the oxidant and the initiator stock is 5 mM catalyst Cu(II) (1,10-phenanthroline)$_3$, created by addition of metal ion (100 μl of 0.5 M CuSO$_4$ in water) and chelator (175 μl of 1 M 1,10-phenanthroline in absolute EtOH) to water (9.725 ml). Dilution of the initiator stock into the sample yields a final catalyst concentration of 0.5 mM Cu(II) (1,10-phenanthroline)$_3$. For studies of transmembrane domains, I_2 chemistry is preferred due to its greater efficiency of disulfide formation in the apolar membrane environment. In this case, the oxidant is I_2 and the initiator stock is 10 mM I_2 in EtOH, created by dissolving iodine crystals in absolute EtOH.

3. At each selected timepoint, 6 μl of the reaction is removed and quenched by addition to a microfuge tube containing 6 μl of quench solution (either 4× nonreducing Laemmli sample buffer for membrane samples to ensure complete solubilization, or 2× non-reducing Laemmli sample buffer for nonmembrane samples, and, in both cases, also containing 10 mM EDTA and 40 mM N-ethyl maleimide to quench the redox chemistry). Immediately after careful mixing, the tube is sealed and incubated at 95° for 2 min to ensure complete quenching. Typically, timepoints of 10, 30, and 300 sec are sufficient to contain the linear initial phase of the reaction timecourse.

4. For each time course, a zero point is generated by diluting the same di-Cys protein to a final concentration of 2 μM in disulfide reaction buffer, yielding a total volume of 9 μl in a microfuge tube. To this sample is added 1 μl of water or EtOH, depending on whether the O_2/Cu(II) (1,10-phenanthroline)$_3$ or I_2 chemistry is utilized, respectively, then 6 μl is removed and quenched as described in step 3. The purpose of this zero point is to determine the level of spontaneous disulfide formation that occurred prior to addition of initiator.

5. The zero and timepoint samples are pelleted in a microfuge to deposit each sample at the bottom of the tube. Then, these samples (10 μl each) are loaded side-by-side on a nonreducing Laemmli SDS-PAGE gel formulated so that the unreacted protein band exhibits maximal resolution from the disulfide-containing product band. The gel is stained with a linear dye such as Coomassie, and the stained bands are imaged by digital photography. Finally, the bands in each sample corresponding to the unreacted protein and the disulfide-containing product protein are quantitated by integration of their dye absorbances.

Note that the migration of the disulfide-containing product band can differ dramatically depending on the nature and location of the disulfide bond. If the disulfide bond is formed between two different polypeptide chains, then the disulfide-containing product band will migrate more slowly than the unreacted protein band, typically in the size range expected for the covalent complex (Falke *et al.*, 1988). If, however, the disulfide bond is formed via an intramolecular reaction between two Cys residues in the same polypeptide chain, then the disulfide-containing product band will migrate more rapidly than the unreacted band (Falke *et al.*, 1988). Moreover, subtle differences in migration are typically seen between different disulfide products of the same class (inter- or intra-chain) when the location of the disulfide bond is changed in the primary structure (Falke *et al.*, 1988). These complex effects on gel migration arise from the covalent crosslink imposed by the disulfide bond, which alters the average conformation and dynamics of the denatured polypeptide chain while it migrates through the gel matrix. If one is uncertain whether a given band is a true disulfide-containing product or a contaminant, disulfide-containing bands can be identified by rerunning the gel samples under reducing conditions.

6. To determine the extent of disulfide bond formation (E) at each timepoint, the integrated absorbances of the unreacted protein (U) and the disulfide-containing product (D) are used to calculate the ratio $E = D/(D + U)$. This ratio represents the fraction of total protein converted to disulfide-containing product.

7. Once the initial disulfide formation time courses have been determined for a range of positions, it may be necessary to adjust the reaction rate. It is useful to optimize the reaction rate so that virtually all the positions examined yield between 5 and 75% disulfide formation at a single timepoint, typically either 10 or 30 sec. This optimization ensures that as many Cys pairs as possible yield sufficient disulfide bond formation to be quantitated, while at the same time the reaction has not proceeded to completion and thus remains in or near its linear range. To decrease or increase the speed of the reaction, one can decrease or increase, respectively, the pH or temperature of the reaction. Alternatively, reducing the glycerol concentration will generally increase the reaction rate, since this protein stabilizer reduces thermal fluctuations, and therefore collision rates and disulfide formation rates (Butler and Falke, 1996). For the $O_2/Cu(II)$ (1,10-phenanthroline)$_3$ chemistry, the concentration of the redox catalyst can also be increased to speed the reaction or decreased to slow the reaction, respectively. Notably, increasing the catalyst to a concentration in excess of EDTA greatly speeds the reaction because the latter chelator no longer buffers the Cu(II) (see step 9). For the I_2 chemistry, the reaction rate is more difficult to adjust, but, in some cases, is moderately sensitive to the final I_2 concentration.

8. After optimization, the disulfide formation reaction is carried out in triplicate for each Cys pair at the chosen timepoint to enable comparison of means and standard deviations. Multiple positions can be tested in a set of simultaneous reactions, with the exact number depending on the timepoint

chosen for sampling the reaction and the preference of the investigator. In each set of reactions, two control reactions are carried out for a selected, well-characterized Cys pair that serves as an internal standard when comparing different sets. For each reaction in a set, a given di-Cys protein is diluted to 2 μM final concentration, as in step 1, except that the total volume is now 9 μl and the newly optimized reaction buffer is utilized. After mixing, spinning, and equilibration to the optimized temperature, 1 μl optimized initiator stock is added as in step 2, and the reaction is allowed to proceed. At the chosen timepoint, a 6 μl aliquot is removed and quenched as in step 3. Then, the quenched samples are analyzed by SDS-PAGE and digital imaging as described in steps 5 and 6, yielding a quantitative extent of disulfide bond formation for each reaction. By running multiple reactions consecutively, a set as large as 20 engineered Cys pairs can be analyzed in triplicate per day. Care should be taken that the control Cys pair included in all sets yields similar extents of disulfide formation in different sets run on the same or different days.

9. After the rapidly forming disulfide bonds have been identified, the effects of these disulfides on activity are tested. First, a stronger oxidation reaction is used to drive disulfide formation to completion by adjusting the pH, temperature, glycerol concentration, or the oxidation components. When the O_2/Cu(II) (1,10-phenanthroline)$_3$ chemistry is used, rapid reactions can often be conveniently driven to completion by decreasing the EDTA concentration of the disulfide reaction buffer (step 1, from 5.0 mM down to 0.8 mM), while increasing the final concentration of Cu(II) (1,10-phenanthroline)$_3$ in the reaction (step 2, from 0.5 mM up to 1.0 mM). These changes yield an excess of redox catalyst relative to EDTA (which, in the standard reaction, buffers most of the Cu(II) and slows the reaction rate for convenient measurement), thereby dramatically increasing the reaction rate. After the reaction is completed, EDTA is added to a final concentration of 2.0 mM to slow further oxidation, or to 10 mM if the activity assay does not require free metal ions such as Mg^{2+} or Ca^{2+} which should be added in excess of EDTA. Finally, the protein activity of this oxidized sample is measured in comparison to an equivalent reduced sample generated by treatment with 20 mM DTT rather than oxidation components for the same length of time.

Disulfide Mapping: Interpretation of Results

Typically, Cys residues that are in close proximity within the native equilibrium structure will exhibit the most rapid rates of disulfide formation, and their reactions will proceed to form nearly 100% disulfide product when driven to completion (Careaga and Falke, 1992). Such rapid disulfide reactions are fast enough to overwhelm the competing reactions that oxidize Cys sulfhydryls to oxyacids incapable of forming disulfide bonds, while slower disulfide reactions exhibit lower final extents of disulfide product even when the reaction is driven to completion, because competing oxidation reactions generate nondisulfide products (Careaga and Falke, 1992). Disulfide bond formation can occur only if, during the transition state of the reaction, the two Cys pairs can approach each other with a maximum distance of 4.6 Å between their α-carbons, and with angular constraints satisfied as well (Careaga and Falke, 1992). However, it has been observed that Cys residues much farther apart in the equilibrium structure can also collide and form a disulfide bond due to protein thermal fluctuations that produce long-range backbone motions. For example, even in the extremely stable galactose binding protein saturated with its ligand, collisions yielding disulfide formation are detected between Cys residues whose α-carbons are separated by 19.8 Å in the crystal structure (Careaga and Falke, 1992). In this study, proximal Cys pairs yielded higher disulfide formation rates than did distal pairs, but the detection of disulfide formation alone was not sufficient to indicate proximity.

To rigorously identify Cys pairs that are nearby in the equilibrium structure, two general approaches are useful. First, by measuring the disulfide formation rates of a large number of Cys pairs under identical conditions, a pattern of rates will often emerge that can be correlated with structural elements to identify multiple proximities between two contact faces. To detect such spatial patterns, it is helpful to classify the disulfide formation rates into three groups—rapid, intermediate, and slow—then to search for structural models in which the distances between the corresponding Cys pairs are short, intermediate, and long. The success of this approach typically depends on the observation of multiple Cys pairs that exhibit high disulfide formation rates and are correlated in space at the same packing interfaces (Bass and Falke, 1999). Second, by measuring the functional effects of disulfide bonds on protein activity, a subset of rapidly formed disulfides that retain normal function, or that trap the protein in a specific native signaling or functional state with a measurable activity, may be identified (Bass and Falke, 1998; Bass et al., 1999; Butler and Falke, 1998; Chervitz and Falke, 1995; Chervitz et al., 1995; Danielson et al., 1997; Winston et al., 2005). Such functional disulfide bonds provide the strongest evidence that the disulfide-crosslinked positions are indeed adjacent in the native, active structure. Moreover, functional disulfide bonds can provide important constraints that are useful in elucidating the molecular mechanisms and conformational changes of proteins that switch among different functional states.

Both the spatial pattern analysis and functional analysis of disulfide bond reactions have been successfully employed in the full-length, membrane-bound aspartate receptor to identify contact surfaces in the periplasmic, transmembrane, and cytoplasmic domains (Bass and Falke, 1998; Bass *et al.*, 1999; Butler and Falke, 1998; Chervitz and Falke, 1995; Chervitz *et al.*, 1995; Danielson *et al.*, 1997; Falke and Koshland, 1987; Falke *et al.*, 1988; Hughson *et al.*, 1997; Lee *et al.*, 1994; Pakula and Simon, 1992; Winston *et al.*, 2005). Moreover, functional analysis of disulfide bonds has played an important role in defining the current mechanistic models for conformational transmembrane and adaptation signals in the receptor (Bass and Falke, 1998; Bass *et al.*, 1999; Butler and Falke, 1998; Chervitz and Falke, 1995; Chervitz *et al.*, 1995; Danielson *et al.*, 1997; Falke and Hazelbauer, 2001; Winston *et al.*, 2005). Together, these studies have identified 22 functional disulfide bonds within the receptor that can be divided into three categories: (1) function-retaining disulfides that retain normal transmembrane signaling and kinase regulation, (2) lock-on disulfides that trap the receptor in its native on-state, and (3) lock-off disulfides that trap the receptor in its native off-state (where the on- and off-states are characterized by multiple, asymmetric activity assays). The information provided by these functional disulfides has directly confirmed contacts between specific pairs of positions in the native 3° and 4° structure of the active receptor in its complex with the kinase CheA. In addition, careful analysis of the functional effects has revealed regions of the receptor that are static during signal transduction, as well as regions that undergo conformational transitions. Furthermore, comparison of the locations of the lock-on and lock-off disulfides in the receptor structure has revealed the piston-type displacement of a specific membrane-spanning helix that is responsible for transmembrane signaling (Chervitz and Falke, 1995, 1996; Chervitz *et al.*, 1995). In short, the disulfide mapping method yields not only 3° and 4° contacts, but also can define conformational transitions by covalently trapping different functional states identified by their signature activities.

Disulfide Trapping of Thermal Backbone and Domain Motions

Overview of Disulfide Trapping

Proteins are highly dynamic structures, but often it is difficult to detect the long-range backbone motions that occur in their native solution and membrane environments. Disulfide trapping uses the formation of a disulfide bond during a collision between two Cys residues to detect and covalently trap the thermal backbone fluctuation that generated the collision (Careaga and Falke, 1992; Falke and Koshland, 1987). Since these collisions are stored and accumulate with time, even rare backbone fluctuations difficult to detect by other methods can be analyzed. Moreover, by examining the disulfide formation patterns and rates of multiple Cys pairs, information can be obtained about the trajectories and frequencies of the observed backbone fluctuations. The approach is most powerful when applied to a protein of known high-resolution structure, thereby facilitating interpretation of the spatial implications of the results, particularly with regard to amplitude and trajectory measurements.

The initial disulfide trapping studies were carried out in the transmembrane aspartate receptor and the aqueous galactose binding protein (Bass and Falke, 1999; Butler and Falke, 1996; Careaga and Falke, 1992; Careaga et al., 1995; Falke and Koshland, 1987; Falke et al., 1988). Like chemical reactivity scanning and disulfide mapping, the disulfide trapping approach can be applied to full-length proteins in their native environment. Dramatic backbone motions were detected in both proteins, including relative sliding motions of adjacent α-helices with amplitudes as large as 15 Å, and hinge-twisting motions yielding rotations of adjacent domains up to 36°. The estimated frequencies of the detected collisions ranged from 10 to 104 per sec. Since the engineered Cys pairs used to trap these motions were introduced at surface sites and had little or no detectable effect on protein stability or activity, the motions appear to be intrinsic features of the native proteins in their native membrane or aqueous environments, respectively.

Disulfide Trapping: General Considerations

The following disulfide trapping procedures were developed for the aspartate receptor and galactose binding protein of bacterial chemotaxis (Bass and Falke, 1999; Butler and Falke, 1996; Careaga and Falke, 1992; Careaga et al., 1995; Falke et al., 1988; Gloor and Falke, unpublished), but are generalizable to other membrane and soluble proteins. As noted for other types of cysteine and disulfide chemistry, the first step in applying disulfide trapping to a new protein system is to use methods described in Chapter 1 of this volume (Bass et al., 2007) to remove intrinsic Cys residues by substitution with Ala or Ser, especially in the domain targeted for disulfide trapping. Intrinsic Cys residues buried in other domains will generally not interfere and need not be removed.

Successful implementation of the disulfide trapping approach requires careful selection of suitable pairs of positions for Cys substitution and subsequent collisional analysis. Since the most powerful application of disulfide trapping is to analyze backbone and domain motions in a protein of known high-resolution structure (Bass and Falke, 1999; Butler and Falke, 1996; Careaga and Falke, 1992; Careaga et al., 1995), the focus here is on such systems. Pairs are chosen at surface positions in the known structure to minimize perturbations of structure and dynamics and to maximize accessibility to the oxidation chemistry. In general, conserved surface residues are avoided because their replacement would perturb activity or stability. To map out the trajectory for a relative motion between two different structural elements, one Cys pair is created with the two Cys residues at adjacent positions, one on each element. Then, other Cys pairs are created keeping one Cys residue, the probe residue, at a fixed position while the other residue is moved to a range of other positions. This strategy enables analysis of the range of positions contacted by the probe residue on an adjacent structural element. After the resulting di-Cys library is created, expressed, and purified, the effects of individual Cys pairs on protein stability and/or activity are measured by appropriate assays, and perturbed proteins are discarded while the unperturbed proteins are subjected to disulfide trapping analysis.

The remaining steps in the disulfide trapping procedure are similar to those outlined for disulfide mapping. In principle, the same two types of chemistry can be used to carry out disulfide trapping reactions: I_2 oxidation for analysis of collisions within transmembrane domains and $O_2/Cu(II)$ (1,10-phenanthroline)$_3$ oxidation for analysis of collisions within aqueous domains and proteins. To date, only the latter chemistry has been tested in disulfide trapping studies, and the procedure has been applied only to soluble proteins and to the aqueous domains of membrane-imbedded proteins (Bass and Falke, 1999; Butler and Falke, 1996; Careaga and Falke, 1992; Careaga et al., 1995; Gloor and Falke, unpublished). The following procedure summarizes how the individual steps of the previously mentioned disulfide mapping procedure are modified for disulfide trapping using the $O_2/Cu(II)$ (1,10-phenanthroline)$_3$ chemistry.

Disulfide Trapping: Detailed Procedure

1. Each selected di-Cys protein of interest is diluted to a final concentration of 2 μM in disulfide reaction buffer, yielding a total volume of 22.5 μl per sample prepared in a microfuge tube, as indicated in step 1 of the previously mentioned disulfide mapping procedure.

2. The Cu(II) (1,10-phenanthroline)$_3$ initiator stock is added as indicated in step 2 of disulfide mapping.

3. At each selected timepoint, typically, 10, 30, and 300 sec, an aliquot of the reaction is removed and quenched as in step 3 of disulfide mapping.

4. For each time course, a zero point is generated as in step 4 of disulfide mapping.

5. The zero and timepoint samples are run side-by-side on a gel. The resulting stained bands are imaged and integrated as in step 5 of disulfide mapping. The bands corresponding to the unreacted protein and disulfide-containing product are identified by their relative migration on the gel,

where an intra-chain disulfide-containing product will typically migrate slightly faster than the unreacted protein. By contrast, an inter-chain disulfide-containing product will migrate slower than the unreacted protein, at about the same R_f expected for the MW of the covalent product (step 5).

6. The extent of disulfide bond formation (E) at each timepoint is calculated as described in step 6 of disulfide mapping.

7. If necessary, the reaction rate is adjusted for convenient measurement of disulfide reaction time courses, as described in step 7 of disulfide mapping.

8. After optimization, the disulfide formation time course is measured, at least in duplicate, for each Cys pair. Multiple positions can be tested in a set of simultaneous time courses, with the exact number depending on the timepoints chosen. In each set of time courses, one control time course is carried out for a selected, well-characterized Cys pair that serves as an internal standard when comparing different sets.

9. Each time course is best-fit via nonlinear least squares analysis to the following equation for a two-component reaction scheme that includes both disulfide formation and a competing oxidation reaction:

$$F_{ss}(t) = \frac{k_{ss}}{k_{ss} + k_2} = \left[1 - e^{-(k_{ss}+k_2)t}\right] \tag{1}$$

where $F_{ss}(t)$ is the fraction of the di-Cys protein converted to disulfide-containing product, k_{ss} is the rate constant for the disulfide formation reaction, and k_2 is the rate constant for the competing reaction. This analysis separates out the desired rate constant for the disulfide formation reaction (k_{ss}) and enables direct comparison of this rate constant for different Cys pairs reacted under the same oxidation conditions.

Disulfide Trapping: Interpretation of Results

Analysis of disulfide formation time courses for an array of Cys pairs differing in the distance between the probe and test positions determines both the amplitude and kinetics of the observed thermal backbone fluctuations. Each Cys pair that yields measurable disulfide bond formation detects a thermal collision produced by the observed fluctuation. These colliding Cys pairs are mapped on the surface of the known protein structure, revealing the spatial pattern of the detected collisions and defining both the amplitude and the trajectory of the fluctuation. If desired, the frequency of each collision can be estimated by determining the fraction of Cys pair collisions successfully converted to a disulfide, as described in detail elsewhere. Overall, the disulfide trapping method provides the most powerful approach currently available for analyzing the amplitudes, trajectories, and frequencies of rare, long-range backbone and domain motions present within many proteins in their native environment. Systematic application of the approach to a diverse array of proteins from different structural and functional families could significantly further the molecular understanding of protein thermal dynamics.

Acknowledgments

The authors gratefully acknowledge support by NIH R01 grant GM040731 to JJF.

References

Akabas, M. H., Stauffer, D. A., Xu, M., and Karlin, A. (1992). Acetylcholine receptor channel structure probed in cysteine-substitution mutants. *Science* **258**, 307–310.

Ames, P., Studdert, C. A., Reiser, R. H., and Parkinson, J. S. (2002). Collaborative signaling by mixed chemoreceptor teams in *Escherichia coli*. *Proc. Natl. Acad. Sci. USA* **99**, 7060–7065.

Baker, M. D., Wolanin, P. M., and Stock, J. B. (2006). Signal transduction in bacterial chemotaxis. *Bioessays* **28**, 9–22.

Bass, R. B., Coleman, M. D., and Falke, J. J. (1999). Signaling domain of the aspartate receptor is a helical hairpin with a localized kinase docking surface: Cysteine and disulfide scanning studies. *Biochemistry* **38**, 9317–9327.

Bass, R. B., and Falke, J. J. (1998). Detection of a conserved alpha-helix in the kinase-docking region of the aspartate receptor by cysteine and disulfide scanning. *J. Biol. Chem.* **273**, 25006–25014.

Bass, R. B., and Falke, J. J. (1999). The aspartate receptor cytoplasmic domain: *In situ* chemical analysis of structure, mechanism, and dynamics. *Struct. Fold Des.* **7**, 829–840.

Bass, R. B., Miller, A. S., Gloor, S. L., and Falke, J. J. (2007) The PICM chemical scanning method for identifying domain–domain and protein–protein interfaces: Applications to the core signaling complex of bacterial chemotaxis. *Methods Enzymol.* **1**, 3–24.

Butler, S. L., and Falke, J. J. (1996). Effects of protein stabilizing agents on thermal backbone motions: A disulfide trapping study. *Biochemistry* **35**, 10595–10600.

Butler, S. L., and Falke, J. J. (1998). Cysteine and disulfide scanning reveals two amphiphilic helices in the linker region of the aspartate chemoreceptor. *Biochemistry* **37**, 10746–10756.

Careaga, C. L., and Falke, J. J. (1992). Structure and dynamics of *Escherichia coli* chemosensory receptors. Engineered sulfhydryl studies. *Biophys. J.* **62**, 209–216; discussion 217–219.

Careaga, C. L., and Falke, J. J. (1995). Thermal motions of surface alpha-helices in the D-galactose chemosensory receptor. Detection by disulfide trapping. *J. Mol. Biol.* **226**, 1219–1235.

Careaga, C. L., Sutherland, J., Sabeti, J., and Falke, J. J. (1995). Large amplitude twisting motions of an interdomain hinge: A disulfide trapping study of the galactose–glucose binding protein. *Biochemistry* **34**, 3048–3055.

Chervitz, S. A., and Falke, J. J. (1995). Lock on / off disulfides identify the transmembrane signaling helix of the aspartate receptor. *J Biol. Chem.* **270**, 24043–24053.

Chervitz, S. A., and Falke, J. J. (1996). Molecular mechanism of transmembrane signaling by the aspartate receptor: A model. *Proc. Natl. Acad. Sci. USA* **93**, 2545–2550.

Chervitz, S. A., Lin, C. M., and Falke, J. J. (1995). Transmembrane signaling by the aspartate receptor: Engineered disulfides reveal static regions of the subunit interface. *Biochemistry* **34,** 9722–9733.

Danielson, M. A., Bass, R. B., and Falke, J. J. (1997). Cysteine and disulfide scanning reveals a regulatory alpha-helix in the cytoplasmic domain of the aspartate receptor. *J. Biol. Chem.* **272,** 32878–32888.

Draheim, R. R., Bormans, A. F., Lai, R. Z., and Manson, M. D. (2005). Tryptophan residues flanking the second transmembrane helix (TM2) set the signaling state of the Tar chemoreceptor. *Biochemistry* **44,** 1268–1277.

Draheim, R. R., Bormans, A. F., Lai, R. Z., and Manson, M. D. (2006). Tuning a bacterial chemoreceptor with protein-membrane interactions. *Biochemistry* **45,** 14655–14664.

Falke, J. J., Bass, R. B., Butler, S. L., Chervitz, S. A., and Danielson, M. A. (1987). The two-component signaling pathway of bacterial chemotaxis: A molecular view of signal transduction by receptors, kinases, and adaptation enzymes. *Annu. Rev. Cell. Dev. Biol.* **13,** 457–512.

Falke, J. J., Dernburg, A. F., Sternberg, D. A., Zalkin, N., Milligan, D. L., and Koshland, D. E., Jr. (1988). Structure of a bacterial sensory receptor. A site-directed sulfhydryl study. *J. Biol. Chem.* **263,** 14850–14858.

Falke, J. J., and Hazelbauer, G. L. (2001). Transmembrane signaling in bacterial chemoreceptors. *Trends Biochem. Sci.* **26,** 257–265.

Falke, J. J., and Kim, S. H. (2000). Structure of a conserved receptor domain that regulates kinase activity: The cytoplasmic domain of bacterial taxis receptors. *Curr. Opin. Struct. Biol.* **10,** 462–469.

Falke, J. J., and Koshland, D. E., Jr. (1987). Global flexibility in a sensory receptor: A site-directed cross-linking approach. *Science* **237,** 1596–1600.

Falke, J. J., Sternberg, D. E., and Koshland, D. E., Jr. (1986). Site-directed sulfhydryl chemistry and spectroscopy: Applications in the aspartate receptor system. *Biophys. J.* **49,** 20a.

Hughson, A. G., and Hazelbauer, G. L. (1996). Detecting the conformational change of transmembrane signaling in a bacterial chemoreceptor by measuring effects on disulfide cross-linking *in vivo. Proc. Natl. Acad. Sci. USA* **93,** 11546–11551.

Hughson, A. G., Lee, G. F., and Hazelbauer, G. L. (1997). Analysis of protein structure in intact cells: Crosslinking *in vivo* between introduced cysteines in the transmembrane domain of a bacterial chemoreceptor. *Protein Sci.* **6,** 315–322.

Hulko, M., Berndt, F., Gruber, M., Linder, J. U., Truffault, V., Schultz, A., Martin, J., Schultz, J. E., Lupas, A. N., and Coles, M. (2006). The HAMP domain structure implies helix rotation in transmembrane signaling. *Cell* **126,** 929–940.

Kim, K. K., Yokota, H., and Kim, S. H. (1999). Four-helical-bundle structure of the cytoplasmic domain of a serine chemotaxis receptor. *Nature* **400,** 787–792.

Lai, W. C., Beel, B. D., and Hazelbauer, G. L. (2006). Adaptational modification and ligand occupancy have opposite effects on positioning of the transmembrane signaling helix of a chemoreceptor. *Mol. Microbiol.* **61,** 1081–1090.

Lee, G. F., Burrows, G. G., Lebert, M. R., Dutton, D. P., and Hazelbauer, G. L. (1994). Deducing the organization of a transmembrane domain by disulfide cross-linking. The bacterial chemoreceptor Trg. *J. Biol. Chem.* **269,** 29920–29927.

Milburn, M. V., Prive, G. G., Milligan, D. L., Scott, W. G., Yeh, J., Jancarik, J., Koshland, D. E., Jr., and Kim, S. H. (1991). Three-dimensional structures of the ligand-binding domain of the bacterial aspartate receptor with and without a ligand. *Science* **254,** 1342–1347.

Miller, A. S., and Falke, J. J. (2004). Side chains at the membrane–water interface modulate the signaling state of a transmembrane receptor. *Biochemistry* **43,** 1763–1770.

Miller, A. S., Kohout, S. C., Gilman, K. A., and Falke, J. J. (2006). CheA kinase of bacterial chemotaxis: Chemical mapping of four essential docking sites. *Biochemistry* **45,** 8699–8711.

Ottemann, K. M., Xiao, W., Shin, Y. K., and Koshland, D. E., Jr. (1999). A piston model for transmembrane signaling of the aspartate receptor [see comments]. *Science* **285,** 1751–1754.

Pakula, A. A., and Simon, M. I. (1992). Determination of transmembrane protein structure by disulfide cross-linking: The *Escherichia coli* Tar receptor. *Proc. Natl. Acad. Sci. USA* **89,** 4144–4148.

Park, S. Y., Borbat, P. P., Gonzalez-Bonet, G., Bhatnagar, J., Pollard, A. M., Freed, J. H., Bilwes, A. M., and Crane, B. R. (2006). Reconstruction of the chemotaxis receptor-kinase assembly. *Nat. Struct. Mol. Biol.* **13,** 400–407.

Sahin-Toth, M., and Kaback, H. R. (1993). Cysteine scanning mutagenesis of putative transmembrane helices IX and X in the lactose permease of *Escherichia coli. Protein Sci.* **2,** 1024–1033.

Starrett, D. J., and Falke, J. J. (2005). Adaptation mechanism of the aspartate receptor: Electrostatics of the adaptation subdomain play a key role in modulating kinase activity. *Biochemistry* **44,** 1550–1560.

Studdert, C. A., and Parkinson, J. S. (2004). Crosslinking snapshots of bacterial chemoreceptor squads. *Proc. Natl. Acad. Sci. USA* **101,** 2117–2122.

Todd, A. P., Cong, J., Levinthal, F., Levinthal, C., and Hubbell, W. L. (1989). Site-directed mutagenesis of colicin E1 provides specific attachment sites for spin labels whose spectra are sensitive to local conformation. *Proteins* **6,** 294–305.

van Iwaarden, P. R., Pastore, J. C., Konings, W. N., and Kaback, H. R. (1991). Construction of a functional lactose permease devoid of cysteine residues. *Biochemistry* **30,** 9595–9600.

Vyas, N. K., Vyas, M. N., and Quiocho, F. A. (1988). Sugar and signal-transducer binding sites of the *Escherichia coli* galactose chemoreceptor protein. *Science* **242,** 1290–1295.

Winston, S. E., Mehan, R., and Falke, J. J. (2005). Evidence that the adaptation region of the aspartate receptor is a dynamic four-helix bundle: Cysteine and disulfide scanning studies. *Biochemistry* **44,** 12655–12666.

Wu, J., Voss, J., Hubbell, W. L., and Kaback, H. R. (1996). Site-directed spin labeling and chemical crosslinking demonstrate that helix V is close to helices VII and VIII in the lactose permease of *Escherichia coli. Proc. Natl. Acad. Sci. USA* **93,** 10123–10127.

[3] Measuring Distances by Pulsed Dipolar ESR Spectroscopy: Spin-Labeled Histidine Kinases

By PETER P. BORBAT and JACK H. FREED

Abstract

Applications of dipolar ESR spectroscopy to structural biology are rapidly expanding, and it has become a useful method that is aimed at resolving protein structure and functional mechanisms. The method of pulsed dipolar ESR spectroscopy (PDS) is outlined in the first half of the chapter, and it illustrates the simplicity and potential of this developing technology with applications to various biological systems. A more detailed description is presented of the implementation of PDS to reconstruct the ternary structure of a large dimeric protein complex from *Thermotoga maritima*, formed by the histidine kinase CheA and the coupling protein CheW. This protein complex is a building block of an extensive array composed of coupled supramolecular structures assembled from CheA/ CheW proteins and transmembrane signaling chemoreceptors, which make up a sensor that is key to controlling the motility in bacterial chemotaxis. The reconstruction of the CheA/CheW complex has employed several techniques, including X-ray crystallography and pulsed ESR. Emphasis is on the role of PDS, which is part of a larger effort to reconstruct the entire signaling complex, including chemoreceptor, by means of PDS structural mapping. In order to precisely establish the mode of coupling of CheW to CheA and to globally map the complex, approximately 70 distances have already been determined and processed into molecular coordinates by readily available methods of distance geometry constraints.

Introduction

An understanding of the intricate machinery of biology depends, among other things, on the knowledge of the structure and internal organization of biomolecules, cells, and tissues. The primary sources of structure at atomic resolution are, of course, X-ray crystallography and nuclear magnetic resonance (NMR), even though they are laborious and require special conditions, (as is also the case with most of the other methods). A number of other methods such as electron microscopy (EM) provide information on a coarser scale. These techniques can lead to useful insights into molecular structure, particularly when study by crystallography or NMR fails or is inapplicable. Many biomolecules are not amenable to study by NMR or crystallography for

METHODS IN ENZYMOLOGY, VOL. 423
0076-6879/07 $35.00
DOI: 10.1016/S0076-6879(07)23003-4

reasons such as insufficient quantities, inability to grow diffraction-quality crystals, large molecular weight, poor solubility, or lack of stability. The labor involved also limits the throughput. Currently, determining the structure of a relatively small membrane protein is a challenge for both NMR and crystal-lography; therefore, less precise methods are widely applied to gain insight into the structure and functional mechanisms. Among them, to name a few, are FRET, chemical cross-linking, ESR nitroxide scan, and cryo-EM.

Since the late 1990s, applications of both pulsed and continuous-wave (cw) Electron Spin Resonance (ESR) techniques to structure determina-tion have grown (Banham et al., 2006; Bennati et al., 2005; Biglino et al., 2006; Borbat et al., 2002, 2004, 2006; Borovykh et al., 2006; Cai et al., 2006; Denysenkov et al., 2006; Dzikovski et al., 2004; Fafarman et al., 2007; Fu et al., 2003; Hilger et al., 2005; Jeschke et al., 2004c; Milov et al., 1999, 2003a, 2005; Ottemann et al., 1999; Park et al., 2006; Schiemann et al., 2004; Xiao et al., 2001; Xu et al., 2006; Zhou et al., 2005). This has followed the development of the site-directed spin-labeling (SDSL) methodology (Altenbach et al., 1989, 1990; Cornish et al., 1994; Farahbakhsh et al., 1992; Hubbell and Altenbach, 1994), wherein nitroxide labels are intro-duced at the desired location in proteins, as well as efforts by leading research groups to develop, perfect, and disseminate the modern ESR techniques. The application of ESR methods has also benefited from the commercialization of pulse ESR instrumentation.

The nitroxide labels typically serve as the reporter groups, providing insights about their environments such as their polarity and their solvent, and oxygen accessibility. Most important in the context of this chapter is measurement of distances between the spin labels. The SDSL-based ESR distance measurement techniques resemble FRET, but have several impor-tant advantages, such as the relatively small label size and the less perturbing nature of the nitroxide side chains (cf. Fig. 1). Furthermore, the labels are relatively easy to introduce and they provide reasonably accurate distance constraints. The notable virtues of ESR-based methods compared to X-ray and NMR methods are that the former require only tiny amounts (nano- to picomole; Klug et al., 2005) of proteins (or other biomolecules), and they can be studied in a variety of environments, such as dilute solutions, micelles, lipid vesicles, native membranes, and supported lipid bilayers. There is no need to grow crystals or be concerned with long-term protein stability at high concentrations. Large biomolecules or complexes that are beyond the range of NMR or X-ray methods are not a major limitation; even unstable or transient biomolecules can be captured and studied.[1]

---

[1] Although solid state NMR, in particular, enhanced by DNP, should also be appreciated in this context.

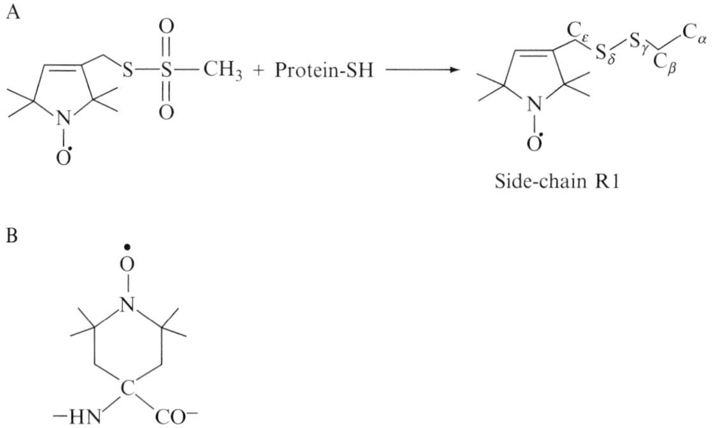

A

Side-chain R1

B

FIG. 1. (A) Protein spin labeling with MTSSL; (B) TOAC spin label.

SDSL combined with cw ESR has been routinely used for nitroxide scans (Altenbach *et al.*, 1989; Crane *et al.*, 2005; Cuello *et al.*, 2004; Dong *et al.*, 2005; Hubbell and Altenbach, 1994), providing insights into the structure and functional mechanisms of water-soluble and membrane proteins and their complexes. Double-labeling combined with pulsed ESR is currently able to readily deliver accurate long-distance constraints in a distance range of 10 to 80 Å. Such constraints may then be used to orient and dock proteins, yielding useful insights into the structure of a protein or a protein complex. They can also aid in refinement of NMR data. We refer to this emerging methodology as "pulsed dipolar (ESR) spectroscopy," or PDS for short. It is the subject of this chapter.

Another notable advantage of ESR is its ability to deal with membrane proteins in their natural environments and to accommodate large protein or protein–RNA complexes composed of several proteins or RNAs. Long-distance constraints hold promise in oligonucleotide study, where they could be used in conjunction with NOE constraints and RDC to refine the structure by reducing the error that accumulates from structure determination based on a large number of short-range constraints (Borbat *et al.*, 2004).

One should, of course, note that FRET provides distances over a comparable range. Its very high sensitivity, access to longer distances, and ability to operate at biological temperatures makes it a very potent tool, but PDS has its distinct virtues. It has now become routine to express, purify, and spin-label dozens of mutants for nitroxide scan (Crane *et al.*, 2005; Cuello *et al.*, 2004; Dong *et al.*, 2005) or to produce and label a set of cysteine double-mutants for distance measurements. The distance between nitroxides is more accurately determined than between chromophores, since it is directly obtained from a simple frequency measurement, and there are no uncertainties in κ^2 as there are in FRET. The reporter group, which is often a methanethiosulfonate spin label (MTSSL), in most cases, introduces only a small (if any) perturbation to the protein structure and functions. Since the nitroxide side-chains are smaller in size than most fluorescent labels, the uncertainty of their positions relative to the backbone is less. A drawback of PDS, as well as of FRET, is that a limited number of constraints, which are themselves the distances between the reporter groups rather than the backbone C_α carbons, may provide only limited insights into the structure. However, the detailed 3D structure is not always required to elucidate the functional mechanism. (Just the fact that the proteins are interacting or how they interact may be all that is sought in many cases.) But the fact that the distances are measured between the reporter groups does lead to a challenge in translating them into distances between the C_α carbons at the labeled sites. Modeling efforts to overcome this are in early stages of development (Bowers *et al.*, 2000); however, future developments of PDS and software tools may improve this situation.

Dipolar ESR Spectroscopy

Background

Both cw and pulsed ESR have been extensively applied to biological problems in the context of molecular dynamics (Borbat *et al.*, 2001; Columbus and Hubbell, 2002; Fanucci and Cafiso, 2006; Freed, 2000) and are now increasingly applied to distance measurements. The method of distance measurements by pulsed double electron–electron resonance (DEER, also known as PELDOR) (Larsen and Singel, 1993; Milov *et al.*, 1981, 1998; Pfannebecker *et al.*, 1996) was introduced more than two decades ago to circumvent multiple problems met in efforts to isolate weak electron–electron dipolar couplings from electron–spin–echo decays, which are usually dominated by relaxation and nuclear modulation effects (Raitsimring and Salikhov, 1985; Salikhov *et al.*, 1981). Since then, several other pulsed methods of distance measurements were introduced (Borbat and Freed, 1999, 2000; Jeschke *et al.*, 2000; Kulik *et al.*, 2001, 2002; Kurshev *et al.*, 1989; Raitsimring *et al.*, 1992); most notable is double-quantum coherence (DQC ESR, or DQC for short) (Borbat and Freed, 1999, 2000). Applications of DEER and DQC to structural problems in biology have rapidly grown in number and scope in the last few years (Borbat *et al.*, 2006; Cai *et al.*, 2006; Fafarman *et al.*, 2007; Fajer, 2005; Fanucci and Cafiso, 2006;

Jeschke *et al.*, 2004c; Milov *et al.*, 2000a, 2001; Park *et al.*, 2006; Schiemann *et al.*, 2004; Steinhoff, 2004). Whereas there are several reviews outlining ESR methods tailored to distance measurements (Berliner *et al.*, 2000; Borbat and Freed, 2000; Columbus and Hubbell, 2002; Dzuba, 2005; Fanucci and Cafiso, 2006; Freed, 2000; Jeschke, 2002; Jeschke and Spiess, 2006; Milov *et al.*, 1998; Prisner *et al.*, 2001), we have made this chapter self-contained, both providing background and emphasizing the latest developments, with a focus on illustrating the methodology through examples taken from our laboratory.

Distance Measurements by ESR

The ESR distance measurements described in this chapter are conducted in low-temperature frozen solutions with the use of nitroxide spin labels. They are based on determining the magnitude of the static dipole–dipole couplings between the spins of unpaired electrons localized on p-π orbitals of the NO groups of the nitroxides. The magnetic moments $\mathbf{m}_{1,2}$ of two electron spins 1 and 2, separated by the distance $r = |\mathbf{R}_{12}|$, interact through space via the electron spin dipole–dipole interaction

$$H_{dd} = \frac{\mathbf{m}_1 \cdot \mathbf{m}_2}{r^3} - \frac{3(\mathbf{m}_1 \cdot \mathbf{R}_{12})(\mathbf{m}_2 \cdot \mathbf{R}_{12})}{r^5} \qquad (1)$$

The electron spin magnetic moment $\mathbf{m}_i$ is given by:

$$\mathbf{m}_i = \gamma_e \hbar \mathbf{S}_i \qquad (2)$$

Where $\mathbf{S}_i$ is the electron spin operator for the ith spin, γ_e is the gyromagnetic ratio of an electron spin, and $\hbar$ is Planck's constant divided by 2π. Eq. (1) may be rewritten as:

$$H_{dd}/\hbar = \frac{\gamma_e^2 \hbar}{r^3}(3\cos^2\theta - 1)[S_{1z}S_{2z} - \frac{1}{4}(S_1^+ S_2^- + S_1^- S_2^+)] \qquad (3)$$

which is valid in high magnetic fields, where the nonsecular terms (not shown) are unimportant (Abragam, 1961). One usually uses the point dipole approximation in employing Eq. (3), that is, the electron spins are far enough apart that their distributions (in, e.g., nitroxide p-π orbitals) are unimportant ($r > 5$ Å for nitroxides).[2] In Eq. (3), θ is the angle between the direction of the large dc magnetic field $\mathbf{B}_0$ and $\mathbf{R}_{12}$ (cf. Fig. 2). The term in $S_{1z}S_{2z}$ in Eq. (3) is known as the secular term, and that in the $S_1^\pm S_2^\pm$

---

[2] An asymmetry parameter may be necessary in the case of delocalized spin density, for example, for closely situated spatially confined tyrosyl radicals, giving rise to a slightly rhombic spectral shape (Jeschke and Spiess, 2006).

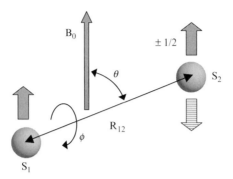

FIG. 2. A pair of electron spins S_1 and S_2 coupled via electron spin dipole–dipole interaction. Vector R_{12} connecting spins is directed along z-axis in the molecular frame of reference. In this molecular frame, the direction of the static magnetic field B_0 is determined by Euler angles $(0, \theta, \varphi)$. In DEER, spin 2 (the B-spin) is selectively flipped by the pumping pulse, changing the sign of its magnetic interaction with spin 1 (A-spin).

pseudosecular term. If, in the absence of the dipolar coupling of Eq. (3), the two electron spins have resonance frequencies ω_1 and ω_2, then the dipolar coupling in frequency units is written as

$$A(r, \theta) = \omega_d(1 - 3\cos^2\theta) \qquad (4)$$

with

$$\omega_d = \gamma_e^2 \hbar / r^3 \qquad (5)$$

For the case of unlike spins, such as $\omega_d \ll |\omega_1 - \omega_2|$, the resonant frequency of each spin is split into a doublet separated by $|A|$; the precise value of A depends on the angle θ, yielding a range of values of A from $-2\omega_d$ to $+\omega_d$. The pulsed dipolar spectrum provides this splitting, which is shown in Fig. 3C as a function of the angle θ, obtained from a macroscopically aligned frozen sample. In the more typical case of an isotropic frozen sample, one observes an average over θ, which yields a distinct dipolar spectrum, known as a Pake doublet[3] (Pake, 1948), (cf. Fig. 3A). It shows a prominent splitting of ω_d, corresponding to $\theta = 90°$, and another splitting of $2\omega_d$, corresponding to $\theta = 0°$. The distance r is immediately and accurately obtained from a measurement of ω_d.

---

[3] Note that the use of spin echoes cancels the effects of hyperfine and g-tensors, so even when the two nitroxides in a given bilabeled molecule resonate at two different frequencies because of different orientations and/or magnetic quantum numbers, they still yield a single Pake doublet, resulting from their common dipolar interaction.

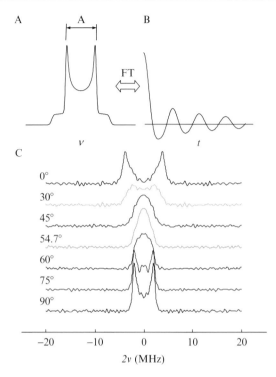

FIG. 3. (A) A dipolar spectrum in isotropic media (Pake doublet) obtained by Fourier transformation of the time-domain (B) dipolar spectrum; (C) An experimental dipolar spectrum of spin-labeled Gramicidin A (Dzikovski *et al.*, unpublished data) obtained by 4-pulse DEER at several orientations in a macroscopically aligned lipid membrane bilayer of DMPC.

The case of unlike spins corresponds to considering only the secular term in Eq. (3) and dropping the pseudosecular term. In the case of like spins, that is, $\omega_d \gg |\omega_1 - \omega_2|$, then the pseudosecular terms become important and Eq. (5) becomes $\omega_d = 3\gamma_e^2\hbar/2r^3$. Otherwise, the results (cf. Fig. 3) are equivalent.

The intermediate case of $\omega_d \sim |\omega_1 - \omega_2|$ is more complex, and is handled by careful simulation using Eq. (3), including both secular and pseudosecular terms. In the case of nitroxide spin labels, the two nitroxide spins in a given molecule usually have their ω_1 and ω_2 substantially different. This arises from their different orientations with respect to the $\mathbf{B}_0$ field, so their effective hyperfine (hf) and g values (arising from their hf and g tensors) are different. At typical ESR frequencies, this means that the unlike spin limit is reached for approximately 20 Å (9–17 GHz ESR).

If ω_d is sufficiently large, it can be determined from the broadening of the nitroxide cw ESR spectrum (Hustedt *et al.*, 1997), but this is likely to fall into the regime where pseudosecular terms are significant. Smaller couplings, ω_d require using pulse ESR methods, as we will discuss. In all cases, accurate values of distances are produced from the measured dipolar couplings.

Applications and Modalities of Dipolar ESR Spectroscopy

Both cw (Altenbach *et al.*, 2001; Hubbell *et al.*, 2000; Hustedt *et al.*, 1997; Koteiche and Mchaourab, 1999; Mchaourab *et al.*, 1997; McNulty *et al.*, 2001) and pulsed (Banham *et al.*, 2006; Bennati *et al.*, 2005; Biglino *et al.*, 2006; Borbat and Freed, 1999; Borbat *et al.*, 2002, 2006; Dzikovski *et al.*, 2004; Fafarman *et al.*, 2007; Hilger *et al.*, 2005; Jeschke *et al.*, 2004c; Milov *et al.*, 2005; Park *et al.*, 2006; Sale *et al.*, 2005; Xu *et al.*, 2006; Zhou *et al.*, 2005) ESR methods are used to measure distances between paramagnetic species, which are usually nitroxide spin-labels. However, PDS is not limited to nitroxides; distances between radical cofactors, nitroxides, and transition metal ions, have been measured in all possible combinations (Astashkin *et al.*, 1994; Becker and Saxena, 2005; Bennati *et al.*, 2003, 2005; Biglino *et al.*, 2006; Borovykh *et al.*, 2006; Codd *et al.*, 2002; Denysenkov *et al.*, 2006; Elsaesser *et al.*, 2002; Kay *et al.*, 2006; Narr *et al.*, 2002). Taken together, cw and pulsed ESR enable the measurement of distances over the range from approximately 5 to 10 Å to nearly 80 Å, with only the shorter distances accessible to cw ESR.

Cw ESR has been most often applied to nitroxides, whose powder spectra are dominated by the inhomogeneous broadenings from nitrogen *hf* and *g*-tensors, and unresolved proton *hf* couplings. One has to extract what usually is a small broadening effect introduced by the dipole–dipole interactions between the spin labels to the nitroxide powder spectra, which is usually accomplished by spectral deconvolution (Rabenstein and Shin, 1995) or a multiple-parameter fit (Hustedt *et al.*, 1997). This requires the spectra from singly labeled species as a reference for the background broadening, which is a complication and not always an option. Incomplete spin labeling makes the task more complex (Persson *et al.*, 2001). For distances less than 15 Å, the dipolar coupling approaches other inhomogeneous spectral broadenings and then can be more easily inferred from cw ESR spectra. The case of strong dipolar coupling has been extensively utilized in cw ESR (Altenbach *et al.*, 2001; Hanson *et al.*, 1996, 1998; Hubbell *et al.*, 2000; Hustedt *et al.*, 1997; Koteiche and Mchaourab, 1999; Mchaourab *et al.*, 1997; McNulty *et al.*, 2001; Rabenstein and Shin, 1995, 1996; Xiao *et al.*, 2001), both in establishing proximity and in providing quantitative distances (Altenbach *et al.*, 2001; Hanson *et al.*, 1996, 1998; Hustedt *et al.*, 1997; McNulty *et al.*, 2001; Rabenstein and Shin, 1995, 1996; Xiao *et al.*, 2001). Cw ESR is thus practical for short distances up to a maximum of approximately 15 to 20 Å, with the values for distances under 15 Å being more reliable.

Pulsed ESR is based on detecting a spin-echo, wherein the inhomogeneous spectral broadening cancels. Spin echo temporal evolution is governed by the weaker effects of spin relaxation, electron-electron dipolar and exchange couplings, Zeeman electron-nuclear superhyperfine and nuclear and quadrupole couplings. The dipolar and exchange coupling can be isolated from the rest by means of a suitable pulse sequence, which also helps to alleviate the problem caused by the presence of single labeled molecules. The direct signal from them is filtered out in PDS, but they do contribute to the background intermolecular dipolar signal, which is best suppressed by working at low concentrations. PDS is routinely used for distances longer than 15 Å (Banham *et al.*, 2006; Borbat *et al.*, 2002, 2004, 2006; Cai *et al.*, 2006; Jeschke, 2002; Park *et al.*, 2006), and it works well all the way down to 10 Å (Fafarman *et al.*, 2007), thus significantly overlapping with the cw ESR range, but it is much less affected by inefficient labeling and can readily yield distance distributions.

Implications of Nitroxide Label Geometry

Even though there is a rigid amino acid spin bearing label (TOAC; cf. Fig. 1), which is being used in peptide studies (McNulty *et al.*, 2001; Milov *et al.*, 2000a, 2001), currently there is no convenient way to incorporate it into proteins; consequently, a variety of cysteine-selective spin labels are in common use (Columbus *et al.*, 2001; Mchaourab *et al.*, 1999). Nitroxide label side chains are flexible and their conformational dynamics (Hustedt *et al.*, 2006; Langen *et al.*, 2000), and the volume they sample, depend on the label type and the details of the protein landscape. They usually reside on the protein surface, since it is difficult to provide efficient labeling to achieve a sizable fraction of double-labeled protein for sites that are deeply buried or in the protein core. Since the distances are measured between nitroxide NO groups rather than between backbone carbons, the side-chain length of about 7 Å causes considerable uncertainty in the $C_\alpha - C_\alpha$ distances accessible by cw ESR. Several studies have attempted refinement of the side-chain geometry to improve the correlation of inter-nitroxide distances with the distances between the respective alpha-carbons (Sale *et al.*, 2002, 2005). The larger distances in the typical PDS range of 20 to 70 Å are relatively more accurate (Sale *et al.*, 2005), but efficient methods of generating backbone constraints from a substantial set of ESR-derived distance restraints are just being developed.

Also, one observes a distribution in the distances between the two NO groups. It depends on the conformational space that the protein samples, the flexibility of the nitroxide side-chain at the particular site, and the relative orientations of the two nitroxide side-chains. Solvent-exposed sites often exhibit wide distance distributions (Borbat *et al.*, 2002; Hustedt *et al.*, 2006; Sale *et al.*, 2002). In any case, the first moments of distance distributions (the average distances) obtained from the time-domain data usually are accurate to 5 Å or better, 1 to 3 Å being typical. The distance distributions between fully or partially buried nitroxides are generally more restricted. Distances between buried radical cofactors reflect their immobilized status and the uncertainty in distance is often a fraction of an Angstrom (Bennati *et al.*, 2003; Kay *et al.*, 2006).

3-Pulse DEER

DEER[4] in its original 3-pulse form (Milov *et al.*, 1981) (also dubbed PELDOR), depicted in Fig. 4, is based on the two-pulse primary spin-echo $\pi/2$-τ-π-τ-*echo* sequence to which a third pumping π-pulse is added. The $\pi/2$ and π pulses, separated by time interval τ, are applied to spins resonating at

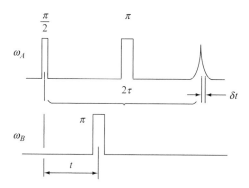

FIG. 4. The original 3-pulse form of DEER (Milov *et al.*, 1981). Primary echo is formed by $\pi/2$ and π pulse sequence at the frequency of A-spins. The pumping pulse at ω_B is applied at a variable time t to probe the dipolar coupling between A and B spins. The spectral excitations at both frequencies should not overlap, thus the pulses are made selective.

---

[4] Both acronyms PELDOR and DEER do not indicate the fact that they are solely concerned with dipolar couplings rather than dynamics. For this reason, we prefer to use PDS, to make explicit the function of the method and make the distinction with classic cw and pulsed ELDOR techniques.

the frequency ω_A, to form the primary echo at the time 2τ after the $\pi/2$ pulse. These spins are commonly referred to as A spins. The third (pumping) pulse is applied at the resonant frequency ω_B (at a variable time t) sufficiently different from ω_B that it does not have any direct effect on the A spins but instead inverts the spins resonating at ω_B, that is, the B spins.[5] The B spins, at a distance r from the A spins, yield the electron dipolar coupling A (cf. Eq. (4)), which splits the resonant line at ω_A into a doublet.[6] Thus, flipping the B spin inverts sign of the coupling[7] sensed by the A spin. This results in the instant shift of the Larmor precession frequency of spins A; it was shown in Milov *et al.* (1981) that the effect manifests itself as a modulation of the spin-echo amplitude, $V(t)$[8]:

$$V(t) = V_0(1 - p(1 - \cos A(r, \theta)t)) \qquad \text{for } 0 < t < \tau. \qquad (6)$$

Here, V_0 is the echo amplitude in the absence of the pumping pulse, p is the probability of flipping spin B, and $A(r, \theta)$ is given by Eq. (4). Powder averaging of $V(t)$ over an isotropic distribution of orientations of $\mathbf{R}_{12}$, under the simplifying assumption of random orientation of the magnetic tensors of the A and B spins relative to $\mathbf{R}_{12}$, produces a decaying oscillatory signal (cf. Fig. 3B):

$$V(t) = V_0(1 - p(1 - v(\omega_d t))) \qquad (7)$$

where

$$v(\omega_d t) = \int_0^{\pi/2} \cos[\omega_d(1 - 3\cos^2\theta)t]d(\cos\theta) \qquad (8)$$

and the frequency of oscillation, $v_d = \omega_d/2\pi$, from which r is calculated as $r[\text{Å}] = 10(52.04/v_d[\text{MHz}])^{1/3}$. Cosine Fourier transformation of $v(\omega_d t)$ versus $2t$ yields the dipolar spectrum with the shape of a Pake doublet (cf. Fig. 3A). Note that, in PDS, it is customary to perform the FT versus t (with the splitting

---

[5] There can be more subtle effects on spin A arising from the dipolar interaction during the pulse at ω_B (Maryasov and Tsvetkov, 2000).

[6] We exclude electron exchange coupling for brevity; it is insignificant for nitroxide labels separated by $r > 15\text{Å}$.

[7] Of course, the A spin also splits the resonant frequency of the B spin into a doublet. A detailed consideration of the spin dynamics for a coupled spin pair (Maryasov and Tsvetkov, 2000) shows that both components of the dipolar doublet are required to be flipped by the π pulse; thus, sufficient amplitude of the microwave magnetic field acting on coupled spins be applied.

[8] The dipolar signal for $\tau < t < 2\tau$ is a repeat of the signal for $t < \tau$. Therefore, in the sequel, we shall assume that t varies only from 0 to τ.

between the singularities in the plot being twice the dipolar splitting), and often only one-half of the dipolar (symmetric) spectrum is plotted versus dipolar frequency, v.

Eqs. (6–7) should be considered as a reasonable approximation for DEER, which is suitable for the majority of cases encountered in biological applications of PDS. In reality, a number of factors affect the signal, and their effects usually cannot be written in closed form or are unwieldy. Some will be discussed later. What is significant is that DEER achieves a good separation of the dipolar coupling from relaxation effects because the time between the $\pi/2$ and π spin-echo pulses at ω_A is constant, (i.e., τ in Fig. 4 is constant in the experiment; this is referred to as a constant time pulse sequence), and relaxation effects introduced by the pumping pulse can normally be ignored.[9] Nuclear ESEEM is also considerably suppressed but still could be an issue when p is not small.

The Newer Methods

4-Pulse DEER

The methods of 4-pulse DEER (Pannier *et al.*, 2000) and 6-pulse DQC (Borbat and Freed, 1999, 2000; Borbat *et al.*, 2001, 2002; Freed, 2000) are illustrated in Fig. 5. The 4-pulse DEER sequence is an improvement over 3-pulse DEER. It is based on the 3-pulse spin- echo sequence $\pi/2$-τ'-π-$(\tau + \tau')$-π-τ-*echo*, which refocuses the primary echo formed by the first two pulses. The additional pumping pulse at ω_B is varied in time between the π pulses at ω_A. Both τ and τ' are fixed; thus, relaxation does not modify the signal envelope recorded versus position of the pumping pulse. The signal is described by Eqs. (6–7) at the same level of approximation as 3-pulse DEER. This pulse sequence substantially simplifies its technical implementation, which has permitted commercial implementation. The pulses do not need to overlap or even come close (cf. the Some Technical Aspects of DEER and DQC section, p. 86), thereby avoiding some small but significant dead times effects in 3-pulse DEER. Also, 4-pulse DEER permits using larger B_1's than does 3-pulse DEER, at optimal settings of ω_A and ω_B, which provides greater sensitivity.

DQC

The 6-pulse DQC pulse sequence $\pi/2$-t_p-π-t_p-$\pi/2$-t_d-π-t_d-$\pi/2$-(t_m-t_p)-π-(t_m-t_p)-*echo* is based on a different principle. All pulses are applied at the same frequency ω_A, and it is important that they all be intense in order to excite the

---

[9] If the flip-flop rate of B spins is low, they do not introduce significant relaxation effects.

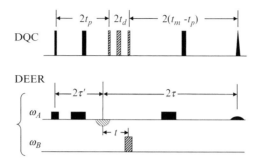

FIG. 5. 6-pulse DQC (top) and 4-pulse DEER (bottom) sequences: The DQC 6-pulse sequence (Borbat and Freed, 1999, 2000) is based on intense pulses in order to probe the dipolar coupling between (nearly) all intramolecular pairs of nitroxide spins. The first part of the sequence $\pi/2$-t_p-π-t_p is a preparation period, at the end of which an echo is formed (not observable) from the anti-phase single coherence between the two coupled spins. The third and fifth pulses ($\pi/2$) convert this coherence into double quantum coherence and then back into anti-phase single-quantum coherence (with the fourth pulse (π) refocusing the spins). This anti-phase single-quantum coherence then develops into the observable single spin coherence after the $2(t_m-t_p)$ time period. The sixth pulse (π) is applied to form an echo of this coherence. This echo is selected by phase cycling of the signal that passes through the double quantum filter (hatched). The time t_d of the 3-pulse double quantum filter is kept short and constant. The time t_m is also kept constant to minimize phase relaxation effects; and it defines the time available for dipolar evolution. The relevant time variable for observing the dipolar signal is $t_\xi \equiv t_m - 2t_p$, which is zero when $t_p = t_m/2$. The pulse sequence is thus dead-time free. The reference point $t_\xi = 0$ is well-defined due to the very short pulses used in DQC.

The 4-pulse form of DEER (Jeschke, 2002; Jeschke and Spiess, 2006) is a modification of its 3-pulse predecessor. It is based on detecting the refocused primary echo formed by $\pi/2$-τ'-π-$(\tau + \tau')$-π-τ-$echo$ pulse sequence at the frequency, ω_A of A-spins. The time variable t is referenced to the point where the primary echo from the first two pulses is formed (but is not detected). At $t = 0$, the dipolar phase is zero for all A spins (the precise $t = 0$ is limited by the width of the pulses). Shifting the starting point for dipolar evolution away from the second pulse by τ' makes this pulse sequence dead-time-free with respect to dipolar evolution and eases its technical implementation.

whole spectral distribution of spins, that is, all the spins are regarded as A spins (cf. Fig. 6). The first interval, $2t_p$, is used to let the normal single-quantum coherence with spin character $S_{1y} + S_{2y}$ evolve into what is known as anti-phase single-quantum coherence between the coupled spins with spin character $S_{1x}S_{2z} + S_{2x}S_{1z}$. Then, the $\pi/2$-t_d-π-t_d-$\pi/2$ pulse "sandwich" (hatched bars in Fig. 5) converts this coherence into double-quantum coherence with spin character $S_{1x}S_{2y} + S_{1y}S_{2x}$ (by means of the first $\pi/2$ pulse), then refocuses it by means of the π-pulse, only to convert it back to (unobservable) anti-phase coherence (by means of the last $\pi/2$ pulse), which evolves back into the observable coherence $S_{1y} + S_{2y}$, giving rise to the echo. Both spins participate equally in the process. The first and the last π-pulses of the 6-pulse sequence

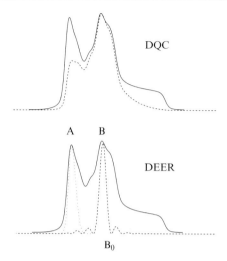

FIG. 6. Excitation of the nitroxide spectrum at 17.3 GHz in the microwave Ku band for DQC (top) and 3-pulse DEER (bottom). The ^{14}N nitroxide ESR spectrum is plotted as a solid line and the spectral excitation profiles are plotted as dashed lines. The detection frequency in DEER is set at the low field edge of the spectrum (A) and the pump pulse frequency corresponds to positioning it at the center (B). The pumping pulse is 4 G (45 ns π pulse) in DEER. The DQC excitation profile corresponds to a 48 G (3.7 ns) π pulse.

are used to refocus in-phase and anti-phase coherences, thereby respectively enhancing the effectiveness of the double-quantum sandwich and producing the echo at time $2t_m + 2t_d$. The signal in the ideal limiting case of intense and nonselective pulses can be written as seen in Eq. (9) (Borbat and Freed, 1999, 2000).

$$
\begin{aligned}
V &= -V_0 \left[\sin A(r,\theta)t_p \right] \sin[A(r,\theta)(t_m - t_p)] \\
&= \frac{V_0}{2} \left[\cos A(r,\theta)t_m - \cos A(r,\theta)t_\xi \right]
\end{aligned}
\tag{9}
$$

The signal is recorded versus $t_\xi - t_m - 2t_p$, with t_m kept constant in order to keep relaxation effects (which decay exponentially in time) constant. Powder averaging gives

$$
V = \frac{V_0}{2} \left[v(\omega_d, t_m) - v(\omega_d, t_\xi) \right]
\tag{10}
$$

with $v(\omega_d, t_\xi)$ as given by Eq. (8). For large $\omega_d t_m$, the first term in Eq. (10), which is constant in t_ξ, is close to zero.

The important feature of the double quantum coherence sandwich is that it very effectively filters out the single quantum signal arising from the individual spins, and only passes the signal from the interacting part of the two spins, which contain only the dipolar oscillations. The only background that can develop is from the double quantum coherence signal that originates from the bath of surrounding spins, that is, from intermolecular dipolar interactions with other doubly labeled molecules (and singly labeled molecules when they are present). The signal envelope $V(t_\xi)$ is symmetric with respect to $t_\xi = 0$ (cf. Fig. 7). This is referred to as being dead-time free, since the dipolar oscillations are a maximum at $t_\xi = 0$ (cf. cosine term in Eq. (9)). This also means that it is sufficient to collect the data points for $t_\xi \geq 0$. There is, however, an apparent dead-time, which is determined by the pulse width; it typically is a few nanoseconds.

Relaxation effects that decay exponentially but nonlinearly in time in the exponent (Borbat *et al.*, 2002), or substantial differences in T_2's from the two spins, can modify the signal. The 6-pulse sequence generates a number of echoes, but with the proper phase cycling, only the dipolar modulation of the double-quantum filtered echo is detected. The details can be found in Borbat and Freed (2000).

The DQC experiment maintains phase coherence between the two coupled spins and treats them equally, whereas in DEER, phase coherence between the two coupled spins is of no importance. The independence

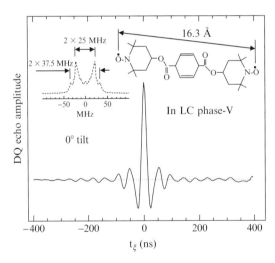

FIG. 7. 17.4 GHz DQC from a 16.3 Å rigid biradical aligned in LC phase-V (Borbat and Freed, 2000). The LC director is oriented parallel to $\mathbf{B_0}$. The dipolar coupling is ~25 MHz with maximum outer splitting of ~37.5 MHz due to pseudosecular term.

of tuning of the pulse conditions at both frequencies, as well as its applica-bility to widely separated spectra, makes the DEER sequence quite flexi-ble. Nevertheless, it can be shown that the dipolar signal recorded in DEER is based on the same type of evolution of in-phase and anti-phase coher-ences as in DQC (Borbat and Freed, 2000). This is also the case with other related pulse sequences (Borbat and Freed, 2000). Although it may look complex, the DQC experiment, once it is set up, is rather simple to use. The similarity in DQC and DEER means that the maximum useful time of the experiment $(2t_m)$ in DQC and 2τ in DEER will be comparable, except for respective differences in signal-to-noise ratio (SNR), as will be discussed.

DQC and DEER have proven to be the most useful methods, and together they address a wide range of applications. (Figs. 9, 10, 11, 14, 16, and 21 show examples of DQC and 4-pulse DEER signals.)

Other Methods

Several other pulse sequences for PDS with useful features have been introduced (Borbat and Freed, 2000; Jeschke *et al.*, 2000; Kulik *et al.*, 2001; Kurshev *et al.*, 1989; Raitsimring *et al.*, 1992). They are related in one way or another to DEER or DQC, since they are all based on dipolar evolution of single quantum in-phase and anti-phase coherence, and some try to minimize relaxation effects based on constant time pulse sequences. They have not been extensively used because of various shortcomings. Addi-tional methods are based on the dipolar contribution to spin relaxation, so they are not as able to provide accurate distances, but they can be useful. The reader is referred to Raitsimring and Salikhov, 1985; Rakowsky *et al.*, 1998; Seiter *et al.*, 1998.

Intermolecular Effects, Clusters, Oligomers, and Spin-Counting

Eqs. (6) and (7) describe the dipolar signal in DEER originating from a pair of spins A and B. The signal from the A and B spins in each (doubly labeled) molecule is usually the signal of interest. All the other A (resonat-ing at ω_A) and B (resonating at ω_B) spins in the sample also contribute to the DEER signal. For example, they represent the intermolecular dipolar interactions. The simplest intermolecular case that can be represented in closed form is the case of uniform spatial spin distribution over an isotropic magnetically dilute sample. Since the dipolar interaction in this case is weak, it can be represented by the secular term in Eq. (3). Then, the effect of all the other B spins on the ith spin, A, is multiplicative, given by the product obtained in Eq. (11) (Milov *et al.*, 1984).

$$V_{i,inter}(t) = \left\langle \prod_{\substack{j \neq i}}^{N-1} [1 - p_{ij}(1 - \cos A(\mathbf{r}_{ij})t)] \right\rangle. \tag{11}$$

Here, N is the number of spins in the sample. Angular brackets denote averaging over all possible configurations of N spins $\{\mathbf{r}_{ij},...,\mathbf{r}_{Nj}\}$. Averaging by the Markov method (Chandrasekar, 1943) leads to a simple exponential decay

$$V_{i,inter}(t) = \exp(-kt) \tag{12}$$

with

$$k^{-1} = 1.0027 \frac{10^{-3}}{pC} \tag{13}$$

where C is the molar concentration and p is the probability of flipping B spins by the pumping pulse (typically, 0.1–0.3). The dipolar signal from the spin pair of interest (cf. Eq. (7)) is then modified by multiplication by this decaying factor, (cf. Figs. 8, 9, and 13a for additional examples). A similar mechanism works among A spins, and is known as the instantaneous diffusion (ID) mechanism (Klauder and Anderson, 1962; Nevzorov and Freed, 2001a; Raitsimring et al., 1974; Salikhov et al., 1981), which, unlike ordinary relaxation mechanisms, can be partially refocused (Slichter, 1990).

A similar approach can be applied to an isotropic uniform distribution in space with fractal dimensionality, where a closed-form solution can be

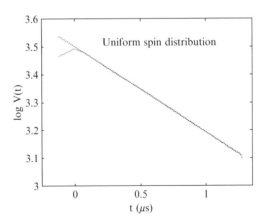

FIG. 8. An example of 4-pulse DEER (solid line) from a uniform distribution of spins in an isotropic sample illustrates the intermolecular signal given by Eq. (12). Dashed line is a fit to the straight line in the logarithmic plot. To achieve the uniform spin distribution, 0.01 mole percent of spin-labeled alamethicin was magnetically diluted with WT by a factor of 20 to avoid effects of its aggregation (unpublished, this lab). (See color insert.)

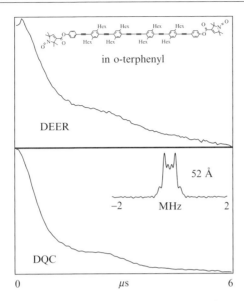

FIG. 9. Long rigid biradical in o-terphenyl glass at 50 K. DQC (bottom) shows a stronger signal than DEER (top) due to strong pulses and partial suppression of nuclear spin diffusion of DQC. The signal profile in DQC is, however, more affected by spin relaxation that decays according to a t^2 dependence in the exponent (Borbat *et al.*, 2002). (Just the oscillating part of the full DEER signal is shown for better comparison of the two signals.) (Unpublished, rigid rod biradical, courtesy of G. Jeschke).

written (Milov and Tsvetkov, 1997). Practical examples of lower dimension are the 2D case of unilamellar lipid membranes or the 1D case of self-avoiding polymer chains.[10] Note that we can distinguish two types of heterogeneous sample—microscopic or macroscopic (Jeschke and Schlick, 2006). This can be understood by realizing that spins beyond a certain radius, call it $R_{inter}(t_m)$, make a negligible contribution to Eq. (11) or (12) (Jeschke *et al.*, 2002). Therefore, such a length scale, R_{inter} can be used to separate micro- and macroscopic domains. Macroscopic heterogeneity represents variations over length scale greater than R_{inter} in concentration or composition throughout the sample (and it also includes pulse amplitude variation over the sample). The signal given by Eqs. (12) and (13) is simply averaged over the sample. Micro-heterogeneous systems such as lipid membranes or clusters, which have characteristic microscopic order, are usually not amenable to simple analytic solutions, and their signals should

---

[10] These are not exactly fractal cases and they are not described as in Milov and Tsvetkov (1997), but have similar time dependence.

be derived based on the appropriate averaging of Eq. (11) for the particular case. Good approximations are possible but are beyond the scope of this chapter. In general, the dipolar signal is modified, and there can be a large nonlinear background, which needs to be accounted for, or else removed to isolate the informative part of the signal $V(Wd,t)$ (cf. Eq. 8) (Borbat et al., 2002; Jeschke, 2002; Maryasov et al., 1998).

The case of a small group of spins (clusters) has been considered in the literature (Milov et al., 1984, 2000b, 2003b; Raitsimring and Salikhov, 1985; Ruthstein et al., 2005; Salikhov et al., 1981). This case requires numerical treatment based on Eq. (11), typically by the Monte Carlo method, although simplified approaches exist and were used to roughly estimate the number of spins in a cluster (Milov et al., 1984). In fact, an accurate treatment is rarely justified in such cases, since there are too many unknown parameters to fit and realistic data permit determining one or two parameters at most. In addition, one must have a priori knowledge about the system in order to model it properly.

We note that a generalization of DQC methods to provide multiple-quantum coherence selective pulses is, in principle, possible (Borbat and Freed, 2000). Such a methodology would be very useful for spin counting, but it has not yet been developed for practical use in ESR.

From the standpoint of PDS, the intermolecular term is usually an unwanted complication, requiring that the intramolecular signal of interest be separated from the intermolecular contribution to the signal. Clearly, the best approach is to minimize the latter by sample dilution, whenever it is an option and sensitivity permits.

For the sophisticated pulse sequence of DQC, there is no rigorous theory for the intermolecular dipolar effects in DQC. However, when pulses are strong and the system is sufficiently dilute, the intramolecular dipolar signal is relatively less affected than in DEER by other spins (cf. Fig. 7). Therefore, at low concentrations, DQC has the advantages of better sensitivity due to all or nearly all the spins participating and weaker effects of surrounding spins. In cases of high local concentrations (lipid vesicles, protein oligomers, or peptide clusters), DEER is able to produce the same (or sometimes even better) sensitivity than DQC because of reduced ID from the weaker DEER pulses. Some examples of dilute samples where DQC works better are shown in Figs. 9 and 10.

Data Processing in DQC and DEER

An example of a typical DEER signal from a spin-labeled protein is shown in Fig. 11 and that for a case of a small cluster in Fig. 12. DQC signals are shown in Fig. 7 (cf. also Figs. 14, 16B, and 21A). For DQC, first, the

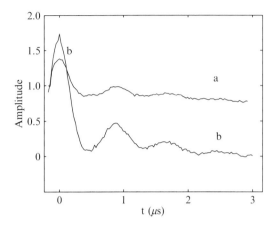

FIG. 10. Comparison of the DQC signal versus t_ξ (b) and the 4-pulse DEER versus t (a) operating at 17.4 GHz for spin-label at position 340 in the cytoplasmic domain of band 3 protein (Zhou *et al.*, 2005). The same resonator and sample was used in both cases, data collection time was 25 min, T was 70 K. In DQC 9 ns π pulses (20 Gauss B_1) were used; 16/32/32 ns observing pulses and 28 ns pumping pulse were used in DEER. SNR of DQC is 142, in DEER it is 43. The DQC SNR may be improved by using shorter π pulses. An additional advantage of DQC was due to its partial cancellation of nuclear spin diffusion (cf. Sensitivity of PDS section, p. 79). [Current operating performance for these conditions (and for the data of Fig. 9) yields SNRs that are greater by a factor of 2.5.] (unpublished data; the protein courtesy of Zheng Zhou).

intermolecular background signal is removed by means of least square polynomial fitting in the time-domain of the latter part of the signal; then, this is extrapolated back to the earlier part of the signal and subtracted out (Borbat *et al.*, 2002). In the case of DEER, the removal of background signal often is performed by fitting the latter part of the signal to a straight line in a log plot under the assumption of an exponent that is linear in time in Eq. (12). When this is not the case, a low-degree polynomial can be used instead. Another way of accounting for the intermolecular background is to use methods of signal reconstruction with simultaneous baseline fitting (cf. the Distance Distributions section, p. 73). This method separates out the part of the signal governed by the intramolecular kernel (see the following text). Using a more dilute sample is recommended when possible since it reduces background and also helps to arrive at a more linear background. Spectral overlap of pulse excitations at the two frequencies in DEER, which can arise when using a stronger pumping pulse to increase p and thereby to enhance the signal, complicates the signal and its analysis due to unwanted effects (cf. Fig. 21B). In the case of a heterogeneous system, one could first study a reference sample with the same spin

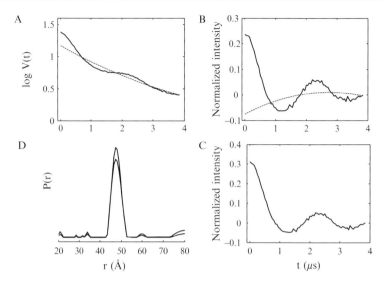

FIG. 11. Data processing of Ku band time-domain DEER signal for nitroxide labeled mono-amine oxidase reconstituted in detergent micelles. (A) The intermolecular background is removed by first fitting the data from 1 to 4 μs to a second-degree polynomial (rather than to a straight line, as relevant to this case) in the log plot, followed by subtracting it out. (B) Dipolar signal after removal of background. Dashed line shows the correction for the background that was generated in the process of MEM reconstruction (Chiang *et al.*, 2005b). (C) Corrected dipolar signal generated by fitting (A) to a linear background signal; it is indistinguishable from (B), indicating the capability of MEM to separate out the inter- and intramolecular contributions to the dipolar signal in this example. (D) $P(r)$'s produced from data from (B) (upper curve) and (C). The very small peaks are caused by noise and signal distortions (mostly caused by using a short pump pulse of 20 ns). (Unpublished, this lab; protein sample provided by A. Upadhyay.)

concentration but with singly labeled molecules or spin probes in order to determine the shape of the intermolecular signal. (But this method has limited applicability and polynomial fitting is usually the method of choice.)

Extracting Distance and Spin-Count Information

The average distance can to good accuracy be extracted by inverse reconstruction (see the Distance Distributions section, p. 73) (Bowman *et al.*, 2004; Chiang *et al.*, 2005a,b; Jeschke *et al.*, 2002, 2004b) of the intramolecular signal obtained from the experimental data (Borbat *et al.*, 2002; Milov *et al.*, 1999); or from analyzing the signal with parameterized geometrical modeling (Borbat *et al.*, 2002); or from the singularities or the half-width of the dipolar spectrum (Park *et al.*, 2006); or simply by an estimate based on temporal envelope (Milov *et al.*, 1999; Park *et al.*, 2006).

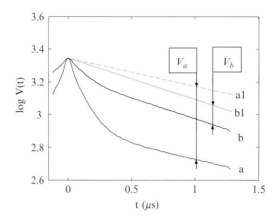

FIG. 12. The DEER signal in the case of clusters for the system of Fig. 8 but less magnetically diluted (b) than in Fig. 8, and without dilution (a). (a1, b1) are straight line fits to the
asymptotic parts of (a, b). (a) is typical for a spin cluster; in this case, single-labeled alamethicin
molecules are organized into small clusters with expected constant number of monomers.
(b) represents the same spin concentration but magnetically diluted by a factor of 5 with
unlabeled peptide, indicating that this signal indeed originates from a spin cluster. The asymptotic DEER amplitudes (V_a, V_b) can be immediately analyzed to yield an estimate of how
many peptide molecules, N, are in the cluster (Milov $et\ al.$, 1984), given that the fraction of
peptides in clusters is known. Based on Milov $et\ al.$ (1984) $\ln V_a = (N-1)\ln(1-p)$, where
p (cf. Eq. (7)) was 0.2. This yields four peptide molecules per cluster. (Unpublished, this lab.)

The latter includes using the period of oscillation or half-width of the initial
decay (ca. $2\pi/5\omega_d$). For very long distances, when it is known a priori that
there should be a reasonably well-defined distance, access to a fraction
of the dipolar oscillation period suffices but requires prediction of the
baseline or knowledge of p (in the case of DEER) and spin labeling
efficiency. The error in distance, even with such crude methods, is relatively
small ($vide\ infra$) due to the inverse cubic dependence of the ω_d on the
distance, so it does not normally exceed the uncertainties introduced by
the nitroxide side-chains.

    For clusters, controlled magnetic dilution proved useful to detect aggre-
gation and evaluate the size of clusters and number of spins (Milov $et\ al.$,
1999). We illustrate in Fig. 12 the practical implementation of the method,
with some additional details given in Milov $et\ al.$ (1984).

Distance Distributions

    Several approaches to determining distance distributions of para-
magnetic centers in solids were utilized in the early applications of DEER
and related methods (Milov $et\ al.$, 1981; Pusep and Shokhirev, 1984;

Raitsimring and Salikhov, 1985). Such methods have been improved (Bowman *et al.*, 2004; Chiang *et al.*, 2005a,b; Jeschke *et al.*, 2002, 2004b) and the Tikhonov regularization method is now a workhorse for extracting distance distributions from the raw or preprocessed data.

The time-domain dipolar signal may generally be viewed as $V_{intra} A_{inter} + B_{inter}$ (B_{inter} originates from singly labeled molecules and, for uniform spin distributions in the sample, its time dependence is given by A_{inter});[11] the A and B terms are removed to the extent possible; and then, what is taken to be a reasonably accurate representation of V_{intra} is subject to inverse reconstruction by Tikhonov regularization or related methods. The problem can be represented by a Fredholm integral equation of the first kind,

$$V_{intra}(t) = V_0 \int_0^\infty P(r)K(r,t)dr \qquad (14)$$

with the kernel $K(r,t)$ for an isotropic sample (cf. Eqs. (4–5)) given by

$$K(r,t) = \int_0^1 \cos\left[\omega_d t(1 - 3u^2)\right]du \qquad (15)$$

The inversion of the signal V_{intra} given by Eq. (14) to obtain $P(r)$, the distance distribution, is, in principle, achievable by standard numerical methods, such as singular value decomposition (SVD), but it is an ill-posed problem that requires regularization methods in order to arrive at a stable solution for $P(r)$. In the practical implementation, the data are discrete and available over a limited time interval, and the actual form of the kernel $K(r,t)$ may differ from the ideal form given by Eq. (15).

Tikhonov regularization (Chiang *et al.*, 2005a,b; Jeschke *et al.*, 2004b) recovers the full distribution in distance, $P(r)$. It is based on seeking an optimum, $P(r)$, which tries to minimize the residual norm of the fit to the data while also trying to maximize the stability of $P(r)$ (to reduce its oscillations). The relative importance of both is determined by the regularization parameter, λ. The L-curve method for optimizing λ is computationally very efficient and the most reliable to date. In the Tikhonov method, the regularization removes the contributions of the small singular values, σ_i in the SVD that are corrupted by the noise by introducing the filter function,

---

[11] But B_{inter} is not, in general, the same as A_{inter} for the case of micro-heterogeneity, wherein the local spin concentration determines B_{inter}.

$$f_i \equiv \frac{\sigma_i^2}{\sigma_i^2 + \lambda^2} \qquad (16)$$

which filters out those contributions for which $\sigma_i^2 \ll \lambda^2$. Further refinement of the $P(r)$ can be performed by means of the maximum entropy method (MEM) (Chiang et al., 2005b), although it is computationally more time-consuming. The latest versions of MEM and Tikhonov regularization permit one to simultaneously fit and remove the effects of A_{inter} and/or B_{inter} while optimizing the $P(r)$ from raw experimental data[12] (Chiang et al., 2005b).

Experimental artifacts, signal distortions, and residual baseline make signal recovery somewhat less accurate than what has been demonstrated on model data that were generated using the ideal kernel of Eq. (15). The test examples of Fig. 13 demonstrate the accuracy of recovery of average distances and distribution widths when the signal is free of artifacts. It is clear that with a good SNR, average distances of the order of 80 Å can be obtained. Very long distances can be recovered, given undistorted and good SNR data. Figure 11 demonstrates the application to real data with baseline correction by MEM.

Relaxation

The amplitude of the primary echo V_0 decays with increasing pulse separation, τ (cf. Fig. 4) due to phase relaxation. Therefore, the maximum dipolar evolution time interval, t_{max} available for recording $V(t)$ is ultimately limited by the phase memory time, T_m (or T_2). In the simplest case, $V(t) = V_0 \propto \exp(-2t/T_m)$. This limits the maximum distance, r_{max} that one can measure, over a reasonable period of signal averaging. Depending on the signal strength, t_{max} is approximately 1 to 3 T_m and cannot be extended much further. Here, t_{max} is essentially $2t_m$ in DQC and $2(\tau' + \tau)$ in DEER (cf. Fig. 5). The largest measurable distance r_{max} is proportional to $t_{max}^{1/3}$ in order to recover the dipolar oscillation (Borbat and Freed, 2000). Thus, only a minor increase in r_{max} could be made by increasing t_{max}, and this would necessarily require a large increase in signal averaging. For nitroxide-labeled proteins, T_m is largely determined by the dynamics of the nearby protons (Huber et al., 2001; Lindgren et al., 1997; Zecevic et al., 1998), especially those from methyl groups, leading to the simple exponential decay expressed above with T_m in the range of 1 to 2 μs for buried or partially buried labels. Such relaxation times are typical for hydrophobic environments that are encountered in lipid membranes (Bartucci et al., 2006) and the protein interior (Lindgren et al., 1997). This permits an r_{max}

---

[12] Available for download through the ACERT web page www.acert.cornell.edu.

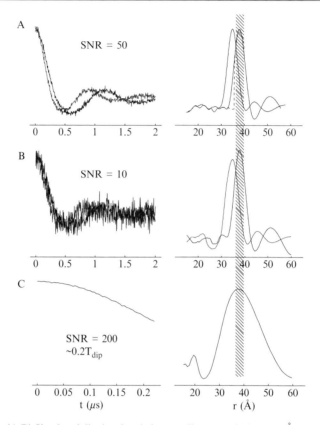

FIG. 13. (A,B) Simulated dipolar signals for two distances of 35 and 38 Å, both rms widths of 3.6 Å, for different levels of noise. Left panels show time-domain signals, and the right panels show distance distributions, $P(r)$ reconstructed using L-curve Tikhonov regularization. Note that just a 3 Å difference produces distinct time-domain signals and distance distributions with correct average distances and widths in case (A). The exact distribution in distance is shown (dashed) for 38 Å distance in (A), and it is nearly coincident with the recovered profile. The poorer SNR in (B) still enables accurate inverse reconstructions. (C) Limited time-domain data (approximately 20% of the period of the dipolar oscillation) with good SNR of 200, which is often achievable, does result in a good estimate of distance, but at the expense of broadening of $P(r)$. By rescaling the curve in (C) to 2 μs, for 20% of the period of the dipolar oscillation, an average r of about 82 Å is obtained in the $P(r)$ of panel (C). The time period of 2 μs in (A,B) is typical for nitroxide labels for proteins in most environments, but is sometimes as long as 3 to 5 μs, which permits one to record approximately a half period of dipolar oscillation for 70 – 80 Å, thus yielding a more reliable estimate of distance.

of typically 50 Å. For water-exposed labels, relaxation at longer τ is dominated by $\exp[-(2\tau/T_m)^\kappa]$ with $\kappa \sim 1.5$–2.5 and $T_m \sim 3$ to 4 μs (Lindgren et al., 1997). This quadratic term in the exponent is governed by the nuclear spin diffusion mechanism (Milov et al., 1973; Nevzorov and Freed, 2001b).

A larger κ may indicate spectral diffusion (Klauder and Anderson, 1962; Raitsimring et al., 1974). This permits an r_{max} of typically ~55 to 60 Å (or ~70 – 75 Å with low accuracy).[13] Such types of relaxation could be partially suppressed by multiple refocusing and/or using deuterated solvent (Borbat et al., 2004; Borbat and Freed, 2000; Jeschke et al., 2004a; Milov and Tsvetkov, 1997). This could extend t_{max} to approximately 6 to 8 μs in favorable cases (Huber et al., 2001), that is, much less than in $D_2O/$ glycerol-$d8$, since there still is a bath of protons of the protein itself (Huber et al., 2001). Using 6-pulse DQC helps to extend t_{max} when T_m is dominated by nuclear spin diffusion (Borbat and Freed, 2000; Borbat et al., 2004). This permits a more accurate estimate of r_{max} to about 70 Å. Further improvement would require much greater effort, such as partial or complete protein deuteration, and this might extend r_{max} to 100 to 130 Å and make distances up to 80 Å much more accurate. Since such enrichment also benefits high-resolution NMR (Hamel and Dahlquist, 2005; Horst et al., 2005; Venters et al., 1996), one could hope that this technology may become, in the future, a standard way to improve the accuracy of distances in the 50 to 80 Å range, which are currently accessible, and to increase the sensitivity dramatically, bringing it to the micromolar level (see next section on Distance Range). This is of particular value for the difficult case of membrane proteins.

The longitudinal relaxation time, T_1, determines how frequently the pulse sequence can be repeated (usually, no more than $1.5/T_1$) and, consequently, the rate at which the data can be averaged. Both T_1 and T_2 are temperature dependent, as is the signal amplitude, which depends on the Boltzmann factor for spins in the dc magnetic field. The combined effect of all these aspects is such that for proteins in water solution or in membranes, the optimal temperature as a rule is in the range of 50 to 70 K. The presence of paramagnetic impurities with short relaxation times shortens both T_1 and T_2. This would require conducting experiments at even lower temperatures.

Distance Range

Long Distances

The ability to measure very long distances is limited by the phase memory time, T_m (see Relaxation and Sensitivity of PDS sections, pp. 75 and 79) and, for proteins, 65 to 75 Å is about the upper limit with current technology. Also, distances measured in this range are typically not very

---

[13] r_{max} is chosen to correspond to t_{max}, which we take as equal to one period of the dipolar oscillation $T_{dip} \equiv 2\pi/\omega_d$ (cf. Eq. (5)). Using $T_{dip}/2$ is often possible, but the accuracy in distance is less (cf. Sensitivity of PDS section, p. 79).

accurate. Modified standard methods have been shown to bring some level of improvement (Borbat *et al.*, 2004; Jeschke *et al.*, 2004a). This situation could be radically improved by protein deuteration. Alternatively, with a good spin labeling strategy, such long distances could often be avoided.

Short Distances

The π-pulse excites a spectral extent (in Gauss) of about B_1. It is necessary to excite both components of the Pake doublet in DEER, which normally uses π-pulses longer than 20 ns (B_1 of $\sim$9 G). This provides a lower limit to DEER of approximately 15 to 20 Å (Fig. 14). However, π-pulses of 30 to 60 ns width are typical, since they provide a cleaner implementation of the method, which requires that the pump pulse and observing pulses do not overlap in spectral extent. This tends to limit DEER to 20 Å and greater. The sensitivity to shorter distances decreases significantly because the coupling increases and both components of the Pake doublet can no longer be adequately excited (Milov *et al.*, 2004). Also, account must be taken of strong dipolar coupling during these long pulses (Maryasov and Tsvetkov, 2000).

DQC uses intense pulses with B_1 of 30 G or greater; hence, it can access distances as short as about 10 Å (Fafarman *et al.*, 2007) (Fig. 14). In this case, the pseudosecular part of the dipolar term in the spin-Hamiltonian (cf. Eq. (3)) cannot be neglected (Fig. 7), but this can be accounted for in rigorous numerical simulations (Borbat and Freed, 2000). Note that for the nitroxide biradical that is aligned in a liquid crystal of Fig. 7, wherein the orientation of $\mathbf{R}_{12}$ is parallel to $\mathbf{B}_0$, the dipolar splitting is 23 MHz; (this corresponds to a 12.7 Å distance for the powder) and the $3v_d/2$ frequency from the pseudosecular terms (37.5 MHz) is also readily excited. The appearance of the outer splittings due to the pseudosecular term tends to broaden out the Pake doublet, thereby reducing the signal amplitude by typically a factor of ≤ 2 (in the case of short distances), which is not a significant issue since the useful signal evolution period is then usually shorter than T_m. For short distances under 20 Å, submicromolar concentrations can suffice (see next section). Also, any intermolecular contribution from singly labeled molecules in the sample is of little concern in this distance range, because the signal acquisition period is short enough that effects given by Eqs. (12–13) are small.

Thus, pulse methods could be applied to most practical cases arising in protein distance mapping. The short distance range is more appropriate, however, for organic biradicals, buried spin labels, or radical cofactors, TOAC, and similar cases, when radicals are substantially immobilized and their geometry is known or can be deduced. This range is less relevant for typical nitroxide labels with long tethers, with uncertain geometry.

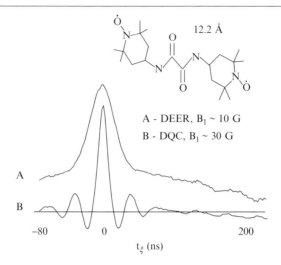

FIG. 14. The challenges of short distances. DQC and DEER were applied to a rigid 12.2 Å nitroxide biradical. Detection pulses in DEER were 16/32/32 ns, the pumping pulse was 18 ns ($B_1 \sim 10$ G). This is found to be insufficient to properly excite the dipolar spectrum. DQC using 6.2 ns π-pulse ($B_1 \sim 30$ G) develops the ~ 30 MHz oscillations very cleanly, similar to the case of an aligned biradical (Fig. 7). The longer pulses of DEER lead to a spread in the refocusing point of different spin packets and the weaker B_1; both smear out the high-frequency dipolar oscillations. (The biradical courtesy of R. G. Griffin.)

Optimal Range of Distances

In our experience, an optimal range of distances for the purposes of PDS is within 20 to 50 Å (45 Å for membrane proteins, whose T_m's could be in the range of 0.5–1 μs), even though larger distances can be measured with a longer period of signal averaging, but usually with reduced accuracy. Distances shorter than 20 Å introduce a relatively larger uncertainty in estimating the $C_\alpha - C_\alpha$ distances. Measurement of distances in the optimal range are fast and accurate, in most cases. The labeling sites and distance network should thus be chosen such that they provide optimal conditions for PDS, by increasing the relative number of optimal distances, as needed. Optimal conditions are not readily available for oligomeric proteins due to multiple labels and their typically large size. For an unknown structure, a preliminary scanning by several trial measurements may be very helpful.

Sensitivity of PDS

The sensitivity of pulse ESR spectroscopy is more difficult to define than for cw ESR, wherein strict criteria were established. In pulse ESR, similar criteria are harder to set, because relaxation times, which are the major determinants of the outcome of a pulse experiment, vary over a wide range among the systems studied. For this reason, often the single-shot SNR for a standard sample (e.g., gamma-irradiated vitreous silica) can be used to calibrate sensitivity. Due to variations in pulsed ESR techniques and samples, the capacity for a meaningful experiment based on considerations of its sensitivity should be decided on a case-to-case basis (Borbat et al., 1997), with all relevant parameters considered. The sensitivity of PDS techniques, specifically DQC and DEER, has been discussed in Borbat and Freed (2000), where the main criterion for sensitivity was based on the ability to perform a successful experiment (of reliably measuring a distance) in a reasonable period of time. It was chosen to correspond to an acceptable SNR, nominally taken as a S_{acc} of 10, which has to be attained in an acceptable time of experiment nominally taken as 8 h of signal averaging. Such an SNR would make it possible to obtain the distance (Fig. 13B), given a sufficient length of, t_{max} (cf. Relaxation section, p. 75), which, conservatively, should be at least one period of the dipolar oscillation. [A relaxed criterion, based on a shorter period, or even half of that, would still enable a less accurate estimate of the distance, depending on the specifics of the signal and given higher SNR than 10. This may include a priori knowledge of spin concentration and labeling efficiency or whether the distance is distributed over a narrow or broad range.] However, an S_{acc} of 10 is a bare minimum, and we usually require an SNR of at least 50, but preferably 100 to 200, to enable reliable distance distribution analysis (Chiang et al., 2005b).

Even though it is possible to estimate sensitivity from first principles (Rinard et al., 1999), we prefer to use an experimental calibration in the spirit of Borbat and Freed (2000), so the following approach has been chosen to give the estimates of sensitivity in distance measurements. First of all, a simple and standard experiment, such as a single-shot amplitude measurement of the primary echo, is performed under conditions when relaxation and other complications can be ignored. Then, the sensitivity of the single-shot experiment of a more complex method is deduced from this, based on the known theory of the method. Within such an approach, it suffices to measure the spin echo amplitude at a selected point of the nitroxide ESR spectrum with a two-pulse primary echo (PE) sequence, applied at a low repetition rate and with a short interpulse spacing. Such an experiment provides the SNR for a single-shot, $S_1(PE)$, which we refer to as per unit of concentration (1 μM) or per the number of spins (1 picomole), whichever is needed. Subsequently, the S_1 for the more complex experiment is estimated from $S_1(PE)$. Due to the limited capacity of simulating the outcome of a complex pulse sequence, such an estimate has limited accuracy, but it should be a reasonable predictor of the actual signal measurement. Finally, all the other major factors that influence the outcome of the actual experiment, such as relaxation, temperature dependence of the signal, instantaneous diffusion, or pulse sequence repetition rate, must be determined and their values used to estimate their effect on the SNR for a given distance and its range of uncertainty.

The calibration of DQC and DEER has been conducted for our pulse ESR spectrometer (Park *et al.*, 2006) at the working frequency of 17.35 GHz on a nitroxide sample of 4-hydroxy TEMPO in a vitrified solution of 50% w/v glycerol in H_2O with a 20 μM spin concentration in a 10 μl sample volume at 70 K, where most PDS measurements are performed. The DEER calibration used a primary echo (Mims, 1965)[14] generated by $\pi/2$-π pulses (π pulse of 32 ns) separated by 80 ns, with the pulses applied at the low-field edge of the nitroxide spectrum. A similar DQC calibration was based on $\pi/2$-π pulses with a 6 ns π pulse, and the same separation as in DEER, but pulses were applied in the middle of the spectrum. For the two measurements, the ratio of the echo amplitudes (DQC versus DEER) was about 6.5 and the ratio of SNRs of the single-shot signals at the condition of optimal signal reception (given by the integration of the spin echo in the time window defined by the time points corresponding to 0.7 of the echo amplitude) was about 3.0, that is, $S_1 \approx 0.42$ μM^{-1} (DEER) and $S_1 \approx 1.25$ μM^{-1} (DQC).

Based on these numbers, the estimates of the dipolar signals for the two methods, according to the analyses given in Borbat and Freed (1999, 2000), are summarized as follows. For 4-pulse DEER with 16/32/32 ns pulses in the detection mode and a 32 ns pump pulse, S_1 is 0.084 μM^{-1}, and for DQC based on a 3/6/3/6/3/6 ns pulse sequence, S_1 is 0.3 μM^{-1}, that is, it is greater for DQC by a factor of 3.6. This ratio is supported by our experimental observations, (cf. Figs. 9–10). Using the sensitivity analysis of Borbat and Freed (2000), we estimate the SNR of the raw data[15] of the full PDS experiment as

$$SNR = 2S_1 x^2 C \eta_c K(f, T_1)(ft_{exp}/N)^{1/2} \exp(-\frac{2t_{max}}{T_m} - 2kxCGt_{max}). \quad (17)$$

---

[14] The classic analysis of the SNR of a primary echo has been given by Mims, and the sensitivity in all PDS is directly related to that of a primary echo.

[15] Note that the factor of $N^{1/2}$ in Eq. (17) accounts for the effective averaging of each data point. But the raw signal can be processed in several ways in order to determine distances and the distributions in distances, when possible. In Borbat and Freed (2000), the number of points was not included in the expression for the SNR, because their sensitivity analysis was conducted within the context of the maximum measurable distances. In that case, based on consideration of spectral analysis (by FT), there should be at least $N_{min} = 4t_{max}/T_{dip}$ sampling points in order to satisfy the Nyquist criterion for the highest dipolar frequency of the Pake doublet, $2\omega_d$ (and just 2 for $t_{max} = T_{dip}/2$). It is this N_{min} that should be used as N in Eq. (17) to estimate the SNR for the dipolar spectrum in the frequency domain. Oversampling does not degrade the SNR, which is determined by the total number of signal samples (ft_{exp}) and N_{min}, but it helps to reduce aliasing in the spectrum and may have other positive effects. For reliable recovery of distributions in distances by Tikhonov analysis, 50 to 100 data points are desirable with the SNR in the data record of at least 30 (Chiang, *et al.* 2005a,b), but as has been demonstrated (cf. Fig. 13B), an SNR of 10 suffices for the case of a single distance. Equation (17) thus gives a conservative estimate.

Here, t_{exp} is the duration of the experimental data acquisition; f is the pulse sequence repetition frequency; N is the number of data points in the record; C is the doubly labeled protein concentration (μM); η_c is the ratio of the sample volume ($\leq 15 \mu l$) to that used in the calibration ($10 \mu l$). The terms in the exponent are consistent with those given in Borbat and Freed (2000), namely, the first accounts for the phase relaxation (but we use $\kappa = 1$ in Eq. $(17)^{16}$) and the second, for instantaneous diffusion. G is method-specific (Borbat and Freed, 2000) (with its definition provided later in reference to Eqs. (18) and (18a), and for the pulse sequences defined previously, it is approximately 0.14 in DEER and approximately 0.52 in DQC. We also include the spin-labeling efficiency, x, which modifies the fraction of both spins that need to be flipped in PDS, showing its strong effect on the outcome of an experiment. In the following text, we assume complete labeling for convenience in the discussion (i.e., $x = 1$). $K(f,T_1) = [1-\exp(-1/fT_1)]$ gives the effect of incomplete spin-lattice relaxation for a given relaxation time, T_1, and repetition rate, f. (K is 0.72 for the optimal repetition rate, when $fT_1 = 0.79$ and is unity when $fT_1 \ll 1$.) As an illustration of the capability of PDS in various regimes, we consider the following examples:

Short Distances, Low Concentrations

For a short distance of 20 Å ($T_{dip} = 154$ ns), we set $t_{max} = 0.48 \mu s \approx 3T_{dip}$ in order to provide very good resolution of distance; T_m is taken as 1.0 μs, that is, the shortest within its typical range (cf. Relaxation section, p. 75); 8 ns steps in t yielding 60 data points are taken as producing the signal record; a pulse repetition frequency f of 1 kHz should be optimal for a spin-labeled protein at 70 K. One finds from Eq. (17) that just $t_{exp} \cong 4$ min of signal averaging of the DQC signal provides an SNR of 10 for a C of 1 μM. DEER will require nearly 1 h (50 min) to achieve this result. Note that this concentration corresponds to just 10 picomoles of protein. A high SNR of 100 for DQC could be attained in 6.5 h for the same amount of protein.

Long Distances

We assume $t_{max} = 4 \mu s$, a typical T_m of approximately 2 μs, and the steps in t are taken to be 50 ns. Then, an SNR of 10 will be reached in 8 h for a C of 2.1 μM for DQC (while for DEER, it would be 104 h). By using one period of T_{dip}, we find $R_{max} = 59$ Å; for half of the period, R_{max} is 75 Å. (Longer distances cannot be estimated reliably with this SNR.) An accurate analysis of the distance distribution requires a higher concentration of at

---

[16] When $\kappa > 1$, such as for relaxation effects from nuclear spin diffusion, its partial refocusing in the DQC experiment provides an improved SNR (Borbat et al., 2004).

least 10 μM in order to provide a SNR of at least 50 (Chiang *et al.*, 2005a,b), under otherwise similar conditions.

Distances in the Optimal PDS Range

We consider 50 Å as an upper limit for the "optimal" PDS distance range (cf. Distance Range section, p. 77). T_{dip} is then 2.4 μs, therefore a t_{max} of 2.4 μs suffices to provide the distance sufficiently accurate for a structure constraint. We assume the rather challenging case of $T_m = 1.5$ μs; steps in t are taken to be of 32 ns; f is 1 kHz, C is taken as 25 μM; but now we require a good SNR of 50. Such a SNR will be achieved in 16 min by DQC. DEER will require nearly 3.5 hours to achieve the same result, or else the concentration must be increased (by a factor of 2–4). Shorter distances of 20 to 45 Å are measured faster, or else yield a better SNR or resolution.

Actual Case

A recent 4-pulse DEER experiment (using a 32 ns pump pulse) conducted on 10 μM of homodimeric protein CheAΔ289 (see Distance Range section, p. 77) mutant Q545C at 60K yielded a SNR of 50 in 1.2 h of signal averaging at an f of 1.2 kHz. The phase relaxation time, T_m, was about 3.3 μs (yielding a factor of 3.7 loss of signal due to relaxation). This SNR fully supports the estimates made previously. The measured distance was approximately 40 Å, although 50 Å ($T_{dip} = 2.4$ us) is also readily accessible. Considerably longer distances could be measured for residues that are more solvent-exposed or if a deuterated buffer is used, when relaxation times could be as long as 4 to 6 μs, and consequently, a t_{max} of up to 6 μs to 8 μs can be used. We also find that, in this case, by using a t_{max} of 3.5 μs (with 32 ns steps), an SNR of 10 could be achieved for 0.5 μM of this protein in 17 h by DEER but in less than 1.5 h by DQC. This corresponds to just 5 picomoles (0.8 μg) of a 160 kDa protein in the sample, which amounts to very high absolute sensitivity.

Absolute spin sensitivity is closely related to the concentration sensitivity; however, it does increase rapidly with an increase of the working frequency due to the smaller volume of a resonator used at a higher frequency; for example, at Ku band 25 to 250 picomoles of protein are routinely used in the optimal distance range. The smaller amounts are better suited for DQC. These amounts can be reduced by about an order of magnitude using smaller resonators than we currently employ, but by an even greater factor at a higher working frequency.

We remind the reader that the previous estimates relate to our 17.3 GHz spectrometer; lower estimates of sensitivity—in particular, absolute sensitivity—would apply to the typical pulse spectrometers that operate at 9 GHz.

The practical cases encountered in the biological structure applications of PDS encompass the following ranges of experimental parameters:

(1) Average double-labeled species concentration (5–100 μM); (2) Phase relaxation times (1–4 μs); (3) Distance of interest (10–80 Å); (4) Error in distance (0.5–10 Å); (5) Acceptable SNR (10–200); (6) Data collection time (0.5–24 h). These parameters are not independent, and the requirements of (6) would not necessarily be met for any arbitrary combination of 1 through 5. However, these examples show that PDS could address most of the cases that usually occur.

Other Aspects

Although the simple form of Eqs. (7) and (10) is well-suited for practical implementation of PDS, in reality, there are many details that are of interest perhaps only to the ESR spectroscopist. Even though subtle features do appear quite often, they are usually not important for the goal of obtaining a sufficient number of distance constraints of reasonable accuracy from nitroxide-based PDS. Those that are encountered most often are noted in the following:

Orientational Selection

As was pointed out in Larsen and Singel (1993), the anisotropy of the nitroxide magnetic tensors in the general case leads to orientational selection and should not, in general, be ignored. This is especially true for DEER, in which the excitations are selective. The problem has been studied for X-band DEER by numerical methods (Maryasov *et al.*, 1998), indicating that multiple sources of broadening tend to reduce effects of correlations between the orientations of the magnetic tensors being selected by the pulses and the orientation of the interspin radial vector.

Eqs. (6) and (10) can be modified to account for orientational selection. Powder averaging then takes the form for DEER:

$$V(t) = V_0 \langle G(\Omega, \Omega_1)[1 - p(\Omega, \Omega_2)(1 - \cos\omega_d(\theta, r)t]\rangle_{\Omega,\Omega_1,\Omega_2} \qquad (18)$$

and for DQC:

$$V(t) = \frac{1}{2}V_0 \langle G(\Omega, \Omega_1)H(\Omega, \Omega_2)[\cos\omega_d(\theta, r)t_\xi - \cos\omega_d(\theta, r)t_m]\rangle_{\Omega,\Omega_1,\Omega_2} \qquad (18a)$$

Here, $\Omega = (\theta, \varphi)$ defines the orientation of $\mathbf{B}_0$ in the molecular frame. The z-axis of the molecular frame is selected along $\mathbf{R}_{12}$, the vector connecting the two nitroxide NO moieties, and the Euler angles $\Omega_k = (\alpha_k, \beta_k, \gamma_k)$ ($k = 1,2$) define the orientations of the *hf* and g-tensors in the molecular frame for the two nitroxides (Fig. 15). G and H (or equivalently p) describe the extent of spectral excitation of spins 1 and 2, respectively[17] (Borbat and

---

[17] Equations 18 and 18a show only one of the two terms; the second term is obtained by permuting the subscripts 1 and 2.

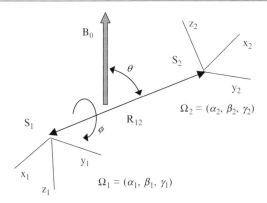

FIG. 15. The orientations of the nitroxides' magnetic tensor frames relative to the molecular frame, which is determined by the interspin vector $\mathbf{R}_{12}$.

Freed, 2000; Milov *et al.*, 2003b). The outcome of orientational selectivity is a distorted dipolar spectrum in DEER, which could result in incorrect distances emerging in the analysis based on Eqs. (6, 10, 15). DQC with its hard pulses is much less sensitive to orientational selectivity, especially when pseudosecular terms are insignificant. When desired, it can reveal the orientational correlations in considerable detail in a 2D mode of signal acquisition versus $t_\xi = t_m - 2t_p$ (Fig. 5) and versus the evolution of the spin echo time, t_2 (Borbat and Freed, 2000).

The issue of orientational selectivity is most pertinent to rigid conformations of nitroxide side-chains where $\Omega_{1,2}$ are distinct. In the case of distinct orientations (at X or Ku band), only Ω is typically used in the averaging implied by Eq. (18), and it suffices to average only over θ and to consider just β_1 and β_2 as the significant parameters (Milov *et al.*, 2003b). A particularly useful case arises for TOAC with its fixed orientation versus backbone, and it was explored in a peptide study (Milov *et al.*, 2003b). In PDS based on SDSL at working frequencies in X or Ku band, the typical side-chain flexibility, aided by unresolved inhomogeneous broadening, considerably decreases correlation effects due to partial averaging of G and H over $\Omega_{1,2}$. This is rather typical for flexible side-chain spin labels; hence, the effect hardly matters in such PDS studies. So far, in our experience, we have rarely encountered effects of orientational selectivity even for DEER for when the MTSSL is conformationally constrained.

Electron Spin-Echo Envelope Modulation (ESEEM)

Neclear ESEEM could also lead to a distorted dipolar spectrum, and in some cases interfere enough that distance determination is no longer feasible. DEER has been designed from the start to avoid ESEEM effects from magnetic nuclei. In the 3-pulse version, based on bimodal resonators and relatively soft pulses, ESEEM from surrounding protons is virtually absent, since the excitation and detection regions of the nitroxide spectrum are well separated. In a typical implementation of 4-pulse DEER utilizing a single power amplifier at X-band and a low-Q dielectric resonator (DR) or split-ring resonator, especially when short pulses of 15 to 20 ns are used, ESEEM cannot be discounted and has to be dealt with by using suppression techniques, which have been developed for both DQC and DEER (Bonora et al., 2004; Borbat et al., 2002; Jeschke et al., 2004a). Standard ESEEM suppression techniques are based on summing up several data records for a set of t_m in DQC (four collections in which t_m is stepped out by a half period of the nuclear modulation frequency), or τ' in DEER, which can be achieved with almost no loss in sensitivity. This, however, is rarely a necessity for protons in Ku-band DQC but sometimes desirable for X-band 4-pulse DEER. For Ku-band DEER, it is harder to avoid proton ESEEM due to the greater proton Zeeman frequency of $\sim$26 MHz, but its depth is smaller by a factor of $\sim$4 compared to X-band, and it can therefore be ignored or filtered out numerically. Deuterium modulation, however, is of greater concern in DQC even at Ku band, unless electron spin concentrations are low or t_m is sufficiently large ($>$4 μs).

Some Technical Aspects of DEER and DQC

A preferred setup for 3-pulse DEER is based on using two independent power amplifiers for the two frequencies (Milov et al., 1981). The amplifiers should be well isolated from each other in order to avoid unwanted interference by the pulse-forming networks. Overlap of the spectral excitations induced by the two frequency sources is undesirable. This requires limiting B_1 or increasing the frequency separation; both will reduce sensitivity. It is natural to use bimodal resonators in this scheme in order to optimize sensitivity and reduce overlap between excitation profiles of pulses in pumping and detection modes. Three-pulse DEER can be conducted with a single amplifier, but this necessitates using a traveling wave tube amplifier (TWTA) in its linear regime, which is some 10 to 12 dB below the preferred saturated mode of operation. But in the linear regime, there is still some "cross-talk" between the pulses, even when separated in time (for example, due to memory effects via the beam current). Simultaneous application of bichromatic irradiation may also contribute a problem.

Four-pulse DEER, however, can be readily set up with a single amplifier, since the pulse of the pumping mode does not coincide with the detection pulses, and thus stronger pulses can be produced. Pulse interaction is not entirely removed but becomes less of a problem if the distance between the first two pulses is not too short. Any residual interaction can be removed by using two independent amplifiers, and both can always be operated in their optimal regimes. Note that, in both forms of DEER, the apparent dead time (time resolution) is limited by the pulse widths, and thus is considerably longer than in DQC, which uses pulses as short as a few nanoseconds. DEER can be used without phase cycling or even with incoherent pulses. However, DEER requires high instrument stability in order to maintain gain, field, phase, etc. since all small drifts directly affect the echo amplitude, leading to low-frequency noise that could limit SNR. This requires state-of-the-art pulse generation and signal detection paths with low noise and drifts, which are difficult to achieve in a home-built instrument, unless it is designed and built with the care given by commercial equipment vendors. Figure 16 compares 3-, 4-pulse DEER, and DQC carried out in the same setup on the same sample with a single TWTA mode of operation.

A key virtue of DQC is the suppression of the large background signal (baseline) by means of its extensive phase-cycling, in particular, its use of the double-quantum filter. Unwanted modulation of the signal due to low-frequency noise and drifts in phase or gain becomes less important, thereby simplifying implementation and use. This also helps to reduce nuclear ESEEM effects, which are due mostly to modulation of the background signal from the single order coherence signals. The basic requirement is to provide reasonably accurate quadrature phase-cycling and sufficient B_1, which requires a more powerful and thus more expensive TWTA. Once these requirements are met, DQC is easy to set up and work with. Originally, the DQC experiment was conducted with a 2D-FT ESR spectrometer designed to achieve less than 30 ns dead-times with full 2 kW power in microwave pulses at a high repetition rate. This was a major instrumental challenge from the receiver protection standpoint. Since the instrumental dead-time does not need to be so short in the zero dead-time DQC technique, and repetition rates usually are low in low-temperature solids, the receiver protection is less of an issue and can be addressed by using readily available not-very-fast high-power limiters and switches. (Three- and 4-pulse DEER also have the advantage of remote echo detection and are not concerned with hard-to-obtain fast-acting receiver protection components, if the spectrometer is to be used exclusively for DEER.)

Case Study: PDS Reconstruction of Histidine Kinases
    Signaling Complex

Bacterial Chemotaxis

Motile bacteria move through a medium toward nutrients and away from repellents. The process that controls the motility is known as chemotaxis. In bacteria, the mechanism is based on the presence at the poles (the ends) of the bacterial cells, discoidal sensors composed of thousands of

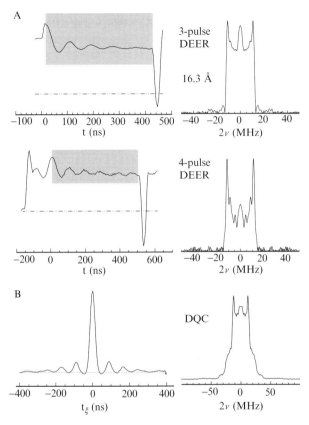

FIG. 16. (A) 3- and 4-pulse DEER, and (B) DQC (Borbat and Freed, 2000) are compared for a 16.3 Å rigid biradical in LC phase V, rapidly frozen from the isotropic phase; at −80° and 17.4 GHz. DEER was set up with a single power amplifier working in the linear regime at 10 dB below saturated output level. A low-Q dielectric resonator was used to accommodate the pulses at both DEER frequencies separated by ∼100 MHz. $\pi/2$ and π pulses were 10 and 20 ns in DEER and 3.2 and 6.2 ns in DQC. The pumping pulse was positioned at the low-field portion of the nitroxide spectrum. The informative parts of the signal traces in DEER are shaded. In 4-pulse DEER, the maximum of the signal is shifted in time as in DQC, so both 4-pulse DEER and DQC are zero dead-time pulse sequences. The outer turnover points of the Pake doublet are missing in the dipolar spectrum from the DEER signals. The DQC signal is considerably stronger and cleaner but decays somewhat faster due to spectral broadening caused by the pseudosecular term of the dipolar coupling. (DEER results unpublished, this lab.)

transmembrane chemical receptors of several types. They relay the effects of substrate binding over the distance of ∼250 Å to the catalytic part of the signaling complexes (Fig. 17), which are formed by the histidine

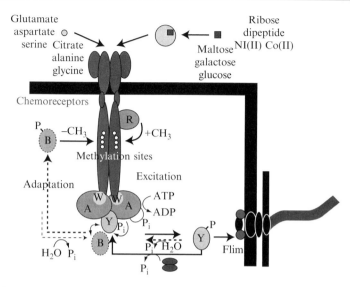

Fig. 17. The signaling bacterial receptor responds to chemical stimulants by activating histidine kinase CheA and, consequently, invoking a phosphorylation cascade that involves several other proteins. This ultimately affects the sense of rotation of the flagella motors and thus the swimming behavior. Receptors are coupled to the CheA dimers via the coupling protein CheW. A small conformational change, caused by stimulant binding at the periplasmic side of a ~300 Å long transmembrane receptor, is relayed over ~250 Å distance to the kinase at its distal cytoplasmic tip. The regulation is provided by enzymes, which (de)methylate or deamidate glutamate residues on the cytoplasmic part of the receptors. This may affect kinase activity via conformational changes of the signaling complex, cooperativity between receptors, or both, thereby changing the catalytic activity of the entire receptor array by orders of magnitude. (Adapted form Bilwes *et al.*, 2003.) (See color insert.)

kinase CheA and the receptor-coupling protein CheW that are attached to the membrane distal ends of the cytoplasmic side of the receptors. CheW-deficient mutants exhibit disrupted chemotaxis *in vivo*. The signaling complex interacts with several soluble proteins to establish control over the sense of rotation of the flagellum motor, thus changing the swimming behavior from tumbling to direct motion. In addition, it realizes feedback control of the gain by regulating the chemotactic response to stimuli by the means of adapting the receptor complex to the concentration of attractant, thereby achieving high sensitivity and wide dynamic range (Blair, 1995; Wolanin *et al.*, 2006). Individual signaling complexes are assembled into arrays and interact with each other, enhancing the capacity of an individual signaling complex.

Objectives

Even though the bacterial chemotaxis system is, in general, well understood, the essential details of the exact functional mechanisms of its main components are still missing. The structure and functional mechanism of the entire chemotaxis receptor complex, which consists of the transmembrane receptor, histidine kinase CheA, and the coupling protein CheW, is yet to be learned, and it was our goal (Park *et al.*, 2006) to shed some light on it. There is no crystal structure of the full-length CheA, nor of CheA in a dimeric complex with CheW. CheA is a homodimer with each monomer organized into five domains P1–P5, separated by linkers of various lengths (Fig. 18). The P1 and P2 deleted protein, CheAΔ289 has a known X-ray derived structure (Bilwes *et al.*, 1999). The dimerization domain, P3, is folded into two anti-parallel α-helices. The P3 domains from the two protomers form a 4-helix bundle, which holds together the dimer. The P1 and P2 domains, separated by long flexible linkers have known X-ray structures. They are believed to have few structural constraints, although NMR work

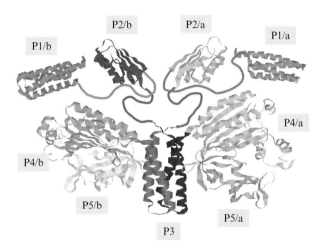

Fig. 18. A collage of the histidine kinase CheA. The whole kinase assembles as a dimer with the monomer composed of 5 domains P1–P5, connected by flexible links of various lengths. CheAΔ289 (lacking P1 and P2 domains) from *Thermotoga maritima* in the center was crystallized; P1 and P2 crystallized separately (PDB codes: 1TQG, 1U0S) were added to the figure to provide a complete view of the protein. Two P3 dimerization domains assemble into a 4-helix bundle; P4 is the catalytic domain, which phosphorylates conserved histidine of the P1 domain; P2 binds CheY or CheB, which, in turn, are phosphorylated by P1. Phosphorylated CheY(B) dissociates and controls flagella and receptor, respectively. P5 is a regulatory domain, which binds CheW and receptor and thus mediates regulation of kinase phosphotransfer activity. (Protein structure was rendered using Chime.) (See color insert.)

in 2005 (Hamel and Dahlquist, 2005) indicates the possibility of P1 association with P4. The NMR method has provided the structure of CheW
(Griswold *et al.*, 2002). Two CheW molecules bind a CheA dimer with
high affinity ($K_d \sim 10$ nM), and they provide binding sites for the receptors.
The structures of the cytoplasmic and periplasmic parts of several types of
transmembrane receptors are available for *E. coli* and *T. maritima*.

Given the potential of accurate long-distance constraints from PDS and
a "triangulation" approach (discussed in the next section), this methodology has been applied to the problem of bacterial chemotaxis for the first
time. Initially, we had the modest goal of confirming the location of the
CheW binding site, according to common belief. This implied obtaining a
few distances. Nevertheless, it was decided to apply the full triangulation
approach and we sought a scaffold of constraints that will orient and dock
CheW with high confidence (cf. following text). Our PDS efforts have,
however, developed to the point of establishing the structure of the whole
signaling complex, based on our findings on the structure of the CheA/
CheW complex in dilute solutions. Our efforts, synergistically combining
X-ray crystallography and PDS with other biochemical methods, were
reported in Park *et al.*, 2006.

Triangulation

The "triangulation" approach to protein mapping (Borbat *et al.*, 2002) is
based on obtaining a network of distance constraints from a set of spin-
labeled sites such that they uniquely define the coordinates of all (or most)
of the sites. This task can be accomplished by making a sufficient number
of double mutations and then measuring the distances between the respective pairs of spin labels in a "one-at-a-time" manner. It is not feasible, in
general, to obtain distances simultaneously among several spin labels due to
the flexibility of the side-chains and the structural heterogeneity of proteins,
which yield fairly broad distributions in each distance. However, there can
be favorable cases (Bennati *et al.*, 2005; Chiang *et al.*, 2005b) (cf. Fig. 21A).
For a monomeric protein, a convenient set of labeling sites should be
selected, and then a number of double mutations for this set would be
made in order to produce a network of distances by PDS. A sufficiently
large rigid distance network (scaffold) based on tetrahedrons (Borbat *et al.*,
2002) will strongly restrain the loci of spin labels and thereby the possible
conformations of the protein (cf. Fig. 19). Building such a rigid network
resembles "triangulating" the protein landscape. Such constraints can be
used to solve the protein structure at a low resolution of 5 to 10 Å. When a
very rough structure or the oligomeric state of a protein complex is of
interest, a few distances may suffice (Banham *et al.*, 2006). Obtaining an

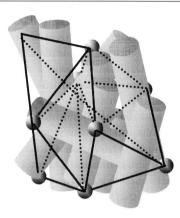

Fig. 19. A cartoon of a rigid triangulation grid based on tetrahedrons. The symbolic protein is encased into a "cocoon" of constraints, which uniquely define its shape and limit possible scenarios of its folds. The known secondary structure aided by homology modeling could produce a fairly accurate structure of a single protein or for it being a part of a larger complex.

accurate structure could be considerably more involved, since it could require obtaining several dozen distances in order to significantly restrain the possible conformations. The sites should be accessible for the spin-labeling reagent, and they should not alter protein structure or function; this may limit their selection. The task of site selection could be facilitated by a knowledge of the secondary structure. Site selection could be aided by nitroxide scan and/or chemical cross-linking data, which gives the information on the secondary or tertiary structure and residue exposure to solvent, thus helping to select the sites for triangulation. However, fewer than a dozen sites can produce an adequately large number of constraints (up to $N(N-1)/2$, where N is the number of sites). For this reason, there should be little concern about site availability.

Another application of this method is to determine the structure of a protein complex, (e.g., that comprised of two proteins, A and B). In this case, the triangulation grid can be based on the individual triangulation grids for each protein (grid A and grid B) and the interprotein distances (grid AB). Intraprotein grids A (and B) are obtained by the triangulation of doubly labeled proteins A (and B), whereas the interprotein grid AB is obtained using singly labeled A and B proteins. Possible structural changes of A and B, when they form the complex, could be elucidated by obtaining the grids A (and B) in the complex with the respective unlabeled "wild-type" (WT) partner. This is the approach that has been applied to the problem of solving the binding structure of CheW with the P5 domain of CheAΔ289. The three grids are then used together to solve the structure of the AB complex. Clearly, this approach can be extended to multiprotein complexes and protein–RNA complexes. The task is much easier when individual protein (or their subdomains) structures are known from crystallography or NMR. However, triangulating the CheAΔ289/CheW complex is an important special case of oligomeric proteins that requires additional considerations, as will be discussed.

CheAΔ289, which is a homodimer, binds two CheWs. Its known X-ray structure indicates (an imperfect) C2 symmetry. Triangulation of a homodimeric (or, in general, oligomeric) protein is a less straightforward task than that for monomeric proteins or complexes, as has been discussed. Double mutations of protomers making up a dimer will result in having four spin labels in the dimer. Thus, four distances are possible in a dimer in the case of the C2 symmetry, and six distances are possible in general. Due to the limited capacity of PDS to resolve multiple distances (except in a few favorable cases; cf. Fig. 21A), the strategy of spin labeling or constructing the dimer should be adjusted to overcome the complications from multiple labels. One solution to this problem is to engineer dimers (or oligomers) such that they contain only one doubly labeled protomer, with the rest being WT (cf. also discussion in the Using Heterodimers, p. 104). This approach was exercised in only a few cases (cf. Using Heterodimers, p. 104) due to the difficulties of making heterodimers. The second approach, which is not necessarily applicable to all oligomeric proteins, is to select the labeling sites such that the distances between the sites on different protomers are considerably different from the intra- or interprotein distances within the protomer, thus making possible their separation.

Four mutation sites, N553, S568, D579, E646, were selected for CheA (cf. Fig. 20) on the distal end of the P5 domain surrounding the putative binding site, based on indirect studies (Bilwes *et al.*, 1999; Boukhvalova *et al.*, 2002; Hamel and Dahlquist, 2005; Shimizu *et al.*, 2000). Site selection was greatly facilitated by a knowledge of the X-ray structure of CheAΔ289. First of all, these four sites on P5 form the vertices of a sufficiently large tetrahedron with respect to which the triangle, made by the three chosen label sites (S15, S72, S80) on CheW is oriented in space, once all necessary interprotein distance measurements are made between labels on CheAΔ289 and CheW. Figure 20C shows the network of interlabel distances that we anticipated at the outset of the study. This triangulation network fixes the vertices, but it does allow for the mirror image. The knowledge of the P5 X-ray structure makes it possible to select between the two solutions. Next, we note from the known CheAΔ289 structure that

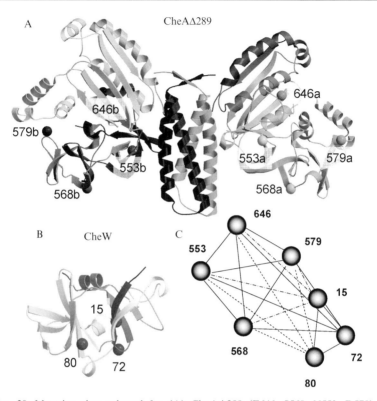

FIG. 20. Mutation sites selected for (A) CheAΔ289 (E646, S568, N553, D579) and (B) CheW (S15, S72, S80). Most of the sites are separated by more than 60Å, thus minimizing problems associated with these multi-spin cases. Additional sites were mutated at a later stage of the study in order to assess the global protein structure after two different conformations of CheAΔ354/CheW were provided by crystallography to confirm ESR-derived structure. (C) A suggested network of restraints for triangulation. Note that the putative binding site on the distal part of P5 domain detected by X-ray crystallography (Bilwes *et al.*, 1999) was under consideration here. (The structures were rendered using Mol Script.) (See color insert.)

the distance between labeled sites on P5 of the two protomers is considerably greater (>60 Å) than that between the sites within each P5 (<40 Å), as schematically illustrated in Fig. 20A. The dipolar signal for a shorter distance exhibits much faster time evolution than that for a considerably longer distance (due to the $1/r^3$ dependence of the dipolar oscillation; cf. Eq. (5)). Consequently, they could be readily separated. Last, these four P5 sites made it possible to detect and quantify possible conformational changes in P5, as has been outlined.

All six double mutants of CheAΔ289 and all three of CheW were engineered to obtain the intraprotein set of distances (within P5 and CheW). This was carried out with and without binding of the respective wild type partner to address the possibility of change in the structure of P5 or CheW. The interprotein measurements employed the same set of mutated sites and encompassed the measurement of 12 possible distances between singly labeled CheAΔ289 and CheW. Single mutations of either CheAΔ289 or CheW were used to probe, at a later stage of study, the interdomain distances between the symmetry-related sites in the CheAΔ289 dimer and the CheAΔ289/CheW complex, with the only problem encountered being the large protein size. Most distances were expected to be in the 40 to 100 Å range. Although initially we regarded distances exceeding 60 Å as practically unmeasurable, we did find that fairly good estimates could be obtained for distances in the 60 to 80 Å range, based on reasonable assumptions regarding the time envelope of the dipolar signal and given a good SNR (cf. Fig. 13C).

Spin-Labeling and Sample Preparation

Certainly, advance knowledge of the structure of individual subunits of a large protein or complex greatly simplifies the choices of the sites for spin-labeling, as we discussed in the previous section. For a protein of unknown structure, several trial-and-error attempts would be necessary in order to establish an idea of its folds and then to refine it by further PDS measurements.

Nitroxide side chain flexibility permits a number of rotamers (Tombolato *et al.*, 2006), which can strongly deviate from the "tether-in-cone model" (Hustedt *et al.*, 2006) as was evidenced by the X-ray study of spin-labeled T4 lysozyme (Langen *et al.*, 2000). This would make it more difficult to predict the locations of the respective backbone carbons. Using different types of spin labels may help to identify the sites with nontypical conformations of side-chains; using site insensitive ones (Cai *et al.*, 2006) could help to eliminate the problem (of unusual rotamers) altogether at the expense of greater side-chain flexibility. In some cases, it is desirable to have a spatially restricted nitroxide side-chain to produce better defined distances especially for shorter distances, or else a solvent exposed site, which is easier to label and which will permit access to larger distances due to longer T_2's for such sites (Huber *et al.*, 2001). All these considerations have been appreciated in planning the study of the CheA/CheW complex. Exposed residues were selected for CheA. For CheW, the selected serine residues were distant from the conserved residues implicated in binding to CheA and they also had limited mobility, according to the NMR structure for CheW (PDB code: 1K0S).

All proteins were prepared as described elsewhere (Park *et al.*, 2006). CheAΔ289 was used for the PDS studies; the full-length CheA (which is more difficult to express) was used, in some cases, to confirm that deletion of P1 and P2 had no significant effect on the structure of the complex. (Note that for PDS, the full-length CheA does not introduce any challenge compared to the case of deletion and distances between all 9 domains could be measured.) Proteins were labeled for 4 h at 4° with 5 to 10 mM (1-oxyl-2,2,5,5-tetraethylpyrollinyl-3-methyl)-methanethiosulfonate (MTSSL, Toronto Research Chemicals, Toronto) in-gel filtration buffer while the His-tagged proteins were bound to nickel-nitrilotriacetic acid agarose beads. The proteins were eluted after 6 to 12 h in the presence of thrombin. Spin labeling sites are shown in Fig. 20A, as discussed in the previous section. Protein concentrations were typically in the range of 10 to 150 μM. They could not be increased considerably due to the onset of what we believe is a tendency of CheA to reversibly assemble into supramolecular constructs of unknown nature. Concentrations in the range we used usually suffice to yield clean dipolar signals for homodimeric protein, given the quantitative spin-labeling. This, however, was rarely the case, as high spin-labeling efficiency was not always achieved, especially at the early stages of the study. As a result, the concentration of double-labeled protein was a factor of 1.5 to 4 less than ideal, depending on labeling site.

As we described in the previous section, the complexes frequently involved four spin-labeled sites. For such cases, the extent of quantitative spin labeling is a complication due to an admixture of complexes bearing 3 or 4 spins, which leads to a more complex time-domain signal, especially when distances are not well separated. Magnetic dilution can sometimes be useful for oligomeric proteins, since it leads to a simpler case of mostly doubly labeled complexes (Klug and Feix, 2005; Zhou *et al.*, 2005).

In all the protein solutions, 30% w/v of glycerol was added to render the protein distribution more uniform throughout the frozen sample and to minimize possible damage to protein by water crystallization. For cryoprotection, sucrose can be used if there is any adverse reaction of protein with glycerol. But sucrose is less convenient, and we did not see any significant effect of glycerol on kinase activity or distances. The 10 to 20 μl of solutions was loaded into an 1.8 mm i.d. suprasil sample tube, plunge-frozen in liquid nitrogen, and transferred into the dielectric-resonator-based ESR probe, where it was cooled down to the working temperature in the 50 to 70 K range. Data collection times were from 10 min to 12 h, depending on concentration and distances, but were typically 30 to 120 min. Long-distance (>50 Å) measurements required using exposed sites (at least one site in the pair) deuterated solvent, and 2 to 12 h of signal averaging.

Technical Aspects

The 4-pulse DEER and 6-pulse DQC ESR methods outlined in The Newer Methods section (p. 63) were used in all cases at a frequency of 17.35 ± 0.05 GHz (corresponding to Ku band) in the two traveling-wave tube amplifier (TWTA) configuration for DEER, with only a single TWTA needed for DQC. The existing 2D-FT ESR spectrometer (Borbat *et al.*, 1997) has been modified to enable DEER by the addition of the second TWTA in order not to significantly alter its existing modes of operation and additionally to minimize interaction between the two channels. A stand-alone microwave pulse-forming channel and signal receiver for the pumping frequency had also been added, and the output pulses from the second amplifier (after passing through a rotary-vane attenuator) were simply combined with the main output by means of the 10-dB directional coupler.

A dielectric resonator coupled to a Ku-band waveguide was installed into CF935 helium flow cryostat (Oxford Instruments, Ltd.) and over-coupled to accommodate short pulses in DQC or frequency-separated pulses in DEER. $\pi/2$ and π pulses were 3.2 and 6 ns in DQC and, typically, 16 and 32 ns in DEER. The pumping pulse in DEER was in the range of 20 to 32 ns. Typically, the pumping pulse frequency, ω_B, was offset by −65 MHz from the detection frequency ω_A in order to pump at the center maximum of the nitroxide spectrum (cf. Fig. 6). At this working frequency, ESEEM from matrix protons is barely visible in DEER and is not a factor in DQC either. However, for a deuterated solvent, a weak 4 MHz ESEEM is detectable in DEER in the case of short pumping pulses and could be problematic for DQC; since a few t_m were suitable to provide blind spots (Borbat and Freed, 2000). Since spin concentrations were not very low, the sensitivity rarely was a concern, so we used DEER in most measurements, which also reduced the effects of the deuterium modulation. However, for short distances, DQC was used, and in the case of low spin concentrations, DQC often improved the SNR by a factor of 2 to 4 (cf. Fig. 10).

DEER/DQC Data and Analysis

The DQC and DEER signals were of very good quality (Fig. 21A–D). Dipolar signals from the sites on the different protomers could be distinguished and they constituted a slowly varying "background." They were removed by fitting them to a second- or third-degree polynomial in the same fashion as removal of intermolecular dipolar contribution (cf. Data Processing in DQC and DEER, p. 73, and Fig. 21C). The signals isolated from the background were then Fourier transformed and, initially, the distances were estimated based on the dipolar splitting in the Pake doublets

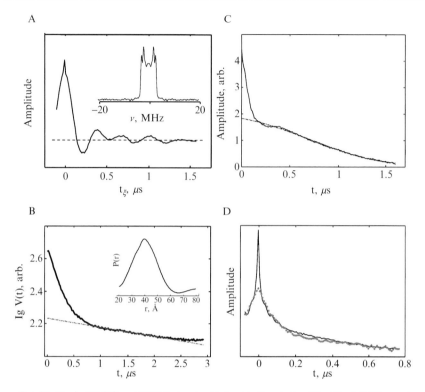

FIG. 21. Typical DQC and DEER signals obtained in the study of the CheAΔ289 complex with CheW. (A) The time-domain DQC signal for the full-length CheA labeled at S318C site on P3. 6-pulse DQC sequence was used at 17.4 GHz. Protein concentration was approximately 80 mM. Estimated doubly labeled protein concentration was ∼30 mM. Data collection time was 45 min. Inset: the dipolar spectrum is the real part of FT of the time-domain data after removing small linear baseline. Two well-defined distances of 26.3 and 28.7 Å are inferred from the two resolved singularities. They probably correspond to two distinct conformations of the P3 domain or entire CheA (Bilwes *et al.*, 1999). (B) Time-domain DEER signal for CheW complex with CheAΔ289-Q545C. Inset shows distance distribution. Concentration of CheAΔ289 was 25 μM, signal averaging time was 8 h 20 min. The latter part of the signal deviates from a straight line due to overlap of pulse excitation at the two frequencies in DEER. The signal after subtracting the background was apodized prior to L-curve Tikhonov regularization. (C) Time-domain DQC signal (solid line) for CheW80 complex with CheAΔ289-N553C. The rapidly decaying component of the signal corresponds to the distance between labels on the same monomer of CheAΔ289; the slowly decaying component of the signal originates from long distances between spin labels on different monomers. It was fitted to a third-order polynomial (dashed line) and removed (cf. DEER/DQC Data and Analysis section, p. 97). The 6-pulse DQC sequence was used at 17.4 GHz, data collection time was 22 min. (D) Time-domain DQC (solid line) and DEER (dotted line) signals from heterodimer of CheAΔ289 labeled at positions 318 and 545 in one of two protomers. The difference in signal shape is due to a greater sensitivity of the first method to short distances. Otherwise, after proper scaling, signal envelopes follow each other closely (not considering a large dc offset present in DEER data), with both reflecting rather broad distributions in distances similarly to (B).

TABLE I

INTRA- AND INTER-DOMAIN DISTANCES IN THE TRIANGULATING CHEΔ289/CHEW
(PARK *ET AL.*, 2006)

Mutant	S15C	S72C	S80C	N553C	S568C	D579C	E646C
S15C		27&29[a]	18.2	37	54.5	61	43.7
S72C	X		24.5 & 30[a]	27	49	46	32.5
S80C	X	X		26	47	54.5	39.5
N553C	X	X	X		23.5	34.5	32
S568C	X	X	X	X		32.5	35.5
D579C	X	X	X	X	X		28
E646C	X	X	X	X	X	X	

[a] The second minor conformation with better defined position is likely due to an immobilized rotamer at the exposed site 72. The main component shows flexibility consistent with the type of the site.

for well-defined distances or by using half-widths of the dipolar spectra in unresolved cases. [Later, the regularization methods were used (cf. Distance Distributions section, p. 75)]. The distances between P5 and CheW and within CheW or P5 are summarized in Table I. The errors were estimated to not exceed ±2 Å, in most cases. A few of the largest distances (>50 Å) were possibly slightly biased by unwanted contributions from long distances between the sites on different protomers. The intra-protein distances were consistent with the expectations based on the CheAΔ289 X-ray structure, CheW NMR structures, and the expected geometries of the nitroxide side chains.

Complexes of double-labeled CheAΔ289 with WT CheW or vice versa probed how the complex formation perturbed the local structure. All nine possible double mutants of CheAΔ289 (6) and CheW (3) were tested. Only subtle changes in widths of distance distribution were observed in a few cases. Thus, perturbations to the local structure were smaller than the spatial resolution of the triangulation method. This justified the use of a rigid-body approach to dock CheW to P5.

Metric Matrix Distance Geometry and Rigid-Body Refinement

A set of triangulation constraints for N nitroxides can be transformed into coordinates by the metric matrix distance geometry method (Crippen and Havel, 1988), at least to determine whether the set is embeddable. One starts with the set of all possible distances d_{ik} between nitroxides i and k and constructs a rank $N-1$ metric matrix $\mathbf{G}$, first, by placing one nitroxide at the origin, as

$$g_{ij} = (d_{i0}^2 + d_{j0}^2 - d_{ij}^2)/2 \tag{19}$$

The Gram matrix $\mathbf{G}$ is then numerically diagonalized, to give the eigensystem $(\mathbf{w}_i, \lambda_i)$, with λ_i being the eigenvalues and $\mathbf{w}_i$ eigenvectors, respectively. The matrix $\mathbf{G}$ is

$$g_{ij} = \sum_{k=x,y,z} x_{ik} x_{jk} = \sum_{k=x,y,z} w_{ik} w_{jk} \lambda_k \tag{20}$$

with x_{ik} being Cartesian coordinates of the i-th atom. The metric matrix is positive semi-definite of rank, at most, 3. Practically, the three largest eigenvalues are used to calculate the coordinates of $N-1$ nitroxides according to

$$x_{jk} = \pm \lambda_k^{1/2} w_{jk} \tag{21}$$

where the sign is the same for all coordinates, reflecting the two mirror image-related solutions.

The method was applied with the distance information from the sites at the four positions on CheAΔ289 (N553C, D568C, E646C, S579C—all on the P5 domain) and the three CheW positions (S15C, S72C, S80C). All 12 interprotein distances and 9 intraprotein distances between 7 nitroxides (cf. Table I) were used. The CheW labeled site S15C was placed at the origin; subsequently, the metric matrix of rank 6 was generated from the other measured distances and diagonalized with a MATLAB program. The eigenvalues are (9326.1, 906.2, 509.9, 208.5, 125.9, −46.5), with the 3 smallest values being nonzero due only to measurement uncertainty. The three largest eigenvalues were used to calculate the coordinates of the four label sites on CheAΔ289 and the two remaining label sites on CheW; two possible mirror image-related coordinate sets were generated. Of the two structures, one, consistent with the relative positions of the four sites on P5, was selected. After regenerating the distance matrix from the coordinates, we found that the new distances were within 2.4 Å bounds (1.2 Å rmsd) from the experimental distances, with three distances showing larger errors (5.2 Å rmsd). A possible reason is a displacement of the E646C residue upon CheW binding and larger errors for distances above 45 Å. The resulting triangulation grid is shown in Fig. 22, and it is clear that it uniquely defines the docking of CheW.[18]

This method, in its generic form, requires all distances between the sites, and it does not consider experimental errors. There is, however, a better

---

[18] In Park et al. (2006), site 568 was not used in an otherwise similar calculation and the remaining efforts were as follows. Least square fitting among the three site coordinates calculated from distance geometry and the C_α atoms of the labeled residues on the CheAΔ289 structure and the three CheW coordinates and their corresponding positions on a rigid CheW structure (taken from the NMR ensemble) produced the complex.

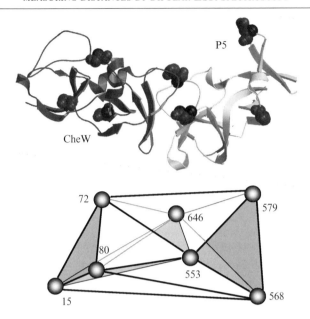

FIG. 22. (top) X-ray structure of P4/P5/CheW complex (P4 is not shown). Residues mutated to nitroxides for PDS are shown in a space-fill representation. (bottom) - Positions of nitroxide moieties found from PDS constraints correspond to that in the top figure. CheW appears to be slightly tilted and rotated about its long axis compared to the X-ray structure. The difference between X-ray and PDS is not large but cannot be discounted. Several reasons can be given: (1) the mutated residues should be replaced by nitroxide side-chains in their site-specific conformations to arrive at a better correspondence of PDS and X-ray derived structures; (2) Couplings between symmetrically positioned residues biased the long-distance constraints; (3) P4/P5/CheW was crystallized in the absence of P3, with which both CheW and P5 are expected to interact (Park *et al.*, 2006); (4) mutations at N553, and especially E646, might have had a small effect on the binding interface. Mutated sites were selected to optimize constraints in the originally presumed case of binding to the distal binding site of P5 (cf. Fig. 18), which turned out not to be the case. Thus, some chosen distances were outside the optimal range. (The ribbon structure was rendered with Mol Script; Delauney triangulation generated and rendered by MATLAB.) (See color insert.)

method (cf. Bhatnagar *et al.*, 2007) to dock domains by rigid modeling based on the CNS software package (Brunger *et al.*, 1998), which implements the distance geometry in a more powerful way. By virtue of the implementation of distance geometry algorithms in CNS, experimental distance constraints and their uncertainties (obtained from the inverse reconstruction of time-domain data) needs to be entered. Any missing constraints could be estimated but with large errors, and thus have only minor impact on the outcome. The coordinates of the P5–CheW complex, determined by distance geometry, have now been optimized by refinement with CNS (cf. Bhatnagar *et al.*, 2007). The PDS measured distances were treated as NMR NOE restraints. The structure was refined until convergence was achieved. To test convergence, CheW was displaced manually in various directions and the complex refined. Within rigid-body displacements of ~15 Å and rotations of ~30 degrees, the same unique solution emerged. When these methods were applied to the CheA dimer under the PDS restraints of the intersubunit P5 distances, they suggested that, in solution, P5 is slightly closer to P3, as compared to the X-ray structure of CheAΔ289. The binding site of CheW inferred from the ESR-derived structure was then confirmed by solving X-ray structure of the CheAΔ354 (Park *et al.*, 2006), which is a monomeric complex of P4–P5, with CheW.

Interdomain and Interprotein Distances in the CheW/CheAΔ289 Complex

After having solved the CheW docking issue, the global topology of CheA/CheW in dilute solution was explored. The following additional single mutations (Fig. 23) S318C (in P3), Q545C, D508C (in P5), E387C, K496C (in P4), and several double mutants shown in Table II were made for this purpose. A set of interdomain, intersubunit, and interprotein distances was generated (Table II) to explore the structure of CheAΔ289 and the CheAΔ289/CheW complex. Heterodimers were constructed in a few cases, so that intrasubunit and intersubunit distances between nonsymmetry-related

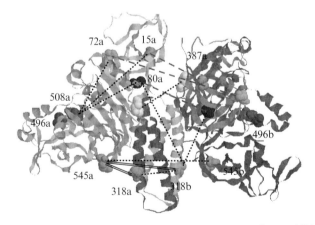

FIG. 23. Distances between several CheW mutants were measured to establish its position on the "top" of CheAΔ289. Heterodimers of P3 (S318C) and P5 (Q545C) mutants were used to establish the fixed position of the P5 domain, stabilized by interaction with P3. (Rendered with Chime.) (See color insert.)

TABLE II

INTER-DOMAIN DISTANCES (Å) IN CHEA, CHEAΔ289, AND CHEAΔ289/CHEW COMPLEX
(PARK *ET AL.*, 2006)

Domains	Residue pair[a]	Proteins in sample	Model, $C_\alpha - C_\alpha$	ESR average	ESR range
P4-P4	A387- B387	CheA[b]	42	44	30%
	A387- B387	CheA + CheW	42	46	30%
	A387- B387	CheA, full length	42	45	30%
	A496- B496	CheA	90	68[c]	30%
	A496- B496	CheA + CheW	90	N/A	
P3-P3	A318- B318	CheA	22	28	26/30[d]
	A318- B318	CheA + CheW	22	28	26/30[d]
P3-CheW	318- W80		35/44	43	30%
P4-CheW	387- W80		16/43	40–50	30%
	387- W15		17/44	40–50	30%
	387- W72		28/50	40–50	30%
P4-P5	508–646		15/65	25	22–38
P5-P5	A568- B568	CheA	76	49[c]	30%
	A568- B568	CheA + CheW	76	57[c]	30%
	A545- B545	CheA	40	41	32–50
	A545- B545	CheA + CheW	40	38	32–50
	A646- B646	CheA	62	N/A	
	A646- B646	CheA + CheW	62	61	30%
	A553- B553	CheA	63	64	30%
	A553- B553	CheA + CheW	63	N/A	
CheW-CheW	W15- W15	CheA + CheW	49	~60[c]	30%
	W80- W80	CheA + CheW	51	59–60[c]	30%
	W72- W72	CheA + CheW	70	67–70[c]	30%

[a] A and B refer to two protomers.
[b] Data for CheA289.
[c] Possible interference by aggregation.
[d] Two well-defined distances.

labels could be determined in the absence of unwanted signals from symmetry-related labels.

It was a time-consuming project, because many large distances (>50Å) were also not well defined due, in part, to the apparent freedom of the P4 and P5 domains to move about their hinges. In particular, P4 appears to exhibit substantial flexibility, as evidenced by the 387/387 data. This was not the result of deletion of P1 and P2 since similar data were obtained for full-length CheA. The P5 domain was found to be somewhat less flexible.

The broad distance distributions in Table II may have been caused, to some extent, by association of CheA dimers, which was not originally anticipated. But increasing the concentration above 500 μM indicated a network or aggregation of unknown nature. That is, the local (as opposed to the average) spin concentration, as indicated by the intermolecular part of the DEER signal, was about triple that expected from a uniform solution and no long distances could be reliably measured. At 50 to 200 μM, a moderate flexibility of the P4 and P5 domains was observed that is probably within the amplitude of difference between subunits found in the asymmetric crystal structure (Bilwes *et al.*, 1999). Association of CheA dimers with each other via the P5 domains seen in the crystal (Bilwes *et al.*, 1999) is also supported by our PDS experiments in solutions as the protein concentration is increased. No efforts to eliminate CheA association were made other than keeping concentrations in the 50 to 100 μM range, where these additional interactions were minimized. Nevertheless, the tendency of the proteins studied to associate possibly had an impact (rather weak but non-negligible) on measured interdomain distances, in particular for those exceeding 50 Å.

The intersubunit distances between labels at residues 387 on P4, 318 on P3, and 553, 545 or 646 on P5 confirmed that the orientations of the domains of the CheAΔ289-CheW complex in solution agree well with those found in the crystal structure of CheAΔ289. Of the two structures of CheW-P5-P4 (PDB code: 2CH4), only one was supported by PDS as relevant for CheAΔ289-CheW complex, with the second possibly present as a minor conformation. The binding of two molecules of CheW to one CheAΔ289 dimer was verified by measuring the inter-CheW distances between spin labels attached to sites 15, 72, 80, and 139 in the presence of wild type CheAΔ289. All measured distances were consistent with the expectations for the reconstructed CheAΔ289-CheW complex.

An interesting finding was that the ESR distances suggest that, in solution, P5 (in both the free CheAΔ289 dimer and the CheAΔ289-CheW complex) assumes an average position relative to P3 that is slightly different but within ±10 Å of that predicted by the CheAΔ289 crystal structure. The position of P5 may well be affected by P5/P5 contacts in oligomers of CheA dimers, subject to further study.

Using Heterodimers

The position of the P3 domain relative to P5 was verified by measuring distances within CheA heterodimers in which P3 residue S318 and P5 residue Q545 were either labeled only in the same subunit or in opposite subunits. The ESR-measured intrasubunit and intersubunit distances between 318 and 545 correlated well with separations in the CheAΔ289 dimer measured by X-rays (Table III). Heterodimers offer a useful approach for detailed ESR mapping of oligomeric proteins. Ideally, one constructs the

TABLE III
INTER-DOMAIN DISTANCES (Å) IN HETERODIMERS OF CheAΔ289 AND IN THE COMPLEX WITH CheW
(PARK *ET AL.*, 2006)

Domains	Residue pair	Proteins in sample	Model, $C_\alpha - C_\alpha$	ESR average	ESR range
P5-P3	A545- A318	CheAHD	12	14	12–30
	A545- A318	CheAHD + CheW	12	14	12–30
	A545- B318	CheAHD	30	30	

complex to have only two labels (e.g., both spin labels on only one proto-mer, or one spin label each on two protomers in well-defined locations in the complex). Heterodimers can be expressed as tandem constructs (Liu *et al.*, 2001); however, this methodology is not simple, and it may fail to express or fold the protein properly. Alternatively, one has to dissociate the protein and reassemble it into the heterodimer when this is at all possible. CheAΔ289 does dissociate at elevated temperatures, and this property was applied to engineer heterodimers with spin labels on the same or opposite subunits. To do so, CheAΔ289 mutants, double-labeled at 318 and 545 sites (Fig. 23), after incubation at 65° with the wild type carrying the histidine tag were recombined and labeled on a nickel affinity column. Magnetic dilution is sometimes a useful approach to studying oligomeric proteins (Klug and Feix, 2005). Spin labeling of different subunits required some level of magnetic dilution (by a factor of 3 to 5) of histidine-tagged mutants with WT to keep the fraction of multi-spin complexes much lower than that of complexes with two spins. This led to increased fraction of single labeled complex.[19]

Discussion and Perspective

We have described a successful application of PDS to the study of bacterial chemotaxis. It has led to a new perspective into the possible organization of the receptor signaling array (Fig. 24), which is substantially different from what had previously been suggested for the *E. coli* receptor (Shimizu *et al.*, 2000). This new perspective remains to be tested. The outcome of the structural study on just the CheAΔ289/CheW complex (Park *et al.*, 2006) by PDS led to a new proposal for the entire signaling complex. The capabilities of PDS are currently being directed to the

[19] This procedure could have been carried out rigorously only by using a multiple affinity approach (tandem purification based on having different tags of opposite protomers in the dimer), which was not attempted.

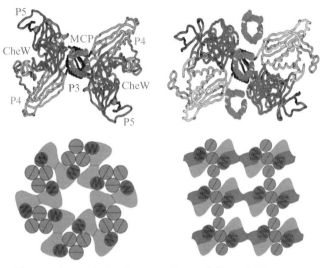

Hexagonal array in *E.coli*   Rows of dimers in *T. maritima*

Fig. 24. Receptor–kinase interactions. In *T. maritima*, the methyl-accepting domain of the receptor forms dimers that are 225 Å long, with a diameter of ~20Å (Park *et al.*, 2006). The canyon formed by two CheW molecules sitting on top of a CheAΔ289 dimer is wide enough to accommodate one receptor dimer. Two additional receptor dimers can be positioned on either side of the center receptor giving a stoichiometry of three receptor dimers to one CheA dimer and two CheWs. The bottom figures compare two possible type of receptor arrays. (Adapted from Weis, 2006.) (See color insert.)

reconstruction of the entire signaling complex of the receptor, which will test this proposal.

In most PDS studies conducted thus far, just a few distances were typically obtained, often with the goal of detecting an important structural change or establishing the oligomerization state. On the other hand, cw ESR routinely employs extensive protein scans (Crane *et al.*, 2005; Cuello *et al.*, 2004; Dong *et al.*, 2005) to elucidate aspects of secondary and tertiary structures. PDS is certainly capable of extensive protein mapping, as we have demonstrated (Borbat *et al.*, 2006; Park *et al.*, 2006). In all, at least 70 distances (including those using WT proteins) have been obtained as the work on CheA/CheW progressed. In retrospect, such a massive effort of PDS could be compressed into 2 to 4 weeks of continuous operation, given that all the samples are prepared. This is, by any measure, quite a short period of time. With good labeling strategy, extensive mapping based on 20 to 50 distances could be completed in about a week, that is, as fast as an

ESR nitroxide scan. The method can be applied to a variety of structural problems that are currently difficult to address by other methods. The main hurdles for PDS are the conformational heterogeneity of proteins; flexibility of the tether of the nitroxide spin label; and oligomeric proteins yielding multiple spin systems, for which there are workarounds, but they remain to be developed into standard techniques. PDS offers spatial resolution that is intermediate between cryo EM and atomic-level resolution methods, although it currently lacks the capacity of sorting out structurally heterogeneous objects as single-particle cryo EM is able to (Stark and Lührmann, 2006; Sigworth, 2007). Functional or conserved cysteines constitute another problem needing solution.

Although PDS may successfully be performed with nanomole or even picomole (at higher working frequency) amount of protein (in the microgram range), extensive study may consume a larger amount. This includes the need to make several different mutants and to account for losses during purification and spin-labeling of the target biomolecule. However, this is still far less than is usually needed to obtain a complete 3D structure by the major methods, although modern NMR is constantly improving its detection limits (as is also true of crystallography), but it often requires expensive isotopic substitution. Also, the procedure of SDSL is well developed and is relatively fast and inexpensive. In the future, it also could be fully automated. After protein purification and nitroxide labeling, the sample is cryopreserved, thereby mitigating difficulties arising from limited protein stability. There is also substantial potential to increase the throughput by perfecting the PDS techniques.

The minuscule amounts of protein that are needed for PDS may well be in line with expected future developments of incorporating unnatural amino-acid targets for selective labeling by suitable (novel) spin-labeling reagents. Attainable improvements in sample preparation (particularly labeling) will allow PDS to address small scales and protein samples that suffer from low-level expression. This could help to reduce the overall costs of using deuterated solvents, lipids, detergents, and proteins that benefit the method. In addition, expected further improvements in pulse ESR instrumentation directed toward enhancing sensitivity to enable the study of even smaller amounts of labeled protein and to increase throughput should further benefit PDS.

At present, protein structure can be reasonably accurately evaluated using just self-consistent nitroxide side-chain modeling (as has been noted) and by structure refinement by CNS for a sufficiently large set of ESR distance constraints (cf. Ch. 4). One could anticipate that future developments will enable ESR distance restraints combined with homology modeling, nitroxide side-chain geometry simulation, and structure prediction to be applied to generate detailed 3D structures of large proteins and their complexes.

Concluding Remarks

PDS has so far been successfully applied on a relatively small scale to a variety of systems in the context of structure and function, and it has the potential to address a wider range of issues. The successful application of PDS triangulation to determine the ternary structure of the CheA/CheW complex of *T. maritima* demonstrates the viability of the method and sets the stage for its future applications in this category. A similar mapping effort has focused on the helix topology of α-Synuclein (Borbat *et al.*, 2006).

In this chapter, we have stressed the point that PDS is a rather straightforward technique in its principles and implementation, and is not overburdened with complexities. We have tried to convey our enthusiasm that PDS will develop into a standard technique for structure determination, given that it does have several virtues, which should lead to its wider acceptance.

Acknowledgments

We thank Brian Crane for a critical reading of the manuscript, Alexandrine Bilwes and Jaya Bhatnagar for help with figures and data processing, and Sang Park for developing the methods for CheA labeling and for metric matrix distance geometry. We also thank Boris Dzikovski, Robert G. Griffin, Igor Grigoriev, Gunnar Jeschke, Anup Upadhyay, and Zheng Zhou for kindly providing samples for PDS. We thank Diane Patzer and Joanne Trutko for their extensive help with the manuscript preparation. This work was supported by grants from the NIH/NCRR P41-RR016292 and NIH/NIBIB EB03150.

References

Abragam, A. (1961). "The Principles of Nuclear Magnetism," pp. 103–105. Clarendon Press, Oxford, UK.

Altenbach, C. A., Flitsch, S. L., Khorana, H. G., and Hubbell, W. L. (1989). Structural studies on transmembrane proteins. 2. Spin labeling of bacteriorhodopsin mutants at unique cysteines. *Biochemistry* **28,** 7806–7812.

Altenbach, C. A., Marti, T., Khorana, H. G., and Hubbell, W. L. (1990). Transmembrane protein structure: Spin labeling of bacteriorhodopsin mutants. *Science* **248,** 1088.

Altenbach, C. A., Oh, K.-J., Trabanino, R. J., Hideg, K., and Hubbell, W. L. (2001). Estimation of inter-residue distances in spin labeled proteins at physiological temperatures: Experimental strategies and practical limitations. *Biochemistry* **40,** 15471–15482.

Astashkin, A. V., Kodera, Y., and Kawamori, A. (1994). Distance between tyrosines Z+ and D+ in plant Photosystem II as determined by pulsed EPR. *Biochim. Biophys. Acta* **1187,** 89–93.

Banham, J. E., Timmel, C. R., Abbott, R. J. M., Lea, S. M., and Jeschke, G. (2006). The characterization of weak protein–protein interactions: Evidence from DEER for the trimerization of a von Willebrand factor A domain in solution. *Angewandte Chemie, International Edition* **45**, 1058–1061.

Bartucci, R., Erilov, D. A., Guzzi, R., Sportelli, L., Dzuba, S. A., and Marsh, D. (2006). Time-resolved electron spin resonance studies of spin-labelled lipids in membranes. *Chem. Phys. Lipids* **141**, 142–157.

Becker, J. S., and Saxena, S. (2005). Double quantum coherence electron spin resonance on coupled Cu(II)–Cu(II) electron spins. *Chem. Phys. Lett.* **414**, 248–252.

Bennati, M., Robblee, J. H., Mugnaini, V., Stubbe, J., Freed, J. H., and Borbat, P. P. (2005). EPR distance measurements support a model for long-range radical initiation in *E. coli* ribonucleotide reductase. *J. Am. Chem. Soc.* **127**, 15014–15015.

Bennati, M., Weber, A., Antonic, J., Perlstein, D. L., Robblee, J., and Stubbe, J. (2003). Pulsed ELDOR spectroscopy measures the distance between the two tyrosyl radicals in the r2 subunit of the *E. coli* ribonucleotide reductase. *J. Am. Chem. Soc.* **125**, 14988–14989.

Berliner, L. J., Eaton, G. R., and Eaton, S. S. (2000). "Distance Measurements in Biological Systems by EPR." Kluwer Academic, NY.

Bhatnagar, J., Freed, J. H., and Crane, B. R. (2006). Rigid body refinement of protein complexes with long-range distance restraints from pulsed dipolar ESR. *Methods in Enzymology* **423**, 117–133.

Biglino, D., Schmidt, P. P., Reijerse, E. J., and Lubitz, W. (2007). PELDOR study on the tyrosyl radicals in the R2 protein of mouse ribonucleotide reductase. *Chem. Phys. Chem.* **8**, 58–62.

Bilwes, A. M., Alex, L. A., Crane, B. R., and Simon, M. I. (1999). Structure of CheA, a signal-transducing histidine kinase. *Cell* **96**, 131–141.

Bilwes, A. M., Park, S. Y., Quezada, C. M., Simon, M. I., and Crane, B. R. (2003). *In* "Structure and Functions of CheA, the Histidine Kinase Central to Bacterial Chemotaxis in Histidine Kinase in Signal Transduction." (M. Inoue and R. Dutta, eds.), Chapter 4, pp. 47–72. Academic Press, NY.

Blair, D. F. (1995). How bacteria sense and swim. *Ann. Rev. Microbiol.* **49**, 489–522.

Bonora, M., Becker, J., and Saxena, S. (2004). Suppression of electron spin-echo envelope modulation peaks in double quantum coherence electron spin resonance. *J. Magn. Reson.* **170**, 278–283.

Borbat, P., Ramlall, T. F., Freed, J. H., and Eliezer, D. (2006). Inter-helix distances in lysophospholipid micelle-bound a-synuclein from pulsed ESR measurements. *J. Am. Chem. Soc.* **128**, 10004–10005.

Borbat, P. P., Crepeau, R. H., and Freed, J. H. (1997). Multifrequency two-dimensional Fourier transform ESR: An X/Ku-band spectrometer. *J. Magn. Reson.* **127**, 155–167.

Borbat, P. P., da Costa-Filho, A. J., Earle, K. A., Moscicki, J. K., and Freed, J. H. (2001). Electron spin resonance in studies of membranes and proteins. *Science* **291**, 266–269.

Borbat, P. P., Davis, J. H., Butcher, S. E., and Freed, J. H. (2004). Measurement of large distances in biomolecules using double-quantum filtered refocused electron spin-echoes. *J. Am. Chem. Soc.* **126**, 7746–7747.

Borbat, P. P., and Freed, J. H. (1999). Multiple-quantum ESR and distance measurements. *Chem. Phys. Lett.* **313**, 145–154.

Borbat, P. P., and Freed, J. H. (2000). *In* "Biological Magnetic Resonance" (L. J. Berliner, G. R. Eaton, and S. S. Eaton, eds.), Vol. 19, pp. 383–459. Kluwer Academic, NY.

Borbat, P. P., Mchaourab, H. S., and Freed, J. H. (2002). Protein structure determination using long-distance constraints from double-quantum coherence ESR: Study of T4-lysozyme. *J. Am. Chem. Soc.* **124**, 5304–5314.

Borovykh, I. V., Ceola, S., Gajula, P., Gast, P., Steinhoff, H.-J., and Huber, M. (2006). Distance between a native cofactor and a spin label in the reaction center of Rhodobacter sphaeroides by a two-frequency pulsed electron paramagnetic resonance method and molecular dynamics simulations. *J. Magn. Reson.* **180,** 178–185.

Boukhvalova, M. S., Dahlquist, F. W., and Stewart, R. C. (2002). CheW binding interactions with CheA and Tar: Importance for chemotaxis signaling in *Escherichia coli. J. Biol. Chem.* **277,** 22251–22259.

Bowers, P. M., Strauss, C. E. M., and Baker, D. (2000). *De novo* protein structure determination using sparse NMR data. *J. Biomol. NMR* **19,** 311–318.

Bowman, M. K., Maryasov, A. G., Kim, N., and deRose, V. J. (2004). Visualization of distance distribution from pulsed double electron–electron resonance data. *App. Magn. Reson.* sb: volume-nr>26, , 23–39.

Brunger, A. T., Adams, P. D., Clore, G. M., DeLano, W. L., Gros, P., Grosse-Kunstleve, R. W., Jiang, J.-S., Kuszewski, J., Nilges, M., Pannu, N. S., Read, R. J., Rice, L. M., *et al.* (1998). Crystallography and NMR System: A new software suite for macromolecular structure determination. *Acta Crystallographica, Section D: Biological Crystallography* **D54,** 905–921.

Cai, Q., Kusnetzow, A. K., Hubbell, W. L., Haworth, I. S., Gacho, G. P. C., Van Eps, N., Hideg, K., Chambers, E. J., and Qin, P. Z. (2006). Site-directed spin labeling measurements of nanometer distances in nucleic acids using a sequence-independent nitroxide probe. *Nucleic Acids Res.* **34,** 4722–4730.

Chandrasekhar, S. (1943). Stochastic problems in physics and astronomy. *Rev. Modern Phys.* **15,** 1.

Chiang, Y.-W., Borbat, P. P., and Freed, J. H. (2005a). The determination of pair distance distributions by pulsed ESR using Tikhonov regularization. *J. Magn. Reson.* **172,** 279–295.

Chiang, Y.-W., Borbat, P. P., and Freed, J. H. (2005b). Maximum entropy: A complement to Tikhonov regularization for determination of pair distance distributions by pulsed ESR. *J. Magn. Reson.* **177,** 184–196.

Codd, R., Astashkin, A. V., Pacheco, A., Raitsimring, A. M., and Enemark, J. H. (2002). Pulsed ELDOR spectroscopy of the Mo(V)/Fe(III) state of sulfite oxidase prepared by one-electron reduction with Ti(III) citrate. *JBIC, J. Biol. Inorg. Chem.* **7,** 338–350.

Columbus, L., and Hubbell, W. L. (2002). A new spin on protein dynamics. *Trends Biochem. Sci.* **27,** 288–295.

Columbus, L., Kalai, T., Jekoe, J., Hideg, K., and Hubbell, W. L. (2001). Molecular motion of spin labeled side chains in a-helices: Analysis by variation of side chain structure. *Biochemistry* **40,** 3828–3846.

Cornish, V. W., Benson, D. R., Altenbach, C. A., Hideg, K., Hubbell, W. L., and Schultz, P. G. (1994). Site-specific incorporation of biophysical probes into proteins. *Proc. Natl. Acad. Sci. USA* **91,** 2910–2914.

Crane, J. M., Mao, C., Lilly, A. A., Smith, V. F., Suo, Y., Hubbell, W. L., and Randall, L. L. (2005). Mapping of the docking of SecA onto the chaperone SecB by site-directed spin labeling: Insight into the mechanism of ligand transfer during protein export. *J. Mol. Biol.* **353,** 295–307.

Crippen, G. M., and Havel, T. F. (1988). "Distance Geometry and Molecular Conformation." John Wiley & Sons, New York, NY.

Cuello, L. G., Cortes, D. M., and Perozo, E. (2004). Molecular architecture of the KvAP voltage-dependent K+ channel in a lipid bilayer. *Science (Washington, DC, United States)* **306,** 491–495.

Denysenkov, V. P., Prisner, T. F., Stubbe, J., and Bennati, M. (2006). High-field pulsed electron-electron double resonance spectroscopy to determine the orientation of the tyrosyl radicals in ribonucleotide reductase. *Proc. Natl. Acad. Sci. USA* **103,** 13386–13390.

Dong, J., Yang, G., and Mchaourab, H. S. (2005). Structural basis of energy transduction in the transport cycle of MsbA. *Science (Washington, DC, United States)* **308**, 1023–1028.

Dzikovski, B. G., Borbat, P. P., and Freed, J. H. (2004). Spin-labeled gramicidin A: Channel formation and dissociation. *Biophys. J.* **87**, 3504–3517.

Dzuba, S. A. (2005). Pulsed EPR structural studies in the nanometer range of distances. *Russ. Chem. Rev.* **74**, 619–637.

Elsaesser, C., Brecht, M., and Bittl, R. (2002). Pulsed electron-electron double resonance on multinuclear metal clusters: Assignment of spin projection factors based on the dipolar interaction. *J. Am. Chem. Soc.* **124**, 12606–12611.

Fafarman, A. T., Borbat, P. P., Freed, J. H., and Kirshenbaum, K. (2007). Characterizing the structure and dynamics of folded oligomers: Pulsed ESR studies of peptoid helices. *Chemical Communications (Cambridge, United Kingdom)* **4**, 377–379.

Fajer, P. G. (2005). Site-directed spin labeling and pulsed dipolar electron paramagnetic resonance (double electron-electron resonance) of force activation in muscle. *J. Phys. Condens. Matter* **17**, S1459–S1469.

Fanucci, G. E., and Cafiso, D. S. (2006). Recent advances and applications of site-directed spin labeling. *Curr. Opin. Struct. Biol.* **16**, 644–653.

Farahbakhsh, Z. T., Altenbach, C. A., and Hubbell, W. L. (1992). Spin labeled cysteines as sensors for protein–lipid interaction and conformation in rhodopsin. *Photochem. Photobiol.* **56**, 1019–1033.

Freed, J. H. (2000). New technologies in electron spin resonance. *Annu. Rev. Phys. Chem.* **51**, 655–689.

Fu, Z., Aronoff-Spencer, E., Backer, J. M., and Gerfen, G. J. (2003). The structure of the inter-SH2 domain of class IA phosphoinositide 3-kinase determined by site-directed spin labeling EPR and homology modeling. *Proc. Natl. Acad. Sci. USA* **100**, 3275–3280.

Griswold, I. J., Zhou, H., Matison, M., Swanson, R. V., McIntosh, L. P., Simon, M. I., and Dahlquist, F. W. (2002). The solution structure and interactions of CheW from *Thermotoga maritima. Nature Struct. Biol.* **9**, 121–125.

Hamel, D. J., and Dahlquist, F. W. (2005). The contact interface of a 120 kD CheA-CheW complex by methyl TROSY interaction spectroscopy. *J. Am. Chem. Soc.* **127**, 9676–9677.

Hanson, P., Anderson, D. J., Martinez, G., Millhauser, G., Formaggio, F., Crisma, M., Toniolo, C., and Vita, C. (1998). Electron spin resonance and structural analysis of water soluble, alanine-rich peptides incorporating TOAC. *Mol. Phys.* **95**, 957–966.

Hanson, P., Millhauser, G., Formaggio, F., Crisma, M., and Toniolo, C. (1996). ESR characterization of hexameric, helical peptides using double TOAC spin labeling. *J. Am. Chem. Soc.* **118**, 7618–7625.

Hilger, D., Jung, H., Padan, E., Wegener, C., Vogel, K.-P., Steinhoff, H.-J., and Jeschke, G. (2005). Assessing oligomerization of membrane proteins by four-pulse DEER: pH-dependent dimerization of NhaA Na+/H+ antiporter of *E. coli. Biophys. J.* **89**, 1328–1338.

Horst, R., Bertelsen, E. B., Fiaux, J., Wider, G., Horwich, A. L., and Wuthrich, K. (2005). Direct NMR observation of a substrate protein bound to the chaperonin GroEL. *Proc. Natl. Acad. Sci. USA* **102**, 12748–12753.

Hubbell, W. L., and Altenbach, C. (1994). Investigation of structure and dynamics in membrane proteins using site-directed spin labeling. *Curr. Opin. Struct. Biol.* **4**, 566–573.

Hubbell, W. L., Cafiso, D. S., and Altenbach, C. (2000). Identifying conformational changes with site-directed spin labeling. *Nature Struct. Biol.* **7**, 735–739.

Huber, M., Lindgren, M., Hammarstrom, P., Martensson, L.-G., Carlsson, U., Eaton, G. R., and Eaton, S. S. (2001). Phase memory relaxation times of spin labels in human carbonic anhydrase II: Pulsed EPR to determine spin label location. *Biophys. Chem.* **94**, 245–256.

Hustedt, E. J., Smirnov, A. I., Laub, C. F., Cobb, C. E., and Beth, A. H. (1997). Molecular distances from dipolar coupled spin-labels: The global analysis of multifrequency continuous wave electron paramagnetic resonance data. *Biophys. J.* **72**, 1861–1877.

Hustedt, E. J., Stein, R. A., Sethaphong, L., Brandon, S., Zhou, Z., and DeSensi, S. C. (2006). Dipolar coupling between nitroxide spin labels: The development and application of a tether-in-a-cone model. *Biophys. J.* **90**, 340–356.

Jeschke, G. (2002). Distance measurements in the nanometer range by pulse EPR. *Chemphyschem. Europ. J. Chem. Phys. Phys. Chem.* **3**, 927–932.

Jeschke, G., Bender, A., Paulsen, H., Zimmermann, H., and Godt, A. (2004a). Sensitivity enhancement in pulse EPR distance measurements. *J. Magn. Reson.* **169**, 1–12.

Jeschke, G., Koch, A., Jonas, U., and Godt, A. (2002). Direct conversion of EPR dipolar time evolution data to distance distributions. *J. Magn. Reson.* **155**, 72–82.

Jeschke, G., Panek, G., Godt, A., Bender, A., and Paulsen, H. (2004b). Data analysis procedures for pulse ELDOR measurements of broad distance distributions. *Appl. Magn. Reson.* **26**, 223–244.

Jeschke, G., Pannier, M., Godt, A., and Spiess, H. W. (2000). Dipolar spectroscopy and spin alignment in electron paramagnetic resonance. *Chem. Phys. Lett.* **331**, 243–252.

Jeschke, G., and Schlick, S. (2006). Spatial distribution of stabilizer-derived nitroxide radicals during thermal degradation of poly(acrylonitrile-butadiene-styrene) copolymers: A unified picture from pulsed ELDOR and ESR imaging. *Phys. Chem. Chem. Phys.* **8**, 4095–4103.

Jeschke, G., and Spiess, H. W. (2006). Distance measurements in solid-state NMR and EPR spectroscopy. *In* "Novel NMR and EPR Techniques," pp. 21–63. Springer, Berlin.

Jeschke, G., Wegener, C., Nietschke, M., Jung, H., and Steinhoff, H.-J. (2004c). Interresidual distance determination by four-pulse double electron-electron resonance in an integral membrane protein: The Na+/proline transporter PutP of *Escherichia coli*. *Biophys. J.* **86**, 2551–2557.

Kay, C. W. M., Elsaesser, C., Bittl, R., Farrell, S. R., and Thorpe, C. (2006). Determination of the distance between the two neutral flavin radicals in augmenter of liver regeneration by pulsed ELDOR. *J. Am. Chem. Soc.* **128**, 76–77.

Klauder, J. R., and Anderson, P. W. (1962). Spectral diffusion decay in spin resonance experiments. *Phys. Rev.* **125**, 912–932.

Klug, C. S., and Feix, J. B. (2005). SDSL: A survey of biological applications. Chapter 10. *In* "Biological Magnetic Resonance" (S. S. Eaton, G. R. Eaton, and L. J. Berliner, eds.), Vol. 24, pp. 269–308. Kluwer, NY.

Klug, C. S., Camenisch, T. G., Hubbell, W. L., and Hyde, J. S. (2005). Multiquantum EPR spectroscopy of spin-labeled arrestin K267C at 35 GHz. *Biophys. J.* **88**, 3641–3647.

Koteiche, H. A., and Mchaourab, H. S. (1999). Folding pattern of the a-crystallin domain in aA-crystallin determined by site-directed spin labeling. *J. Mol. Biol.* **294**, 561–577.

Kulik, L. V., Dzuba, S. A., Grigoryev, I. A., and Tsvetkov, Y. D. (2001). Electron dipole–dipole interaction in ESEEM of nitroxide biradicals. *Chem. Phys. Lett.* **343**, 315–324.

Kulik, L. V., Grishin, Y. A., Dzuba, S. A., Grigoryev, I. A., Klyatskaya, S. V., Vasilevsky, S. F., and Tsvetkov, Y. D. (2002). Electron dipole-dipole ESEEM in field-step ELDOR of nitroxide biradicals. *J. Magn. Reson.* **157**, 61–68.

Kurshev, V. V., Raitsimring, A. M., and Tsvetkov, Y. D. (1989). Selection of dipolar interaction by the "2 + 1" pulse train ESE. *J. Magn. Reson. (1969–1992)* **81**, 441–454.

Langen, R., Oh, K. J., Cascio, D., and Hubbell, W. L. (2000). Crystal structures of spin labeled T4 lysozyme mutants: Implications for the interpretation of EPR spectra in terms of structure. *Biochemistry* **39**, 8396–8405.

Larsen, R. G., and Singel, D. J. (1993). Double electron-electron resonance spin-echo modulation: Spectroscopic measurement of electron spin pair separations in orientationally disordered solids. *J. Chem. Phys.* **98**, 5134–5146.

Lindgren, M., Eaton, G. R., Eaton, S. S., Jonsson, B.-H., Hammarstrom, P., Svensson, M., and Carlsson, U. (1997). Electron spin echo decay as a probe of aminoxyl environment in spin-labeled mutants of human carbonic anhydrase II. *Journal of the Chemical Society, Perkin Transactions* **2**, 2549–2554.

Liu, Y. S., Somponpisut, P., and Perozo, E. (2001). Structure of the KcsA channel intracellular gate in the open state. *Nat. Struct. Biol.* **8**, 883–887.

Maryasov, A. G., and Tsvetkov, Y. D. (2000). Formation of the pulsed electron-electron double resonance signal in the case of a finite amplitude of microwave fields. *Appl. Magn. Reson.* **18**, 583–605.

Maryasov, A. G., Tsvetkov, Y. D., and Raap, J. (1998). Weakly coupled radical pairs in solids. ELDOR in ESE structure studies. *Appl. Magn. Reson.* **14**, 101–113.

Mchaourab, H. S., Kalai, T., Hideg, K., and Hubbell, W. L. (1999). Motion of spin-labeled side chains in T4 lysozyme: Effect of side chain structure. *Biochemistry* **38**, 2947–2955.

Mchaourab, H. S., Oh, K. J., Fang, C. J., and Hubbell, W. L. (1997). Conformation of T4 lysozyme in solution. Hinge-bending motion and the substrate-induced conformational transition studied by site-directed spin labeling. *Biochemistry* **36**, 307–316.

McNulty, J. C., Silapie, J. L., Carnevali, M., Farrar, C. T., Griffin, R. G., Formaggio, F., Crisma, M., Toniolo, C., and Millhauser, G. L. (2001). Electron spin resonance of TOAC labeled peptides: Folding transitions and high frequency spectroscopy. *Biopolymers* **55**, 479–485.

Milov, A. D., Erilov, D. A., Salnikov, E. S., Tsvetkov, Y. D., Formaggio, F., Toniolo, C., and Raap, J. (2005). Structure and spatial distribution of the spin-labeled lipopeptide trichogin GA IV in a phospholipid membrane studied by pulsed electron-electron double resonance (PELDOR). *Phys. Chem. Chem. Phys.* **7**, 1794–1799.

Milov, A. D., Naumov, B. D., and Tsvetkov, Y. D. (2004). The effect of microwave pulse duration on the distance distribution function between spin labels obtained by PELDOR data analysis. *Appl. Magn. Reson.* **26**, 587–599.

Milov, A. D., Maryasov, A. G., and Tsvetkov, Y. D. (1998). Pulsed electron double resonance (PELDOR) and its applications in free-radicals research. *Appl. Magn. Reson.* **15**, 107–143.

Milov, A. D., Maryasov, A. G., Tsvetkov, Y. D., and Raap, J. (1999). Pulsed ELDOR in spin-labeled polypeptides. *Chem. Phys. Lett.* **303**, 135–143.

Milov, A. D., Ponomarev, A. B., and Tsvetkov, Y. D. (1984). Electron-electron double resonance in electron spin echo: Model biradical systems and the sensitized photolysis of decalin. *Chem. Phys. Lett.* **110**, 67–72.

Milov, A. D., Salikhov, K. M., and Shirov, M. D. (1981). Application of the double resonance method to electron spin echo in a study of the spatial distribution of paramagnetic centers in solids. *Soviet Physics-Solid State* **23**, 565–569.

Milov, A. D., Salikhov, K. M., and Tsvetkov, Y. D. (1973). Phase relaxation of hydrogen atoms stabilized in an amorphous matrix. *Soviet Physics–Solid State* **15**, 802–806.

Milov, A. D., and Tsvetkov, Y. D. (1997). Double electron-electron resonance in electron spin echo. Conformations of spin-labeled poly-4-vinylpyridine in glassy solutions. *Appl. Magn. Reson.* **12**, 495–504.

Milov, A. D., Tsvetkov, Y. D., Formaggio, F., Crisma, M., Toniolo, C., and Raap, J. (2000a). Self-assembling properties of membrane-modifying peptides studied by PELDOR and CW-ESR spectroscopies. *J. Am. Chem. Soc.* **122**, 3843–3848.

Milov, A. D., Tsvetkov, Y. D., Formaggio, F., Crisma, M., Toniolo, C., and Raap, J. (2001). The secondary structure of a membrane-modifying peptide in a supramolecular assembly studied by PELDOR and CW-ESR spectroscopies. *J. Am. Chem. Soc.* **123,** 3784–3789.

Milov, A. D., Tsvetkov, Y. D., Formaggio, F., Crisma, M., Toniolo, C., and Raap, J. (2003a). Self-assembling and membrane modifying properties of a lipopeptaibol studied by CW-ESR and PELDOR spectroscopies. *J. Peptide Sci.* **9,** 690–700.

Milov, A. D., Tsvetkov, Y. D., Formaggio, F., Oancea, S., Toniolo, C., and Raap, J. (2003b). Aggregation of spin labeled trichogin GA IV dimers: Distance distribution between spin labels in frozen solutions by PELDOR data. *J. Phy. Chem. B* **107,** 13719–13727.

Milov, A. D., Tsvetkov, Y. D., and Raap, J. (2000b). Aggregation of trichogin analogs in weakly polar solvents: PELDOR and ESR studies. *Appl. Magn. Reson.* **19,** 215–226.

Mims, W. B. (1965). Electron echo methods in spin resonance spectrometry. *Rev. Sci. Instrum.* **36,** 1472–1479.

Narr, E., Godt, A., and Jeschke, G. (2002). Selective measurements of a nitroxide–nitroxide separation of 5 nm and a nitroxide–copper separation of 2.5 nm in a terpyridine-based copper(II) complex by pulse EPR spectroscopy. *Angewandte Chemie, International Edition* **41,** 3907–3910.

Nevzorov, A. A., and Freed, J. H. (2001a). Direct-product formalism for calculating magnetic resonance signals in many-body systems of interacting spins. *J. Chem. Phys.* **115,** 2401–2415.

Nevzorov, A. A., and Freed, J. H. (2001b). A many-body analysis of the effects of the matrix protons and their diffusional motion on electron spin resonance line shapes and electron spin echoes. *J. Chem. Phys.* **115,** 2416–2429.

Ottemann, K. M., Xiao, W., Shin, Y.-K., and Koshland, D. E., Jr. (1999). A piston model for transmembrane signaling of the aspartate receptor. *Science (Washington, D. C.)* **285,** 1751–1754.

Pake, G. E. (1948). Nuclear resonance absorption in hydrated crystals: Fine structure of the proton line. *J. Chem. Phys.* **16,** 327–336.

Pannier, M., Veit, S., Godt, A., Jeschke, G., and Spiess, H. W. (2000). Dead-time free measurement of dipole–dipole interactions between electron spins. *J. Magn. Reson.* **142,** 331–340.

Park, S.-Y., Borbat, P. P., Gonzalez-Bonet, G., Bhatnagar, J., Pollard, A. M., Freed, J. H., Bilwes, A. M., and Crane, B. R. (2006). Reconstruction of the chemotaxis receptor–kinase assembly. *Nature Struct. Mol. Biol.* **13,** 400–407.

Persson, M., Harbridge, J. R., Hammarstrom, P., Mitri, R., Martensson, L.-G., Carlsson, U., Eaton, G. R., and Eaton, S. S. (2001). Comparison of electron paramagnetic resonance methods to determine distances between spin labels on human carbonic anhydrase II. *Biophys. J.* **80,** 2886–2897.

Pfannebecker, V., Klos, H., Hubrich, M., Volkmer, T., Heuer, A., Wiesner, U., and Spiess, H. W. (1996). Determination of end-to-end distances in oligomers by pulsed EPR. *J. Phys. Chem.* **100,** 13428–13432.

Prisner, T., Rohrer, M., and MacMillan, F. (2001). Pulsed EPR spectroscopy: Biological applications. *Annu. Rev. Phys. Chem.* **52,** 279–313.

Pusep, A. Y., and Shokhirev, N. V. (1984). Application of a singular expansion in the analysis of spectroscopic inverse problems. *Optika I spectroscopia* **57,** 792–798.

Rabenstein, M. D., and Shin, Y.-K. (1995). Determination of the distance between two spin labels attached to a macromolecule. *Proc. Natl. Acad. Sci. USA* **92,** 8239–8243.

Rabenstein, M. D., and Shin, Y.-K. (1996). HIV-1 gp41 tertiary structure studied by EPR spectroscopy. *Biochemistry* **35,** 13922–13928.

Raitsimring, A. M., Peisach, J., Lee, H. C., and Chen, X. (1992). Measurement of distance distribution between spin labels in spin-labeled hemoglobin using an electron spin echo method. *J. Phys. Chem.* **96,** 3526–3531.

Raitsimring, A. M., and Salikhov, K. M. (1985). Electron spin echo method as used to analyze the spatial distribution of paramagnetic centers. *Bull. Magn. Reson.* **7,** 184–217.

Raitsimring, A. M., Salikhov, K. M., Umanskii, B. A., and Tsvetkov, Y. D. (1974). Instantaneous diffusion in the electron spin echo of paramagnetic centers stabilized in solid matrixes. *Fizika Tverdogo Tela (Sankt-Peterburg)* **16,** 756–766.

Rakowsky, M. H., Zecevic, A., Eaton, G. R., and Eaton, S. S. (1998). Determination of high-spin iron(III)-nitroxyl distances in spin-labeled porphyrins by time-domain EPR. *J. Magn. Reson.* **131,** 97–110.

Rinard, G. A., Quine, R. W., Song, R., Eaton, G. R., and Eaton, S. S. (1999). Absolute EPR spin echo and noise intensities. *J. Magn. Reson.* **140,** 69–83.

Ruthstein, S., Potapov, A., Raitsimring, A. M., and Goldfarb, D. (2005). Double electron-electron resonance as a method for characterization of micelles. *J. Phys. Chem. B* **109,** 22843–22851.

Sale, K., Sar, C., Sharp, K. A., Hideg, K., and Fajer, P. G. (2002). Structural determination of spin label immobilization and orientation: A Monte Carlo minimization approach. *J. Magn. Reson.* **156,** 104–112.

Sale, K., Song, L., Liu, Y.-S., Perozo, E., and Fajer, P. (2005). Explicit treatment of spin labels in modeling of distance constraints from dipolar EPR and DEER. *J. Am. Chem. Soc.* **127,** 9334–9335.

Salikhov, K. M., Dzyuba, S. A., and Raitsimring, A. (1981). The theory of electron spin-echo signal decay resulting from dipole–dipole interactions between paramagnetic centers in solids. *J. Magn. Reson.* **42,** 255–276.

Schiemann, O., Piton, N., Mu, Y., Stock, G., Engels, J. W., and Prisner, T. F. (2004). A PELDOR-based nanometer distance ruler for oligonucleotides. *J. Am. Chem. Soc.* **126,** 5722–5729.

Seiter, M., Budker, V., Du, J.-L., Eaton, G. R., and Eaton, S. S. (1998). Interspin distances determined by time domain EPR of spin-labeled high-spin methemoglobin. *Inorg. Chim. Acta* **273,** 354–366.

Shimizu, T. S., Le Novere, N., Levin, M. D., Beavil, A. J., Sutton, B. J., and Bray, D. (2000). Molecular model of a lattice of signaling proteins involved in bacterial chemotaxis. *Nature Cell Biol.* **2,** 792–796.

Sigworth, F. J. (2007). From cryo-EM multiple protein stuctures in one shot. *Nature methods* **4,** 20–21.

Slichter, C. P. (1990). "Principles of Magnetic Resonance." Springer-Verlag, Berlin-Heidelberg-New York.

Stark, H., and Lührmann, R. (2006). Cyro-electron microscopy of spliceosomal components. *Annu. Rev. Biophys. Biomol. Struct.* **35,** 435–457.

Steinhoff, H.-J. (2004). Inter- and intra-molecular distances determined by EPR spectroscopy and site-directed spin labeling reveal protein–protein and protein–oligonucleotide interaction. *Biol. Chem.* **385,** 913–920.

Tombolato, F., Ferrarini, A., and Freed, J. H. (2006). Modeling the effects of structure and dynamics of the nitroxide side chain on the ESR spectra of spin-labeled proteins. *J. Phys. Chem. B* **110,** 26260–26271.

Venters, R. A., Farmer, B. T., II, Fierke, C. A., and Spicer, L. D. (1996). Characterizing the use of perdeuteration in NMR studies of large proteins: 13C, 15N, and 1H assignments of human carbonic anhydrase II. *J. Mol. Biol.* **264,** 1101–1116.

Weis, R. M. (2006). Inch by inch, row by row. *Nat. Struct. Mol. Biol.* **13,** 382–384.

Wolanin, P. M., Baker, M. D., Francis, N. R., Thomas, D. R., DeRosier, D. J., and Stock, J. B. (2006). Self-assembly of receptor/signaling complexes in bacterial chemotaxis. *Proc. Natl. Acad. Sci. USA* **103,** 14313–14318.

Xiao, W., Poirier, M. A., Bennett, M. K., and Shin, Y.-K. (2001). The neuronal t-SNARE complex is a parallel four-helix bundle. *Nature Struct. Biol.* **8,** 308–311.

Xu, Q., Ellena, J. F., Kim, M., and Cafiso, D. S. (2006). Substrate-dependent unfolding of the energy coupling motif of a membrane transport protein determined by double electron-electron resonance. *Biochemistry* **45,** 10847–10854.

Zecevic, A., Eaton, G. R., Eaton, S. S., and Lindgren, M. (1998). Dephasing of electron spin echoes for nitroxyl radicals in glassy solvents by non-methyl and methyl protons. *Mol. Phys.* **95,** 1255–1263.

Zhou, Z., DeSensi, S. C., Stein, R. A., Brandon, S., Dixit, M., McArdle, E. J., Warren, E. M., Kroh, H. K., Song, L., Cobb, C. E., Hustedt, E. J., and Beth, A. H. (2005). Solution structure of the cytoplasmic domain of erythrocyte membrane band 3 determined by site-directed spin labeling. *Biochemistry* **44,** 15115–15128.

[4]   Rigid Body Refinement of Protein Complexes
with Long-Range Distance Restraints from
Pulsed Dipolar ESR

By JAYA BHATNAGAR, JACK H. FREED, and BRIAN R. CRANE

Abstract

   The modeling of protein–protein complexes greatly benefits from the incorporation of experimental distance restraints. Pulsed dipolar electron spin resonance spectroscopy is one such powerful technique for obtaining long-range distance restraints in protein complexes. Measurements of the dipolar interaction between two spins placed specifically within a protein complex give information about the spin–spin separation distance. We have developed a convenient method to incorporate such long-range distance information in the modeling of protein–protein complexes that is based on rigid body refinement of the protein components with the software Crystallography and NMR System (CNS). Factors affecting convergence such as number of restraints, error allocation scheme, and number and position of spin labeling sites were investigated with real and simulated data. The use of 4 to 5 different labeling sites on each protein component was found to provide sufficient coverage for producing accuracies limited by the uncertainty in the spin-label conformation within the complex. With an asymmetric scheme of allocating this uncertainty, addition of simulated restraints revealed the importance of longer distances within a limited set of total restraints. We present two case studies: (1) refinement of the complex formed between the histidine kinase CheA and its coupling protein CheW, and (2) refinement of intra-helical separations in the protein a-synuclein bound to micelles.

Introduction

   Elucidation of the structures of protein complexes is often critical for understanding molecular mechanism and function. This is no more evident than for two-component signaling systems where transient associations of proteins mediate the propagation of information. Despite numerous successes, the structure determination of complexes remains a challenge because of the difficulty in growing crystals for X-ray crystallography or in obtaining enough suitable small distance and orientation restraints by NMR. Techniques such as electron microscopy (Frank, 1996), small-angle X-ray scattering (Glatter and Kratky, 1982), and small-angle neutron scattering (Chen and

METHODS IN ENZYMOLOGY, VOL. 423   0076-6879/07 $35.00
                  DOI: 10.1016/S0076-6879(07)23004-6

Bendedouch, 1986) can provide molecular envelopes for complexes but the results suffer from lack of contrast and resolution. In addition, a number of useful approaches map molecular interfaces by measuring perturbations to interfacial residues, such as changes in cross-linking reactivity, accessibility, or NMR chemical shifts. While these methods provide points of contact between partners, relative orientations can be difficult to discern. Finally, distance measurements between specifically labeled positions on associating molecules are possible with FRET and ESR. The former relies on resonance energy transfer between a donor excited state and an accepter ground state; the latter relies on the direct dipolar coupling between two spins. In each case, probe positioning is often achieved with site-directed cysteine substitution, but whereas FRET usually employs two different types of labels, ESR requires only one, usually a nitroxide derivative. Also, ESR provides the distance directly, since it does not require calibrations nor does it have uncertain parameters. In addition, the distribution in distance, P(r), can readily be obtained. Pulsed ESR techniques, such as Double Electron Electron Resonance (DEER) and Double Quantum Resonance (DQC), are capable of measuring biologically relevant distances in the range of 1 to 8 nm between spin labels (Borbat and Freed, 2007; Chiang *et al.*, 2005).

Such long-range pairwise distance restraints can, in principle, be processed to formulate precise structures. Related methodologies have already been applied to FRET-derived distances (Knight *et al.*, 2005; Mukhopadhyay *et al.*, 2004). In the present study, we have developed a simple and convenient method for modeling the structure of a binary complex by rigid body refinement of known substructures, using as restraints the intermolecular distances derived from pulsed ESR. Also, by testing simulated restraints, we produced a set of guidelines to optimize spin label location, number of labels, and measurement error schemes for achieving reasonable model accuracies. Our method is general enough to be applied to any type of distance restraints provided a reasonable estimate of uncertainty associated with the particular measurement is known.

Method

Rigid Body Minimization with CNS

The software package Crystallography and NMR System (CNS) developed by Brunger *et al.* (1998) is primarily designed for structure determination using data from X-ray crystallography or nuclear magnetic resonance (NMR) spectroscopy. However, the optimization algorithms can readily be applied to other types of structural restraints. Our incorporation of ESR-measured distances into CNS is quite similar to that for distances derived from Nuclear Overhauser Effect (NOE) data in structure refinement. Distances (d) were input in the form of a table, each line of which specifies the pair of atoms between which the d has been measured and the error limits, d_{minus} and d_{plus}, which represent the minimum and maximum allowed distances associated with that measurement. CNS provides six possible restraining functions associated with NOE-derived distances: biharmonic function, square-well function, soft-square function, symmetry function, 3D NOE-NOE function, and high dimensional function. We have applied the soft square potential (Brunger et al., 1998) to generate the energy term (E_{ESR}), which is then minimized by conjugate gradient refinement based on the agreement between measured distances "d" and model distance "R." Taking default values for most of the constants[1], a simplified version of the function has the form:

$$E_{ESR} = S \times \begin{cases} a + b/\Delta + \Delta; & R > d + d_{plus} + r_{sw} \\ \Delta^2; & R < d + d_{plus} + r_{sw} \end{cases} \qquad (1)$$

$$\text{where } \Delta = \begin{cases} R - (d + d_{plus}); & d + d_{plus} < R \\ 0; & d - d_{minus} < R < d + d_{plus} \\ d - d_{minus} - R; & R < d - d_{minus} \end{cases}$$

We assign "d" as the ESR-measured distance between C_β atoms of the corresponding amino acid residues at which the spin label is attached, R is the corresponding distance in the model, d_{plus} and d_{minus} are the positive and negative errors associated with each distance, r_{sw} is a constant with default value 0.5 Å, a and b are determined by the program such that E_{ESR} is a smooth function at point $R = d + d_{plus} + r_{sw}$. S is a scale factor that weights the ESR energy relative to the van der Waals energy. A similar soft square potential has also been used to model constraints for a system of transmembrane helices (Sale et al., 2004) and is analogous to a global penalty function developed by Knight et al. (2005) for modeling FRET-derived restraints. However, an additional property of the restraining function in CNS is that it becomes linear for large deviations between experimental and model restraints. This allowance maintains numerical stabilities (Brunger et al., 1999).

Initial Conformation of the Complex

Gradient-descent optimization methods such as conjugate gradient minimization converge to a global minimum of the system if the starting conformation is not very different from the correct structure. Various computational

---

[1] Softexp = 1; Exp = 2,C = 1; c = 1; d_{off} = 0.

procedures have been developed to model initial conformations for refinement, provided distance restraints are available (Faulon *et al.*, 2003; Sale *et al.*, 2004). For our first case study of the complex between chemotaxis proteins CheA and CheW, we test initial conformations determined randomly with those generated with matrix distance geometry from both X-ray crystallography and pulsed ESR, as discussed in our previous work (Borbat and Freed, 2007; Park *et al.*, 2006). With our second case, the protein alpha-synuclein (aS), we compare refinements beginning with either the NMR-determined structure or random orientations of the two synuclein helices.

Evaluation Criterion

In case 1, the separate structures of CheW and the P5 domain of CheA were taken from the crystal structure of CheW and the CheA domains P4-P5, where CheW predominantly binds to the CheA domain, P5 (Park *et al.*, 2006). The final conformation of the complex after rigid body minimization was evaluated by comparing the ESR-refined complex to the coordinates of the P4:P5:CheW crystal structure. The tight binding between CheW and CheAΔ289 (domains P3,P4,P5 collectively called CheAΔ289; $K_b = 100$ nM) makes it unlikely that crystal packing forces significantly alter the association mode of the complex (Park *et al.*, 2004). The measure of agreement was assigned as the root-mean-square deviation (RMSD) in the position of C_α atoms of CheW in the final refined structure with respect to the crystal structure after least square fitting of the P5 domains from both structures (McRee, 1999).

In case 2, we aimed to reproduce the orientation of two anti parallel helices of α-synuclein when bound to micelles, for which an NMR structure has been determined (Bussell *et al.*, 2005; Ulmer *et al.*, 2005). Interhelical distances measured by ESR give information about relative orientation of the helices, which cannot be determined with certainty from NMR data alone. For comparison, the quality of the ESR-refined structure was evaluated by superimposing one of the two α-synuclein helices with the NMR structure of the molecule bound to micelles (Ulmer *et al.*, 2005) and then calculating the RMSD between the ESR refined second helix and that from the NMR structure.

Results

Case Study 1

CheA:CheW complex. CheW forms a complex with the histidine kinase CheA that is necessary for assembly with chemorecepters. To construct the structure of the CheA:CheW complex, 12 intermolecular distances were measured between nitroxide spin labels on four residues (N553C, S568C, E646C, and D579C) of the P5 domain of *T. maritima* CheAΔ289 (which contains domains P3-P4-P5) and three residues (S15C, S72C and S80C) on *T. maritima* CheW (Fig. 1). Positioning of the labels was achieved by site-directed cysteine mutagenesis followed by reaction with (1-oxyl-2,2,5,5-tetramethylpyrolinyl-3-methyl)-methanethiosulfonate (MTSSL). An initial conformation of the complex was predicted by using a matrix distance geometry method, and it was found to agree with a root mean square deviation (RMSD) in C_α atom positions of about 16 Å when compared with the crystal structure of the complex. Rigid-body refinement using CNS reduced the RMSD to 11 Å. The total energy function in the refinement of CNS is a sum of $E_{EMPIRICAL}$ and $E_{EFFECTIVE}$ terms (Brunger *et al.*, 1998). This force field is similar to the Bundler penalty function used to model transmembrane helices against sparse distance constraints (Sale *et al.*, 2004). $E_{EMPIRICAL}$ describes the energy of the molecule as a function of atomic coordinates (energy associated with bonds, angles, dihedral angles, etc.), whereas $E_{EFFECTIVE}$ refers to restraining energy terms associated with agreement of the model to the ESR data, that is, it equals E_{ESR} given by Equation (1). In rigid body refinement, only energy terms that reflect van der Waals contacts contribute to $E_{EMPIRICAL}$.

The convergence was tested by randomly orienting CheW in various positions and evaluating the refined complex. Within rigid body displacements of 15 Å and rotations of 30°, the same final conformation was found (within an RMSD difference of ∼3 Å). In the following sections, we

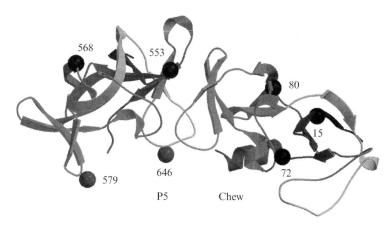

FIG. 1. Crystal structure of CheW-P5 complex showing positions of spin label sites (balls) along the polypeptide. Both proteins shown as ribbon representations colored blue to red from N to C terminus. Sites producing the most aberrant ESR restraints compared to the crystal structure shown in red.(See color insert.)

investigate how parameterization of the refinement affects the quality of the final solution. By adding restraints comprising distances taken from the crystal structure with errors derived from the standard deviations observed in the ESR measurements, we also explore how the number and nature of distance restraints affect the modeling results. In particular, we present guidelines to aid selection of potential spin labeling sites on the protein components within a general complex.

Error Allocation Scheme

A pulsed ESR experiment with a pair of nitroxide spin labels measures the separation between the nitroxyl groups of the spin labels, which can have considerable orientational freedom with respect to the protein backbone and with respect to each other because of their flexible tethers. In the absence of information about the spin label orientation, we have assigned the ESR experimental distance to coincide with the C_β position of the native amino acid residue. If the spin label tethers point away from each other in the complex, the model distances will underestimate the nitroxide separations. In fact, the ESR measured distances are almost always larger than those predicted by C_β separations (Table I). In contrast, if the spin label tethers project toward each other in the complex, then the spin–spin separation will be overestimated by C_β separations. However, when globular domains associate, there is a bias against facing labels because they tend to reside on protein surfaces that participate in the interface.

TABLE I

COMPARISON OF ESR-MEASURED DISTANCES TO $C_{\beta\beta}$ SEPARATIONS BETWEEN CORRESPONDING RESIDUES IN THE P5-CHeW CRYSTAL STRUCTURE

Residue P5-CheW	Distance between C_β atoms in crystal structure(R_{crys}) (Å)	ESR measured distances (R_{esr}) (Å)	$R_{esr} - R_{Crys}$ (Å)
553–15	34.9	37	2.1
646–15	31.8	43.7	11.9
568–15	55.4	54.5	−0.9
579–15	52	61	9
553–72	28.3	27	−1.3
646–72	27.5	32.5	5
568–72	47.9	49	1.1
579–72	41	46	5
553–80	23.6	26	2.4
646–80	26.8	39.5	12.7
568–80	44.5	47	2.5
579–80	44.5	54.5	10

To compensate for overall longer experimental distances, on average, we have found that an asymmetric uncertainty model is effective. In previous work (Park et al., 2006), we presented a distance-dependent error allocation scheme. However, better results are obtained by setting $d_{minus} = 5$ Å and $d_{plus} = 1$ Å for all restraints which are the boundaries within which most of the experimental distances are over- or underestimated by C_β separations (Table I). Similar magnitudes in error are consistent with other spin label-ing studies (Faulon et al., 2003; Rabenstein and Shin, 1995) which may also benefit from asymmetric error boundaries. However, 4 out of 12 distances do not meet these criteria due to reasons related to the location of spin label site on the protein surface. The reasons for such inaccuracies in distance measurements are discussed in a later section.

Weighting Scheme for Contact Parameters

For ESR restraints to determine the final configuration, $E_{EFFECTIVE}$ must account for a considerable percentage of the total energy. This can be achieved by simply increasing the scale factor (S) in the input file associated with E_{NOE}, or in our case, E_{ESR}. With the ceiling constant assigned to 10^5, the scale factor was increased from 75 to 75,000 in steps of 100 and the RMSD in C_α positions were evaluated. Predictably, the convergence improves progressively as the scale factor increases (S = 75 yields an RMSD = 16.38 Å; S = 75,000, an RMSD = 11.06 Å). Above S = 75,000, there is no further improvement.

Type and Number of Restraints

Applying all 12 experimental intermolecular distance restraints be-tween CheW and P5 domain, while setting $d_{minus} = 5$ Å and $d_{plus} = 1$ Å, the best structure that could be achieved has an RMSD on C_a positions of 11.06 Å compared to the crystal structure. To evaluate the effect of addi-tional arbitrarily chosen distance restraints, four new label sites on CheW were successively added to the refinement. Each new site generated four new distances to the P5 labels. The standard deviation of the parameter (R_{esr}–R_{crys}) as defined in Table I is 5 Å. In order for the new distances to mimic the experimental ones, the standard deviation obtained previously was added to the C_β separations. Then, for each successive addition of a label site on CheW, the RMSD in positions of the C_α atoms in the final structure was calculated, and the results were plotted against total number of restraints. Two error schemes, $d_{minus} = 5$ Å, $d_{plus} = 1$ Å (Fig. 2A) and $d_{minus} = 5$ Å, $d_{plus} = 5$ Å (Fig. 2B), were used for comparison. The procedure was also repeated for five different initial conformations of the complex. The results indicate that irrespective of the initial conformation

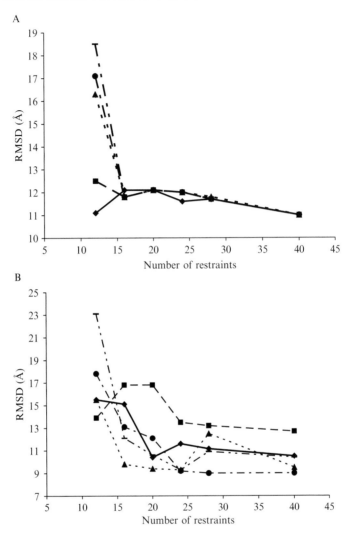

FIG. 2. The effects of different error schemes and simulated restraints on refinement accuracy. RMSDs for the refined CheW/P5 complex are shown for five different initial conformations of the complex ■, ◆, ▲, ●, _ . Two different error schemes: (A) $d_{minus} = 5$ Å, $d_{plus} = 1$ Å and (B) $d_{minus} = 5$ Å, $d_{plus} = 5$ Å.

prior to refinement, addition of random distance restraints leads to an improved RMSD of 8 to 12 Å but beyond 20 and 24 restraints with $d_{minus} = 5$ Å $d_{plus} = 1$ Å and $d_{minus} = 5$ Å $d_{plus} = 5$ Å, respectively, there is no improvement. It is interesting to note that Knight *et al.* (2005) also reported that model accuracies of only 10 Å RMSD can be obtained with 20 or more FRET restraints.

Addition of More Accurate Restraints

As has been illustrated, about 20 distance restraints with standard deviation of 5 Å from the crystal-structure derived distances are sufficient to produce model accuracies of about 10 Å. The inability of additional restraints to obtain better results suggested that convergence is limited by inaccuracies in the experimental distances. With the initial configuration taken from distance geometry, even the addition of 28 accurate crystal-derived distances (setting $d_{minus} = 1$ Å and $d_{plus} = 1$ Å) to 12 experimental distances (setting $d_{minus} = 5$ Å and $d_{plus} = 5$ Å) only improved the final agreement to a limited degree (from RMSD 15.4 to 10.5 Å); thus, a few distances with large inconsistencies appear to dominate the more accurate restraints. Comparison of 12 experimental distances with crystal separations revealed that 2 of the distances were highly skewed, with average deviations up to 12.7 Å (Table I). If we take the same set of 28 crystal-structure derived distances and 12 experimental distances, and the observed ESR distances are deleted two at a time, beginning with the most deviant ones, the RMSD drastically reduces from 10.5 to 6.2 Å and then becomes constant at 2.6 Å (Fig. 3). Adding only the two highly skewed measurements

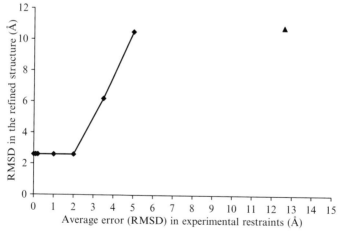

FIG. 3. The effect of aberrant measurements on refinement accuracy in the presence of additional restraints derived from the crystal structure. From a set of 28 crystal distances with ±1 Å and experimental distances with ±5 Å, the most deviant ESR distances were deleted two at a time (◆). Addition of only the two most deviant experimental distances to the defined crystal distances give a slightly higher RMSD compared to the overall set of experimental distances (▲).

to the crystal-structure derived restraints produces a worse RMSD than the entire set of experimental restraints, emphasizing the deleterious effects of these aberrant measurements.

However, in the absence of the simulated restraints, the deletion of the 2 most deviant distances from the set of 12 experimental distances increased the RMSD from 11 to 16 Å. This is probably because the refinement now suffers from underdetermination. Alternatively, since the total energy associated with distance restraints is the sum of individual contributions, improved convergence may result from a weighting scheme based on experimental-to-model agreement that adjusts on successive iterations to reduce the weight of the contribution of aberrant measurements. Simply, if the difference between a measurement and its predicted distance by the refined complex deviates by more than two standard deviations, as given by the distribution of residuals from all the measurements, then the measurement should be removed and the refinement repeated. As we will discuss, due to surface site mobility, interference of labels with complex formation, and other conformational effects, it is reasonable to encounter some outliers in these experiments.

If experimental restraints are deleted successively in the absence of simulated restraints, the RMSD increases as expected. However, the additional increase in RMSD is more sensitive to removal of the shortest, rather than the longest, distance (Fig. 4). This suggests that longer experimental distances in the CheA:CheW system are more inaccurate than shorter ones.

Effect of Spin Label Position

Site-directed spin labeling (SDSL) is a convenient method to attach ESR probes to cysteine residues on proteins (Hubbell and Altenbach, 1994); however, it is unclear how the pattern of sites affects the refinement, apart from the considerations that a solvent-exposed residue is more likely to react with the spin label, and that spin labels in the interfacial region may disrupt complex formation.

To test the effect of label position on predicting the CheA:CheW solution complex, CheW was broadly divided into three sections relative to the CheA interface—front, middle, and back—and from each of these sections, one amino acid residue was randomly selected as a label site. Additional distance restraints from these sites to P5 were measured as before while setting $d_{minus} = 5$ Å and $d_{plus} = 1$ Å. This procedure was repeated seven more times, selecting a random site in each section each time, and finally the RMSD on C_α positions after refinement was averaged for all eight cases per section. The trend in RMSD values

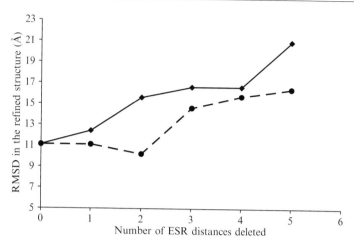

FIG. 4. Long vs short restraints in refinement of CheA/CheW complex. Successive deletions of distances beginning with the shortest (◆) or the longest (●) restraints.

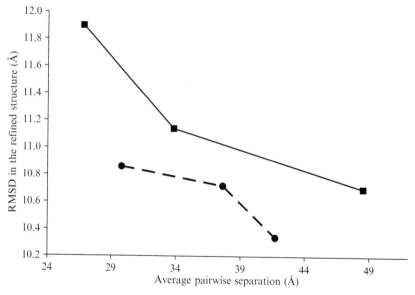

FIG. 5. Variation of RMSD with average pairwise separation of the label sites with addition of one new spin label (■) and four spin labels on CheW (●).

(Fig. 5) showed a slight preference for locating the new sites in the middle and back sections of the protein (from 11.99 to 11.14 to 10.7 Å, respectively).

We also considered the effect of adding four more CheW sites (16 new restraints). The selection of sites was organized the following six ways:

1. All from front section (residues I60, S45, N54, S37)
2. All from middle section (E90, K67, D139, I34)
3. All from back section (V101, K123, N107, N113)
4. Two from front and two from middle section
5. Two from middle and two from back section
6. Two from back and two from front section

For the first three cases, the RMSD from the refined structure shows slightly better agreement with the crystal structure when sites in the distal end of CheW are selected compared to sites closer to the P5 interface. For scenarios 4 through 6, the selection of two residues from each section of CheW was done in eight different ways and the final RMSD was averaged for all eight cases. Plotting the final RMSD in the final structure versus the average pairwise separation of each new label from sites on P5 demonstrates that only minimal improvement in RMSD is seen, no matter how the sites are chosen (Fig. 5).

However, we can conclude that more restraints result in a lower RMSD, and longer restraints play a crucial role only when the total number of restraints is fewer than 16. As more restraints are added, the locations of sites on the surface of CheW have little effect on the refined complex.

Additional considerations affecting the choice of labeling sites are discussed in Chapter 3 of this volume.

Case Study 2

Helix Orientations of α-Synuclein Bound to Micelles. NMR studies on the protein alpha-synuclein (aS) have shown that when bound to sodium dodecyl sulphate (SDS) micelles, the protein adopts a conformation of two separate anti-parallel helices (helix 1 residues: 3–37; helix 2 residues: 45–92) connected by an ordered linker (Bussell *et al.*, 2005; Ulmer *et al.*, 2005). Pulsed ESR has been used to determine interhelix distances between spin labels at various positions on the two helices when the protein is bound to both SDS and lyso-1-palmitoylphosphotidylglycerol (LPPG) micelles (Borbat *et al.*, 2006). In total, 13 interhelical dipolar couplings were measured and from them the average distance (R_{avg}) and its root mean square deviation (RMSD) were evaluated. We tested the ability of our refinement procedure to orient the two helices relative to each other under the assumption that each helix behaves as a rigid body. To generate two rigid bodies, the helices were separated between residues 40 and 41 in the linker. To account for the covalent bonding between residues 40 to 41, additional restraints were added between residues 40 to 41, 39 to 41, and 40 to 42 (C_i-N_k, $C_{\alpha i}$-$C_{\alpha k}$, C_i-C_k, $C_{\alpha i}$-N_k, N_i-N_k and C_i-$C_{\alpha k}$, for i = 40; k = i + 1 and k = i + 2, for i = 39; k = i + 2). In this scheme, the restraints were calculated by summing the bond lengths connecting the two atoms of interest and d_{plus} was set to 0 because any distance measured through space is shorter than that measured along the summed bond lengths. For a hypothetical case where no information is available regarding the conformation of the turn residues, the d_{minus} error was given more flexibility by assigning $d_{minus} = 1$ Å for distances between adjacent residues and $d_{minus} = 8$ Å for distances between non-adjacent residues.

For SDS bound α-synuclein (with the exception of two distances, between V3C/E61C and E13C/H50C), 11 interhelical ESR distances, taken as their reported R_{max} values, were incorporated into the refinement. The RMSDs in label position obtained from P(r) measurements were taken as estimates for the d_{minus} error. As the ESR measurements likely overestimate R, as in the CheA/CheW case, d_{plus} was set to a smaller value, but was increased to reflect changes in d_{min} ($d_{plus} = 1$ for 5 Å < $d_{minus} = 8$ Å, $d_{plus} = 2$ for 9 Å < $d_{minus} = 15$ Å and $d_{plus} = 5$ Å for $d_{minus} = 15$ Å). Combining all the restraints, and starting with what was available from the NMR structure, the refinement places the ends of the two rigid helices close to each other (the length of amide bond C_{40} – N_{41} is 2.2 Å compared to ideal bond length 1.3 Å). When the helical fragment from 1 to 40 is superimposed on its position in the NMR structure, the anti-parallel partner helix (residues 41–103) is rotated by an angle $\Phi \sim 30°$ with respect to its position in the NMR structure (Fig. 6). However, the angle separating the two helical axes (θ) is better determined. Thus, the ESR refinement is unable to distinguish which sides of the helices face each other, and this generates inaccuracy in Φ. This is not surprising, since the errors in the spin label position are larger than the width of a helix. We noted that the absolute orientation of the two helices can be determined if precise restraints on the conformation of linker residues are known by other means. If rigid restraints are added for the conformation of residues within the loops, the agreement with the NMR structure is excellent.

Discussion

In this chapter, we have described a simple and readily implemented method for refining association modes of protein complexes from ESR restraints. Agreement with crystal data improves with number of ESR restraints until approximately 20 restraints are available; additional restraints beyond this number result in little further improvement due to errors associated with the knowledge of the label position. A 2005 study

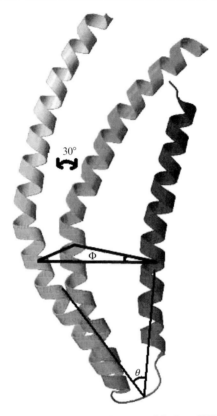

FIG. 6. A comparison of helix orientations in α-synuclein from ESR-refinement and NMR. The orientation of the two anti-parallel α-synuclein helixes (residues 3–34 and 44–94) as derived from NMR is shown in blue. Superposition of N terminal of helix from the rigid body refined structure, places the second helix rotated by angle of 30° with respect to the NMR structure. (See color insert.)

reported agreement between C_β to C_β distances and ESR distance restraints with mean errors up to 6 Å (Sale *et al.*, 2005). These errors are similar in magnitude to those accounted for by our asymmetric error scheme in CheA/CheW case study. Molecular modeling approaches such as Monte Carlo simulations and molecular dynamics have been found to be useful in lowering the uncertainty associated with spin-label positions (Borbat *et al.*, 2002; Sale *et al.*, 2005; Schiemann *et al.*, 2004).

Type of Restraints

We investigated how positioning of the spin labels influences convergence of varying accuracies. Addition of longer simulated restraints appeared most effective in driving convergence to the target model, provided the total number of restraints was fewer than 16. However, removal of the shortest experimental distances has more deleterious effect in the absence of simulated restraints. This apparent contradiction may derive from the longer experimental restraints being unusually aberrant due to conformational properties of these sites in the CheA:CheW system. In addition, any real differences between the solution and crystal complex would be expected to be greatest at sites farthest from the high-affinity interface. Nonetheless, our combination of studies suggests that spin label-ing 4 or 5 sites on each protein at positions distributed as far apart as possible on the structures of the individual components is a reasonable strategy for covering the distance space.

Inaccuracies in Distance Measurement

Apart from the technical limitations of the experimental method in measuring accurate distances (cf. Borat and Freed, 2007), local conforma-tional changes in the protein structure, backbone dynamics, and the flexi-bility of the spin label lead to ambiguity in measurements. The two most deviant intermolecular distances in the CheW/P5 complex were those measured from site 646 on P5 domain (P5/CheW: 646-15, 646-80). In the crystal structure of the complex, P5-646 is very close to the binding interface with CheW, and thus the label conformation may be unusually perturbed in the complex. In addition, the 646 site resides in a loop, which may impart more than usual flexibility (Fig. 1). Aberrant distances involving P5 site 579 may also be caused because this residue resides in a loop with few neighbor contacts and, hence, may be more mobile.

Spatial Resolution of ESR-Derived Structures

Case 2 demonstrates that this method as implemented is less effective at orienting secondary structure elements within a protein than at defining association modes within the complex. It follows that, even with a large number of measurements, it may be difficult to precisely define conforma-tional changes involving small to medium amplitude shifts in secondary structure positions. This limitation could be overcome by more rigid spin labels whose positions on the protein surface are fixed and well defined. In this regard, metal complexes may be an attractive alternative to nitroxide-based labels (Rodriguez-Castaneda *et al.*, 2006).

In conclusion, pulsed dipolar ESR, combined with site-directed spin labeling, can reconstitute structures of protein–protein complexes with reasonable accuracies provided structures of the individual components are well defined. CNS-based rigid-body refinement is a straightforward and accessible method for generating complexes from the distance restraints. Further improvements may be possible with a weighting scheme that identifies and adjusts the contribution of outliers during the course of refinement.

Acknowledgments

This work has been supported by grants from the National Institutes of Health GM: R01066775 (to B.R.C), NCRR: P41-RR016292 and NIBIB: R01-EB03150 (to J.H.F).

References

Borbat, P., Ramlall, T. F., Freed, J. H., and Eliezer, D. (2006). Inter-helix distances in lysophospholipid micelle-bound alpha-synuclein from pulsed ESR measurements. *J. Am. Chem. Soc.* **128,** 10004–10005.

Borbat, P. P., and Freed, J. H. (2007). Measuring distances by pulsed dipolar ESR spectroscopy: Spin-labeled histidine kinases. *Methods Enzymol.* **423,** 52–116.

Borbat, P. P., Mchaourab, H. S., and Freed, J. H. (2002). Protein structure determination using long-distance constraints from double-quantum coherence ESR: Study of T4 lysozyme. *J. Am. Chem. Soc.* **124,** 5304–5314.

Brunger, A. T., Adams, P. D., Clore, G. M., DeLano, W. L., Gros, P., Grosse-Kunstleve, R. W., Jiang, J. S., Kuszewski, J., Nilges, M., Pannu, N. S., Read, R. J., Rice, L. M., Simonson, T., and Warren, G. L. (1998). Crystallography and NMR system: A new software suite for macromolecular structure determination. *Acta Crystallogr. , Sect. D: Biol. Crystallogr.* **54,** 905–921.

Brunger, A. T., Adams, P. D., and Rice, L. M. (1999). Annealing in crystallography: A powerful optimization tool. *Prog. Biophys. Mol. Biol.* **72,** 135–155.

Bussell, R., Ramlall, T. F., and Eliezer, D. (2005). Helix periodicity, topology, and dynamics of membrane-associated alpha-Synuclein. *Protein Sci.* **14,** 862–872.

Chen, S. H., and Bendedouch, D. (1986). Structure and interactions of proteins in solution studied by small-angle neutron scattering. *Methods Enzymol.* **130,** 79–116.

Chiang, Y. W., Borbat, P. P., and Freed, J. H. (2005). The determination of pair distance distributions by pulsed ESR using Tikhonov regularization. *J. Magn. Reson.* **172,** 279–295.

Faulon, J. L., Sale, K., and Young, M. (2003). Exploring the conformational space of membrane protein folds matching distance constraints. *Protein Sci.* **12,** 1750–1761.

Frank, J. (1996). "Three-Dimensional Electron Microscopy of Macromolecular Assemblies." Academic Press, San Diego, CA.

Glatter, D., and Kratky, O. (1982). "Small Angle X-ray Scattering." Academic Press, London, UK.

Hubbell, W. L., and Altenbach, C. (1994). Investigation of structure and dynamics in membrane-proteins using site-directed spin-labeling. *Curr. Opin. Struct. Biol.* **4,** 566–573.

Knight, J. L., Mekler, V., Mukhopadhyay, J., Ebright, R. H., and Levy, R. M. (2005). Distance-restrained docking of rifampicin and rifamycin SV to RNA polymerase using systematic FRET measurements: Developing benchmarks of model quality and reliability. *Biophys. J.* **88,** 925–938.

McRee, D. E. (1999). XtalView Xfit—A versatile program for manipulating atomic coordinates and electron density. *J. Struct. Biol.* **125,** 156–165.

Mukhopadhyay, J., Sineva, E., Knight, J., Levy, R. M., and Ebright, R. H. (2004). Antibacterial peptide microcin J25 inhibits transcription by binding within and obstructing the RNA polymerase secondary channel. *Mol. Cell* **14,** 739–751.

Park, S. Y., Quezada, C. M., Bilwes, A. M., and Crane, B. R. (2004). Subunit exchange by CheA histidine kinases from the mesophile *Escherichia coli* and the thermophile *Thermotoga maritima*. *Biochemistry* **43,** 2228–2240.

Park, S. Y., Borbat, P. P., Gonzalez-Bonet, G., Bhatnagar, J., Pollard, A. M., Freed, J. H., Bilwes, A. M., and Crane, B. R. (2006). Reconstruction of the chemotaxis receptor–kinase assembly. *Nat. Struct. Mol. Biol.* **13,** 400–407.

Rabenstein, M. D., and Shin, Y. K. (1995). Determination of the distance between 2 spin labels attached to a macromolecule. *Proc. Natl. Acad. Sci. USA* **92,** 8239–8243.

Rodriguez-Castaneda, F., Haberz, P., Leonov, A., and Griesinger, C. (2006). Paramagnetic tagging of diamagnetic proteins for solution NMR. *Magn. Reson. Chem.* **44,** S10–S16.

Sale, K., Faulon, J. L., Gray, G. A., Schoeniger, J. S., and Young, M. M. (2004). Optimal bundling of transmembrane helices using sparse distance constraints. *Protein Sci.* **13,** 2613–2627.

Sale, K., Song, L. K., Liu, Y. S., Perozo, E., and Fajer, P. (2005). Explicit treatment of spin labels in modeling of distance constraints from dipolar EPR and DEER. *J. Am. Chem. Soc.* **127,** 9334–9335.

Schiemann, O., Piton, N., Mu, Y. G., Stock, G., Engels, J. W., and Prisner, T. F. (2004). A PELDOR-based nanometer distance ruler for oligonucleotides. *J. Am. Chem. Soc.* **126,** 5722–5729.

Ulmer, T. S., Bax, A., Cole, N. B., and Nussbaum, R. L. (2005). Structure and dynamics of micelle-bound human alpha-synuclein. *J. Biol. Chem.* **280,** 9595–9603.

[5] TonB/TolA Amino-Terminal Domain Modeling

By KIMBERLY L. KELLER, KERRY K. BRINKMAN, and RAY A. LARSEN

Abstract

TonB and TolA proteins are energy transducers that couple the ion electrochemical gradient of the cytoplasmic membrane to support energy-dependent processes in the outer membrane of gram-negative bacteria. Energization of these proteins involves specific interactions with multiprotein cytoplasmic membrane energy harvesting complexes. The specific mechanisms by which these energy transfers occur remain unclear, but the evidence to date indicates that the amino-terminally located signal anchors of TonB and TolA play essential roles in the process. Mutant hunts have identified one motif in this region, common to both TonB and TolA, as important for energization. Because TonB and TolA each have a "preferred" energy-harvesting complex, it is clear that additional motifs, not shared between TonB and TolA, are involved in interactions with energy harvesting complexes. We have adopted a strategy of examining derivatives with multiple-residue substitutions to identify such regions. This involves the characterization of specific TonB derivatives generated by two similar approaches: the block substitutions in TonB by alanyl residues and the exchange of short regions between TonB and TolA. The methods by which these derivatives are generated are described, with an illustrative example for each.

Introduction

A dual membrane envelope forms the interface between gram-negative bacteria and their surroundings. The inner layer of this envelope provides a permeability barrier in the form of a cytoplasmic membrane (CM)—a phospholipid bilayer rich in proteins that create and harvest ion gradients to drive essential transport processes. While enclosing the cell proper, the CM is itself enfolded by an aqueous compartment, the periplasmic space. A viscous mix of osmoregulatory polysaccharides and proteins that support nutrient transport and envelope biogenesis, the periplasm also houses a thin corset of peptidoglycan—a concentric mesh of oligopeptide-crosslinked glycan that confers rigidity to the cell. Trussed to this network and enclosing the periplasm is a final layer—a diffusion barrier in the form of an outer membrane (OM). This membrane is uniquely asymmetric, with an inner

METHODS IN ENZYMOLOGY, VOL. 423
0076-6879/07 $35.00
DOI: 10.1016/S0076-6879(07)23005-8

phospholipid face and an external surface rich in anionic lipid-anchored oligopolysaccharides. Small hydrophilic nutrients can passively traverse this barrier via the aqueous channels of resident porin proteins; however, the polar surface of the OM hinders the passage of various detergents and other hydrophobic toxins that commonly corrupt bacterial surroundings.

The barrier functions of the OM allow gram-negative bacteria to compete in a variety of niches, yet this fortification also poses complications. One major consequence of envelope architecture is that the OM is spatially removed from any significant source of energy. Specifically, the aqueous channels that afford the entry of hydrophilic nutrients also preclude the formation of energetically useful ion gradients; further, the hydrolytic resident proteins of the periplasmic space are not compatible with those phosphorylated molecules that cells favor as energy currency. Thus, any energy-dependent processes that occur at the OM must rely upon imported energy.

An obvious energy need of the OM is simply to support upkeep. It is thermodynamically unlikely that a structure as complex as the OM can be maintained without an input of energy. How such energy is actually used remains ill defined; indeed, the details of OM biogenesis are themselves far from resolved (Ruiz *et al.*, 2006). One set of proteins implicated in OM maintenance is the Tol system. The core components of this system include two cytoplasmic membrane proteins (TolQ and TolR) that appear to form a heteromultimeric complex with a third protein (TolA) that spans the periplasmic space to interact with a fourth protein (TolB) that is peripherally associated with the OM (reviewed in Lazzaroni *et al.*, 1999). Tol mutations disrupt OM integrity, with resultant hypersensitivity to bile salts, detergents, and certain antibiotics, leakage of periplasmic contents (Lazzaroni *et al.*, 1989; Lloubes *et al.*, 2001) and shedding of OM vesicles (Bernadac *et al.*, 1998). The possibility that the Tol system provides for the transfer of energy to the OM was first inferred from the ability of TolQ and TolR to partially replace two paralogous proteins (ExbB and ExbD, respectively) from a system known to support energy-dependent OM processes—the TonB system (Braun and Herrrmann, 1993).

The TonB system supports a different, more well-defined OM energy need, facilitating the active transport of specific nutrients across the outer membrane. Primarily serving to fuel high-affinity transporters of iron siderophores and host iron sequestration proteins (reviewed in Perkins-Balding *et al.*, 2004; Postle and Kadner, 2003; Weiner, 2005), TonB also energizes the high-affinity transport of cobalamin, with recent evidence indicating that TonB might support the transport of a more diverse set of ligands in some species (Neugebauer *et al.*, 2005). Owing in part to its well-defined function and relative amenability to quantitative analysis, the TonB system provides the paradigm for the transduction of CM energy to the OM.

Like the Tol system, the core components of the TonB system include two cytoplasmic membrane proteins (ExbB and ExbD) that form a heteromultimeric complex with a third protein (TonB) that traverses the periplasmic space (Held and Postle, 2002; Higgs et al., 1998; Skare et al., 1993). Lacking a TolB analog, TonB instead interacts directly with TonB-dependent OM receptors, facilitating the transport of ligand into the periplasmic space (reviewed in Weiner, 2005). TonB-dependent transport can be blocked by protonophores (Hancock and Braun, 1976; Reynolds et al., 1980) and, in unc strains (which lack CM-bound ATP synthase) by cyanide (Bradbeer, 1993), suggesting that the proton gradient of the CM (i.e., protonmotive force: pmf) is the energy source. Subsequent demonstration that TonB undergoes pmf-dependent conformational changes and that this phenomenon requires the ability to interact with ExbB and ExbD suggests that ExbB and ExbD together function to harvest the potential energy of the CM proton gradient, which is then stored conformationally in TonB (Larsen et al., 1999). Further support for a role as the energy-harvesting complex comes from the apparent homology of ExbB and ExbD (and their paralogs TolQ and TolR) with the proton-harvesting MotA and MotB proteins of the bacterial flagellar motor (Kojima and Blair, 2001).

The preponderance of data suggests that the ability of TonB to interact productively with the ExbB/D energy-harvesting complex is dependent upon the amino-terminal region of TonB. Originally predicted to contain a single transmembrane domain (residues 12–32) on the basis of hydrophobicity (Postle and Good, 1983), this region was subsequently shown to mediate the Sec-dependent partitioning of TonB to the CM (Postle and Skare, 1988). Topological analyses indicate the bulk of TonB localizes to the periplasmic space, confirming a single amino-terminal anchorage in the CM (Hannavy et al., 1990; Roof et al., 1991). Essential for function (Jaskula et al., 1994; Karlsson et al., 1993), specific interactions between this signal anchor domain and the energy-harvesting ExbB/ExbD complex provide for the conversion of TonB to the energized state (Larsen et al., 1999). In the absence of ExbB and ExbD, TonB function is impaired, but not absent (Eick-Helmerich and Braun, 1989), with the residual TonB activity reflecting the ability of the ExbB/D paralogs TolQ/R to energize TonB (Braun and Herrmann, 1993).

When modeled as an α-helix, the TonB signal anchor presents a face of four residues (Ser_{16}, His_{20}, Leu_{27}, and Ser_{31}) shared by the TonB paralog TolA (Koebnik, 1993) and conserved among various gram-negative enteric bacteria (reviewed in Postle and Larsen, 2004). Because this motif marks the sole sequence identity between TonB and TolA, it is likely that these residues contribute to the ability of these energy transducers to (albeit imperfectly) be energized by each other's energy-harvesting complexes.

Clearly, two of the residues in this motif are important for the activity of TonB (Ser$_{16}$ and His$_{20}$) and TolA (Ser$_{18}$ and His$_{22}$), with substitutions of either residue (and, at least in the case of TonB, alterations in the spacing between the two residues) rendering the proteins unable to support OM functions (Germon *et al.*, 1998; Larsen and Postle, 2001; Larsen *et al.*, 1994, 1999). In the case of TonB, the mutant phenotype can be suppressed by second site mutations that map to the gene encoding the energy-harvesting complex protein ExbB (Larsen *et al.*, 1994, 1999); similar suppressors of TolA mutations map to the gene encoding the ExbB paralog TolQ (Germon *et al.*, 1998). Interestingly, in the presence of extragenic suppression, these TonB mutants are unstable, with significantly shorter physical half-lives—but only in cells exposed to transportable ligand. This suggests that substitutions at Ser$_{16}$ and His$_{20}$ alter features required for efficient recycling of TonB following the release of conformationally stored energy (Larsen *et al.*, 1999).

Thus, beyond simply tethering the protein to the energy source, the amino termini of TonB and TolA make specific contributions to their function as energy transducers. First, this region participates in the acquisition of energy from the energy-harvesting complex (ExbB/D or TolQ/R). Second, at least in the case of TonB, it contributes to the efficient recycling of TonB following the transfer of energy to an OM recipient. It is likely that each of these roles involves the specific recognition of structural motifs by components of the energy-harvesting complexes. As has been noted, the Ser-His couple appears to contribute one such motif. Scanning single-residue deletion mutagenesis of the amino-terminal portion of the TonB signal anchor (residues 10–24) found only residues 16 through 20 crucial for function. In that study, the replacement of the residues at positions 17 through 19 by three consecutive alanyl residues did not alter TonB function, indicating that these residues served solely to maintain the relative positioning of Ser$_{16}$ to His$_{20}$ (Larsen and Postle, 2001).

Mutagenesis studies involving single-residue changes have not provided for the identification of additional regions that participate in the interaction of TonB and TolA with their respective energy-harvesting complexes. However, because the efficiency with which TonB and TolA are energized varies depending upon which energy-harvesting complex is available, there must be features in addition to the shared Ser-His motif that dictate this specificity. To identify such amino-terminal regions, we have adopted a strategy wherein blocks, rather than individual residues, are altered. Specifically, we have used two approaches. In the first approach, consecutive alanyl residues replace blocks of wild type residues, creating a set of TonB derivatives with an increasingly "generic" signal anchor. In the second approach, specific TolA sections are exchanged for the corresponding TonB region, producing a set of chimeric proteins for evaluation. Both sets of derivatives are generated by adaptations of polymerase chain reaction (PCR), as described and illustrated with the examples that follow.

Alanyl Replacement

Two previously constructed plasmids (Larsen *et al.*, 1999) provide the foundation for constructing *tonB* derivatives that encode proteins with multiple alanyl substitutions. The first of these is pKP315, a pBAD18-based construct (Guzman *et al.*, 1995) in which the *tonB* gene (including its transcriptional start and rho-independent terminator) is placed under control of the bidirectional *araBAD* promoter, providing for the arabinose-regulated expression of wild type TonB. The second is pKP325, in which this *tonB* assemblage and the flanking *araC* gene have been moved into the plasmid pACYC184. In both plasmids, a pair of *BamHI* restriction sites flanks the *tonB* gene to facilitate the recovery of engineered derivatives (Fig. 1A). Site-directed mutagenesis is performed by a modification of a method described by Michael (1994) in which a third, 5'-phosphorylated oligonucleotide primer carrying the desired base changes is included in a modified polymerase chain reaction that uses a thermostabile ligase to

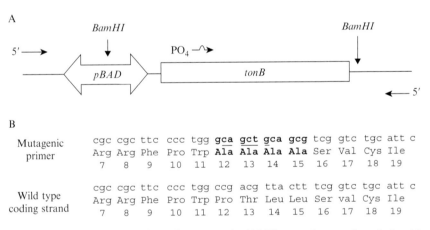

FIG. 1. Three primer site-directed mutagenesis. (A) The template portion of plasmids pKP315 and pKP325 is depicted, including the *araBAD* promoter (*pBAD*) and the *tonB* gene. Primers used for amplification are indicated at each edge of the region, and the relative position of the 5' phosphorylated mutagenic primer is shown. The *BamHI* sites used for cloning are indicated. (B) The mutagenic primer and corresponding wild type coding strand are shown. Substitutions introduced by the mutagenic primer are indicated in bold, with the introduced *PvuII* site used for screening indicated by the underlined sequence.

incorporate this mutagenic primer in the extending strand (Larsen and Postle, 2001). Using nonmutagenic primers that flank the *BamHI* sites, an amplimer is generated that, upon restriction with *BamHI*, can be inserted as a cassette into a similarly restricted parent vector. By designing mutagenic primers to include (or, as possible, exclude) a given restriction site, products bearing the desired mutations can be identified by restriction mapping, with identities then verified by nucleotide sequence determination. Initially used to make silent mutations that introduce restriction sites and to make single-residue substitutions and deletions, we ultimately used this approach to generate a larger mutation, replacing TonB residues 17 through 19 with alanyl residues (Larsen and Postle, 2001). We have subsequently used this approach to replace blocks of four to seven residues with alanyl residues. To illustrate this approach, the replacement of TonB residues 12 through 15 by four consecutive alanyl residues is described.

To replace TonB residues 12 through 15, a mutagenic primer was designed (Fig. 1B) in which the four-codon change was nested between five upstream and four downstream codons identical to the wild type coding strand. The mutagenic codons also introduced a *PvuII* restriction site (CAGCTG – as underlined in Fig. 1B) for screening purposes. Reactions were set up in volumes of 50 μl as follows:

1. 50 pmol of each flanking primer
2. 5 pmol of mutagenic primer
3. 200 pmol each dNTP
4. 5 μl 10$\times$ *Taq* DNA thermoligase buffer (200 mM Tris-Cl [pH 7.6], 250 mM potassium acetate, 100 mM magnesium acetate, 100 mM dithiolthritol, 10 mM NAD, 1% triton X-100; New England Biolabs, Inc., Beverly, MA)
5. 10 units *Taq* DNA ligase (New England Biolabs, Inc.)
6. 2 units Deep Vent *Taq* DNA polymerase (New England Biolabs, Inc.)
7. 0.01 μg pKP315 (as template)

Controls lacking either the mutagenic primer or the template were included. Mixtures were subjected to 35 cycles of melting (94° for 30 sec), annealing (52° for 30 sec), and extension (67° for 240 sec). Note that extensions are allowed to run longer than for normal polymerase chain reactions because our reaction conditions, optimized for the ligation reaction, are not optimal for the polymerase itself. Also, note that these reactions will generate three products—two will be full length (in this case, 963 bp), the majority of which will have incorporated the mutagenic primer and are distinguished from the wild type by having a new *PvuII* site, and one will be shorter (in this case, 780 bp), produced from the mutagenic and downstream primers. If the mutagenic primer is used in a 1:1 molar ratio with the flanking primers, virtually all of the final product will be present in this shorter form; thus, we use 10-fold less mutagenic primer relative to the flanking primers, and under these circumstances, the full-length amplimers are generally the dominant reaction products.

Resultant products are purified using a Qiaquick PCR purification kit (Qiagen Inc., Valencia, CA), then digested with BamHI and resolved on a 1% agarose gel. In this example, digestion of the full-length amplimer (963 bp) resulted in a band of about 900 bp, readily distinguished from the secondary, ~780 bp band generated by the mutagenic and the downstream primers. The full-length amplimer was excised from the gel and recovered using a Qiagen gel extraction kit (Qiagen, Inc.), then inserted into a BamHI restricted, dephosphorylated pKP325 vector in a standard ligation reaction. Products were electroporated into E. coli DH5α cells and selected on standard LB plates with 34 μg ml^{-1} chloramphenicol. Plasmids were recovered from transformants by alkaline lysis and restriction mapped for the presence and orientation of insert, as well as the presence of a new PvuII site (indicative of the mutation). In this case, two of the twelve transformants evaluated met all of the previously stated criteria, with subsequent sequence determination confirming the identity of each.

Having generated the mutation, it is important to verify that the construct will indeed make protein. Thus, the two derivative-bearing plasmids created previously were transformed into the $\Delta tonB$ strain KP1344. These transformants and an isogenic strain encoding wild type TonB from the native chromosomal promoter were grown with aeration at 37° in supplemented M9 minimal salts containing 34 μg ml^{-1} chloramphenicol and various levels of L-arabinose. Cells were harvested at an A$_{550}$ of 0.4 (as determined with a Spectronic 20 spectrophotometer with a path length of 1.5 cm) by precipitation with 10% w/v trichloroacetic acid (TCA), resolved on SDS 11% polyacrylamide gels, immunoblotted, and visualized as previously described (Larsen et al., 1999). Visual comparisons of the ECL results identified concentrations of L-arabinose that provided levels of full-length TonB for each derivative similar to that of chromosomally encoded TonB. For these particular constructs, a concentration of 0.001% L-arabinose produced TonB levels similar to that of wild type cells grown under the same conditions. One of the derivatives was thus selected for further characterization and named "pRA022."

To determine the relative stability of this derivative, cells bearing either pRA022 or the wild type pKP325 plasmid were grown as previously described in supplemented M9 minimal salts containing 34 μg ml^{-1} chloramphenicol and 0.001% L-arabinose. When cells reached an A$_{550}$ of 0.4, the synthesis of new TonB was halted by the addition of D-fucose and D-glucose-6-phosphate (to 12 and 7 mM, respectively) to repress the araBAD promoter, and spectinomycin (to 50 μg ml^{-1}) to halt protein synthesis. Samples

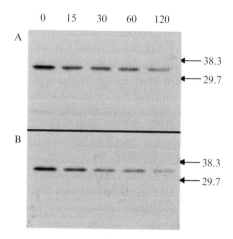

FIG. 2. Alanyl substitution of TonB residues 12 through 15 does not reduce the chemical stability of the protein. A Δ*tonB* strain expressing either wild type (A) or the alanyl-substituted TonB (B) were grown, with the chemical stability of individual derivatives monitored by sampling cultures for immunoblot analysis at the indicated time points (in minutes) following the cessation of protein synthesis, as described in the text. The position of molecular mass standards, and their apparent mass (in kilo Daltons) is indicated at the right of each panel.

were harvested at 0, 15, 30, 60, and 120 min following additions, precipitated in 10% v/w TCA and processed as previously described for immunoblot analysis. Consistent with previous studies (Skare and Postle, 1991), the chemical half-life of wild type TonB under these culture conditions is greater than one hour (Fig. 2A). A similar degree of stability is evident for the TonB derivative encoded by pRA022 (Fig. 2B), indicating that the replacement of four residues in the amino-terminal portion of the signal anchor does not significantly alter the stability of this protein. Further, *in vivo* chemical cross-linking studies indicate that this derivative is properly trafficked to the CM, indicating that the substitutions have not disrupted signal function; similarly, this derivative is able to support iron transport, indicating this region is not essential for energy transduction (data not shown).

Using this approach, we have made alanyl replacements throughout the signal anchor of TonB, with second- and third-generation derivatives created using plasmids that already bear substitutions as templates. Characterization of these derivatives is ongoing.

TonB/TolA Chimeras

Early studies in the Higgins laboratory (Karlsson *et al.*, 1993) suggested that the amino terminal regions of TonB and TolA contributed to the ability of these energy transducers to discriminate between their respective energy-harvesting complexes. Specifically, they found that replacement of the first 32 residues of TonB by the first 34 residues of TolA produced a derivative that retained activity (as measured by sensitivity to the TonB-dependent bacteriophage $\phi80$) and appeared to be more active in the presence of TolQ/R than in the presence of ExbB/D. Because these regions bear little similarity save for the shared Ser/His motif (Fig. 3), these findings suggest that a similar strategy, focusing on smaller portions of the amino-terminal region, might identify motifs that contribute to the recognition of specific energy-harvesting complexes. To build such chimeras, we have adapted a sequence overlap extension (SOEing) PCR procedure (Lee *et al.*, 2004) to replace specific regions of TonB with the analogous regions of TolA. Here, we describe the production of one such construct, in which the cytoplasmic domain of the extreme amino terminus of TolA replaces the corresponding region of TonB.

   The basic approach for assembling chimeric genes is depicted in Fig. 4. To provide a source of *tolA*, the plasmid pRA004 was constructed by inserting a PCR-derived *tolA* amplimer into a pBAD24 plasmid (Guzman *et al.*, 1995) under control of the bidirectional *araBAD* promoter, using the

1	2	3	4	5	6	7	8	9	10	11	12	13	14
M	T	L	D	L	P	R	R	F	P	W	P	T	L
atg	acc	ctt	gat	tta	cct	cgc	cgc	ttc	ccc	tgg	ccg	acg	tta
gtg	tca	aag	gca	acc	gaa	caa	aac	gac	aag	ctc	aag	cgg	gcg	ata	att
V	S	K	A	T	E	Q	N	D	K	L	K	R	A	I	I
1	2	3	4	5	6	7	8	9	10	11	12	13	14	15	16

15	16	17	18	19	20	21	22	23	24	25	26	27	28	29	30
L	S	V	C	I	H	G	A	V	V	A	G	L	L	Y	T
ctt	tcg	gtc	tgc	att	cat	ggt	gct	gtt	gtg	gcg	ggt	ctg	ctc	tat	acc
att	tca	gca	gtg	ctg	cat	gtc	atc	tta	ttt	gcg	gcg	ctg	atc	tgg	agt
I	S	A	V	L	H	V	I	L	F	A	A	L	I	W	S
17	18	19	20	21	22	23	24	25	26	27	28	29	30	31	32

31	32	33	34	35	36	37	38	39	40	41	42	43	44	45	46
S	V	H	Q	V	I	E	L	P	A	P	A	Q	P	I	S
tcg	gta	cat	cag	gtt	att	gaa	cta	cct	gcg	cct	gcg	cag	ccg	att	tct
tcg	ttc	gat	gag	aat	ata	gaa	gct	tca	gcc	gga	ggc	ggc	ggt	ggt	tcg
S	F	D	E	N	I	E	A	S	A	G	G	G	G	G	S
33	34	35	36	37	38	39	40	41	42	43	44	45	46	47	48

FIG. 3. The amino-terminal regions of TonB and TolA. An alignment of the coding strands of TonB and TolA (in bold), with the corresponding amino acid residues and codon numbers indicated. The predicted signal anchor-encoding region (TonB codons 12–32) is highlighted. Boxes identify the conserved Ser and His residues. The *Fsp*I restriction site used in the construction of pRA0027 is underlined.

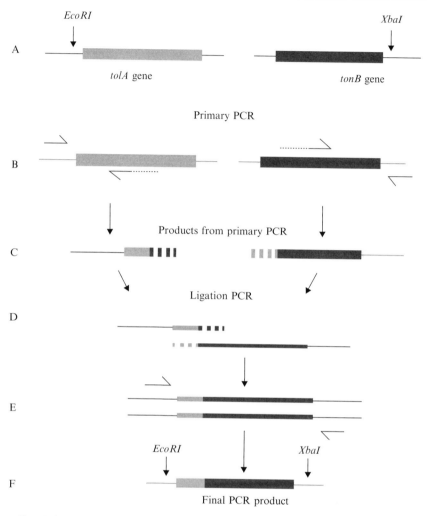

FIG. 4. Strategy for construction of TolA/TonB chimeric proteins. Graphic summary of the sequence overlap extension (SOEing) PCR strategy as described in the text. The beginning templates and the relative positions of restriction sites used in subsequent cloning of products are depicted in (A). Two primary polymerase chain reactions are performed (B) to amplify portions of the two genes using a standard flanking primer and a SOEing primer designed with a 9 residue overhang (dashed line). Amplified products are purified (C) and used in the ligation PCR in which (D) the overlapping sequences are allowed to anneal and elongate in the absence of primers. A pair of flanking primers is then added (E) and the reaction continued to produce the final chimeric PCR product (F).

plasmid-provided ribosome binding site and start codon. To either side of this insert are a number of unique restriction sites, including a 5' located *EcoRI* site and a 3' located *XbaI* site (the latter of which is also present in pKP315) that allow for the subsequent cassetting of PCR-derived constructs into an expression vector (much as *BamHI* sites were used in recovering alanyl-substituted derivatives previously). As with the production of alanyl-substituted derivatives, the plasmid pKP315 served as a template for *tonB* DNA. Primary PCR reactions were run on these templates (Fig. 4B), each pairing a normal flanking primer with an internal "SOEing" primer—designed to contain a nine-base overhang (the dotted portion of the primers in Fig. 4B) complementary to a region of the gene to which the fusion will ultimately be made. Thus, the two primary PCR runs produce a "front" and a "back" amplimer, each containing an 18 base homologous region consisting of the nine residues from each gene that will flank the eventual fusion site (Fig. 4C). In a subsequent "ligation" PCR, the front and back amplimers are paired, with the homologous regions annealing to provide for an extension reaction (Fig. 4D) that produces a fused template that can now be amplified by conventional PCR to produce a final fused product (Fig. 4E,F). It should be noted that initial experiments using the same flanking primers as used to generate the initial front and back amplimers yielded very little full-length product in this final amplification—this problem was readily resolved by using a second, recessed set of flanking primers.

To construct a fusion in which the first 11 codons of the *tonB* gene are replaced by the first 13 codons of the *tolA* gene, reactions were performed as follows:

The front amplimer (a 488 bp product), produced from the *tolA* template pRA004 and back amplimer (a 898 bp product), produced from the *tonB* template pKP315, were generated in standard 50 μl reactions:

1. 50 pmol of each flanking primer
2. 200 pmol of each dNTP
3. 5 μl 10× ThermoPol Reaction buffer (200 mM Tris-HCl [pH 8.8 @ 25°], 100 mM potassium chloride, 100 mM ammonium sulfate, 20 mM magnesium sulfate, 1% Triton X-100; New England Biolabs, Inc.)
4. 2 units Deep Vent *Taq* DNA polymerase (New England Biolabs, Inc.)
5. 0.01 μg pKP315 or pRA004 (as template)

Mixtures were subjected to 35 cycles of melting (94° for 30 sec), annealing (60° for 30 sec), and extension (72° for 70 sec). Products were resolved on 1% agarose gels, with the appropriate products excised and recovered using a gel extraction kit (Qiagen Inc.). Purified front and back amplimers were then used for ligation PCR as follows:

Reactions were initiated in a 48 μl volume using $\sim$0.08 pmol each of front and back amplimers, with 200 pmol of each dNTP and 2 units of Deep Vent DNA polymerase, in a 1× ThermoPol Reaction buffer with additional amounts (0, 1 mM, or 2 mM) of magnesium sulfate. Mixes were subjected to 3 cycles of melting (94° for 30 sec), annealing (60° for 150 sec), and extension (72° for 150 sec). Following the third cycle, products were heated to 94° and 50 pmol of each recessed flanking primer was added. Samples were then subjected to 32 cycles of melting (94° for 30 sec), annealing (60° for 30 sec), and extension (72° for 90 sec). Reactions were resolved on 1% agarose gels and examined for the presence of the predicted 963 bp product. The ligation reaction containing the predicted 963 bp product and the least by-products (in this case, the reaction not supplemented with additional magnesium sulfate) was purified with a Qiaquick PCR purification kit (Qiagen, Inc.). The purified amplimer was then digested with *EcoR*I and *Xba*I and inserted into a similarly restricted (and additionally dephosphorylated) pRA004 in a standard ligation reaction. Products were transformed into chemically competent DH5α cells and selected on standard LB plates with 100 μg ml^{-1} ampicillin. Plasmids were recovered from transformants by alkaline lysis and restriction mapped for the presence of insert. Twelve plasmids with putative *tolA*$_{1-13}$/*tonB*$_{12-239}$ inserts were transformed into the Δ*tonB* strain KP1229, induced with 0.0001% L-arabinose, and their ability to encode a protein detectible by a TonB-specific monoclonal antibody was subsequently verified by immunblot analysis (data not shown). Three positive isolates were chosen and the identity of each construct was confirmed by sequence determination. Interestingly, each of the selected constructs had an unintended point mutation downstream from the fusion site that resulted in a residue substitution (Ser$_{46}$Ala, Glu$_{74}$Lys, and Gln$_{107}$Arg). To obtain a derivative free of unintended substitutions, an *Fsp*I site (at codons 41–43; underlined in Fig. 3) and a *BstE*II site in the vector 5′ to the construct (not shown) were used to move the fusion region into pKP325, replacing the corresponding 5′ portion of *tonB* with the *tolA/tonB* fusion. The resultant construct was recovered as previously described, screened by restriction mapping, confirmed by sequence analysis, and named "pRA027."

The strategy used to screen the initial fusion products included an immunoblot verification that confirmed these constructs did encode protein. The final derivative, now on a pKP325 framework similar to that used for the alanyl substitution derivatives, was similarly shown to express a product upon induction with L-arabinose, the chemical half life of which was similar to that of a wild type TonB (data not shown). To determine whether this chimeric protein retained activity, its ability to support siderophore-mediated iron transport was examined. Iron transport assays were performed using *aroB, ΔtonB* strains as previously described (Larsen and Postle, 2001), here bearing plasmids either encoding wild-type TonB (pKP368, a pKP325 derivative carrying a silent mutation that introduces a novel restriction site irrelevant to the current experiment), the TolA/TonB fusion (pRA027), or a vector control (pACYC184). In these preliminary studies, the TolA/TonB fusion protein provided for the transport of [^{55}Fe]-ferrichrome at rates only slightly less than those achieved with wild type TonB (Fig. 5).

In the previous experiment, both the ExbB/D and the TolQ/R energy-harvesting complexes are present. Future studies with this and additional chimeric constructs will use iron transport and other assays with cells lacking one or the other energy-harvesting complex to determine which regions of TonB and TolA provide for specific interactions with the energy-harvesting complexes. Together with the ongoing characterization of alanyl-substituted TonB derivatives, these studies will afford dissection of the molecular interactions that provide for the coupling of the CM electrochemical potential to the activation of these energy transduction proteins.

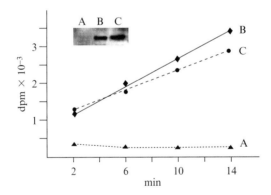

FIG. 5. The TolA/TonB chimera supports TonB-dependent transport of (FeIII)-sidero-phores. The ΔtonB, aroB strain RA1023 carrying plasmids that either express no TonB (A), wild type TonB (B), or the TolA/TonB chimera (C) were grown and assayed for the uptake of [^{55}Fe]-ferrichrome as described later. To verify levels of TonB expression, samples were harvested at the onset of the experiment (at 0 min) and processed for immunoblot analysis (inset). Briefly, cells were grown with aeration at 37° in supplemented M9 minimal salts (containing 34 μg ml^{-1} chloramphenicol and 0.001% L-arabinose) to an A$_{550}$ of 0.4, centrifuged, and then suspended to 2 × 10^8 colony-forming units ml^{-1} in M9 minimal salts containing 0.1 mM nitrilotriacetate and 0.2% w/v$_D$-glucose. Cells were equilibrated for 5 min at 30°, then transport initiated with the addition of 150 pmol of [^{55}Fe]-ferrichrome. Samples were harvested at indicated time points by filtration onto Whatman GF/C filters, washed three times with 5 ml of 0.1 M LiCl, then air dried. Incorporated [^{55}Fe] was determined by liquid scintillation counting, and recorded as dpm per 10^8 colony-forming units. All experiments were performed in triplicate.

Acknowledgments

The studies reported herein adhered to the chemical and radiation safety guidelines overseen by Bowling Green State University and to the appropriate NIH guidelines for the generation and experimental use of recombinant DNA molecules. This research was supported by an award to R.L. from the National Science Foundation (MCB0315983).

References

Bernadac, A., Gavioli, M., Lazzaroni, J.-C., Reina, S., and Lloubes, R. (1998). *Escherichia coli tol-pal* mutants from outer membrane vesicles. *J. Bacteriol.* **180**, 4872–4878.

Bradbeer, C. (1993). The proton motive force drives the outer membrane transport of cobalamin in *Escherichia coli*. *J. Bacteriol.* **175**, 3146–3150.

Braun, V., and Herrmann, C. (1993). Evolutionary relationship of uptake systems for biopolymers in *Escherichia coli*: Cross-complementation between the TonB-ExbB-ExbD and TolA-TolQ-TolR proteins. *Mol. Microbiol.* **8**, 261–268.

Eick-Helmerich, K., and Braun, V. (1989). Import of biopolymers into *Escherichia coli*: Nucleotide sequences of the *exbB* and *exbD* genes are homologous to those of the *tolQ* and *tolR* genes, respectively. *J. Bacteriol.* **171**, 5117–5126.

Germon, P., Clavel, T., Vianney, A., Portalier, R., and Lazzaroni, J. C. (1998). Mutational analysis of the *Escherichia coli* K-12 TolA N-terminal region and characterization of its TolQ-interacting domain by genetic suppression. *J. Bacteriol.* **180**, 6433–6439.

Guzman, L. M., Belin, D., Carson, M. J., and Beckwith, J. (1995). Tight regulation, modulation, and high-level expression by vectors containing the arabinose P-BAD promoter. *J. Bacteriol.* **177**, 4121–4130.

Hancock, R. E. W., and Braun, V. (1976). Nature of the energy requirement for the irreversible adsorption of bacteriophages T1 and ϕ80 to *Escherichia coli*. *J. Bacteriol.* **125**, 409–416.

Hannavy, K., Barr, G. C., Dorman, C. J., Adamson, J., Mazengera, L. R., Gallagher, M. P., Evans, J. S., Levine, B. A., Trayer, I. P., and Higgins, C. F. (1990). TonB protein of *Salmonella typhimurium*: A model for signal transduction between membranes. *J. Mol. Biol.* **216**, 897–910.

Held, K. G., and Postle, K. (2002). ExbB and ExbD do not function independently in TonB-dependent energy transduction. *J. Bacteriol.* **184**, 5170–5173.

Higgs, P. I., Myers, P. S., and Postle, K. (1998). Interactions in the TonB-dependent energy transduction complex: ExbB and ExbD from homomultimers. *J. Bacteriol.* **180**, 6031–6038.

Jaskula, J. C., Letain, T. E., Roof, S. K., Skare, J. T., and Postle, K. (1994). Role of the TonB amino terminus in energy transduction between membranes. *J. Bacteriol.* **176**, 2326–2338.

Karlsson, M., Hannavy, K., and Higgins, C. F. (1993). A sequence-specific function for the amino-terminal signal-like sequence of the TonB protein. *Mol. Microbiol.* **8**, 379–388.

Koebnik, R. (1993). Microcorrespondence: The molecular interaction between components of the TonB-ExbBD-dependent and of the TolQRA-dependent bacterial uptake systems. *Mol. Microbiol.* **9**, 219.

Kojima, S., and Blair, D. F. (2001). Conformational change in the stator of the bacterial flagellar motor. *Biochemistry* **40**, 13041–13050.

Larsen, R. A., and Postle, K. (2001). Conserved residues Ser(16) and His(20) and their relative positioning are essential for TonB activity, cross-linking of TonB with ExbB, and the ability of TonB to respond to proton motive force. *J. Biol. Chem.* **276**, 8111–8117.

Larsen, R. A., Thomas, M. G., Wood, G. E., and Postle, K. (1994). Partial suppression of an *Escherichia coli* TonB transmembrane domain mutation (ΔV17) by a missense mutation in ExbB. *Mol. Microbiol.* **13,** 627–640.

Larsen, R. A., Thomas, M. G., and Postle, K. (1999). Protonmotive force, ExbB, and ligand-bound FepA drive conformational changes in TonB. *Mol. Microbiol.* **31,** 1809–1824.

Lazzaroni, J.-C., Fognini-Lefebvre, N., and Portalier, R. (1989). Cloning of *excC* and *excD* genes involved in the release of periplasmic proteins by *Escherichia coli* K12. *Mol. Gen. Genet.* **218,** 460–464.

Lazzaroni, J.-C., Germon, P., Ray, M.-C., and Vianney, A. (1999). The Tol proteins of *Escherichia coli* and their involvement in the uptake of biomolecules and outer membrane stability. *FEMS Microbiol. Lett.* **177,** 191–197.

Lee, J., Lee, H.-J., Shin, M.-K., and Ryu, W.-S. (2004). Versatile PCR-mediated insertion or deletion mutagenesis. *BioTechniques* **36,** 398–400.

Lloubes, R., Cascales, E., Walburger, A., Bouveret, E., Lazdunski, J.-C., Bernadac, A., and Journet, L. (2001). The Tol-Pal proteins of the *Escherichia coli* cell envelope: An energized system required for outer membrane integrity? *Res. Microbiol.* **152,** 523–529.

Michael, S. F. (1994). Mutagenesis by incorporation of a phosphorylated oligo during PCR amplification. *BioTechniques* **16,** 410–412.

Neugebauer, H., Herrmann, C., Kammer, W., Schwarz, G., Nordheim, A., and Braun, V. (2005). ExbBD-dependent transport of maltodextrins through the novel MalA protein across the outer membrane of *Caulobacter crescentus. J. Bacteriol.* **187,** 8300–8311.

Perkins-Balding, D., Ratliff-Griffin, M., and Stojiljkovic, I. (2004). Iron transport systems in *Neisseria meningitidis. Microbiol. Mol. Biol. Rev.* **68,** 154–171.

Postle, K., and Good, R. F. (1983). DNA sequence of the *Escherichia coli tonB* gene. *Proc. Natl. Acad. Sci.* **80,** 5235–5239.

Postle, K., and Skare, J. T. (1988). *Escherichia coli* TonB protein is exported from the cytoplasm without proteolytic cleavage of its amino terminus. *J. Biol. Chem.* **263,** 11000–11007.

Postle, K., and Kadner, R. J. (2003). Touch and go: Tying TonB to transport. *Mol. Microbiol.* **49,** 869–882.

Postle, K., and Larsen, R. A. (2004). The TonB, ExbB, and ExbD proteins. *In* "Iron Transport in Bacteria" (J. H. Crosa, A. R. Mey, and S. M. Payne, eds.), pp. 96–112. ASM Press, Washington, DC.

Reynolds, P. R., Mottur, G. P., and Bradbeer, C. (1980). Transport of vitamin B12 in *Escherichia coli.* Some observations on the role of the gene products of *btuC* and *tonB. J. Biol. Chem.* **255,** 4313–4319.

Roof, S. K., Allard, J. D., Bertrand, K. P., and Postle, K. (1991). Analysis of *Escherichia coli* TonB membrane topology by use of PhoA fusions. *J. Bacteriol.* **173,** 5554–5557.

Ruiz, N., Kahne, D., and Silhavy, T. J. (2006). Advances in understanding bacterial outer-membrane biogenesis. *Nat. Rev. Microbiol.* **4,** 57–66.

Skare, J. T., and Postle, K. (1991). Evidence for a TonB-dependent energy transduction complex in *Escherichia coli. Mol. Microbiol.* **5,** 11000–11007.

Skare, J. T., Ahmer, B. M. M., Seachord, C. L., Darveau, R. P., and Postle, K. (1993). Energy transduction between membranes. TonB, a cytoplasmic membrane protein, can be chemically cross-linked *in vivo* to the outer membrane receptor FepA. *J. Biol. Chem.* **268,** 16302–16308.

Weiner, M. C. (2005). TonB-dependent outer membrane transport: Going for Baroque? *Curr. Opin. Struct. Biol.* **15,** 394–400.

[6] Functional Dynamics of Response Regulators Using NMR Relaxation Techniques

By ALEXANDRA K. GARDINO and DOROTHEE KERN

Abstract

A fundamental concept of phosphorylation-mediated signaling is the precise switching between discrete functional conformations. According to the traditional view, phosphorylation induces a new, active conformation. In this chapter, a series of NMR experiments performed on a response regulator are described that challenge this traditional notion. The combination of NMR relaxation experiments with chemical shift data and the linkage to structure/function reveals a fundamentally different activation mechanism. The NMR data for the response regulator NtrC provide kinetic (rates of interconversion), thermodynamic (relative populations), and structural (chemical shift) information for the conformational exchange process. The results demonstrate that both the inactive and active states are present before phosphorylation, and activation occurs via a shift of this preexisting equilibrium. This concept is in accordance with the energy landscape view of proteins that embraces the existence of conformational substates. We conjecture that this population-shift mechanism is a general paradigm for response regulator activation and possibly more universal for phosphorylation-mediated signaling.

Introduction

Phosphorylation-mediated signaling is a common mechanism of signal transduction enabling an organism to respond appropriately to its environment. Signaling events in bacteria are dominated by "two-component" systems, a basic module consisting of a histidine protein kinase and a response regulator (reviewed in Parkinson and Kofoid, 1992; Stock *et al.*, 2000; West and Stock, 2001). Response regulators can be single or multidomain proteins. In all cases, the receiver domain serves as a phosphorylation-driven molecular switch. The corresponding cellular response is regulated through interactions of the switch domain with different downstream effector domains or target proteins.

While the three dimensional structures of several receiver domains have been solved in their inactive (or nonphosphorylated) (Feher *et al.*, 1997; Gouet *et al.*, 1999; Lewis *et al.*, 2000; Meyer *et al.*, 2001; Volkman *et al.*, 1995; Volz and Matsumura, 1991) and activated forms (Bachhawat

METHODS IN ENZYMOLOGY, VOL. 423
0076-6879/07 $35.00
DOI: 10.1016/S0076-6879(07)23006-X

et al., 2005; Birck *et al.*, 1999; Cho *et al.*, 2000; Gardino *et al.*, 2003; Halkides *et al.*, 2000; Hastings *et al.*, 2003; Kern *et al.*, 1999; Lewis *et al.*, 1999; Park *et al.*, 2002), the mechanism of activation in these molecular switch domains cannot be elucidated from the static structures of receiver domains in different functional states alone. It is well accepted that ligand-mediated activation can act through two different mechanisms: (1) ligand binding induces a new, active structure otherwise known as the induced-fit or Koshland-Nemethy-Filmer (KNF) mechanism (Koshland *et al.*, 1966) or (2) ligand binding shifts a pre-existing equilibrium between the inactive and active states toward that of the active form otherwise known as the population-shift or Monod-Wyman-Changeux (MWC) mechanism (Monod *et al.*, 1965). This second mechanism of allosteric activation is in accordance with a "new view" of the energy landscape of proteins that embraces the existence of conformational ensembles that are kinetically distinct but similar in free energies (Austin *et al.*, 1975; Dill and Chan, 1997; Elber and Karplus, 1987; Frauenfelder *et al.*, 1988). According to this energy landscape view, activation can be achieved by ligand-binding to kinetically accessible but lowly populated states, a state that most likely is not seen in the crystal structure or represented in the NMR structure that is based on bulk, ensemble-averaged macroscopic measurements.

NMR spectroscopy is currently the most powerful biophysical tool to study ensemble averages of molecules in solution at atomic resolution and under equilibrium conditions. NMR has the advantage to go beyond static structure determination and incorporate the fourth dimension or "time domain" to examine molecular motions spanning 15 orders of magnitude (10^{-12}–10^3 sec) (reviewed in Akke, 2002; Palmer, 2004; Palmer *et al.*, 2001). This chapter focuses on the efforts made toward deciphering the mechanism of activation in response regulators using NMR spectroscopy as a probe of both structural rearrangements and protein dynamics. The experiments discussed are sensitive to motions on the microsecond to millisecond timescales, the time regime on which many biological processes such as allosteric regulation and ligand binding occur (Mulder *et al.*, 2001; Palmer *et al.*, 2001; Volkman *et al.*, 2001). Motions on this timescale capture transitions between kinetically distinct substates in contrast to faster motions (picoseconds–nanoseconds) that describe the ruggedness of the energy landscape.

In an NMR experiment, kinetic substates are characterized by different chemical shifts (ω) for nuclei in different electronic environments. Conformational exchange between two substates, A and B (also referred to as chemical exchange), results in accelerated dephasing of coherence in the transverse plane, which manifests as additional line broadening in solution. This additional dephasing from chemical exchange, called R_{ex}, contributes additively to the transverse relaxation rate in the absence of exchange, R_{20}, to yield the overall observed transverse relaxation rate (R_2^{eff}):

$$R_2^{eff} = R_{20} + R_{ex} \qquad (1)$$

Besides a qualitative identification of nuclei experiencing conformational exchange, transverse relaxation experiments can yield information about the kinetic parameters (rate of interconversion between two states, $k_{ex} = k_f + k_r$), thermodynamic parameters (the relative populations of each state, $p_A = 1-p_B$), and structural information (chemical shift difference between states, $\Delta\omega = \omega_A - \omega_B$). These parameters are encoded in R_{ex} through the following relationship assuming a two-state exchange process that is fast on the NMR chemical shift timescale (Palmer et al., 2001):

$$R_{ex} \sim p_A p_B * \Delta\omega^2 / k_{ex} \qquad (2)$$

The experimental approach that allows the extraction of these parameters is described in the last section of this chapter.

The Experimental Setup

To shed light on the question of whether response regulators are activated via an induced fit mechanism versus a shift of a preexisting equilibrium, we chose NtrC (Nitrogen regulatory protein C from *Salmonella typhimurium*) as a model system. NtrC is a multidomain response regulator (Weiss et al., 1991). The N-terminal "switch" or receiver domain (residues 1–124, termed NtrC^r) is activated by phosphorylation of the conserved Asp54. The NMR solution structures of NtrC^r and P-NtrC^r (the fully active phosphorylated form) showed a large conformational rearrangement in roughly half of the protein upon phosphorylation (termed the "3445 face," which stands for helix 3, strand 4, helix 4, and strand 5) exposing a hydrophobic surface (Kern et al., 1999; Volkman et al., 2001) (Fig. 1A). This newly exposed surface has been shown to be the region responsible for signal transmission to the downstream central domain (Hastings et al., 2003; Hwang et al., 1999; Lee et al., 2000) triggering ATPase activity (Weiss et al., 1991), oligomerization (Wyman et al., 1997), and ultimately transcriptional activation.

NMR experiments were performed on the isolated receiver domain since it represents the actual molecular switch. Moreover, the structures representing the on (active) and off (inactive) substates are known. Interestingly, a series of mutations in this domain were identified as having varying levels of activity in the absence of phosphorylation. Consequently, this model system offers the opportunity to link protein dynamics to function by studying the dynamics in different functional states. While the mutants are stable forms of an "active state" (although only partially

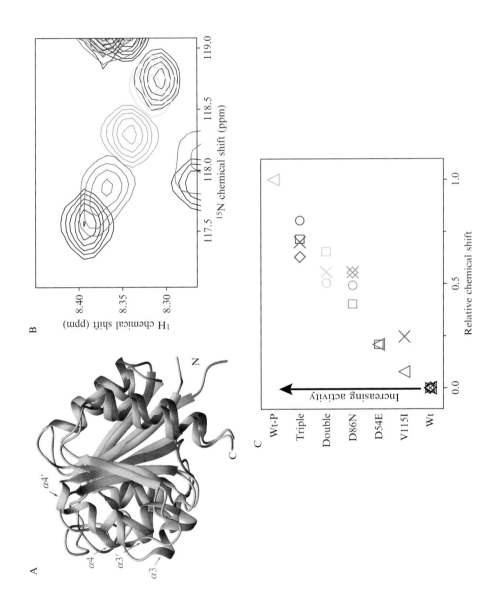

active), the fully active, phosphorylated form has a lifetime (about 2 min) that is too short for data collection in NMR dynamics and/or structural experiments. An experimental approach to overcome this problem is to regenerate the phosphorylated state utilizing NtrC's catalytic activity to phosphorylate D54 from a small molecule phosphor donor (in this case, carbamoyl phosphate) (Kern et al., 1999).

Two-State Allosteric Activation Identified by NMR Chemical Shift Analysis

Chemical shift is a very sensitive marker for small changes in the local environment. As has been discussed, the conformational substates can easily be identified by different chemical shifts in an NMR spectrum. However, distinct resonances are only observed if the interconversion between them is slow on the NMR time scale, meaning that $k_{ex} < \Delta\omega$. Here, the relative populations of the substates can be directly determined from the relative integrals of the peak intensities. Conversely, if $k_{ex} > \Delta\omega$, the interconversion process is fast on the NMR time scale and a single population weighted average resonance is observed (Palmer et al., 2001). Consequently, from the averaged peak position, the equilibrium constant can be read out if the chemical shifts of the two individual substates are known.

Chemical shift analysis using ^{19}F NMR on 4-fluorophenylalanine in CheY was used early on to probe the structural changes associated with activation (Bourret et al., 1993; Drake et al., 1993) in this response regulator. To obtain structural information on a per residue basis, however, chemical shifts of ^{15}N backbone amides were compared for inactive (wild-type), a series of partially active mutants forms and fully active (phosphorylated wild-type and mutant) forms of NtrC^r (Fig. 1B). First, for all forms, single resonances were observed for all amides. Strikingly, for a number of amides in the 3445 face, a linear progression of the chemical shifts was observed that correlated with increased activity (Fig. 1C). This spectroscopic phenomenon can be rationalized by an interconversion between an inactive and active substate that is fast on the NMR time scale resulting in population-averaged chemical shifts. The chemical shift

---

FIG. 1. Two-state model implied by NMR structure and chemical shift analysis. (A) The NMR structures of NtrC^r (blue) and P-NtrC^r (orange) are superimposed. The switch region is highlighted in lighter colors. (B, C) Chemical shift changes for NtrC^r forms with increasing activities: V115I (grey), D54E (red), D86N (green), D86N/A89T (gold), D86N/AS89T/V115I (blue), P-NtrC^r (cyan) with respect to wild-type (black). Changes in peak position are shown for the backbone amide of D88 (B). The same pattern is observed for a number of residues as seen from the normalized chemical shifts (C), therefore reflecting the equilibrium between the two states. Reproduced with permission from Volkman et al., 2001. (See color insert.)

of nonphosphorylated wild-type would represent approximately the chemical shift of the inactive substate, whereas the resonance position in P-NtrC^r would represent the chemical shift of the active substate. From the chemical shift of the individual mutant proteins relative to those "endstates," the equilibrium constant could be directly calculated. This simple model is supported by the fact that all amide resonances with the characteristic linear chemical shift dependence fall at the same position between the two endstates after normalization (Fig. 1C).

We want to address three points regarding this chemical shift analysis: First, this linear progression can only be observed if there is no local perturbation of chemical shift due to local changes such as amino acid substitutions or phosphorylation. Second, the absolute change in chemical shift does not relate to the magnitude of conformational change. Third, the chemical shifts of nonphosphorylated and phosphorylated wild-type NtrC^r do not necessarily represent the true chemical shifts of the inactive and active conformations but should be good approximates due to the fact that the populations should be skewed toward those substates as their structures were solved by NMR-based ensemble-averaged NOE measurements (Kern et al., 1999; Volkman et al., 2001). In addition, phosphorylation of the partially active mutant forms results in chemical shifts identical to phosphorylated wild-type, strengthening the idea of a two-state interconversion and a highly skewed population of the active conformation.

Two-State Allosteric Activation Buttressed by Standard NMR Relaxation Experiments

If our interpretation of the chemical shift behavior is correct, one should be able to measure the conformational exchange process directly by NMR relaxation techniques. For this purpose, longitudinal (R_1) and transverse (R_2) relaxation times, together with heteronuclear NOE experiments (Farrow et al., 1994), were performed on NtrC^r in three distinct functional states—inactive wild-type, partially active mutant, and fully active P-NtrC^r. Sophisticated analysis of this set of relaxation data allows extraction of ps-ns as well as μs-ms timescale motions, termed Model-free analysis (for details, see Mandel et al., 1995, and Lipari and Szabo, 1982). We want to focus here entirely on the results regarding the μs-ms dynamics in the form of an R_{ex} term from such an analysis.

Indeed, many residues in the 3445 face or switch region showed μs-ms dynamics as manifested through R_{ex} contributions for the partially active mutant protein (Fig. 2B). In contrast, residues outside of the signaling face do not have R_{ex}. This result is in agreement with our model that the mutant protein samples both the inactive and active conformation based on the

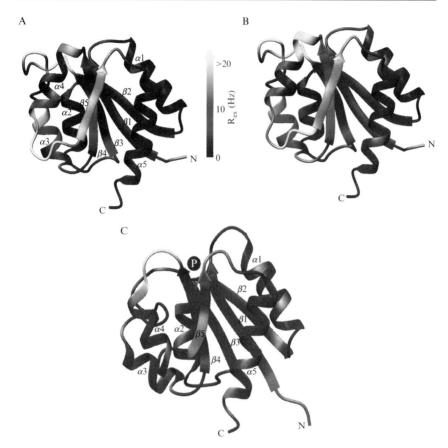

FIG. 2. Conformational exchange detected by standard transverse relaxation measurements. R_{ex} contributions are shown for (A) NtrC^r, (B) NtrC^r (D86N/A89T), and (C) P-NtrC^r as a continuous color scale. Reproduced with permission from Volkman *et al.*, 2001. (See color insert.)

chemical shift analysis (Volkman *et al.*, 2001). Interestingly, inactive wild-type shows the same qualitative pattern (Fig. 2A), suggesting that NtrC^r already samples both the on and off substates before phosphorylation. Phosphorylation could simply shift this preexisting equilibrium toward the active substate. The fact that R_{ex} contributions could no longer be detected in P-NtrC^r (Fig. 2C) suggests that the active substate predominates the equilibrium.

Taken together, both the chemical shift analysis and NMR relaxation techniques support a two-state equilibrium between an active and inactive conformation (both structures were previously determined) (Kern *et al.*, 1999;

Volkman *et al.*, 1995) that is fast on the NMR time scale. Activation by mutation or phosphorylation shifts this preexisting equilibrium.

A New Approach for Quantitative Analysis of Microsecond Protein Dynamics

CPMG Relaxation Dispersion for Millisecond Dynamics

To go beyond a qualitative description of μs-ms dynamics on the basis of R_{ex} only, one would like to characterize the energy landscape by determining the free energy of activation from the rates of interconversion as well as relative free energies of the substates from the relative populations. These kinetic and thermodynamic parameters can, in principle, be obtained by Carr-Purcell-Meiboom-Gill (CPMG) experiments that take advantage of the fact that R_{ex} has a dependence on an externally applied B_1 field (ν_{CPMG}), whereas the intrinsic transverse relaxation rate, R_{20}, does not:

$$R_2^{eff}(\nu_{CPMG}) = R_{20} + R_{ex}(\nu_{CPMG}) \qquad (3)$$

CPMG relaxation dispersion experiments allow for the dissection of the chemical exchange (R_{ex}) contribution to the exchange-free rate of transverse relaxation (R_{20}) by employing a CPMG pulse train of differentially spaced 180° refocusing pulses ($\nu_{CPMG} = 1/4\tau_{CPMG}$) to effectively suppress R_{ex} (Loria *et al.*, 1999a) (Fig. 3). Curve-fitting of relaxation dispersion data to the following equation (Luz and Meiboom, 1963) provides an accurate

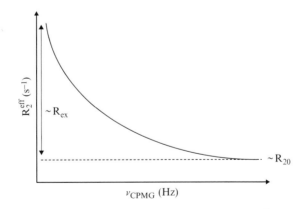

FIG. 3. Dependence of the effective transverse relaxation rate (R_2^{eff}) on applied CPMG field strengths (ν_{CPMG}). The amplitude of the dispersion curve represents the R_{ex} contribution at ν_{CPMG} of 0 Hz while the shape of the dispersion curve is governed by the rate of chemical exchange (k_{ex}), shown here theoretically. At ($\nu_{CPMG} \rightarrow \infty$), R_{ex} is completely suppressed yielding R_{20}.

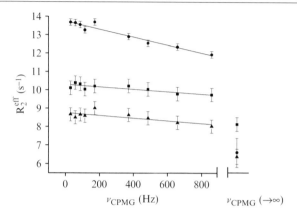

FIG. 4. Quantification of the conformational transition between the inactive and active states of NtrC^r at 25°. ^{15}N CPMG relaxation dispersion data were combined with independently determined values of R$_{20}\alpha$ ($\nu_{CPMG}\rightarrow\infty$) to fit to Eq. (4), yielding a global rate of k$_{ex}$ of 13,580 ± 920 s^{-1}. Data was processed using NMRPipe (Delaglio et al., 1995). Data processing, analysis and error estimation were performed as described (Eisenmesser et al., 2005). (This same error analysis was also applied to the experimental data shown in Fig. 5.)

measurement of the rate of chemical exchange as long as the interconversion rate (k$_{ex}$ ≤ 2,000 s^{-1}) is near that of the physical CPMG pulsing limit (ν_{CPMG} about 2000 Hz maximum):

$$R_2^{eff}(\nu_{CPMG}) = R_{20}(\nu_{CPMG} \rightarrow \infty) + (p_A p_B \Delta\omega^2/k_{ex})$$
$$[1 - (4\nu_{CPMG}/k_{ex})\tanh(k_{ex}/4\nu_{CPMG})] \tag{4}$$

In this time regime, k$_{ex}$/2ν_{CPMG} < 1 and R$_{ex}$ can be fully suppressed (Wang et al., 2001). When this condition is not met, R$_2^{eff}$ does not approach that of the exchange-free transverse relaxation rate at the maximum applied ν_{CPMG} field strength yielding inaccurate values of k$_{ex}$.

^{15}N CPMG relaxation dispersion experiments performed on wild-type NtrC^r revealed that the interconversion rate must be larger than this upper limit because R$_{ex}$ could not be fully suppressed at the maximal field strength of 1000 Hz (Fig. 4). Here, we describe a new approach that allows quantifying chemical exchange on this faster timescale.

CPMG Relaxation Dispersion Combined with Exchange-Free Relaxation for Microsecond Dynamics

The idea of this approach is to combine CPMG dispersion data with an independently determined exchange-free rate of relaxation (R$_{20}$) (Eq. 1). Furthermore, to improve the sensitivity of this approach, particularly for larger proteins, the TROSY component of the ^{15}N doublet is selected (Pervushin *et al.*, 1997) to reduce the rate of exchange-free transverse relaxation which consequently leads to an overall larger contribution from chemical exchange to the measured relaxation rate, R_2^{eff} (Loria *et al.*, 1999b).

The exchange-free relaxation rate in a TROSY-based experiment (R_{20}^{α}) is composed of three components:

$$R_{20}^{\alpha} = R_2^{\text{o}} - \eta_{xy} + R_1^{H}/2 \qquad (5)$$

where R_2^{o} is the auto-relaxation rate for in-phase magnetization, η_{xy} is the cross correlation rate between ^{15}N CSA and ^{15}N-^{1}H dipolar interactions resulting in the differential transverse relaxation rates of the ^{15}N doublet, and R_1^{H} is the amide proton longitudinal relaxation rate of dipolar interactions with nearby protons. The first two components of the R_{20}^{α} rate (Eq. 5), $R_2^{\text{o}} - \eta_{xy}$, were determined using a previously published pulse sequence (Wang *et al.*, 2003) that measures the relaxation rates of the narrow (R_2^{α}) and broad (R_2^{β}) components of the ^{15}N doublet as well as the longitudinal two-spin order relaxation rate (R_1^{2HzNz}). The ratio of peak intensity for I^{β}/I^{α} (Wang *et al.*, 2003) yields the cross-correlation rate, η_{xy}.

R_2^{o} cannot be determined for residues undergoing exchange; however, it can be estimated through the relationship with η_{xy} via the correlation coefficient κ, which is independent of local or global motions (Fushman *et al.*, 1998; Fushman and Cowburn, 1998). In order to determine κ, a trimmed mean approach over all residues has been proposed (Fushman and Cowburn, 1998; Wang *et al.*, 2003). In NtrC^{r}, however, about 20% of the residues suitable for this analysis exhibit chemical exchange. This large amount of R_{ex} contributions causes the mean value of κ over all residues in NtrC^{r} to be artificially high. In practice, taking the trimmed mean of κ over all residues yielded a value of 1.59, which resulted in R_{20}^{α} values that were approximately 0.5 to 1.0 Hz higher than the experimental R_2^{eff} values for non-exchanging residues. However, separation of residues with and without exchange from the ^{15}N CPMG relaxation dispersion data (Fig. 5A,B) allows calculation of a more accurate trimmed mean value of κ by using only non-exchanging residues. The recalculated κ was 1.49.

The final component needed in order to determine the exchange-free relaxation rate of the TROSY component is $R_1^{H}/2$ (Eq. 5). R_1^{H} can be estimated through the following approximation (Wang *et al.*, 2001):

$$R_1^{HzNz} \approx R_1^{H} + R_1^{N} \qquad (6)$$

A TROSY-based two-spin order pulse sequence developed by Kay *et al.* yields the R_1^{HzNz} relaxation rate, and through the subtraction of R_1^{N} rates measured with conventional experiments (Farrow *et al.*, 1994) yields R_1^{H}.

With these exchange-free rates of relaxation in hand (R_{20}^α), we can now fit ^{15}N TROSY-CPMG dispersion data together with these independently determined exchange-free rates to the Luz-Meiboom equation (Fig. 4). Fits of individual residues gave similar rates of k_{ex}. This result, together with previously published chemical shift data discussed earlier, support

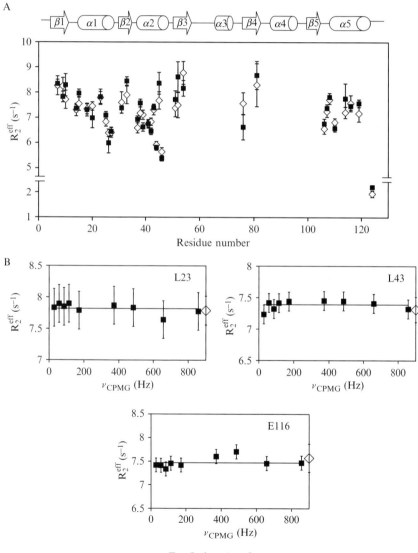

FIG. 5. (*continued*)

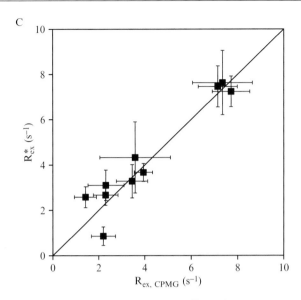

FIG. 5. Validation of the applied method to quantify exchange processes in the micro-second time regime. (A) For non-exchanging residues, R_2^{eff} values obtained from the ^{15}N TROSY-CPMG experiment (■) are compared to the exchange-free R_{20}^{α} calculated from three independent relaxation experiments (◇) described in the text. (B) The full dispersion profiles (■) for representative residues are shown with the exchange-free R_{20}^{α} rate (◇) plotted on the right-hand side of the graph. Lines were drawn at the weighted mean of R_2^{eff}. (C) R_{ex} values extracted from Fig. 4 (R_{ex}, CPMG) are compared to R_{ex} values determined from a published sequence (R_{ex}^{*}) (Wang *et al.*, 2003) designed to measure maximal R_{ex}. All data were collected at 25°.

a global two-state model (Volkman *et al.*, 2001) and justified a global fit of all residues to a k_{ex} value of $13{,}580 \pm 920 \text{ s}^{-1}$. This final determined rate demonstrates that this new approach expands the time regime of CPMG experiments into the microsecond time scale. It is, of course, also apparent that rates faster than the ones observed here will not be accessible by this method since R_{ex} cannot be significant suppressed by a 1000 Hz ν_{CPMG} field strength. For those faster processes, $R_{1\rho}$ experiments are the method of choice (Igumenova and Palmer, 2006).

The accuracy of the approach described above was validated using several controls. First, the exchange-free transverse relaxation rates (R_{20}^{α}) (calculated from the three independent relaxation experiments needed to determine the three parameters in Eq. (5)) were directly compared to relaxation rates from the CPMG data alone for non-exchanging residues (Fig. 5A,B). Evidently, the R_{20}^{α} values (shown as a single value on the right side of the plot) are in very good agreement with the R_2^{eff} values measured across the range of CPMG field strengths (Fig. 5B). This validation ensures that R_{20}^{α} values determined for residues with exchange are accurate for combination with CPMG dispersion data to extract precise rates of conformational exchange.

Second, R_{ex} was determined from the fits as the difference between R_2^{eff} at free precession (ν_{CPMG} is 0 Hz) and the exchange-free rate (Eq. 3). These R_{ex} values were compared to R_{ex} values determined using a previously published procedure by Wang et al. (Wang et al., 2003) that solely allows extraction of accurate R_{ex} values (Fig. 5C). For non-exchanging residues, R_{ex} was distributed around zero using both methods with a standard deviation of $\pm 1.3 \text{ s}^{-1}$. Importantly, for exchanging residues, a very good agreement of R_{ex} values between these two methods is observed (correlation coefficient of 0.94) (Fig. 5C).

In conclusion, three independent relaxation experiments were combined to determine the exchange free relaxation rate of the TROSY component of the ^{15}N doublet. This parameter allows for accurate fitting of ^{15}N TROSY-CPMG dispersion data for a two-state chemical exchange process that is too fast to be suppressed with maximal ν_{CPMG} field strengths due to experimental limitations. The highly sensitive ^{15}N TROSY-CPMG experiment (Loria et al., 1999b) can quickly identify residues that have R_{ex} contributions and provide an estimation of the time regime of exchange. For residues that have exchange processes of $k_{ex} > 2000 \text{ s}^{-1}$, dispersion data can be accurately combined with the newly determined exchange-free R_{20}^{α} values to determine k_{ex} quantitatively. The described method applies a more sensitive and time efficient approach to quantifying fast dynamic processes than $R_{1\rho}$ experiments (Korzhnev et al., 2003) and can be applied to proteins of MW > 20kDa due to the implementation of the TROSY method.

Conclusions

The combination of structural and dynamic data and its relation to function facilitates the distinction between two fundamentally different mechanisms in response to regulator activation, namely, induced fit versus population shift. Detailed characterization of protein dynamics in the response regulator NtrC unambiguously supports the population shift model, in which NtrC^r samples both the inactive and active conformation *before* phosphorylation and phosphorylation of the active site or mutations in the switch region purely shift this preexisting equilibrium toward the active substate. This mechanism has been put forward based on qualitative NMR relaxation measurements on nonphosphorylated SpoOF (Feher and Cavanagh, 1999; Gardino et al., 2003) and functional assays on wild-type and several mutant forms of CheY (Silversmith and Bourret, 1999) as well as OmpR (Ames et al., 1999).

NMR relaxation techniques described in this chapter directly monitor the interconversion between kinetic substates on a per residue basis. It is the combination of this dynamic information with chemical shift data and the linkage to structure/function that revealed the two-state interconversion between the inactive and active conformation in NtrC^r. The two-state model was further buttressed by the quantitative analysis of the interconversion rate yielding a *global* rate of $\sim$13,000 s^{-1}. While this fast rate is unexpected given the complexity of conformational rearrangements in NtrC^r (Volkman *et al.*, 2001), it suggests that this corresponding low energy barrier of the structural transition guarantees fast sampling of both states at thermal equilibrium and implies that this step does not limit cellular responses. Importantly, the described new NMR relaxation techniques not only identify residues with conformational exchange, but also allow characterization of the nature of the transition and its barrier height. While usually only the low-energy major substate is detected with standard structural biology methods, excursions to higher energy states (consequently lowly populated) can be characterized by NMR CPMG relaxation dispersion experiments (Beach *et al.*, 2005; Eisenmesser *et al.*, 2005; Mulder *et al.*, 2001; Palmer, 2004; Tollinger *et al.*, 2001).

From the quantitative analysis performed on NtrC, together with both functional and dynamic information on other response regulators (Ames *et al.*, 1999; Feher and Cavanagh, 1999; Gardino *et al.*, 2003; Silversmith and Bourret, 1999; Volkman *et al.*, 2001), an emerging view arises that the mechanism of activation in response regulators occurs via a shift of a preexisting equilibrium. We conjecture that this population-shift mechanism is a general paradigm of phosphorylation-mediated signaling, or even more universally, of ligand binding.

Acknowledgments

We thank Dr. Chunyu Wang at Columbia University for providing us with a pulse sequence to measure R_{ex} in large proteins and for helpful discussions, Dr. Lewis E. Kay at the University of Toronto for the R_1^{HzNz} pulse sequence, and Dr. Jack Skalicky at the NHMFL at Florida with support from NSF for NMR time. This work was supported by NIH grants GM067963 and GM62117 and by a DOE grant DEFG0205ER15699 to D.K., HHMI, and by instrumentation grants by the NSF and the Keck foundation to D.K.

References

Akke, M. (2002). NMR methods for characterizing microsecond to millisecond dynamics in recognition and catalysis. *Curr. Opin. Struct. Biol.* **12,** 642–647.
Ames, S. K., Frankema, N., and Kenney, L. J. (1999). C-terminal DNA binding stimulates N-terminal phosphorylation of the outer membrane protein regulator OmpR from *Escherichia coli. Proc. Natl. Acad. Sci. USA* **96,** 11792–11797.

Austin, R. H., Beeson, K. W., Eisenstein, L., Frauenfelder, H., and Gunsalus, I. C. (1975). Dynamics of ligand binding to myoglobin. *Biochemistry* **14**, 5355–5373.

Bachhawat, P., Swapna, G. V., Montelione, G. T., and Stock, A. M. (2005). Mechanism of activation for transcription factor PhoB suggested by different modes of dimerization in the inactive and active states. *Structure* **13**, 1353–1363.

Beach, H., Cole, R., Gill, M. L., and Loria, J. P. (2005). Conservation of mus-ms enzyme motions in the apo- and substrate-mimicked state. *J. Am. Chem. Soc.* **127**, 9167–9176.

Birck, C., Mourey, L., Gouet, P., Fabry, B., Schumacher, J., Rousseau, P., Kahn, D., and Samama, J. P. (1999). Conformational changes induced by phosphorylation of the FixJ receiver domain. *Structure* **7**, 1505–1515.

Bourret, R. B., Drake, S. K., Chervitz, S. A., Simon, M. I., and Falke, J. J. (1993). Activation of the phosphosignaling protein CheY. II. Analysis of activated mutants by 19F NMR and protein engineering. *J. Biol. Chem.* **268**, 13089–13096.

Cho, H. S., Lee, S. Y., Yan, D., Pan, X., Parkinson, J. S., Kustu, S., Wemmer, D. E., and Pelton, J. G. (2000). NMR structure of activated CheY. *J. Mol. Biol.* **297**, 543–551.

Delaglio, F., Grzesiek, S., Vuister, G. W., Zhu, G., Pfeifer, J., and Bax, A. (1995). Nmrpipe—A multidimensional spectral processing system based on Unix pipes. *J. Biomol. NMR* **6**, 277–293.

Dill, K. A., and Chan, H. S. (1997). From Levinthal to pathways to funnels. *Nature Struct. Biol.* **4**, 10–19.

Drake, S. K., Bourret, R. B., Luck, L. A., Simon, M. I., and Falke, J. J. (1993). Activation of the phosphosignaling protein CheY. I. Analysis of the phosphorylated conformation by 19F NMR and protein engineering. *J. Biol. Chem.* **268**, 13081–13088.

Eisenmesser, E. Z., Millet, O., Labeikovsky, W., Korzhnev, D. M., Wolf-Watz, M., Bosco, D. A., Skalicky, J. J., Kay, L. E., and Kern, D. (2005). Intrinsic dynamics of an enzyme underlies catalysis. *Nature* **438**, 117–121.

Elber, R., and Karplus, M. (1987). Multiple conformational states of proteins: A molecular dynamics analysis of myoglobin. *Science* **235**, 318–321.

Farrow, N. A., Muhandiram, R., Singer, A. U., Pascal, S. M., Kay, C. M., Gish, G., Shoelson, S. E., Pawson, T., Forman-Kay, J. D., and Kay, L. E. (1994). Backbone dynamics of a free and phosphopeptide-complexed Src homology 2 domain studied by 15N NMR relaxation. *Biochemistry* **33**, 5984–6003.

Feher, V. A., Zapf, J. W., Hoch, J. A., Whiteley, J. M., McIntosh, L. P., Rance, M., Skelton, N. J., Dahlquist, F. W., and Cavanagh, J. (1997). High-resolution NMR structure and backbone dynamics of the *Bacillus subtilis* response regulator, Spo0F: Implications for phosphorylation and molecular recognition. *Biochemistry* **36**, 10015–10025.

Feher, V. A., and Cavanagh, J. (1999). Millisecond-timescale motions contribute to the function of the bacterial response regulator protein Spo0F. *Nature* **400**, 289–293.

Frauenfelder, H., Parak, F., and Young, R. D. (1988). Conformational substates in proteins. *Annu. Rev. Biophys. Chem.* **17**, 451–479.

Fushman, D., Tjandra, N., and Cowburn, D. (1998). Direct measurement of N-15 chemical shift anisotropy in solution. *J. Am. Chem. Soc.* **120**, 10947–10952.

Fushman, D., and Cowburn, D. (1998). Model-independent analysis of N-15 chemical shift anisotropy from NMR relaxation data. Ubiquitin as a test example. *J. Am. Chem. Soc.* **120**, 7109–7110.

Gardino, A. K., Volkman, B. F., Cho, H. S., Lee, S. Y., Wemmer, D. E., and Kern, D. (2003). The NMR solution structure of BeF(3)(−)-activated Spo0F reveals the conformational switch in a phosphorelay system. *J. Mol. Biol.* **331**, 245–254.

Gouet, P., Fabry, B., Guillet, V., Birck, C., Mourey, L., Kahn, D., and Samama, J. P. (1999). Structural transitions in the FixJ receiver domain. *Structure* **7**, 1517–1526.

Halkides, C. J., McEvoy, M. M., Casper, E., Matsumura, P., Volz, K., and Dahlquist, F. W. (2000). The 1.9 A resolution crystal structure of phosphono-CheY, an analogue of the active form of the response regulator, CheY. *Biochemistry* **39,** 5280–5286.

Hastings, C. A., Lee, S. Y., Cho, H. S., Yan, D., Kustu, S., and Wemmer, D. E. (2003). High-resolution solution structure of the beryllofluoride-activated NtrC receiver domain. *Biochemistry* **42,** 9081–9090.

Hwang, I., Thorgeirsson, T., Lee, J., Kustu, S., and Shin, Y. K. (1999). Physical evidence for a phosphorylation-dependent conformational change in the enhancer-binding protein NtrC. *Proc. Natl. Acad. Sci. USA* **96,** 4880–4885.

Igumenova, T. I., and Palmer, A. G., 3rd. (2006). Off-resonance TROSY-selected R 1rho experiment with improved sensitivity for medium- and high-molecular-weight proteins. *J. Am. Chem. Soc.* **128,** 8110–8111.

Kern, D., Volkman, B. F., Luginbuhl, P., Nohaile, M. J., Kustu, S., and Wemmer, D. E. (1999). Structure of a transiently phosphorylated switch in bacterial signal transduction. *Nature* **402,** 894–898.

Korzhnev, D. M., Orekhov, V. Y., Dahlquist, F. W., and Kay, L. E. (2003). Off-resonance R1rho relaxation outside of the fast exchange limit: An experimental study of a cavity mutant of T4 lysozyme. *J. Biomol. NMR* **26,** 39–48.

Koshland, D. E., Jr., Nemethy, G., and Filmer, D. (1966). Comparison of experimental binding data and theoretical models in proteins containing subunits. *Biochemistry* **5,** 365–385.

Lee, J., Owens, J. T., Hwang, I., Meares, C., and Kustu, S. (2000). Phosphorylation-induced signal propagation in the response regulator ntrC. *J. Bacteriol.* **182,** 5188–5195.

Lewis, R. J., Brannigan, J. A., Muchova, K., Barak, I., and Wilkinson, A. J. (1999). Phosphorylated aspartate in the structure of a response regulator protein. *J. Mol. Biol.* **294,** 9–15.

Lewis, R. J., Muchova, K., Brannigan, J. A., Barak, I., Leonard, G., and Wilkinson, A. J. (2000). Domain swapping in the sporulation response regulator Spo0A. *J. Mol. Biol.* **297,** 757–770.

Lipari, G., and Szabo, A. (1982). Model-free approach to the interpretation of nuclear magnetic resonance in macromolecules. 1. Theory and range of validity. *J. Am. Chem. Soc.* **104,** 4546–4559.

Loria, J. P., Rance, M., and Palmer, A. G. (1999a). A relaxation-compensated Carr-Purcell-Meiboom-Gill sequence for characterizing chemical exchange by NMR spectroscopy. *J. Am. Chem. Soc.* **121,** 2331–2332.

Loria, J. P., Rance, M., and Palmer, A. G., 3rd. (1999b). A TROSY CPMG sequence for characterizing chemical exchange in large proteins. *J. Biomol. NMR* **15,** 151–155.

Luz, Z., and Meiboom, S. (1963). Nuclear magnetic resonance study of protolysis of trimethyl-ammonium ion in aqueous solution—Order of reaction with respect to solvent. *J. Chem. Phys.* **39,** 366–370.

Mandel, A. M., Akke, M., and Palmer, A. G., 3rd. (1995). Backbone dynamics of *Escherichia coli* ribonuclease HI: Correlations with structure and function in an active enzyme. *J. Mol. Biol.* **246,** 144–163.

Meyer, M. G., Park, S., Zeringue, L., Staley, M., McKinstry, M., Kaufman, R. I., Zhang, H., Yan, D., Yennawar, N., Yennawar, H., Farber, G. K., and Nixon, B. T. (2001). A dimeric two-component receiver domain inhibits the sigma54-dependent ATPase in DctD. *FASEB J.* **15,** 1326–1328.

Monod, J., Wyman, J., and Changeux, J. P. (1965). On the nature of allosteric transitions: A plausible model. *J. Mol. Biol.* **12,** 88–118.

Mulder, F. A., Mittermaier, A., Hon, B., Dahlquist, F. W., and Kay, L. E. (2001). Studying excited states of proteins by NMR spectroscopy. *Nat. Struct. Biol.* **8,** 932–935.

Palmer, A. G., 3rd, Kroenke, C. D., and Loria, J. P. (2001). Nuclear magnetic resonance methods for quantifying microsecond-to-millisecond motions in biological macromolecules. *Methods Enzymol.* **339,** 204–238.

Palmer, A. G., 3rd. (2004). NMR characterization of the dynamics of biomacromolecules. *Chem. Rev.* **104,** 3623–3640.

Park, S., Meyer, M., Jones, A. D., Yennawar, H. P., Yennawar, N. H., and Nixon, B. T. (2002). Two-component signaling in the AAA + ATPase DctD: Binding Mg2$^+$ and BeF3$^-$ selects between alternate dimeric states of the receiver domain. *FASEB J.* **16,** 1964–1966.

Parkinson, J. S., and Kofoid, E. C. (1992). Communication modules in bacterial signaling proteins. *Annu. Rev. Genet.* **26,** 71–112.

Pervushin, K., Riek, R., Wider, G., and Wuthrich, K. (1997). Attenuated T2 relaxation by mutual cancellation of dipole–dipole coupling and chemical shift anisotropy indicates an avenue to NMR structures of very large biological macromolecules in solution. *Proc. Natl. Acad. Sci. USA* **94,** 12366–12371.

Silversmith, R. E., and Bourret, R. B. (1999). Throwing the switch in bacterial chemotaxis. *Trends Microbiol.* **7,** 16–22.

Stock, A. M., Robinson, V. L., and Goudreau, P. N. (2000). Two-component signal transduction. *Annu. Rev. Biochem.* **69,** 183–215.

Tollinger, M., Skrynnikov, N. R., Mulder, F. A. A., Forman-Kay, J. D., and Kay, L. E. (2001). Slow dynamics in folded and unfolded states of an SH3 domain. *J. Am. Chem. Soc.* **123,** 11341–11352.

Volkman, B. F., Nohaile, M. J., Amy, N. K., Kustu, S., and Wemmer, D. E. (1995). Three-dimensional solution structure of the N-terminal receiver domain of NTRC. *Biochemistry* **34,** 1413–1424.

Volkman, B. F., Lipson, D., Wemmer, D. E., and Kern, D. (2001). Two-state allosteric behavior in a single-domain signaling protein. *Science* **291,** 2429–2433.

Volz, K., and Matsumura, P. (1991). Crystal structure of *Escherichia coli* CheY refined at 1.7-Å resolution. *J. Biol. Chem.* **266,** 15511–15519.

Wang, C., Grey, M. J., and Palmer, A. G., 3rd. (2001). CPMG sequences with enhanced sensitivity to chemical exchange. *J. Biomol. NMR* **21,** 361–366.

Wang, C. Y., Rance, M., and Palmer, A. G. (2003). Mapping chemical exchange in proteins with MW > 50 kD. *J. Am. Chem. Soc.* **125,** 8968–8969.

Weiss, D. S., Batut, J., Klose, K. E., Keener, J., and Kustu, S. (1991). The phosphorylated form of the enhancer-binding protein NTRC has an ATPase activity that is essential for activation of transcription. *Cell* **67,** 155–167.

West, A. H., and Stock, A. M. (2001). Histidine kinases and response regulator proteins in two-component signaling systems. *Trends Biochem. Sci.* **26,** 369–376.

Wyman, C., Rombel, I., North, A. K., Bustamante, C., and Kustu, S. (1997). Unusual oligomerization required for activity of NtrC, a bacterial enhancer-binding protein. *Science* **275,** 1658–1661.

[7] The Design and Development of Tar-EnvZ Chimeric Receptors

By TAKESHI YOSHIDA, SANGITA PHADTARE, and MASAYORI INOUYE

Abstract

Escherichia coli histidine kinases play an essential role in sensing external environmental changes. Since the majority of these are transmembrane proteins, it is believed that their periplasmic domains function as receptor and transduce a signal through the transmembrane domain to their cytoplasmic enzymatic domains. Therefore, it is important to understand how signal transduction modulates the enzymatic activities of histidine kinase across transmembrane. Osmosensor histidine kinase EnvZ and chemoreceptor Tar are well-characterized signal-transducing proteins; a fusion of these two proteins would prove to be an ideal tool not only for characterization of histidine kinase EnvZ, but also, more importantly, as a general approach for studying the molecular mechanism of signal transduction across transmembranes. Tar-EnvZ chimeric protein served as a useful tool to study how the signal modulates enzymatic activities of EnvZ by using a well-defined chemical, aspartate, as a receptor ligand. As more and more genome sequences are being published, the number of identified histidine kinases is rapidly growing. The analysis of these newly identified histidine kinases revealed that the architecture of their cytoplasmic domains is more complex than was perceived based on *E. coli* histidine kinases. Therefore, chimeric proteins of these histidine kinases with Tar receptor would be helpful to study the mechanism of signal transduction. This chapter describes methods for designing chimeric proteins between a histidine kinase of interest and the Tar receptor and applications of the chimeric protein.

Introduction

Adaptation to environmental changes is essential for survival of *Escherichia coli* under various stress conditions (Forst and Inouye, 1988; Parkinson and Kofoid, 1992). A signal transduction system in *E. coli* is known as a two-component signal transduction, or Histidyl-Aspartyl phosphorelay system, and plays a crucial role in this adaptation (Hoch and Silhavy, 1995; Inouye, 2003; Parkinson, 1993; Parkinson and Kofoid, 1992; Stock *et al.*, 1989, 1990). A typical signal transduction system consists of two

METHODS IN ENZYMOLOGY, VOL. 423
0076-6879/07 $35.00
DOI: 10.1016/S0076-6879(07)23007-1

proteins; one is a histidine protein kinase, which acts as a sensor, and the other is a response regulator, which plays an important role as a general factor in gene regulation or cellular locomotion. The majority of receptor proteins as well as histidine protein kinases are transmembrane proteins with a functional cytoplasmic domain that can be modified and has enzymatic activity. Since binding of a ligand to a periplasmic receptor will modulate the activity of its downstream cytoplasmic domain through signal transduction, it is important to understand the molecular mechanism of how signal transduction occurs across transmembrane. However, identification of a specific ligand for each receptor is critical and can be challenging.

One of the well-characterized Histidyl-Aspartyl phosphorelay systems is the EnvZ-OmpR phosphorelay system (Egger et al., 1997; Forst and Roberts, 1994; Pratt and Silhavy, 1995). This system is responsible for regulation of two major outer membrane porin proteins, OmpF and OmpC, upon medium osmolarity changes (Aiba et al., 1989a; Forst and Inouye, 1988; Inouye et al., 2003). EnvZ is a transmembrane histidine kinase while OmpR is a cognate response regulator for EnvZ and a transcription factor for ompF and ompC genes (Aiba et al., 1989b; Forst et al., 1987, 1989; Igo and Silhavy, 1988; Igo et al., 1990). The cytoplasmic domain of EnvZ has enzymatic activities. EnvZ can autophosphorylate a conserved His residue using its ATP-binding domain and ATP (Roberts et al., 1994). Subsequently, the phosphoryl group on His is transferred to a conserved Asp residue on OmpR to generate phosphorylated OmpR (OmpR-P) (OmpR kinase) (Delgado et al., 1993). EnvZ also has phosphatase activity toward OmpR-P (OmpR-P phosphatase) (Aiba et al., 1989a; Igo et al., 1989). Although EnvZ responds to medium osmolarity changes to regulate OmpF and OmpC expressions, a specific stimuli, or ligand, for EnvZ is as yet unknown. Thus, EnvZ can be used as a model protein to study how a signal modulates the enzymatic activity of its cytoplasmic domain. An ideal approach for this would be to create a chimeric protein consisting of a periplasmic receptor, of which the ligand is known, and the cytoplasmic domain of EnvZ.

Tar, the well-characterized periplasmic receptor domain from one of the methyl-accepting chemotaxis proteins (MCPs), was used to make the first chimeric protein with EnvZ (Utsumi et al., 1989). Tar has architecture similar to that of EnvZ and consists of three parts: the periplasmic receptor domain, the transmembrane domain (TM1 and TM2), and the cytoplasmic domain (Forst et al., 1987; Krikos et al., 1983). Importantly, Asp is known as a ligand for this receptor (Silverman and Simon, 1977; Wang and Koshland, 1980). There is Nde I site in both genes at a similar position after TM2, which is convenient for construction of chimeric protein. Therefore, a chimeric protein consisting of the Tar receptor of Tar and the cytoplasmic enzymatic domain of EnvZ can be constructed without introducing extra amino acid residues. This chimeric protein termed Taz responds to the concentration of Asp in the media and generates more OmpR-P, resulting in increased OmpC expression (Utsumi *et al.*, 1989).

Study of the chimeric protein, Taz, provided important information about the molecular mechanism of signal transduction and function of EnvZ. More specifically, it was demonstrated that Tar and EnvZ share a common mechanism of signal transduction (Utsumi *et al.*, 1989). An important region connecting TM2 and a cytoplasmic functional domain was also identified and was shown to play an important role in signal transduction (Jin and Inouye, 1994). This region, called the linker region, was later identified to be present among wide range of transmembrane receptor proteins involved in signal transduction. Therefore, this linker region is now known as HAMP domain (*H*istidine kinase, *A*denylyl cyclases, *M*ethyl-accepting proteins, and *P*hosphatases) (Aravind and Ponting, 1999). Using another chimeric protein, Tez1A1, which consists of Tar receptor and the cytoplasmic domain of EnvZ including EnvZ HAMP domain, it was demonstrated that EnvZ HAMP domain is able to replace Tar HAMP domain and can transduce a signal from Tar receptor to the cytoplasmic domain of EnvZ, indicating that HAMP domains of these proteins share a common mechanism of signal transduction (Zhu and Inouye, 2003).

Various chimeric proteins containing a Tar receptor or a receptor that has known ligand have been successfully used for elucidating functions of other histidine kinases (Appleman *et al.*, 2003; Baumgartner *et al.*, 1994; Kristich *et al.*, 2003; Kumita *et al.*, 2003; Ward *et al.*, 2006). This suggests that the chimeric protein approach will be extremely useful for characterizing other histidine kinases of interest with unknown stimuli or ligands. In this chapter, we describe creation of a chimeric protein between Tar and EnvZ and its applications.

Construction of Tar-EnvZ Chimeric Protein, Taz

Overview

The construction of the Tar-EnvZ chimeric protein was based on the topological orientation of these two transmembrane proteins and made use of the *Nde* I site present in both proteins downstream of their second transmembrane domains (TM2) (Utsumi *et al.*, 1989). This unique *Nde* I site allows replacing of Tar cytoplasmic domain in Tar with EnvZ cytoplasmic domain. The region between TM2 and *Nde* I has an important role in signal transduction (Jin and Inouye, 1994) and is termed HAMP domain, since it is found in a broad range of signal-transducing protein families such as *H*istidine kinase, *A*denylyl cyclases, *M*ethyl-accepting proteins, and *P*hosphatases (Aravind and Ponting, 1999). This suggests that the role of HAMP domain in signal transduction is conserved in a number of membrane proteins, including EnvZ and Tar. Maintaining the organization of structural elements of HAMP domain is very critical for constructing chimeric protein between EnvZ and Tar, since an Asp-regulatable Tar-EnvZ chimeric protein cannot be created by simply exchanging the HAMP domains from these two proteins (Zhu and Inouye, 2003). In fact, even a single point mutation in HAMP domain of Tar or EnvZ significantly affects its function and the mutated proteins lose their ability to respond to Asp or medium osmolarity changes (Ames and Parkinson, 1988; Park and Inouye, 1997). This aspect will be discussed later in this chapter.

Bacterial Plasmids and Strains

Throughout our study, the low copy number vectors such as pACYC (Cm^r), pBR322 (Am^r) or pIN-III (Am^r or Cm^r) containing *lpp*$^{p-5}$ promoter and *lac*$^{p-o}$, lac promoter-operator from pIN-III (Masui *et al.*, 1983; Utsumi *et al.*, 1989), are used to clone a Tar-EnvZ chimeric protein construct. The plasmid containing a Tar-EnvZ chimeric sequence is transformed into RU1012 [MC4100 *ara*$^+$, Φ (*ompC-lacZ*) 10–25, Δ*envZ*::Km^r] cells.

Asp-Dependent Induction of *ompC-lacZ* Fusion Gene by Taz and OmpR

Overview

A number of two-component signal transduction systems are involved in the regulation of gene expression, in which response regulators play an important role as transcription factors by binding the upstream region of a target gene's promoter. Therefore, to study how a two-component system regulates a gene, one can replace the gene regulated by response regulator of interest with a reporter gene, such as β-galactosidase or green fluorescent protein (GFP). In order to study *in vivo* the mechanism of regulation by Taz, *E. coli* RU1012 [MC4100 *ara*+, Φ (*ompC-lacZ*) 10-25, Δ*envZ*::Km^r] cells that carry chromosomal deletion of the gene encoding EnvZ, but contain the *ompC-lacZ* fusion gene, are used (Utsumi *et al.*, 1989).

Chemosensor Tar is also one of the well-characterized signal-transducing proteins and its receptor is able to respond to Asp directly (Wang and Koshland, 1980) but indirectly to maltose through maltose-binding protein (Hazelbauer, 1975; Koiwai and Hayashi, 1979; Manson *et al.*, 1985). Biochemical analysis of its ligand-binding properties showed

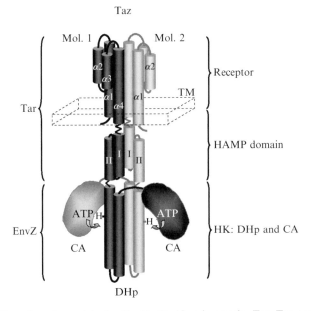

FIG. 1. The schematic model of a Tar-EnvZ chimeric protein, Taz. Taz consists of Tar receptor, Tar HAMP domain, and EnvZ histidine kinase domain. Taz undergoes ATP-dependent trans-autophosphorylation in which the ATP-binding pocket of Mol. 1 (dark gray) phosphorylates the phosphorylation site, H243, of Mol. 2 (light gray). TM, transmembrane; DHp. a dimerization histidine phosphotransfer domain; CA, a catalytic ATP-binding domain.

that it binds to Asp with K_d value of 0.5 to 6 μM (Foster *et al.*, 1985; Russo and Koshland, 1983; Wang and Koshland, 1980), while it exhibits less affinity for maltose (K_d value ~250 μM) (Manson *et al.*, 1985). The induction of β-galactosidase was monitored in the presence of several possible ligands including Asp and maltose using RU1012 cells expressing Taz, and it was found that *ompC-lacZ* can be induced only by Asp and its analog α-methyl-D,L-aspartate (CH$_3$-Asp) (Utsumi *et al.*, 1989).

Taz has a fully functional EnvZ cytoplasmic enzymatic domain and thus can phosphorylate OmpR, a cognate response regulator of EnvZ, to generate phosphorylated OmpR (OmpR-P)(Fig. 1). OmpR-P then functions as a transcription factor for *ompF* and *ompC* genes. Taz also contains a receptor domain from Tar (Fig. 1). The induction of β-galactosidase by Asp suggests that binding of Asp to Tar receptor modulates the enzymatic activities of EnvZ cytoplasmic domain, resulting in higher kinase-to-phosphatase ratio and generating more OmpR-P, which, in turn, activates *ompC* promoter of *ompC-lacZ* fusion gene.

Measurement of Asp-Dependent Induction of ompC-lacZ Fusion Gene by β-Galactosidase Assay

RU1012 cells harboring pIN-III-Taz are grown in M9 minimal medium containing 50 μg/ml each of ampicillin and kanamaycin, and various concentrations (0 to 10 mM) of L-aspartate (with monopotassium salt). Cells are grown to mid-log phase, and β-galactosidase assay is carried out (Miller, 1972). A typical assay of Asp-dependent induction of *ompC-lacZ* fusion gene using RU1012 cells expressing Taz is shown in Fig. 2. It is important to note that in the experiments involving Tar, the concentration of Asp needed to elicit detectable response was in the range of 10^{-6} M, while in the case of experiments involving Taz, the concentration of Asp required to detect the induction of *ompC-lacZ* fusion gene was in the range of 10^{-3} M. This difference in the concentration of Asp possibly arises from the different detection systems used or is due to the Taz construct, which is less sensitive to Asp.

Phenotype Analysis of the Taz Construct

Analysis of Phenotype of the Taz Construct Using MacCONKEY or X-gal Agar Plates

MacCONKEY and X-gal agar plates are used to assess β-galactosidase activity of the Taz constructs. Asp is incorporated in these plates at the concentration of 5 mM. β-galactosidase producing colonies can be identified

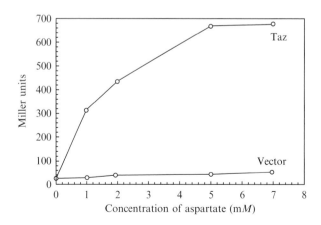

FIG. 2. Asp-dependent induction of *ompC-lacZ* is monitored by β-galactosidase assay. RU1012 cells harboring pIN-III-Taz or pIN-III vector alone are grown in M9 minimal medium containing different concentrations of Asp (0, 1, 2, 5, and 7 mM). β-galactosidase assay is carried out using actively growing cells at mid-log phase.

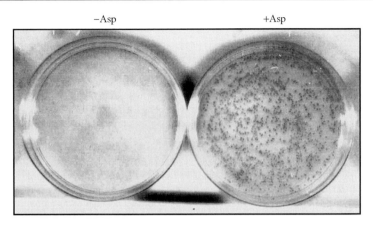

FIG. 3. Asp-dependent induction of *ompC-lacZ* is detected on X-gal plate. RU1012 cells harboring pIN-III-Taz construct are plated on X-gal agar plates with or without 5 m*M* Asp. Blue-colored colonies represent the cells expressing β-galactosidase. (See color insert.)

by change in color from white to red on MacCONKEY agar plates and from white to blue on X-gal containing agar plates. Simplicity of use of this system makes it a useful tool for analyzing Asp-regulatable phenotype or for screening suppressor mutants of Asp-nonregulatable phenotype.

MacCONKEY agar plates (MacCONKEY AGAR CS, DIFCO) or X-gal agar plates (M9 minimal medium agar plate) containing 50 μg/ml each of ampicillin and kanamaycin with or without 5 m*M* L-aspartate (with monopotassium salt) are prepared. For X-gal containing plates, 50 μl of 20 mg/ml 5-bromo-4-chloro-3-indolyl-β-D-galactoside in dimethylformamide is spread on plates before use. RU1012 cells harboring pIN-III-Taz construct are grown in M9 minimal medium without Asp and spread on MacCONKEY or X-gal agar plates with or without 5 m*M* Asp. It is important to allow the formation of well-distributed single colonies to avoid false-positive colonies. Figure 3 represents an X-gal agar plate showing Asp-dependent induction of β-galactosidase using RU1012 cells.

Regulation of Binding of Asp to One of Two Asp-Binding Pockets of Tar Receptor to Study Signal Transduction

Overview

As has been described, a Tar-EnvZ chimeric protein responds to Asp as its ligand and binding of Asp modulates the enzymatic activities of EnvZ cytoplasmic domain. The mutational analysis of the Tar receptor in Tar demonstrated that three Arg residues (Arg64, 69, and 73) and Thr154 play an important role in Asp binding (Lee and Imae, 1990; Mowbray and Koshland, 1990; Wolff and Parkinson, 1988). The X-ray crystal structure of Tar receptor bound to Asp confirmed that these amino acids are part of the Asp-binding pocket and interact with Asp (Milburn *et al.*, 1991; Yeh *et al.*, 1993). Tar receptor forms a dimer and two Asp-binding pockets are formed by the interface of two subunits. It has been shown that binding of Asp to one of two Asp-binding pockets inhibits the binding of second molecule of Asp to the other site, indicating that binding of one Asp molecule to the Tar receptor is sufficient to trigger signal transduction (Milburn *et al.*, 1991; Mowbray and Koshland, 1990; Yeh *et al.*, 1993). We introduced mutations in a Tar-EnvZ chimeric protein, which are known to diminish Asp binding. This allows one to specifically control Asp binding to one of the two Asp-binding pockets and to study how binding of Asp to one site influences the enzymatic activities of the EnvZ cytoplasmic domain.

Key Mutations within a Tar Receptor

Based on earlier mutational analysis in a Tar receptor, four mutations can be considered ideal to abolish Asp binding—Arg64Cys, Arg69His, Arg73Gln, and Thr154Ile. Although single mutations affect Asp binding to a certain extent, two point mutations within a Tar receptor are incorporated in order to effectively block binding of Asp to the Asp-binding pocket. Since Arg64 and Thr154 of each subunit of a Tar dimer contribute to form Asp-binding pockets with Arg69 and Arg73 of the other subunit, we created two pairs of two point mutations; one pair consists Arg64Cys and Thr154Ile mutations (Tarα) and the other pair consists of Arg69His and Arg73Gln mutations (Tarβ).

Asp-Binding Pockets of Tar Receptor are Formed by Dimerization

It is important to note that Asp-binding pockets are formed at the interface of a Tar dimer (Fig. 4A). As shown in Fig. 4, when a homodimer is formed by a Tarα (Fig. 4D) or a Tarβ (Fig. 4E), it does not contain functional Asp-binding pockets. Only a heterodimer between a Tar and a Tarα subunit contains one active Asp-binding pocket formed by Arg69 and Arg73 of a Tarα and Arg64 and Thr154 of a Tar (Fig. 4B) or a heterodimer between a Tar and a Tarβ subunit contains one functional Asp-binding pocket formed by Arg69 and Arg73 of a Tar and Arg64 and Thr154 of a Tarβ (Fig. 4C). Note that when a heterodimer is formed by a Tarα and a Tarβ subunit, regulation by Asp is diminished, even though there is only

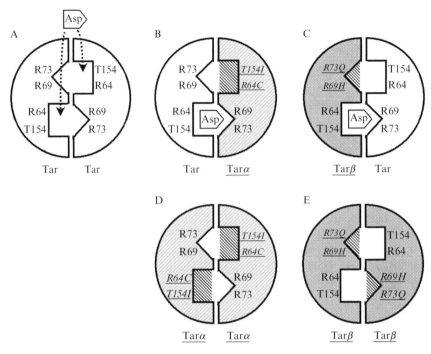

FIG. 4. Rendering one of the two Asp-binding pockets of Tar receptor inactive by Tarα or Tarβ mutation. Tar receptor and mutated Tar receptor with Asp-binding pocket(s) are illustrated. Two Asp-binding pockets are formed at the interface between two subunits in the Tar receptor dimer (A). Two essential amino acid residues for Asp binding are mutated in Tarα receptor (shaded by lines); Arg64Cys and Thr154Ile and in Tarβ receptor (shaded by dots); Arg69His, and Arg73Gln. When one subunit of mutated receptors forms a heterodimer with a subunit of Tar, one complete Asp-binding pocket is formed, as shown in (B) and (C). Binding of Asp to both Asp-binding pockets is inhibited in homodimers of Tarα and Tarβ receptors as shown in (D) and (E), respectively. Mutated residues are shown by underlined and italicized texts.

one Asp-binding pocket, suggesting that creation of these four mutations in the Tar receptor is detrimental to its function (Yang *et al.*, 1993).

Use of Tarα and Tarβ Receptors to Study Signal Transduction

The Tar-EnvZ chimeric protein containing (α) and (β) mutations described previously has been used successfully to study signal transduction upon Asp binding (Yang *et al.*, 1993). For this study, two types of EnvZ mutants are used to create a Tar-EnvZ chimeric protein; one is EnvZ[H243V] mutant, termed EnvZH1, in which the phosphorylation site is eliminated, and the other one is EnvZ[G1] mutant is called EnvZG1, which has three mutations at the conserved G1 box (Gly375Ala, Gly377Ala, and Ala379-Ser). These mutant proteins lack the ability to bind ATP. We also use another set of EnvZ mutants, EnvZH1 and EnvZ[G405A] mutant, called EnvZG2. The mutation at the conserved G2 box of Gly405 to Ala is enough to abolish the binding of EnvZ to ATP (Zhu and Inouye, 2002, 2004). Therefore, one can presume that a similar effect will be observed with the protein containing G2 mutation as that seen with the protein containing G1 mutation. EnvZ, similar to other histidine kinases, forms a dimer to form its active site and to phosphorylate His243. As shown in Fig. 5, when these two EnvZ mutants form a heterodimer, only one active site is formed. The combination of Tar receptor mutation, (α) or (β), and EnvZ cytoplasmic domain mutations, His243Val and G1 or G2, provides a useful tool to study signal transduction regulated by Asp *in vivo*.

Different combinations of Tar and EnvZ mutations have different phenotypes. It has been shown that the heterodimer formed between TazαH1 (Tarα and EnvZH1 mutations) and TazG1 (EnvZG1 mutation) (Fig. 5A) has the phenotype closest to the phenotype of Taz, followed by the phenotype of heterodimer formed between TazH1 (EnvZH1 mutation) and TazβG1 (Tarβ and EnvZG1 mutations) (Fig. 5B) (Yang *et al.*, 1993). This result demonstrates that one Asp-binding pocket and one EnvZ active site is enough to regulate cellular concentration of OmpR-P upon Asp binding. However, in the other two combinations of Asp-binding pocket and EnvZ active site shown in Fig. 5C and D, the Asp-dependent induction of *ompC-lacZ* is lost (Yang *et al.*, 1993). These results seem to be related to a conformational change in the Tar receptor upon Asp binding. The structural studies of the Tar receptor show a conformational change in helix 4, which contains the Thr154 residue (Chervitz and Falke, 1996; Milburn *et al.*, 1991). The helix 4 leads to TM2, which is connected to the cytoplasmic domain. Upon binding of Asp to the active Asp-binding pocket, a conformational change in helix 4 occurs, and this conformational change modulates the downstream region of the EnvZ kinase domain. This Asp-dependent regulation is detected when the His243 residue is located downstream of helix 4 (in the Tar receptor of TazG1 or TazβG1) that is involved in forming a functional Asp-binding pocket (Fig. 5A,B), and its active site is formed by the His243 residue and the ATP-binding pocket of TazαH1 via TM2 (Fig. 5). This shows that by identifying the site to which the Asp molecule binds, one can monitor how signal transduction modulates enzymatic activities of histidine kinase of interest.

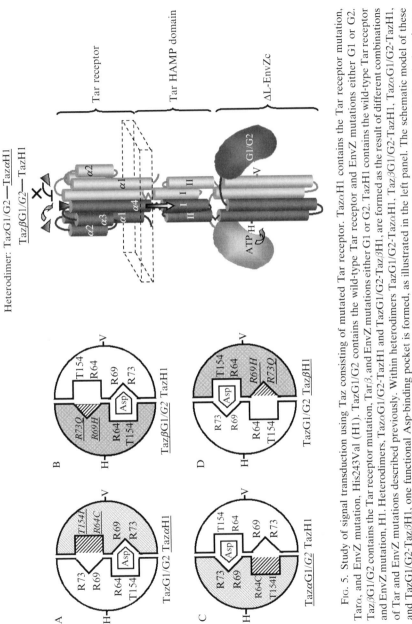

Fig. 5. Study of signal transduction using Taz consisting of mutated Tar receptor. TazαH1 contains the Tar receptor mutation, Tarα, and EnvZ mutation. His243Val (H1). TazG1/G2 contains the wild-type Tar receptor and EnvZ mutations either G1 or G2. TazβG1/G2 contains the Tar receptor mutation, Tarβ, and EnvZ mutations either G1 or G2. TazH1 contains the wild-type Tar receptor and EnvZ mutation, H1. Heterodimers, TazαG1/G2-TazH1 and TazG1/G2-TazβH1, are formed as the result of different combinations of Tar and EnvZ mutations described previously. Within heterodimers TazG1/G2-TazαH1, TazβG1/G2-TazαH1, TazαG1/G2-TazH1, and TazG1/G2-TazβH1, one functional Asp-binding pocket is formed, as illustrated in the left panel. The schematic model of these heterodimers in the right panel illustrates that upon Asp binding, a signal is transduced to and modulates EnvZ cytoplasmic domain as a conformational change occurs in α4 (shown by an arrow). (H) or (V) represents H243 or His243Val mutation, respectively, in EnvZ kinase domain located downstream of each Tar receptor. The details are described in the text.

The Right Configuration of HAMP Domain is Crucial for Proper
Signal Transduction in a Tar-EnvZ Chimeric Protein

Overview

The critical role of the HAMP domain in signal transduction of Tar
and EnvZ was revealed after successful construction of Taz. Therefore, a
Tar-EnvZ chimeric protein was also used to test whether the EnvZ HAMP
domain can replace the role of Tar HAMP domain. As shown in Fig. 6,
amino acid sequences of Tar and EnvZ HAMP domains are not highly
homologous, whereas secondary structure prediction analysis indicates that
they have similar structure organization consisting of helix–turn–helix. The
number of amino acids in these HAMP domains is almost the same: EnvZ
HAMP domain has one amino acid fewer than does the Tar HAMP domain.
Based on our assumption that they have a similar mechanism of signal
transduction, we constructed another Tar-EnvZ chimeric protein called
Tez, consisting of Tar receptor, EnvZ HAMP domain, and EnvZ kinase
domain, without introducing extra amino acids. Experiments with Tez
demonstrated that a simple exchange of HAMP domains was not enough
to create functional Tar-EnvZ chimeric protein, and Tez was unable to

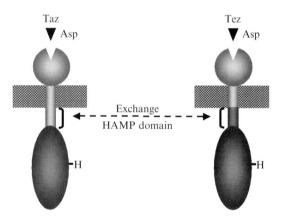

Tar: 213 RRMLLTP⌐LAKIIAHIREIA⌐GGNLANTLTIDGR⌐S⌐EMGDLAQSVSH⌐M 257
EnvZ: 180 •RIQNRP⌐LVDLEHAALQV⌐GKGIIPPPLREYQAS⌐E⌐VRSVTRAFNH⌐M 223
              Helix I   Helix II

FIG. 6. Construction of Tez. By exchanging the Tar HAMP domain of Taz with EnvZ
HAMP domain, Tez is constructed. Tar receptor and HAMP domain are shown in light gray,
while EnvZ HAMP domain and histidine kinase domain are shown in dark gray. H,
phosphorylation site, His243. Amino acid sequences of *E. coli* Tar and *E. coli* EnvZ HAMP
domain are shown and boxed regions are predicted helices.

respond to Asp (Zhu and Inouye, 2003). Extensive mutation analysis of EnvZ HAMP domain of Tez was carried out by deletion, addition, or random mutagenesis. We observed that Tez carrying a point mutation was able to respond to Asp (Zhu and Inouye, 2003). This new construct allowed us to study function of EnvZ HAMP domain as well as to show that the Tar HAMP domain and the EnvZ HAMP domain share a similar molecular mechanism of signal transduction. Importantly, it suggested that the right configuration of HAMP domain is absolutely necessary for proper transduction of signal from Tar receptor upon Asp binding and to modulate enzymatic activities of EnvZ kinase domain.

It is worthwhile to mention that there is a specific interaction between two HAMP domains in Taz and Tez, indicating that it forms a dimer (Zhu and Inouye, 2004). The NMR structure of a HAMP domain from a hyperthermophilic bacterium *Archaeoglobus fulgidus* indeed demonstrated that HAMP domain consisting of a helix–turn–helix structure forms a dimer (Hulko *et al.*, 2006). Interestingly, coexpression of Asp-regulatable Taz and Tez with the combination of EnvZ mutations, H1 and G2, described previously, demonstrated that Asp-dependent signal transduction still occurs through a heterodimeric HAMP domain formed by Tar HAMP domain of Taz subunit and EnvZ HAMP domain of Tez subunit in Taz-Tez heterodimer, resulting in Asp-regulatable phenotype (Zhu and Inouye, 2004).

Modifications in the HAMP Domain That Influence the Asp-Regulatibility of the Tez Construct

Certain chimeric proteins may exhibit nonregulatable phenotype, as seen in the case of the Tez construct. There may be several factors that contribute to the nonregulatable phenotype. Here, we present a screening method for isolating the Asp-regulatable phenotype by mutating the HAMP domain of Tez construct that displays Asp-nonregulatable phenotype. We observed that the expression of β-galactosidase in the RU1012 cells expressing Tez is low, irrespective of the presence or absence of Asp. This suggests that Tez does not respond to Asp and has a typical constitutive "off" phenotype ($ompF^-/ompC^-$). Given the fact that the HAMP domain plays a crucial role in signal transduction and the signal transduced by binding of Asp to the Tar receptor may not be properly transduced to EnvZ kinase domain through the EnvZ HAMP domain in Tez construct, mutagenesis on the EnvZ HAMP domain was carried out to explore the possibility that the Asp-regulatable phenotype can be recovered.

EnvZ HAMP domain has one amino acid fewer than the Tar HAMP domain and is predicted to have two helices, as shown in Fig. 6. Therefore, it may be necessary to add one amino acid to the domain since change in the number of amino acids may result in change in the relative position of helices. As described by Zhu and Inouye (2003), the deletion of one amino acid does not change the phenotype of Tez construct, whereas the addition of two Ala residues at the transmembrane/HAMP domain junction results in changing its phenotype from constitutively "off" to constitutively "on" [$ompF^-/ompC^{c(constitutive)}$]. This clearly indicates that the addition or deletion of amino acids can influence the function of EnvZ HAMP domain to modulate the downstream EnvZ kinase domain. We obtained a Tez construct with Asp-regulatable phenotype by adding one Ala residue before or after Arg180. Importantly, in this construct, Ala is highly specific, as evidenced by the fact that replacing Ala by the other 12 amino acids abolished the Asp-regulatable phenotype, and replacing Ala to Asn or Asp showed partially Asp-regulatable phenotype. The position of the inserted Ala residue is also important.

We also used another approach to isolate Asp-regulatable phenotype of Tez by screening for a suppressor mutant of Tez construct. Random mutagenesis of EnvZ HAMP domain by PCR is carried out as described by Spee *et al.* (1993). In this method, the PCR reaction mixture contains dITP and Mn^{2+} ions to increase errors by the DNA polymerase. The EnvZ HAMP domain in the Tez construct is replaced with the mutagenized HAMP domain sequence and the resultant plasmids are transformed in the RU1012 cells. The cells are plated on MacCONKEY agar plates containing 5 mM Asp. Note that, since RU1012 cells expressing Tez with the Asp nonregulatable phenotype form white colonies on MacCONKEY agar plates with or without 5 mM Asp. Thus, it is expected that if the cells express a mutant Tez protein that has Asp-regulatable phenotype, the color of the colony should be red on these plates in the presence of 5 mM Asp. If the cells expressing a chimeric protein of interest show red color even in the absence of Asp (possibly a kinase$^+$/Phosphatase$^-$ phenotype), use the MacCONKEY agar plates without Asp and screen for white colonies.

The cells from colonies showing red color on MacCONKEY agar plate containing 5 mM Asp are grown in M9 medium with or without 5 mM Asp and β-galactosidase assay is carried out. After Asp-dependent induction of β-galactosidase is confirmed, plasmids are isolated from those cells and retransformed into RU1012 cells. After confirming the Asp-regulatable phenotype, the coding region of EnvZ HAMP domain from these plasmids is sequenced to identify mutations. If a particular construct contains multiple mutations, then the individual mutations are isolated by site-directed mutagenesis and their effect on the Asp-dependent phenotype is tested. Note that while MacCONKEY agar plates are simple to prepare and cells grow faster on them, we observed that, in some cases, it is hard to distinguish the phenotype on MacCONKEY agar plates and, after incubation for longer periods of time, all colonies on the plate become red. In this case,

X-gal agar plates with 5 mM Asp are recommended. Our screening for Asp-regulatable phenotype yielded a plasmid containing two mutations, Pro185Gln and Arg184Gln. We isolated these two mutations as TezP185Q or TezR184Q and observed that Pro185Gln mutation is responsible for Asp-regulatable phenotype. As seen with Ala insertion as previously described, the Pro185Gln mutation is also highly specific and the Pro185 residue cannot be replaced with other amino acids except Ile (Zhu and Inouye, 2003).

These results suggest that the HAMP domain is finely tuned and slight changes influence its function, resulting in regulatable or nonregulatable phenotype of chimeric proteins. If a chimeric protein has a nonregulatable phenotype, one approach to recover the regulatable phenotype is to carefully examine the amino acid residues to be added or deleted from the HAMP domain and to examine the position of intended change. As has been demonstrated, a random mutagenesis approach can also be used to isolate Asp-regulatable phenotypes from a large variety of mutants.

Conclusions

In this chapter we described the design and construction of functional Tar-EnvZ chimeric proteins. A number of successful chimeric proteins have been created not only to study histidine kinases, but also to understand the general molecular mechanism of signal transduction. Through Tar-EnvZ chimeric constructs, we now know that while constructing a functional chimeric protein containing the HAMP domain, it is extremely important to maintain the right phasing of the HAMP domain between TM2 and the kinase domain. This approach may be even more useful for characterization of the regulation of histidine kinases with highly complex architecture.

Acknowledgments

This work was supported by Grant GM076587 from the National Institutes of Health. The authors thank N. E. Severinov for his critical reading of the manuscript.

References

Aiba, H., Nakasai, F., Mizushima, S., and Mizuno, T. (1989a). Evidence for the physiological importance of the phosphotransfer between the two regulatory components, EnvZ and OmpR, in osmoregulation in *Escherichia coli*. *J. Biol. Chem.* **264,** 14090–14094.

Aiba, H., Nakasai, F., Mizushima, S., and Mizuno, T. (1989b). Phosphorylation of a bacterial activator protein, OmpR, by a protein kinase, EnvZ, results in stimulation of its DNA-binding ability. *J. Biochem. (Tokyo)* **106,** 5–7.

Ames, P., and Parkinson, J. S. (1988). Transmembrane signaling by bacterial chemoreceptors: *E. coli* transducers with locked signal output. *Cell* **55,** 817–826.

Appleman, J. A., Chen, L. L., and Stewart, V. (2003). Probing conservation of HAMP linker structure and signal transduction mechanism through analysis of hybrid sensor kinases. *J. Bacteriol.* **185,** 4872–4882.

Aravind, L., and Ponting, C. P. (1999). The cytoplasmic helical linker domain of receptor histidine kinase and methyl-accepting proteins is common to many prokaryotic signaling proteins. *FEMS Microbiol. Lett.* **176,** 111–116.

Baumgartner, J. W., Kim, C., Brissette, R. E., Inouye, M., Park, C., and Hazelbauer, G. L. (1994). Transmembrane signaling by a hybrid protein: Communication from the domain of chemoreceptor Trg that recognizes sugar-binding proteins to the kinase/phosphatase domain of osmosensor EnvZ. *J. Bacteriol.* **176,** 1157–1163.

Chervitz, S. A., and Falke, J. J. (1996). Molecular mechanism of transmembrane signaling by the aspartate receptor: A model. *Proc. Natl. Acad. Sci. USA* **93,** 2545–2550.

Delgado, J., Forst, S., Harlocker, S., and Inouye, M. (1993). Identification of a phosphorylation site and functional analysis of conserved aspartic acid residues of OmpR, a transcriptional activator for *ompF* and *ompC* in *Escherichia coli. Mol. Microbiol.* **10,** 1037–1047.

Egger, L. A., Park, H., and Inouye, M. (1997). Signal transduction via the histidyl-aspartyl phosphorelay. *Genes Cells* **2,** 167–184.

Forst, S., Comeau, D., Norioka, S., and Inouye, M. (1987). Localization and membrane topology of EnvZ, a protein involved in osmoregulation of OmpF and OmpC in *Escherichia coli. J. Biol. Chem.* **262,** 16433–16438.

Forst, S., Delgado, J., and Inouye, M. (1989). Phosphorylation of OmpR by the osmosensor EnvZ modulates expression of the *ompF* and *ompC* genes in *Escherichia coli. Proc. Natl. Acad. Sci. USA* **86,** 6052–6056.

Forst, S., and Inouye, M. (1988). Environmentally regulated gene expression for membrane proteins in *Escherichia coli. Annu. Rev. Cell. Biol.* **4,** 21–42.

Forst, S. A., and Roberts, D. L. (1994). Signal transduction by the EnvZ-OmpR phosphotransfer system in bacteria. *Res. Microbiol.* **145,** 363–373.

Foster, D. L., Mowbray, S. L., Jap, B. K., and Koshland, D. E., Jr. (1985). Purification and characterization of the aspartate chemoreceptor. *J. Biol. Chem.* **260,** 11706–11710.

Hazelbauer, G. L. (1975). Maltose chemoreceptor of *Escherichia coli. J. Bacteriol.* **122,** 206–214.

Hoch, J. A., and Silhavy, T. J. (1995). "Two-Component Signal Transduction." American Society for Microbiology, Washington, DC.

Hulko, M., Berndt, F., Gruber, M., Linder, J. U., Truffault, V., Schultz, A., Martin, J., Schultz, J. E., Lupas, A. N., and Coles, M. (2006). The HAMP domain structure implies helix rotation in transmembrane signaling. *Cell* **126,** 929–940.

Igo, M. M., Ninfa, A. J., Stock, J. B., and Silhavy, T. J. (1989). Phosphorylation and dephosphorylation of a bacterial transcriptional activator by a transmembrane receptor. *Genes Dev.* **3,** 1725–1734.

Igo, M. M., and Silhavy, T. J. (1988). EnvZ, a transmembrane environmental sensor of *Escherichia coli* K-12, is phosphorylated *in vitro. J. Bacteriol.* **170,** 5971–5973.

Igo, M. M., Slauch, J. M., and Silhavy, T. J. (1990). Signal transduction in bacteria: Kinases that control gene expression. *New Biol.* **2,** 5–9.

Inouye, M. (2003). Histidine kinases: Introductory remarks. *In* "Histidine Kinases in Signal Transduction" (M. Inouye and R. Dutta, eds.), pp. 1–24. Academic Press, Inc., San Diego, CA.

Inouye, M., Dutta, R., and Zhu, Y. (2003). Regulation of porins in *Escherichia coli* by the osmosensing histidine kinase/phosphatase EnvZ. *In* "Histidine Kinases in Signal Transduction" (M. Inouye and R. Dutta, eds.), pp. 25–46. Academic Press, Inc., San Diego, CA.

Jin, T., and Inouye, M. (1994). Transmembrane signaling. Mutational analysis of the cytoplasmic linker region of Taz1–1, a Tar-EnvZ chimeric receptor in *Escherichia coli. J. Mol. Biol.* **244,** 477–481.

Koiwai, O., and Hayashi, H. (1979). Studies on bacterial chemotaxis. IV. Interaction of maltose receptor with a membrane-bound chemosensing component. *J. Biochem. (Tokyo)* **86,** 27–34.

Krikos, A., Mutoh, N., Boyd, A., and Simon, M. I. (1983). Sensory transducers of *E. coli* are composed of discrete structural and functional domains. *Cell.* **33,** 615–622.

Kristich, C. J., Glekas, G. D., and Ordal, G. W. (2003). The conserved cytoplasmic module of the transmembrane chemoreceptor McpC mediates carbohydrate chemotaxis in *Bacillus subtilis. Mol. Microbiol.* **47,** 1353–1366.

Kumita, H., Yamada, S., Nakamura, H., and Shiro, Y. (2003). Chimeric sensory kinases containing O2 sensor domain of FixL and histidine kinase domain from thermophile. *Biochim. Biophys. Acta* **1646,** 136–144.

Lee, L., and Imae, Y. (1990). Role of threonine residue 154 in ligand recognition of the tar chemoreceptor in *Escherichia coli. J. Bacteriol.* **172,** 377–382.

Manson, M. D., Boos, W., Bassford, P. J., Jr., and Rasmussen, B. A. (1985). Dependence of maltose transport and chemotaxis on the amount of maltose-binding protein. *J. Biol. Chem.* **260,** 9727–9733.

Masui, Y., Coleman, J., and Inouye, M. (1983). Multipurpose expression of cloning vehicles in *Escherichia coli. In* "Experimental Manipulation of Gene Expressions" (M. Inouye, ed.), pp. 15–32. Academic Press, New York.

Milburn, M. V., Prive, G. G., Milligan, D. L., Scott, W. G., Yeh, J., Jancarik, J., Koshland, D. E., Jr., and Kim, S. H. (1991). Three-dimensional structures of the ligand-binding domain of the bacterial aspartate receptor with and without a ligand. *Science* **254,** 1342–1347.

Miller, J. H. (1972). Assay of β-galactosidase. *In* "Experiments in Molecular Genetics" (J. H. Miller ed.), pp. 352–355. Cold Spring Harbor Laboratory, Cold Spring Harbor, NY.

Mowbray, S. L., and Koshland, D. E., Jr. (1990). Mutations in the aspartate receptor of *Escherichia coli* which affect aspartate binding. *J. Biol. Chem.* **265,** 15638–15643.

Park, H., and Inouye, M. (1997). Mutational analysis of the linker region of EnvZ, an osmosensor in *Escherichia coli. J. Bacteriol.* **179,** 4382–4390.

Parkinson, J. S. (1993). Signal transduction schemes of bacteria. *Cell* **73,** 857–871.

Parkinson, J. S., and Kofoid, E. C. (1992). Communication modules in bacterial signaling proteins. *Annu. Rev. Genet.* **26,** 71–112.

Pratt, L. A., and Silhavy, T. J. (1995). Porin regulon of *Escherichia coli. In* "Two-Component Signal Transduction" (J. A. Hoch and T. J. Silhavy, eds.), pp. 105–127. American Society for Microbiology, Washington, DC.

Roberts, D. L., Bennett, D. W., and Forst, S. A. (1994). Identification of the site of phosphorylation on the osmosensor, EnvZ, of *Escherichia coli. J. Biol. Chem.* **269,** 8728–8733.

Russo, A. F., and Koshland, D. E., Jr. (1983). Separation of signal transduction and adaptation functions of the aspartate receptor in bacterial sensing. *Science* **220,** 1016–1020.

Silverman, M., and Simon, M. (1977). Chemotaxis in *Escherichia coli*: Methylation of che gene products. *Proc. Natl. Acad. Sci. USA* **74,** 3317–3321.

Spee, J. H., de Vos, W. M., and Kuipers, O. P. (1993). Efficient random mutagenesis method with adjustable mutation frequency by use of PCR and dITP. *Nucleic Acids Res.* **21,** 777–778.

Stock, J. B., Ninfa, A. J., and Stock, A. M. (1989). Protein phosphorylation and regulation of adaptive responses in bacteria. *Microbiol. Rev.* **53,** 450–490.

Stock, J. B., Stock, A. M., and Mottonen, J. M. (1990). Signal transduction in bacteria. *Nature* **344,** 395–400.

Utsumi, R., Brissette, R. E., Rampersaud, A., Forst, S. A., Oosawa, K., and Inouye, M. (1989). Activation of bacterial porin gene expression by a chimeric signal transducer in response to aspartate. *Science* **245,** 1246–1249.

Wang, E. A., and Koshland, D. E., Jr. (1980). Receptor structure in the bacterial sensing system. *Proc Natl. Acad. Sci. USA* **77,** 7157–7161.

Ward, S. M., Bormans, A. F., and Manson, M. D. (2006). Mutationally altered signal output in the Nart (NarX-Tar) hybrid chemoreceptor. *J. Bacteriol.* **188,** 3944–3951.

Wolff, C., and Parkinson, J. S. (1988). Aspartate taxis mutants of the *Escherichia coli* tar chemoreceptor. *J. Bacteriol.* **170,** 4509–4515.

Yang, Y., Park, H., and Inouye, M. (1993). Ligand binding induces an asymmetrical transmembrane signal through a receptor dimer. *J. Mol. Biol.* **232,** 493–498.

Yeh, J. I., Biemann, H. P., Pandit, J., Koshland, D. E., and Kim, S. H. (1993). The three-dimensional structure of the ligand-binding domain of a wild-type bacterial chemotaxis receptor. Structural comparison to the cross-linked mutant forms and conformational changes upon ligand binding. *J. Biol. Chem.* **268,** 9787–9792.

Zhu, Y., and Inouye, M. (2002). The role of the G2 box, a conserved motif in the histidine kinase superfamily, in modulating the function of EnvZ. *Mol. Microbiol.* **45,** 653–663.

Zhu, Y., and Inouye, M. (2003). Analysis of the role of the EnvZ linker region in signal transduction using a chimeric Tar/EnvZ receptor protein, Tez1. *J. Biol. Chem.* **278,** 22812–22819.

Zhu, Y., and Inouye, M. (2004). The HAMP linker in histidine kinase dimeric receptors is critical for symmetric transmembrane signal transduction. *J. Biol. Chem.* **279,** 48152–48158.

[8] Functional and Structural Characterization
of EnvZ, an Osmosensing Histidine Kinase
of *E. coli*

By Takeshi Yoshida, Sangita Phadtare, and Masayori Inouye

Abstract

EnvZ is an osmosensing histidine kinase located in the inner membrane, and one of the most extensively studied *Escherichia coli* histidine kinases. Because of its structural complexity, functional and structural studies have been quite challenging. It is a multidomain transmembrane protein consisting of 450 amino acid residues. In addition, it must form a dimer to function as a histidine kinase like all the other histidine kinases. EnvZ consists of the 115-residue periplasmic domain, two transmembrane domains (TM1 and TM2), and the cytoplasmic domain consisting of the 43-residue linker (HAMP) domain and the 228-residue kinase domain. It has been shown that the kinase domain of EnvZ, responsible for its enzymatic activities, contains all of the conserved regions of histidine kinases such as H, F, N, G1, G2, and G3 boxes. Therefore, the 271-residue cytoplasmic domain of EnvZ (termed EnvZc) has been used as a model system to establish fundamental characteristics of histidine kinases. The DNA fragment encoding EnvZc was cloned in pET vector and EnvZc was expressed and purified. It is highly soluble and retains all the enzymatic activities of EnvZ. We demonstrated that it consists of two functional domains, domain A and domain B. NMR spectroscopic studies of these two domains revealed, for the first time, the structure of a histidine kinase. Domain A is responsible for dimerization of EnvZc forming a four-helical bundle containing two α-helical hairpin structures, while domain B is a monomer and has an ATP-binding pocket formed by regions conserved among the histidine kinases. In this chapter, we describe functional and structural studies of EnvZc, which can be applied to characterize other histidine kinases.

Introduction

EnvZ, an osmosensor histidine kinase, is a multidomain transmembrane protein which consists of 450 amino acid residues which obligatorily form a dimer in the inner membrane. Earlier studies on EnvZ showed that it contains a short N-terminal cytoplasmic tail (residues 1–15), two transmembrane segments, TM1 (residues 16–47) and TM2 (residues 163–179) (Forst *et al.*, 1987), a periplasmic receptor domain (residues 48–162)

METHODS IN ENZYMOLOGY, VOL. 423
0076-6879/07 $35.00
DOI: 10.1016/S0076-6879(07)23008-3

(Egger and Inouye, 1997; Khorchid *et al.*, 2005), and a cytoplasmic C-terminal domain (residues 180–450). Biochemical studies demonstrated that using ATP, EnvZ can autophosphorylate its conserved His residue, His243 (Roberts *et al.*, 1994). The phosphoryl group is then transferred to its cognate response regulator, OmpR, a transcription factor of *ompF* and *ompC* genes, at the conserved Asp residue, Asp55 (Delgado *et al.*, 1993). EnvZ can also dephosphorylate phosphorylated OmpR (OmpR-P) (Aiba *et al.*, 1989; Igo *et al.*, 1989).

The cytoplasmic kinase domain of EnvZ (termed EnvZc) consists of the 43-residue (HAMP) domain and the 228-residue kinase domain, and the kinase domain contains regions highly conserved among other histidine kinases (Dutta and Inouye, 2000; Parkinson and Kofoid, 1992). Therefore, EnvZc domain has been used as a model system for functional and structural studies of histidine kinases. Biochemical studies on EnvZc elucidated the properties and roles of conserved regions of histidine kinases. EnvZc also exists as a dimer having all the enzymatic activities of full-length EnvZ (Hidaka *et al.*, 1997; Park and Inouye, 1997). Dimerization is the key feature of all histidine kinases since it was originally demonstrated that EnvZ undergoes *trans*-autophosphorylation; the ATP-binding domain of one subunit phosphorylates His243 on the other subunit within a dimer (Cai and Inouye, 2003; Qin *et al.*, 2000; Yang and Inouye, 1991). Further dissection of EnvZc revealed two functional domains (Park *et al.*, 1998). One domain is a 67-residue domain, domain A, which is responsible for dimerization of EnvZc and which contains the H box with the phosphorylation site, His243. It has been demonstrated that domain A is responsible for not only phosphorylation of OmpR, but also dephosphorylation of OmpR-P and the conserved His residue plays an essential role in these reactions (Zhu *et al.*, 2000). The other domain is a 169-residue domain, domain B, which is a monomer, which contains an ATP-binding site and conserved N, F, G1, G2, and G3 boxes (Dutta and Inouye, 2000). Mutational analysis of these conserved regions in EnvZc demonstrated that these are involved in ATP binding (Dutta and Inouye, 1996; Hsing *et al.*, 1998; Yang and Inouye, 1993) and the G2 box is also involved in phosphatase activity of EnvZ (Zhu and Inouye, 2002).

The structural studies of domain A and domain B revealed the structure–function relationship of these subunits in activity of EnvZc (Tanaka *et al.*, 1998; Tomomori *et al.*, 1999). NMR structure of domain A revealed that it consists of two long antiparallel helices (helix I: 235–255; helix II: 265–286) connected by a 9-residue turn and is a dimer forming a four-helix bundle (Fig. 1A) (Tomomori *et al.*, 1999). Importantly, the His residue, which is the phosphorylation site in helix I, is exposed to the solvent (Fig. 1B). NMR structure of domain B showed a similar folding to that

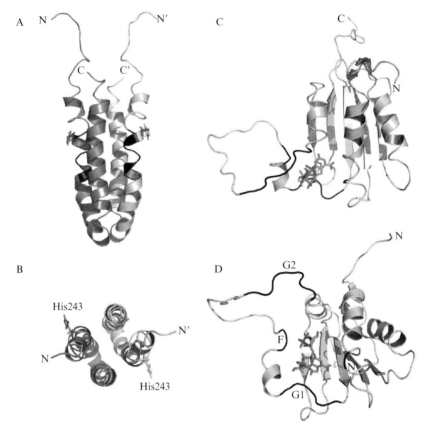

Fig. 1. NMR structures of domain A and domain B of EnvZ. (A) Side view of NMR structure of domain A. The conserved H box is shown in black. (B) Top view of domain A showing the side chain of a phosphorylation site, His243, exposed to the solvent. (C) NMR structure of domain B with ATP analog, AMP-PNP. The conserved N, G1, F, and G2 boxes are indicated and shown in black. AMP-PNP molecule is shown. (D) Bottom view of domain B showing ATP-binding pocket surrounded by the conserved boxes. Structure images are generated by using PyMOL from DeLano Scientific (DeLano, 2002).

seen among other ATPase proteins, such as Hsp90 (heat shock protein), DNA gyrase B (type II topoisomerase), and MutL (DNA mismatch repair protein) (Dutta and Inouye, 2000; Tanaka *et al.*, 1998). It consists of an α/β sandwich fold made by a five-stranded β-sheet and three α-helices and a large flexible loop (Fig. 1C). This flexible loop contains conserved G1, F, and G2 boxes and is important for ATP binding. Based on the structure, it was concluded that all conserved regions are involved in formation of the ATP-binding pocket (Fig. 1D). It should be noted that between TM2 and EnvZc domain there is a 43-residue linker—a HAMP domain—which plays a critical role in signal transduction from the periplasmic receptor domain to the kinase domain. The structure of the HAMP domain identified as a sole cytoplasmic domain from a hyperthermophilic bacterium *Archaeoglobus fulgidus* has been determined by NMR and a possible model for the role of the HAMP domain in signal transduction was proposed (Hulko *et al.*, 2006; Inouye, 2006).

In this chapter we present purification methods for EnvZc, domain A and domain B and the assays to study basic functions of EnvZc. We also describe mutations used for characterizing EnvZc. These studies should provide guidelines for characterization of histidine kinase of interest.

Expression and Purification of EnvZc

E. coli BL21(DE3) cells are transformed with pET11a-EnvZc (Park and Inouye, 1997) and the transformants are grown on an M9 medium plate supplemented with casamino acids and 50 μg/ml ampicillin. A single colony is inoculated into 30 ml of M9 medium with casamino acids and 50 μg/ml ampicillin (6 g dibasic sodium phosphate, 3 g monobasic potassium phosphate, 0.5 g NaCl, 1 g NH$_4$Cl, 0.2% casamino acids, 1 mM MgSO$_4$, 50 mg tryptophan, 4 g glucose, and 2 mg thiamine per liter) and incubated overnight at 37° on a rotary shaker. The overnight culture is then used as an inoculum for 1.5 liter of M9 medium with casamino acids and 50 μg/ml ampicillin in a 4-liter flask and incubated on a rotary shaker at 37° until the culture reaches early logarithmic phase (OD$_{600}$ of ~0.6). The overproduction of EnvZc is induced by the addition of 1 mM IPTG (isopropyl β-D thiogalactopyranoside) for 3 h. Cells are harvested by centrifugation at 4500×g for 20 min at 4° and the cell pellet is washed once with 30 ml of 50 mM Tris-HCl buffer (pH 8.0), containing 50 mM KCl, and 5% glycerol. Cells are collected by centrifugation at 20,000×g for 20 min and the cell pellet is resuspended in 30 ml of 20 mM Tris-HCl buffer (pH 8.0), containing 100 mM NaCl, 5% glycerol, and 1 mM phenylmethylsulfonyl fluoride (PMSF, a serine protease inhibitor). The cells are lysed by 2 to 3 passes through a French Press at 14,000 psi (internal cell pressure), and the lysate is centrifuged at 20,000×g for 20 min to remove cell debris. The supernatant is subjected to ultracentrifugation 163,000×g for 1 h at 4° to remove the membrane fraction. The supernatant is loaded onto a DEAE anion exchange column equilibrated with 20 mM Tris-HCl buffer (pH 8.0), containing 100 mM NaCl, 5 mM β-mercaptoethanol, and 5% glycerol and EnvZc bound to the column is eluted by salt gradient of 100 mM to 500 mM NaCl. Six-milliliter fractions are collected at flow rate of 1.5 ml/min and analyzed by 17.5% SDS-PAGE. The fractions containing EnvZc are pooled and

EnvZc is further purified by 35% ammonium sulfate fractionation. The protein is collected by centrifugation at $20,000 \times g$ for 20 min. The protein pellet is resuspended with 5 ml of 50 mM Tris-HCl buffer (pH 8.0) containing 150 mM KCl and loaded onto a Sephacryl S-100HR gel-filtration column (5 cm $\times$ 80 cm) equilibrated with 50 mM Tris-HCl buffer (pH 8.0) containing 150 mM KCl. The homogeneity of the protein sample is established by SDS-PAGE and Coomassie Blue staining. The protein concentration is determined by Bradford method with BioRad protein assay dye using bovine serum albumin as a standard. The protein sample is aliquoted and stored at $-80°$ with 5% (v/v) glycerol. Typically, 60 to 70 mg of pure EnvZc can be obtained from an 1-liter culture.

Expression and Purification of Domain A and Domain B

Expression of Proteins

E. coli BL21(DE3) cells are transformed with pET11a-domain A or pET11a-domain B (Park *et al.*, 1998) and the trasformants are grown on LB plates with 50 μg/ml ampicillin. A single colony is inoculated into 30 ml of M9 medium supplemented with casamino acids and 50 μg/ml ampicillin and incubated overnight at 37° on a rotary shaker. The overnight culture is used as an inoculum for 1.5 liter of M9 medium with casamino acids and 50 μg/ml ampicillin in a 4-liter flask and is incubated at 37° on a rotary shaker until the culture reaches early logarithmic phase (OD$_{600}$ of $\sim$0.6). The overexpression of domain A or domain B is induced by the addition of 1 mM IPTG for 3 h. Cells are harvested by centrifugation at $4500 \times g$ for 20 min at 4° and the cell pellet is washed once with 30 ml of 50 mM Tris-HCl buffer (pH 8.0), containing 50 mM KCl, and 5% glycerol. Cells are collected by centrifugation at $20,000 \times g$ for 20 min.

Purification of Domain A

The cell pellet is resuspended in 30 ml of 50 mM Tris-HCl buffer (pH 8.0), containing 150 mM KCl, 5 mM β-mercaptoethanol, 5% glycerol, and 1 mM PMSF. The cells are lysed by 2 to 3 passes through French Press at 14,000 psi (internal cell pressure), and the lysate is centrifuged at $20,000 \times g$ for 20 min to remove cell debris. The supernatant is subjected to ultracentrifugation at $163,000 \times g$ for 1 h at 4° to remove the membrane fraction. Domain A is purified by 50% ammonium sulfate fractionation. The protein is collected by centrifugation at $20,000 \times g$ for 20 min. The protein pellet is resuspended with 5 ml of 50 mM Tris-HCl buffer (pH 8.0), containing 150 mM KCl and 5 mM β-mercaptoethanol, and loaded onto a Sephacryl S-100HR gel-filtration column equilibrated with 50 mM Tris-HCl buffer (pH 8.0), containing 150 mM KCl and 5 mM β-mercaptoethanol. The fractions containing domain A protein are further purified by a Q-Sepharose FF anion exchange column equilibrated with 20 mM Tris-HCl buffer (pH 8.0), containing 150 mM KCl, 5 mM β-mercaptoethanol, and 5% glycerol. Domain A protein is eluted by salt gradient of 150 mM to 400 mM KCl. Six-milliliter fractions are collected at flow rate of 1.5 ml/min and analyzed by 20% SDS-PAGE. The fractions containing domain A are pooled and dialyzed against 50 mM Tris-HCl buffer (pH 8.0) containing 50 mM KCl and 5% glycerol. Homogeneity of the samples is established by SDS-PAGE and Coomassie Blue staining. The protein concentration is determined by Bradford method with BioRad protein assay dye using bovine serum albumin as a standard. The protein sample is aliquoted and stored at $-80°$. Typically, 10 mg of pure domain A can be obtained from a 1-liter culture.

Purification of Domain B

Purification of Domain B is carried out using a method similar to that used for purification of domain A, described previously. After cells are lysed by French Press and cell debris and unsoluble membrane fraction are removed by centrifugation, the resultant supernatant is loaded onto a Q-Sepharose FF anion exchange column equilibrated with 20 mM Tris-HCl (pH 8.0) containing 150 mM KCl, 5 mM β-mercaptoethanol, and 5% glycerol and domain B is eluted by the salt gradient from 150 mM to 500 mM KCl. After further purification by 50% ammonium sulfate fractionation, Domain B is purified by a Sephacryl S-100HR gel-filtration column equilibrated with 50 mM Tris-HCl (pH 8.0) containing 150 mM KCl and 5 mM β-mercaptoethanol. The protein sample is aliquoted and stored at $-80°$. Typically, 60 to 70 mg of pure domain B can be obtained from an 1-liter culture.

Characterization of EnvZc

ATP Binding Assay

Using domain B, it has been shown that an ATP molecule can be cross-linked to the ATP-binding pocket using ultraviolet (UV) light (Tanaka *et al.*, 1998). Since this reaction is exothermic, it is recommended to carry out this experiment on ice.

Radioactive, nonhydrolizable ATP, [^{35}S]-ATPγS (800 Ci/mmol, 10 mCi/ml; Perkin-Elmer Life Sciences) is used for this experiment (Zhu and Inouye, 2002). Four μg of EnvZc is incubated with 20 μCi of [^{35}S]-ATPγS in 50 mM Tris-HCl buffer (pH 8.0), containing 50 mM KCl, 5 mM CaCl$_2$ (or MgCl$_2$),

5% glycerol for 15 min on ice (final reaction volume is 25 μl). Note that it is convenient to use 96-well plates for this experiment. The reaction mixture is then exposed to 254-nm (short wavelength) UV light at a distance of approximately 4.5 cm from the sample for 5 min using a UV lamp (model UVG-45, 115V, 60 Hz, 0.16A; UVP). The reaction is stopped by adding 5× SDS loading buffer [100 mM Tri-HCl (pH 6.8), 4% SDS, 200 mM β-mercaptoethanol, 20% glycerol, 0.2% bromophenol blue dye]. The reaction products are then analyzed by 17.5% SDS-PAGE, followed by drying the gel for autoradiography.

Autophosphorylation Assay

EnvZc has an ATP-binding domain and is able to use ATP to autophosphorylate at the conserved His residue, His243. This reaction is divalent metal ion-dependent, requiring Ca^{2+} or Mg^{2+} ions. It has been shown that the rate of autophosphorylation of EnvZc is enhanced in the presence of Ca^{2+} ions compared to the reaction carried out in the presence of Mg^{2+} ions (Rampersaud *et al.*, 1991).

For autophosphorylation of EnvZc, the final concentration of EnvZc used is 2 μM and the reaction mixture includes 50 mM Tris-HCl buffer (pH 8.0) containing 50 mM KCl, 5 mM β-mercaptoethanol, and 5 mM $CaCl_2$ (or $MgCl_2$, if it is preferred). The reaction is carried out at room temperature. An appropriate amount of EnvZc is incubated in 90 μl of the reaction buffer. The reaction is started by adding 10 μl of ATP mixture containing nonradioactive ATP (500 μM) and 0.5 μl of [γ-^{32}P]ATP (3000 Ci/mmol, 10 mCi/ml). Note that, under this condition, by 15 min the reaction will reach the saturation point. Ten-μl aliquots are taken and the reaction is stopped by mixing with 5× SDS loading buffer. The reaction products are then analyzed by 17.5% SDS-PAGE, followed by drying the gel for autoradiography.

Identification of Phosphohistidine

Phosphohistidine is known to be sensitive to acidic pH and resistant to alkaline pH, while phosphoserine, phosphothreonine, and phosphotyrosine have exactly opposite properties (Forst *et al.*, 1989). Therefore, by exposing phosphorylated histidine kinase to acidic or alkaline pH, one can distinguish phosphorylation of His residue from that of Ser, Thr, or Tyr.

The autophosphorylation reaction is carried out in the 50-μl reaction mixture described previously for 15 min. A 14-μl aliquot is removed and mixed with 2 μl of 10% SDS. Denatured phosphorylated EnvZc is exposed to acidic or basic pH by adding 2 μl of 0.7 M HCl or 2 μl of 3.3 M NaOH, respectively, and is incubated at 43° for 60 min. The treated sample is neutralized by adding 1 μl of 3.3 M NaOH or 4 μl of 0.7 M HCl, respectively, to minimize the detrimental effect of unfavorable pH conditions. 5× SDS loading buffer is added and the samples are analyzed by 17.5% SDS-PAGE, followed by drying the gel for autoradiography.

Phosphotransfer Assay

EnvZc is autophosphorylated at His243 and the phosphoryl group is subsequently transferred to Asp55 residue of its cognate response regulator, OmpR. Phosphotransfer assay helps to monitor the generation of phosphorylated OmpR (OmpR-P) as a result of phosphotransfer from EnvZc-P to OmpR. This reaction occurs very rapidly and is completed within a few minutes and is dependent on divalent metal ions. After EnvZc-P phosphorylates OmpR, unphosphorylated EnvZc acts as a phosphatase to dephosphorylate OmpR-P and inorganic phosphate, Pi, is released. Importantly, phosphatase activity of EnvZc toward OmpR-P is differently stimulated in the presence of Ca^{2+} or Mg^{2+} ions; it is much higher in the presence of Mg^{2+} ions than in the presence of Ca^{2+} ions. Therefore, if phosphotransfer reaction is carried out in the presence of Mg^{2+} ions, the band representing OmpR-P will rapidly disappear with concomitant release of Pi.

The phosphotransfer reaction leading to transfer of a phosphoryl group from phosphorylated EnvZc to OmpR can be monitored continuously by adding purified OmpR into the reaction mixture of the autophosphorylation reaction of EnvZc. It is recommended to carry out a time course in minutes and the reaction should be completed by 5 min. In this simple method, ATP can be used by EnvZc to autophosphorylate itself or can be a cofactor to stimulate the phosphatase activity of EnvZc. Therefore, it is recommended to completely remove free ATP associated with phosphorylated EnvZc (EnvZc-P) to specifically analyze the phosphotransfer reaction. Since the phosphoryl group of EnvZc-P is stable and has a long half-life (several hours), free ATP can be removed from EnvZc-P by exchanging the buffer as follows. Fifty μg of EnvZc is first autophosphorylated in 200 μl of the autophosphorylation buffer containing 50 μM ATP and 1 μl of [γ-^{32}P]ATP (3000 Ci/mmol, 10 mCi/ml) for 15 min at room temperature. The reaction mixture is transferred to 0.5-ml centrifugal concentrator (10K NMWL, UFV5BGC00, Millipore). Three hundred μl of 50 mM Tris-HCl buffer (pH 8.0) containing 50 mM KCl, 5 mM β-mercaptoethanol, 1mM EDTA, and 15% glycerol is added and the volume (500 μl) is then reduced by centrifugation to approximately 100 μl. This is then diluted by adding 400 μl of the buffer described previously and the volume is again reduced to approximately 100 μl. This procedure is repeated (10–15 times) until radio-activity from flowthrough is barely detectable by using a Geiger counter. The reaction mixture is then concentrated to 50 μl and it is stored at $-20°$. Phosphotransfer reaction is carried out in the autophosphorylation buffer containing 5 mM CaCl$_2$ using 2 μM of EnvZc-P and 4 μM of OmpR. Since the phosphotransfer reaction occurs very rapidly, a short time course such as 20, 40, 60, 120, and 300 sec is recommended to follow the course of the reaction. Note that if MgCl$_2$ is used instead of CaCl$_2$, dephosphoryl-ation of generated OmpR-P occurs, resulting in disappearance of the band representing OmpR-P at later time points. The reaction mixtures are then analyzed by 17.5% SDS-PAGE, followed by drying the gel for autoradiography.

Phosphatase Assay

EnvZ also has phosphatase activity and dephosphorylates phosphory-lated OmpR (OmpR-P). It is known that ATP, ADP, AMP, or ATP analog, such as AMP-PNP, function as cofactors to significantly stimulate the phosphatase activity of EnvZc. For measuring phosphatase activity of EnvZc, after phosphotransfer reaction in the presence of 5 mM CaCl$_2$ is carried out, ADP is added to this reaction mixture. In this mixture, OmpR is phosphorylated and the major fraction of EnvZc is unphosphorylated. Autophosphorylation is first carried out for 15 min, followed by addition of OmpR to allow phosphotransfer reaction, which is carried out for 5 min. To initiate phosphatase reaction, ADP is added to the phosphotransfer reaction mixture already prepared at a final concentration of 1 mM. Under this condition, the rate of dephosphorylation of OmpR-P is slow. Therefore, it is recommended to carry out a time course between 5 and 30 min. The reaction is stopped by adding 5× SDS loading buffer. In order to compare the phosphatase activity of EnvZc mutant proteins, it is better to use purified OmpR-P. The preparation of OmpR-P is described in the following text.

Preparation of Membrane Fraction Containing Overexpressed EnvZ[T247R] Mutant Protein

The preparation of OmpR-P has been established using the membrane fraction containing overexpressed full-length EnvZ[T247R] mutant (kinase$^+$/ phosphatase$^-$) protein. For the preparation of OmpR-P, first, the mem-brane fraction containing EnvZ[T247R] mutant protein is prepared using an *envZ* deletion strain, RU1012 [MH225, $\Delta envZ$::Km^r] (Utsumi *et al.*, 1989).

EnvZ[T247R] mutant protein is then autophosphorylated. Finally, OmpR-P is prepared by incubating purified OmpR with the membrane fraction containing phosphorylated EnvZ[T247R] protein and subsequently removing the membrane fraction by ultracentrifugation.

The T247R mutation is created in EnvZ by site-directed mutagenesis using the plasmid pDR200, which contains wild-type full-length EnvZ (Utsumi et $al.$, 1989). The construct is cloned in a pINIII vector to create pIN-III-EnvZ[T247R]. RU1012 cells harboring pIN-III-EnvZ[T247R] are grown in 40 ml of LB medium supplemented with 50 μg/ml of ampicillin and 50 μg/ml of kanamycin. At OD$_{600}$ of 0.5, IPTG is added at a final concentration of 1 mM, and the culture is further incubated for one hour to induce EnvZ[T247R] mutant. Cells are harvested at mid-log phase and washed in 100 mM sodium phosphate buffer (pH 7.2) containing 10% glycerol, 1 mM PMSF, 5 mM EDTA, 5 mM 1,10-phenanthroline, and 1 μM leupeptin. Cells are lysed by sonication and cell debris is removed by centrifugation at 10,000×g for 15 min. The supernatant is then subjected to ultracentrifugation at 334,000×g for 20 min at 4°. The membrane fraction thus precipitated contains EnvZ[T247R] mutant protein. The pellet is resuspended by sonication in 200 μl of 20 mM Tris-HCl buffer (pH 8.0) containing 5% glycerol, 1 mM EDTA, 10 mM β-mercaptoethanol, and 2 M KCl. The suspension is centrifuged to collect the membrane fraction. The membrane fraction thus obtained is resuspended in 100 μl of 20 mM Tris-HCl buffer (pH 8.0) containing 5% glycerol, 1 mM EDTA, and 10 mM β-mercaptoethanol, stored at −20°.

Autophosphorylation of EnvZ[T247R] Mutant

It is important to sonicate the membrane fraction every time prior to use. Ten μl of the membrane fraction containing overexpressed EnvZ [T247R] protein is first autophosphorylated with 50 μM ATP and 4 μl of [γ-^{32}P]ATP (3000 Ci/mmol, 10 mCi/ml) in 200 μl of autophosphorylation buffer for 20 min at room temperature. The reaction mixture is transferred to a 1.5-ml ultracentrifugation tube and subjected to ultracentrifugation at 334,000×g for 15 min at 4°. The yellow-brown membrane pellet thus obtained is resuspended in 300 μl of the autophosphorylation buffer and subjected to brief sonication. The resultant lysate is transferred into a new 1.5-ml ultracentrifugation tube and subjected to ultracentrifugation at 334,000×g for 15 min at 4°. Reuse of a tube is not recommended because of its possible breaking and subsequent contamination of the ultracentrifuge. The washing steps are repeated twice. The final pellet is resuspended by sonication in 80 μl of the autophosphorylation buffer.

Preparation of Phosphorylated OmpR by Phosphorylated EnvZ[T247R]
  Mutant Protein

Purified OmpR (50 μg) is mixed with the membrane fraction prepared previously (80 μl) in 200 μl of autophosphorylation buffer and incubated for 20 min at room temperature. The reaction mixture is subjected to ultracentrifugation at 334,000×g for 15 min at 4° to remove the membrane fraction. The supernatant containing phosphorylated OmpR (OmpR-P) is transferred to 0.5-ml centrifugal concentrator (10K NMWL, UFV5BGC00, Millipore). Care should be taken while removing the supernatant to avoid contamination of the membrane fraction. In order to remove Ca^{2+} ions, free ATP, the buffer exchange is carried out with use of 50 mM Tris-HCl buffer (pH 8.0) containing 50 mM KCl, 5 mM β-mercaptoethanol, 1 mM EDTA, 10% glycerol, as described in the "Phosphotransfer Assay" section. The radioactivity of the phosphorylated OmpR sample as detected by Geiger counter should not be significantly reduced during the buffer exchange procedure. Phosphorylated OmpR is concentrated to 50 μl and is stored at −20°. The purity of the sample is checked by subjecting a 1 μl sample to thin layer chromatography (TLC) using a polyethyleneimine-cellulose TLC plate. The plate is processed at room temperature using a TLC chamber saturated with 0.75 M KH$_2$PO4 buffer (pH 3.65) to detect whether the OmpR-P preparation is free of [γ-^{32}P]ATP or inorganic phosphate, ^{32}Pi. The plate is dried and processed for autoradiography. The spotted area should show radioactivity, since ^{32}P-labeled OmpR-P is unable to migrate, while ATP and Pi migrate with the buffer.

Measuring Phosphatase Activity of EnvZc Using Purified OmpR-P

2 μM of OmpR-P is incubated in 50 mM Tris-HCl buffer (pH 8.0) containing 50 mM KCl, 5 mM MgCl$_2$, and 5 mM β-mercaptoethanol. The reaction is initiated by the addition of EnvZc (1 μM) and 1 mM ADP. Note that, under this condition, dephosphorylation of OmpR-P occurs rapidly and is completed within a few minutes. The reaction is stopped by adding 5× SDS loading buffer. The reaction products are then analyzed by 17.5% SDS-PAGE, followed by drying the gel for autoradiography.

Characterization of EnvZ with Help of its Specific Mutants

A number of mutations at the conserved amino acid residues in EnvZc have been mutated in order to study the roles of these residues in EnvZ function. The following mutations are created at highly conserved amino acid residues in EnvZc, and are thus expected to exert a similar effect in other histidine kinases.

H243V Mutation (Kinase$^-$/Phosphatase$^-$)

This mutation eliminating the phosphorylation site of EnvZ abolishes its phosphatase activity, demonstrating that His243 is the phosphorylation site of EnvZ and is required for its enzymatic activities (Forst *et al.*, 1989; Kanamaru *et al.*, 1990). However, the ATP-binding domain of EnvZ[H243V] mutant protein fully retains its function if it forms a heterodimer with another EnvZ construct that has a phosphorylation site, His243 (Yang and Inouye, 1991, 1993; Zhu and Inouye, 2004).

T247R Mutation (Kinase$^+$/Phosphatase$^-$)

This mutant protein is known as EnvZ11 (Hall and Silhavy, 1981; Wandersman *et al.*, 1980). This mutation increases the kinase activity of EnvZ, while abolishing its phosphatase activity (Aiba *et al.*, 1989). Thus, the EnvZ[T247R] mutant protein is known as "super kinase." Thr247 residue is located within the H box, and highly conserved among histidine kinases (Grebe and Stock, 1999). Mutation of Thr247 affected both the autophosphorylation and phosphotransfer activities of EnvZc, but mostly the phosphatase activity was diminished (Dutta *et al.*, 2000). This mutation is ideal for an experiment involving OmpR-P as described in the "Phosphatase Assay" section, since the mutant protein will not dephosphorylate the OmpR-P generated during the phosphotransfer reaction. For instance, phosphorylation of OmpR is required for its binding to DNA. Thus, in order to generate a sufficient amount of fully phosphorylated OmpR for DNA binding studies, one can use EnvZc[T247R] mutant protein.

G405A Mutation (Kinase$^-$/Phosphatase$^-$)

This mutation in EnvZc abolishes its ATP binding and phosphatase activity (Zhu and Inouye, 2002). Gly405 is one of Gly residues in the conserved G2 box that is important for ATP binding (Hsing *et al.*, 1998; Yang and Inouye, 1993). Therefore, EnvZ[G405A] mutant protein can neither autophosphorylate itself nor phosphorylate OmpR. As the active sites of EnvZ are formed by two subunits within a dimer and autophosphorylation occurs in *trans*, complementation between G405A and H243V mutations is extremely useful; as shown in Fig. 2A, heterodimer between EnvZc[H243V] and EnvZc[G405A] mutant proteins will create only one active site containing the phosphorylation site, H243, from EnvZc[G405A] mutant and ATP-binding domain from EnvZc[H243V] mutant, while the other active site is inactive because it consists of the mutated phosphorylation site, H243V, from EnvZc[H243V] mutant and the mutated ATP-binding domain from

EnvZc[G405A]. This enables one to specifically monitor the activity of only one active site in an EnvZ dimer to study how signal transduction regulates the function of EnvZ (or other histidine kinases of interest) or to follow a single phosphoryl group within a hybrid histidine kinase.

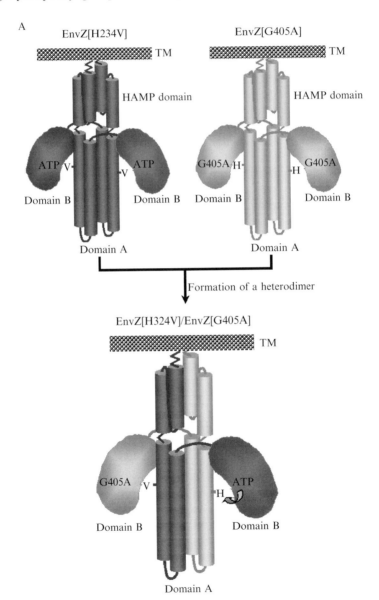

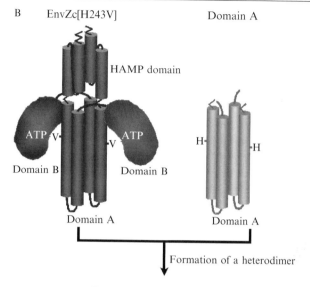

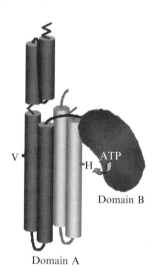

Fig. 2. The scheme of heterodimer formation of EnvZ[H243V] mutant protein. (A) The cytoplasmic domains of full-length EnvZ[H243V] (dark gray) and EnvZ[G405A] (light gray) mutants are illustrated. In a homodimer of EnvZ[H243V] and EnvZ[G405A] mutants, there is no functional active site. As the result of the formation of heterodimer between these two EnvZ mutants, only one active site is formed by His243 of EnvZ[G405A] mutant and ATP-binding pocket of EnvZ[H243V] mutant. (B) By mixing EnvZc[H243V] mutant and domain A, a heterodimer is formed, in which one active site is formed by His243 on domain A and ATP-binding pocket of EnvZc[H243V] mutant.

Phosphorylation of Domain A with EnvZc[H243V] Mutant

Although domain A has a phosphorylation site, it cannot autophosphorylate because it lacks the ATP-binding domain. However, it can be phosphorylated if domain B is added in the presence of ATP (Park *et al.*, 1998). Since domain A is a dimerization domain for EnvZc and autophosphorylation occurs in *trans*, it should be efficiently phosphorylated in the heterodimer using EnvZc[H243V] mutant protein, which lacks a phosphorylation site, but has an intact ATP-binding domain (Fig. 2B) (Park *et al.*, 1998). This approach allows one to study the phosphotransfer from domain A-like subunits of a histidine kinase of interest to downstream components such as a response regulator or a regulatory domain within a hybrid histidine kinase.

For preparation of phosphorylated domain A, domain A and His-tagged EnvZc[H243V] protein are mixed in equimolar ratio and the mixture is incubated in the autophosphorylation buffer for 20 min. His-tagged EnvZc [H243V] protein can then be removed by adding an appropriate amount of Ni-NTA resin into the reaction mixture, followed by filtration using a centrifugal filter (0.45 μm, UFC30HV00, Millipore). This phosphorylated domain A is stable and can be stored at $-20°$.

Creation of a Monomeric Histidine Kinase Using EnvZc

Once two subunits of EnvZc form a dimer, two active sites are formed (Fig. 3A). To investigate how these active sites are formed and how domain A and domain B are topologically oriented within a dimer, one can construct a monomeric EnvZc by connecting two subunits of domain A in a single polypeptide chain, to which domain B is connected at the C-terminal end, as shown in Fig. 3B. A monomeric EnvZc is designed and created to study the active site and topological orientation between domain A and domain B.

To create such as a monomeric EnvZ protein, an extra domain A (residues 223–289) is added at residue 223 of EnvZc with use of two types of flexible spacers, GlyGlySerIle (S) or GlyGlySerGlyGlySerIle (SS). The resultant EnvZc constructs have two domain A and one domain B with a flexible spacer (S) or (SS) and are termed as EnvZ[AsAB] or EnvZ [AssAB], respectively. These two constructs are expressed using pET system and purified. They are fully soluble and exist mostly as a monomer. Importantly, they retain all enzymatic activities of EnvZc. The detail characterization of these two proteins is described by Qin *et al.* (2000). This approach may be also useful in studying the phosphotransfer reaction of a single phosphoryl group, especially within hybrid histidine kinases.

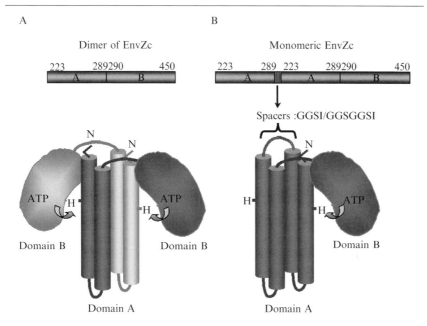

FIG. 3. The schematic model of monomeric EnvZc. (A) Two subunits, shown in dark and light gray, form a functional dimer in EnvZc. (B) A monomeric histidine kinase is designed and created. In this construct, the C-terminal end of first domain A is connected to the N-terminal end of second domain A by two types of short flexible spacers, GlyGlySerIle or GlyGlySer-GlyGlySerIle, as indicated. This second domain A is linked with domain B, which is able to phosphorylate His243 on the first domain A.

NMR Structural Analysis of Domain A and Domain B

By the mass spectroscopic analysis of limited tryptic digested fragments of EnvZc, a fragment from Gln229 to Arg289 containing His243 is found to be a stable domain in EnvZc. Analysis of digestion products of ^{32}P-labeled EnvZc-P also confirmed that this fragment contains a phosphorylation site, His243. We thus constructed EnvZc[223–289] (domain A) and EnvZc [290–450] (domain B). Biochemical studies on these two domains demonstrated that domain A forms a dimer, can be phosphorylated in the presence of ATP by domain B, and is able to phosphorylate OmpR, while domain B is a monomeric protein and has an ATP binding pocket. We used these two domains to solve the first structure of a histidine kinase by NMR.

NMR Study of Domain A and Domain B

^{15}N-, ^{13}C-, or ^{15}N/^{13}C-labeled domain A or domain B is prepared by expressing these proteins in M9 minimal medium containing ^{15}N-ammoniumn chloride and/or ^{13}C$_6$-D-glucose. Labeled domain A or domain B is purified as has been described. Domain A is prepared at the concentration of 1 mM in 95% H_2O/5% ^2H_2O containing 20 mM sodium phosphate buffer (pH 7.2), 50 mM KCl, 5 mM MgCl$_2$, 0.5 mM 4-(2-aminoethyl)-benzensul-fonylfluoride hydrochloride (AEBSF), 5 mM perdeuterated dithiothreitol (DTT), and 50 μM sodium azide. Domain B is prepared at the concentration of 1.0 to 1.5 mM in 95% H_2O/5% ^2H_2O or 99.996% ^2H_2O containing 20 mM sodium phosphate (pH 7.0), 50 mM KCl, 5 mM MgCl$_2$, 0.5 mM 4-(2-aminoethyl)-benzensulfonylfluoride hydrochloride (AEBSF), and 50 μM sodium azide with 5 mM nonlabeled or ^{15}N/^{13}C-labeled AMP-PNP. The details of analysis of NMR studies of these two domains are described by Tanaka *et al.* and Tomomori *et al.*, respectively (Tanaka *et al.*, 1998; Tomomori *et al.*, 1999).

Conclusion

In this chapter, we described general methods for functional and structural characterization of *E. coli* histidine kinase EnvZ. The use of the functional cytoplasmic domain of EnvZ, EnvZc enables us to quite extensively investigate the function and structure of EnvZ, since it is fully soluble, retaining all the enzymatic activities of the full-length EnvZ. The methods described here will serve as a guideline for functional and structural studies of other histidine kinases of interest.

Acknowledgments

This work was supported by Grant GM076587 from the National Institutes of Health. The authors thank Dr. M. Ikura for his contributions to NMR structural studies of domain A and domain B of Env, Z, and Dr. E. Severinov for his critical reading of our manuscript.

References

Aiba, H., Nakasai, F., Mizushima, S., and Mizuno, T. (1989). Evidence for the physiological importance of the phosphotransfer between the two regulatory components, EnvZ and OmpR, in osmoregulation in *Escherichia coli*. *J. Biol. Chem.* **264,** 14090–14094.

Cai, S. J., and Inouye, M. (2003). Spontaneous subunit exchange and biochemical evidence for trans-autophosphorylation in a dimer of *Escherichia coli* histidine kinase (EnvZ). *J. Mol. Biol.* **329,** 495–503.

DeLano, W. L. (2002). "The PyMOL Molecular Graphics System." DeLano Scientific, San Carlos, California.

Delgado, J., Forst, S., Harlocker, S., and Inouye, M. (1993). Identification of a phosphorylation site and functional analysis of conserved aspartic acid residues of OmpR, a transcriptional activator for *ompF* and *ompC* in *Escherichia coli*. *Mol. Microbiol.* **10,** 1037–1047.

Dutta, R., and Inouye, M. (1996). Reverse phosphotransfer from OmpR to EnvZ in a kinase$^-$/phosphatase$^+$ mutant of EnvZ (EnvZ.N347D), a bifunctional signal transducer of *Escherichia coli*. *J. Biol. Chem.* **271,** 1424–1429.

Dutta, R., and Inouye, M. (2000). GHKL, an emergent ATPase/kinase superfamily. *Trends Biochem. Sci.* **25,** 24–28.

Dutta, R., Yoshida, T., and Inouye, M. (2000). The critical role of the conserved Thr247 residue in the functioning of the osmosensor EnvZ, a histidine kinase/phosphatase, in *Escherichia coli. J. Biol. Chem.* **275,** 38645–38653.

Egger, L. A., and Inouye, M. (1997). Purification and characterization of the periplasmic domain of EnvZ osmosensor in *Escherichia coli. Biochem. Biophys. Res. Commun.* **231,** 68–72.

Forst, S., Comeau, D., Norioka, S., and Inouye, M. (1987). Localization and membrane topology of EnvZ, a protein involved in osmoregulation of OmpF and OmpC in *Escherichia coli. J. Biol. Chem.* **262,** 16433–16438.

Forst, S., Delgado, J., and Inouye, M. (1989). Phosphorylation of OmpR by the osmosensor EnvZ modulates expression of the *ompF* and *ompC* genes in *Escherichia coli. Proc. Natl. Acad. Sci. USA* **86,** 6052–6056.

Grebe, T. W., and Stock, J. B. (1999). The histidine protein kinase superfamily. *Adv. Microb. Physiol.* **41,** 139–227.

Hall, M. N., and Silhavy, T. J. (1981). Genetic analysis of the *ompB* locus in *Escherichia coli* K-12. *J. Mol. Biol.* **151,** 1–15.

Hidaka, Y., Park, H., and Inouye, M. (1997). Demonstration of dimer formation of the cytoplasmic domain of a transmembrane osmosensor protein, EnvZ, of *Escherichia coli* using Ni-histidine tag affinity chromatography. *FEBS Lett.* **400,** 238–242.

Hsing, W., Russo, F. D., Bernd, K. K., and Silhavy, T. J. (1998). Mutations that alter the kinase and phosphatase activities of the two-component sensor EnvZ. *J. Bacteriol.* **180,** 4538–4546.

Hulko, M., Berndt, F., Gruber, M., Linder, J. U., Truffault, V., Schultz, A., Martin, J., Schultz, J. E., Lupas, A. N., and Coles, M. (2006). The HAMP domain structure implies helix rotation in transmembrane signaling. *Cell* **126,** 929–940.

Igo, M. M., Ninfa, A. J., Stock, J. B., and Silhavy, T. J. (1989). Phosphorylation and dephosphorylation of a bacterial transcriptional activator by a transmembrane receptor. *Genes Dev.* **3,** 1725–1734.

Inouye, M. (2006). Signaling by transmembrane proteins shifts gears. *Cell* **126,** 829–831.

Kanamaru, K., Aiba, H., and Mizuno, T. (1990). Transmembrane signal transduction and osmoregulation in *Escherichia coli*: I. Analysis by site-directed mutagenesis of the amino acid residues involved in phosphotransfer between the two regulatory components, EnvZ and OmpR. *J. Biochem. (Tokyo)* **108,** 483–487.

Khorchid, A., Inouye, M., and Ikura, M. (2005). Structural characterization of *Escherichia coli* sensor histidine kinase EnvZ: The periplasmic C-terminal core domain is critical for homodimerization. *Biochem. J.* **385,** 255–264.

Park, H., and Inouye, M. (1997). Mutational analysis of the linker region of EnvZ, an osmosensor in *Escherichia coli. J. Bacteriol.* **179,** 4382–4390.

Park, H., Saha, S. K., and Inouye, M. (1998). Two-domain reconstitution of a functional protein histidine kinase. *Proc. Natl. Acad. Sci. USA* **95,** 6728–6732.

Parkinson, J. S., and Kofoid, E. C. (1992). Communication modules in bacterial signaling proteins. *Annu. Rev. Genet.* **26,** 71–112.

Qin, L., Dutta, R., Kurokawa, H., Ikura, M., and Inouye, M. (2000). A monomeric histidine kinase derived from EnvZ, an *Escherichia coli* osmosensor. *Mol. Microbiol.* **36,** 24–32.

Rampersaud, A., Utsumi, R., Delgado, J., Forst, S. A., and Inouye, M. (1991). Ca2(+)-enhanced phosphorylation of a chimeric protein kinase involved with bacterial signal transduction. *J. Biol. Chem.* **266,** 7633–7637.

Roberts, D. L., Bennett, D. W., and Forst, S. A. (1994). Identification of the site of phosphorylation on the osmosensor, EnvZ, of *Escherichia coli. J. Biol. Chem.* **269,** 8728–8733.

Tanaka, T., Saha, S. K., Tomomori, C., Ishima, R., Liu, D., Tong, K. I., Park, H., Dutta, R., Qin, L., Swindells, M. B., Yamazaki, T., Ono, A. M., Kainosho, M., Inouye, M., and Ikura, M. (1998). NMR structure of the histidine kinase domain of the *E. coli* osmosensor EnvZ. *Nature* **396**, 88–92.

Tomomori, C., Tanaka, T., Dutta, R., Park, H., Saha, S. K., Zhu, Y., Ishima, R., Liu, D., Tong, K. I., Kurokawa, H., Qian, H., Inouye, M., and Ikura, M. (1999). Solution structure of the homodimeric core domain of *Escherichia coli* histidine kinase EnvZ. *Nat. Struct. Biol.* **6**, 729–734.

Utsumi, R., Brissette, R. E., Rampersaud, A., Forst, S. A., Oosawa, K., and Inouye, M. (1989). Activation of bacterial porin gene expression by a chimeric signal transducer in response to aspartate. *Science* **245**, 1246–1249.

Wandersman, C., Moreno, F., and Schwartz, M. (1980). Pleiotropic mutations rendering *Escherichia coli* K-12 resistant to bacteriophage TP1. *J. Bacteriol.* **143**, 1374–1383.

Yang, Y., and Inouye, M. (1991). Intermolecular complementation between two defective mutant signal-transducing receptors of *Escherichia coli*. *Proc. Natl. Acad. Sci. USA* **88**, 11057–11061.

Yang, Y., and Inouye, M. (1993). Requirement of both kinase and phosphatase activities of an *Escherichia coli* receptor (Taz1) for ligand-dependent signal transduction. *J. Mol. Biol.* **231**, 335–342.

Zhu, Y., and Inouye, M. (2002). The role of the G2 box, a conserved motif in the histidine kinase superfamily, in modulating the function of EnvZ. *Mol. Microbiol.* **45**, 653–663.

Zhu, Y., and Inouye, M. (2004). The HAMP linker in histidine kinase dimeric receptors is critical for symmetric transmembrane signal transduction. *J. Biol. Chem.* **279**, 48152–48158.

Zhu, Y., Qin, L., Yoshida, T., and Inouye, M. (2000). Phosphatase activity of histidine kinase EnvZ without kinase catalytic domain. *Proc. Natl. Acad. Sci. USA* **97**, 7808–7813.

[9]   Light Modulation of Histidine-Kinase Activity
in Bacterial Phytochromes Monitored by Size
Exclusion Chromatography, Crosslinking,
and Limited Proteolysis

By STEFFI NOACK and TILMAN LAMPARTER

Abstract

Phytochromes are photoreceptors that have been found in plants, bacteria, and fungi. Most bacterial and fungal phytochromes are histidine kinases and, for several bacterial phytochromes, light regulation of kinase activity has been demonstrated. Typical histidine kinases are homodimeric proteins in which one subunit phosphorylates the substrate histidine residue of the other subunit; dimerization is an intrinsic property of the histidine kinase itself. Truncated phytochromes which lack the histidine kinase can also form dimers, but the interaction between subunits is modulated by light. This light-dependent dimerization can give a clue to the intramolecular signal transduction of phytochromes which modulates the histidine kinase activity. Size exclusion chromatography, limited proteolysis, and protein crosslinking can be used to study light-induced conformational changes and the interaction of subunits within the homodimer.

Introduction

The history of the early phytochrome research is closely connected to agricultural studies on seed germination, flower induction, or the greening of chloroplasts. It was found that many light responses of plants follow a common principle of red/far-red photoreversibility. These responses are induced by a pulse of red light, that is, light around 660 nm, and the induction is reverted by a subsequent pulse of far-red light, that is, light around 730 nm. It turned out that this feature is directly related to the spectral property of the photoreceptor, which was termed phytochrome (plant pigment) (Sage, 1992). Phytochrome holoproteins can exist in two spectrally different forms, called Pr and Pfr, for red- and far-red-absorbing phytochrome, respectively. The Pr and Pfr forms of plant phytochromes have an absorption maximum of approximately 665 and 730 nm, respectively. Phytochromes are synthesized as Pr, which converts into Pfr upon light absorption. Light absorption of Pfr leads to the opposite conversion into Pr. The photoconversion quantum yields are usually almost equal for both directions and are wavelength independent (Mancinelli, 1994). The rates of Pr to

METHODS IN ENZYMOLOGY, VOL. 423   0076-6879/07 $35.00

Pfr and Pfr to Pr photoconversion are therefore proportional to the light absorption of Pr and Pfr, respectively. If the sample is irradiated with red light, which is close to the absorption maximum of Pr, photoconversion leads to a high Pfr/Pr ratio. If light of longer wavelengths (far-red) is used, a low Pfr/Pr ratio is obtained. In this way, phytochromes can be switched by light between Pr and Pfr.

Each phytochrome monomer bears one bilin chromophore, which is covalently attached to a cysteine residue during an autocatalytical process. The natural chromophore differs. Plant- and cyanobacterial-phytochromes use phytochromobilin (PΦB) and phycocyanobilin (PCB), respectively, whereas other bacteria and fungi incorporate biliverdin (BV). The chromophore-binding cysteines of PΦB- and PCB-binding phytochromes lie within the GAF domain of the protein, whereas the BV binding site lies close to the N-terminus of a prototypical phytochrome (see Fig. 1). Chromophores other than the natural ones can assemble with phytochromes *in vitro* (Inomata *et al.*, 2005) and *in vivo* (Hanzawa *et al.*, 2002).

Sequence comparisons showed that the C-terminal part of plant phytochromes is homologous to bacterial histidine kinases (Schneider-Poetsch *et al.*, 1991). The upper part of Fig. 1 shows the domain arrangement of a typical plant phytochrome. Due to the homology, it was suggested that plant phytochromes might act as light-regulated histidine kinases. However, the proposed canonical substrate histidine is not conserved in plant phytochromes. Oat phytochrome A, which has the homologous histidine

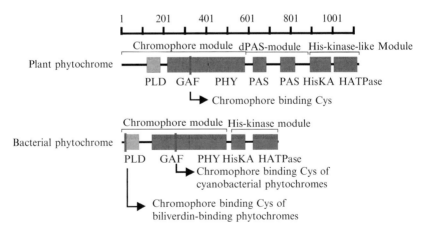

Fig. 1. Domain arrangement of a typical plant and a typical bacterial phytochrome. A ruler for the amino acid numbers is given above. Plants use PΦB as a phytochrome chromophore. The natural chromophore of cyanobacterial phytochromes such as Cph1 is PCB, other bacteria and fungi use BV. The position of chromophore-binding cysteines is indicated by the arrow.

residue, is also physiologically active after mutation of this residue (Boylan and Quail, 1996). Experiments in which deletion mutants were tested showed that the C-terminal part of plant phytochrome, which bears the histidine-kinase homologous part, is dispensable for phytochrome action (Krall and Reed, 2000). A protein in which the N-terminal part of plant phytochrome (chromophore module in Fig. 1) was fused to dimerization and nuclear translocation domains of other proteins is able to induce phytochrome effects (Matsushita *et al.*, 2003). Thus, the histidine kinase function has obviously been lost during the evolution of plant phyto-chromes. Dimerization by the histidine–kinase-like domain is, however, important for phytochrome action (Matsushita *et al.*, 2003). A Ser/Thr kinase activity has been described for several plant phytochromes (Yeh and Lagarias, 1998), and the phosphorylation status of phytochrome is important for adaptation (Ryu *et al.*, 2005).

Many bacterial phytochromes, which have been discovered several dec-ades after the plant phytochromes (Lamparter, 2004), are true histidine kinases. The lower panel of Fig. 1 shows the domain arrangement of such prototypical bacterial phytochromes. Also, some fungal phytochromes have a histidine kinase module (Blumenstein *et al.*, 2005). The cyanobacterial phytochrome Cph1 was the first bacterial phytochrome that has been char-acterized (Hughes *et al.*, 1997; Yeh *et al.*, 1997). It was shown that the histidine kinase of the activity of Cph1 is dependent on the light conditions (Yeh *et al.*, 1997). Quite interestingly, autophosphorylation of Cph1 is stronger in the Pr than in the Pfr form, a finding that contrasts with the dogma that Pfr is the only active form of phytochromes. It is as yet not clear whether the low phosphorylation activity of photoconverted Cph1 is in-duced by residual Pr or whether Pfr itself has a weak kinase activity. It was also shown that Cph1 transfers the phosphate to a cognate response regula-tor, Rcp1, which is encoded in one operon together with Cph1 (Yeh *et al.*, 1997). Further details of the signal transduction initiated by Cph1 are still unclear, since effects that are controlled by this cyanobacterial phytochrome are difficult to analyze (Fiedler *et al.*, 2004; Hübschmann *et al.*, 2005). Other bacterial phytochromes for which histidine autophosphorylation has been demonstrated are CphA and CphB from *Calothrix* SP PCC7601 (Hübschmann *et al.*, 2001), BphP1 of *Pseudomonas syringae* (Bhoo *et al.*, 2001), Agp1 (or BphP1) from *Agrobacterium tumefaciens* (Lamparter *et al.*, 2002) or BphP from *Pseudomonas aeruginosa* (Tasler *et al.*, 2005). The second phytochrome of *Agrobacterium tumefaciens*, termed BphP2 or Agp2, contains a novel type of histidine kinase; autophosphorylation has also been shown for this phytochrome (Karniol and Vierstra, 2003). In most but not all cases, the kinase activity was modulated by light. Phosphotransfer to cognate response regulators has also been shown for some of the bacterial phytochromes listed previously, but further signal transduction steps are also unknown in these cases. A fusion protein that contains the chromophore module of Cph1 and the histidine kinase of the EnvZ osmosensor has been constructed for light regulation of bacterial gene expression (Levskaya *et al.*, 2005).

The regulation of histidine kinase activity may be divided into two steps. In the first step, light absorption of the chromophore leads to protein conformational changes within the chromophore module, that is, the N-terminal part of the protein. In the second step, these changes are transmitted to the C-terminal histidine kinase. The experimental separation of photoconversion from intramolecular signal transduction is possible by using truncation fragments. The autocatalytic incorporation of the chromophore (chromophore assembly) and the photochemistry of phytochrome are unaffected if the protein is expressed without the histidine kinase (Otto *et al.*, 2003; Scheerer *et al.*, 2006; Yeh *et al.*, 1997).

Both photoconversion and intramolecular signal transduction are only partially understood at the molecular level. A structure of the chromophore binding domain (CBD) of a bacterial phytochrome gained insight into the stereochemistry of the chromophore and the protein fold of the Pr form (Wagner *et al.*, 2005). The CBD contains N-terminal amino acids including the GAF domain (see Fig. 1), but not the PHY domain, which is also part of the chromophore module. The PHY domain is required for the formation of spectrally integer Pfr. Structural information about the histidine kinase module is also lacking.

In our group, we have probed protein conformational changes of bacterial phytochromes by size exclusion chromatography, limited proteolysis, and protein crosslinking (Esteban *et al.*, 2005; Noack *et al.*, 2007). Although these techniques belong to the repertoire of classical protein biochemistry, their advantages have not been fully exploited for phytochrome protein conformational changes. By combining these techniques with mutagenesis, mass spectrometry, and autophosphorylation assays, it was possible to obtain further information about protein conformational changes and intramolecular signal transduction. A light-dependent interaction between N-terminal chromophore module subunits within the homodimer might be regarded as an intermediate step in the modulation of kinase activity. The interaction of chromophore-modules could affect the distance between both histidine kinase subunits and thereby lead to a modulation of kinase activity.

Sample Preparation

The bacterial phytochromes of these assays were always expressed as poly-histidine-tagged proteins by recombinant *E. coli* cells. Purification via Ni^{2+}-NTA affinity chromatography followed the standard procedures given by the manufacturer Qiagen and is described in several publications (Lamparter *et al.*, 1997, 2001, 2002; Scheerer *et al.*, 2006). After induction of expression, the *E. coli* cells are kept at 18 to 20° to reduce the fraction of insoluble phytochrome. At higher temperatures, a large portion of the expressed protein becomes insoluble. Following Ni^{2+}-affinity purification, the protein was further purified by preparative size exclusion chromatography in some cases. Since the phytochrome chromophore is not synthesized by *E. coli*, the protein is extracted as apoprotein. The bilin chromophore, which assembles autocatalytically with the protein, can be added at any stage of purification. In the case of cyanobacterial phytochrome Cph1, which incorporates PCB as a natural chromophore, it proved advantageous to add excess PCB directly to the extraction buffer. The tendency to form protein aggregates is reduced if Cph1 is purified as holoprotein. Coexpression of heme oxygenase, an enzyme that catalyzes the conversion of heme into BV, or heme oxygenase and PCB synthase for PCB-binding phytochromes results in the formation of holophytochrome in the cell (Bhoo *et al.*, 2001; Gambetta and Lagarias, 2001; Landgraf *et al.*, 2001; Scheerer *et al.*, 2006). Coexpression allows detection of holophytochrome directly in the host cells, which is an advantage for the screening of many clones after random mutagenesis (Fischer and Lagarias, 2004). There are, however, several restrictions of the coexpression approach: (1) It is possible that the amount of chromophore produced by the cell is not sufficient to saturate all phytochrome apoproteins. In this case, a mixture of apoprotein and holoprotein is extracted. (2) Growth of the cells is reduced, probably as a consequence of heme degradation. In addition, the relative amount of phytochrome is reduced because rather high amounts of enzymes have to be synthesized by the *E. coli* host. (3) It is not possible to compare holo- and apoprotein and to study chromophore assembly.

After purification, the buffer in which the protein is dissolved has to be exchanged. The imidazole-containing buffer used for the elution from the Ni^{2+} column is harmful for the protein and can not be used for long-term storage. A "basic buffer" with 300 m*M* NaCl, 50 m*M* Tris/Cl, 5 m*M* EDTA, pH 7.8, may be used for long-term storage of the protein at −80°, although this buffer might also have to be exchanged afterwards. Ammonium sulfate precipitation is the most convenient method for protein concentration and buffer exchange. For the precipitation of full-length proteins, 50% ammonium sulfate is used. Complete precipitation of shorter fragments is possible with 66 or 75% saturation. After the resuspension of precipitated protein, residual ammonium sulfate will remain in the sample. Desalting columns (e.g., NAP-10 columns from GE healthcare) should be used to for its removal. If the protein concentration is high enough, these columns can also be used for direct buffer exchange (without ammonium sulfate precipitation). Excess free bilin chromophores can also be removed with these columns.

Photoconversion, Experimental Light Conditions,
  Protein Concentration

Chromophore assembly of a typical phytochrome results in the formation of Pr, which remains stable in darkness. There are, however, some examples for bacterial phytochromes, including Agp2, that convert from Pr to Pfr in darkness (Giraud *et al.*, 2002; Karniol and Vierstra, 2003). In addition, phytochromes and phytochrome-like proteins with spectral properties that differ from the classical scheme have been described (Giraud *et al.*, 2005; Wu and Lagarias, 2000). The present descriptions are for phytochromes with spectrally normal Pr and Pfr forms, but might be useful in a more general way. Usually, monochromatic light provided by projectors and interference filters, by light emitting diodes (LEDs) or laser diodes (LDs), is used for photoconversion. A simple custom-built "irradiation unit" that contains an LED or an LD can be placed on top of a cuvette (Fig. 2). In most cases, photoconversion is performed until photoequilibrium is almost reached. At photoequilibrium, the Pr to Pfr and Pfr to Pr conversions are equal and no net photoconversion occurs. If the aim is to convert Pr into Pfr, the "spectral overlap" between the actinic light and the Pr form should be high, whereas the spectral overlap between the actinic light and the Pfr form should be low. The same principle with opposite signs holds for the photoconversion of Pfr to Pr. Thus, the light sources used for Pr to Pfr and Pfr to Pr conversions should have a wavelength lower than the λ_{max} of Pr and higher than the λ_{max} of Pfr, respectively. A selection of light-emitting diodes and laser diodes that are used for photoconversion is given in Table I. It should be noted that the market for these optical

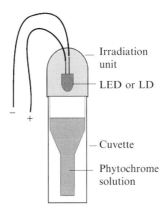

FIG. 2. Cuvette and irradiation unit. The custom-built irradiation unit can be positioned on top of the cuvette for irradiation in the photometer. The unit contains an LED or LD (see Table I), which is connected to an appropriate power supply.

TABLE I
EXAMPLES FOR LIGHT-EMITTING DIODES (LEDs) AND LASER DIODES (LDs) USED FOR
PHOTOCONVERSION OF PHYTOCHROMES

LED or LD	Type (company)	Operating currency, mA	Intensity (μmol m^{-2}s^{-1})	Emission maximum at	Photoconversion
LED	TLRH190P (Toshiba)	20	50	650 nm	Pr to Pfr of Cph1, Agp1, Agp2, and plant phytochromes
LED	QDDH73502 (Quantum Devices)	20	100	730 nm	Pfr to Pr of Cph1 and plant phytochromes
LD	RLD-78MA (Rohm)	40	100	780 nm	Pfr to Pr of Agp1 and Agp2

semiconductors fluctuates rapidly and that similar other devices might have to be chosen. Depending on the duration of the assay, dark conversion might affect the Pr/Pfr ratio significantly. In this case, it might be preferable to irradiate the sample continuously. In our studies with Agp1, which converts from Pfr to Pr in darkness, continuous irradiation was performed during protein crosslinking and limited proteolysis, but not during size exclusion chromatography and autophosphorylation assays. The cyanobacterial phytochrome Cph1 is stable in the Pr and Pfr forms.

Photoconversion should be monitored by UV/vis spectroscopy, at least until irradiation conditions are optimized. To this end, the Pr/Pfr ratio is estimated either by repetitively measuring absorption spectra or by continuously measuring the absorption at a single wavelength, preferentially the λ_{max} of Pr or Pfr. In this way, the time point where no more significant photoconversion occurs can be estimated. Theoretically, the photoequilibrium is only asymptotically approached and therefore never reached, but the small absorbance changes that occur after prolonged irradiation are negligible. It has to be considered that phytochrome itself filters actinic light and that high phytochrome concentrations may lead to a significant reduction of the light intensity when light passes through the cuvette.

The concentration of phytochrome can be judged from the absorption at 280 nm (A_{280nm}) or at the λ_{max} of Pr in the red spectral region ($A_{\lambda max,R}$). A rough estimation for the extinction coefficient $\varepsilon_{280\ nm}$ can be obtained from computer programs that are used for the management of DNA and protein sequences, such as Vector NTI (Invitrogen) or GENtle (http://en. wikibooks.org/wiki/GENtle_manual). It has to be considered that the chromophore, contaminating proteins, and DNA also absorb in that spectral region, and that the extinction coefficient might be modified by the protein fold. Experiments in which A_{280nm} of the purified apoprotein was measured in the native state and after denaturation imply that the "error" caused by the protein fold is in the range of 10%. Measurements at $A_{\lambda max,R}$ are more specific for phytochrome and will give more precise estimations on the concentration of the holoprotein, if the extinction coefficient $\varepsilon_{\lambda max,R}$ is known. Published values $\varepsilon_{max,R}$ for oat phytochrome, Cph1, and Agp1 are 130,000 (Lagarias et al., 1987), 85,000 (Lamparter et al., 2001) and 90,000 $M^{-1}cm^{-1}$ (Lamparter et al., 2002), respectively. Quite interestingly, the ratio between $A_{\varepsilon max,R}$ and A_{280nm} of the purified protein is close to unity in all three phytochromes. With the approximations that ε_{280nm} and $\varepsilon_{max,R}$ are 100,000 $M^{-1}cm^{-1}$ and that the molecular weight is 100 kDa (more accurate typical values are ~85 kDa for typical bacterial and ~125 kDa for plant phytochromes; these values can also be obtained from Vector NTI or GENtle), A_{280nm} of 1 or $A_{\varepsilon max,R}$ of 1 (measured in a 1-cm cuvette) correspond to a protein concentration of 1 mg/ml or 10 μM.

All experiments with phytochrome should be performed in a dark room. Light from blue-green LEDs (e.g., E1L52-YC1A2-03, Marl) with 505 nm emission maximum can be used to build simple safelight devices. This light will not lead to significant photoconversion of phytochrome.

Size Exclusion Chromatography

Size exclusion chromatography is used as an assay for the quaternary structure. From comparisons with marker proteins of known molecular size, a value for the "apparent molecular size" can be obtained. A direct estimation of molecular size is, however, often not possible, because the mobility is also dependent on the shape of the protein. Moreover, the mobility can be affected by an interaction with the gel matrix. To minimize possible interactions with the matrix, SEC is usually performed in the presence of 150 mM NaCl. Another restriction of SEC is that the protein becomes diluted during the assay and that only stable interactions are recognized. Despite these restrictions, SEC has been combined with other techniques to show that phytochromes form stable dimers (Jones and Quail, 1986; Kidd and Lagarias, 1990; Lamparter et al., 2001; Noack et al., 2007), that this stable dimerization is mediated by the C-terminal part of the protein, and that two chromophore module subunits interact with each other in a conformation-dependent way (Esteban et al., 2005; Noack et al., 2006; Strauss et al., 2005). Figure 3 gives an example for Agp1-M15, the chromophore module of Agp1, which was applied as apoprotein, non-irradiated, and irradiated adduct to the SEC column. The mobility differences between

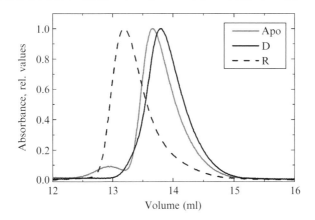

FIG. 3. Size exclusion chromatography with Agp1-M15. The elution profiles are from the apoprotein (Apo, grey line), the non-irradiated (D, black straight line), and the irradiated (R, dashed line) adducts.

the non-irradiated adduct (Pr) and the irradiated adduct (Pfr) are related to a Pfr-specific dimerization.

The minimal system for SEC requires a controllable pump, a column, an injection valve, and a UV/vis detector. We use readymade Superdex 200 10/300 FPLC columns, a "P-500" pump (both from GE Healthcare; formerly Pharmacia) and a "Multichannel Photo Detector L-3000" (Merck) with a 5-mm flow through cuvette. The analog output of the detector is coupled via an A/D converter to a computer to record the values measured continuously. Fractions are either collected manually or with the aid of a fraction collector. An injection loop of 200 μl should be chosen. The separation must be performed in a dark room.

A set of globular proteins from Sigma/Aldrich, which is specifically selected for native SEC, may serve as marker proteins. This selection comprises cytochrome C (12.4 kDa), carbonic anhydrase (29 kDa), bovine serum albumin (66 kDa), alcohol dehydrogenase (150 kDa), β-amylase (299 kDa), and apoferritin (443 kDa). The "void volume" is estimated with blue dextran (2000 kDa). Marker proteins might elute with several peaks from the column, since they can form dimers or oligomers. If degradation and contamination can be excluded, the last peak corresponds to the monomer. Since the shape of the oligomers is different from that of the monomer, the elution parameters of the oligomers cannot be used for calibration. The column (and pump) should be tested routinely with acetone, which elutes with a maximum retention time and is recognized by the ultraviolet (UV) detector.

The following protocol is for SEC with purified full-length Agp1, deletion mutants, or other mutants of Agp1; other phytochromes might be used accordingly:

1. Prepare at least 1 l running buffer (150 mM NaCl, 50 mM Tris/Cl, 5 mM EDTA, pH 7.8).

2. Filter the buffer by using sterile filters of $< 0.6 \, \mu M$ pore size.

3. Equilibrate the Superdex column with at least 100 ml buffer at a flow rate of 0.5 ml/min. If the column buffer contains ethanol, the flow rate has to be reduced to approximately 0.1 ml/min until ethanol is replaced. Then, the 0.5 ml/min flow rate should be re-adjusted. The maximum pressure of the column must not be exceeded; otherwise, the column will irreversibly be damaged. Record the absorption of the elution continuously at 280 nm with a sensitivity of 0.01 or 0.05 AUFS (absorbance units full scale). After equilibration, a stable value around zero should be obtained.

4. Centrifuge a phytochrome apoprotein stock solution at 45,000g for 30 min in order to remove insoluble material. Dilute the supernatant with running buffer to achieve a final concentration of , for example, 1mg/ml (approximately $10 \, \mu M$). For each run, approximately $200 \, \mu$l must be prepared.

5. For holoprotein assays: mix the apoprotein with BV from a 2 mM stock solution (in DMSO or methanol) to obtain a final BV concentration of $15 \, \mu M$. Incubate for 5 min or longer at $20°$. The assembly rates of phytochromes vary drastically. If another phytochrome is used, longer incubations might be required.

6. For assays with the photoproduct: irradiate the holoprotein with red light until no apparent net photoconversion occurs. Irradiation conditions have to be optimized (see preceding text).

7. Make sure that the running buffer is pumped at a flow rate of 0.5 ml/min through the column and that the baseline of the detector is stable. Set the sensitivity of the detector to 0.1 AUFS. Switch the injection valve to the "Load" position. Inject $100 \, \mu$l of the phytochrome into the $200 \, \mu$l injection loop with the use of a Hamilton syringe. For measurements of the photoproduct, the sample should be irradiated in the syringe until the sample is injected. Do not distract the syringe.

8. Switch the valve into the "Inject" position. This will start the separation. Start the recording process simultaneously (or indicate the start time).

9. Optional: collect 0.5 ml (1 min) fractions between 7 ml and 20 ml of elution. These fractions can be analyzed by SDS-PAGE or spectrophotometry.

10. The elution should be continued for 60 min before another sample is injected. Note that free BV and other bilins elute rather late due to interactions with the gel matrix.

11. Start with another sample or switch the system off. For long-term storage of the column, please refer to the manufacturer's manual.

Protein Crosslinking

Crosslinkers form covalent links between interacting proteins or protein subunits so that they do not dissociate upon SDS-denaturation. Therefore, interactions can become detectable by SDS-PAGE. The dimer formation of plant phytochrome was studied by using the crosslinkers glutaraldehyde and dimethylsuberimidate (Jones and Quail, 1986), but until recently, no other crosslinking studies in the context of phytochrome have been published. The previously mentioned limitations of SEC prompted us to use protein crosslinking as an additional tool for studies on the quaternary structure of Agp1. We could show that the full-length protein forms stable dimers and less stable oligomers, that the chromophore module Agp1-M15 also forms dimers and oligomers, that the N-terminal nine amino acids are involved in oligomer formation, and that an intramolecular crosslink between the N-terminus and the so-called PHY domain is formed (Noack *et al.*, 2007). It could also be shown by dilution series that the dimer formation of Agp1-M15 is more pronounced in the Pfr than in the Pr form.

During initial experiments, glutaraldehyde (Sigma/Aldrich), dimethylsuberimidate (Sigma/Aldrich), and bis[sulfosuccinimidyl]suberate (BS^3, Sigma/Aldrich) were tested. Full-length Agp1 was incubated with these crosslinkers and subjected to SDS-PAGE. Whereas we did not obtain crosslinking products with dimethylsuberimidate, both other agents induced the formation of covalent dimers and oligomers (Fig. 4). For routine experiments, we used glutaraldehyde as a crosslinker.

To distinguish between strong and weak interactions, protein dilution series can be carried out. Weak interactions will only be detected when the protein concentration is high, whereas strong interactions will result in the formation of crosslinking products at any protein concentration. An intramolecular crosslink should also be independent on the protein concentration. We also combined crosslinking and SEC for a clear assignment of the number of subunits that elute under each SEC peak (Noack *et al.*, 2007). Based on the positive experience with the crosslinker, we assume that it is possible to identify the (conformation-dependent) contact sites of two subunits by mass spectrometry-based methods (Pearson *et al.*, 2002). Crosslinking might also help to stabilize complexes with interacting proteins.

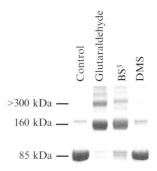

Fig. 4. Crosslinking of Agp1 (full-length holoprotein). Agp1 (final concentration 1 mg/ml) was incubated with 5 mM glutaraldehyde, 17 mM BS^3 or 35 mM DMS for 3 h at 18° and thereafter separated on a NuPage gel (Invitrogen). The apparent molecular sizes of the protein bands are indicated on the left.

Results from dilution series with Agp1-M15 are shown in Fig. 5. At lower protein concentrations, the amount of crosslinked dimers is higher for the irradiated than for the non-irradiated adduct. Thus, two subunits of this construct bind more strongly to each other in the Pfr than in the Pr form.

The following protocol is for crosslinking studies with Agp1 or fragments thereof.

1. Transfer apoprotein and holoprotein into phosphate buffer (20 mM NaP$_i$, pH 7.8); keep buffer and protein solutions at 18°.
2. Prepare a 2 mg/ml stock solution of the apoprotein and the holoprotein.
3. Pipette one dilution series with the apoprotein and two dilution series with the holoprotein; use phosphate buffer for the dilutions. The protein concentration should vary between 2 mg/ml and 0.06 mg/ml. For lower protein concentrations, more sensitive staining methods are required. Each sample should have >50 μl.
4. Prepare a fresh solution of 10 mM glutaraldehyde in phosphate buffer.
5. Photoconvert the holoprotein samples of one of the dilution series with red light (as has been described).
6. Mix 50 μl of each protein sample with 50 μl of 10 mM glutaraldehyde.
7. Keep the irradiated protein in red light.
8. Incubate all samples for 3 h at 18°.
9. Stop the reaction by the addition of 50 μl SDS sample buffer (30% glycerol, 6% SDS, 300 mM DTT, 0.01% bromphenol blue, 240 mM Tris/Cl, pH 6.8). Heat the samples for 5 min at 95° or 15 min at 60°.
10. Subject 10 to 30 μl of each sample to SDS-PAGE. Stain the gel with Coomassie.

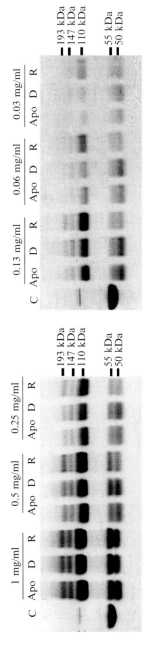

FIG. 5. Crosslinking of Agp1-M15 (chromophore module of Agp1) with glutaraldehyde. Dilution series were prepared with the apoprotein (Apo), the non-irradiated (D), and the irradiated (R) adducts. The protein concentration of each sample is given above its symbol. Samples were incubated for 3 h with 5 mM glutaraldehyde and separated on NuPage gels (Invitrogen). The apparent molecular sizes of the protein bands are indicated on the right side of each subpanel. Reprinted with permission from Biochemistry (Noack et al., 2007). Copyright 2007, American Chemical Society.

We got a better separation if the electrophoresis was performed with NuPage gels (Invitrogen) instead of custom-made standard gels (Laemmli, 1970).

Limited Proteolysis

This method has been applied to plant phytochromes to demonstrate that Pr and Pfr have a different protein conformation (Abe *et al.*, 1989; Grimm *et al.*, 1988; Lagarias and Mercurio, 1985; Park *et al.*, 2000; Schendel *et al.*, 1989). Different proteolysis patterns for Pr and Pfr were also found with the bacterial phytochromes Cph1 (Esteban *et al.*, 2005) and Agp1 (Noack *et al.*, 2007). We used both trypsin and V8 as proteases. Trypsin cleaves C-terminal of Arg and Lys residues, V8 cleaves C-terminal of Glu residues. Despite the different cleavage sites, the proteolysis pattern and the Pr/Pfr specificity of both proteases were similar. In the case of Cph1, the cleavage sites of selected fragments were determined by mass spectrometry. A trypsin cleavage site at Arg 472 was shown to be specifically protected in the Pfr form by subunit interaction within the homodimer (Esteban *et al.*, 2005). Cph1 appeared more stable to trypsin digestion in the Pfr than in the Pr form. The general rule for plant phytochromes and Agp1 is, however, that Pfr is more rapidly degraded than Pr. Chromophore-binding fragments (chromopeptides) can specifically be detected on SDS-PAGE gels by Zn^{2+}-induced fluorescence (Berkelman and Lagarias, 1986). The cleavage sites of chromophore-bearing fragments of Agp1 can be estimated from the mobility on SDS-PAGE, because the chromophore binding site Cys 20 is close to the N-terminus of Agp1 (Lamparter *et al.*, 2002). Chromopeptide fragments have lost not more than 7 or 16 N-terminal amino acids if endoproteinase Glu-C (V8) or trypsin is used, respectively. An example for a Trypsin digest

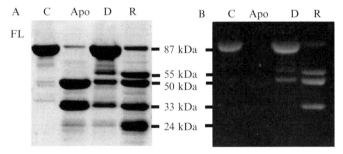

FIG. 6. Trypsin proteolysis of Agp1 (full-length). The apoprotein (Apo), the non-irradiated (D), and the irradiated (R) adducts were incubated with trypsin and the products separated on a NuPage gel (Invitrogen). The same gel was (A) stained with Coomassie and (B) used for Zn^{2+}-induced fluorescence. The apparent molecular sizes of the peptide bands are indicated between the subpanels.

of Agp1 samples is given in Fig. 6. The Zn fluorescence is shown in panel B and the Coomassie stain of the same gel in panel A. Since almost all bacterial BV-binding phytochromes have their chromophore binding site close to the N-terminus, the same principle holds for these proteins as well.

The following protocol is for proteolysis of Agp1 or fragments thereof.

1. Transfer apoprotein and holoprotein into "basic buffer" (300 mM NaCl, 50 mM Tris/Cl, 5 mM EDTA, pH 7.8); keep buffer and protein solutions at 18°.
2. Prepare 50 μl of a 2 mg/ml stock solution with the apoprotein and two 50 μl holoprotein samples.
3. For trypsin digestion: prepare a fresh 1.2 μg/ml trypsin (TPCK-treated; Sigma/Aldrich) solution in basic buffer. For V8 digestion: prepare a fresh 20 μg/ml V8 (Sigma/Aldrich) solution in basic buffer.
4. Photoconvert one of the holoprotein samples with red light.
5. Mix 50 μl of each protein sample with 50 μl of the protease solution.
6. Keep the irradiated protein in red light.
7. Incubate all samples for 30 min (for trypsin) or 3 h (for V8) at 18°.
8. Stop the reaction with 50 μl SDS sample buffer (30% glycerol, 6% SDS, 300 mM DTT, 0.01% bromphenol blue, 240 mM Tris/Cl, pH 6.8). Heat the samples for 5 min at 95° or 30 min at 60°.
9. Separate 10 μl of each sample by SDS-PAGE along with marker proteins and undigested protein. Before start of electrophoresis, add 1 mM Zn^{2+}-acetate from a 1 M aqueous stock solution to the anode buffer. During electrophoresis, the Zn^{2+} ions will migrate into the gel.
10. For Zn^{2+}-induced bilin fluorescence: transfer the SDS gel in water and place it onto the UV transilluminator of a gel documentation system. Any system that is used for the fluorescence of ethidium bromide-stained DNA on electrophoresis gels can be used. Adjust the sensitivity of the system so that the background of the gel is slightly above the dark level. Record the fluorescence of the gel. Irradiation with UV will increase the fluorescence transiently, but prolonged irradiation will lead to a decrease and finally to a loss of bilin fluorescence.
11. Stain the gel with Coomassie. The comparison of both stains shows which fragments contain the chromophore and which are free of chromophore.

Autophosphorylation

Studies on the modulation of kinase activity require the kinase itself to be measured. We adapted a protocol that has been used for the phosphorylation of Cph1 (Yeh et al., 1997). A similar method was used for the phosphorylation of plant phytochromes (Yeh and Lagarias, 1998).

The activity of the histidine kinase is estimated indirectly by measuring the covalent incorporation of radioactive phosphate from $[\gamma\text{-}^{32}P]ATP$. Phosphorylation requires Mg^{2+} or other divalent cations, the level is dependent on the incubation time and temperature, and the level might also be influenced by ions and other compounds. The following protocol is for phosphorylation of Agp1 or Cph1. The reaction should be performed in a dark room with safe light until the phosphorylation is stopped by sample buffer.

1. Prepare the following stock solutions: 500 mM Tris/Cl pH 7.8; 100 mM MgCl$_2$; 80 mM 2-mercaptoethanol; 1 M KCl; 1 mM ATP (not radioactive); 50% ethylene glycol. The phosphorylation reaction should be performed at 20°. We found it unnecessary to incubate the samples at 37° as described in Yeh et al., 1997.

2. Calculate the volumes for the master mix: Let x be the number of phosphorylation reactions that are to be performed in parallel. Increase x by 0.5 or 1. Mix 1 x μl of each stock solution, but 2 x μl of the ethylene glycol solution and add 0.3 x μl $[\gamma\text{-}^{32}P]ATP$ (300 MBq/ml). Fill up with water to achieve a final volume of 15 x μl. The concentrations in the phosphorylation reaction are 25 mM Tris, 5 mM MgCl$_2$, 4 mM 2-mercaptoethanol, 50 mM KCl, 50 $\mu$$M$ ATP, and 5% ethylene glycol. Pipette 15 μl of the master mix into reaction vials (Eppendorf tubes). It should be noted that KCl and nonradioactive ATP can be omitted.

3. Prepare 1.5 mg/ml stock solutions with the apoprotein and the holoprotein.

4. For photoconversion into Pfr, some holoprotein samples must be irradiated. Make sure that the irradiation induces complete photoconversion. Irradiation should be continued in the pipette tip (with a separate irradiation unit) until the protein is given to the phosphorylation solution.

5. Start the first phosphorylation reaction by mixing 5 μl protein into the 15 μl master mix solution. The other reactions should be started in 30 s intervals. It is important that the samples are kept in darkness during the incubation.

6. Stop each phosphorylation 5 min (in the case of Agp1) or 30 min (in the case of Cph1) after the start of the reaction by the addition of 10 μl sample buffer (30% glycerol, 6% SDS, 300 mM DTT, 0.01% bromphenol blue, 240 mM Tris/Cl, pH 6.8). It is not necessary to heat the samples.

7. Subject 10 μl of each sample to SDS-PAGE.

8. Transfer the protein from the gel onto a PVDF membrane (Millipore) with a semi-dry blot apparatus (Biorad).

9. Expose the dried membrane to an X-ray film or a phosphoimager plate. The protein on the membrane can be stained with Coomassie (in 50% methanol, the same methanol concentration is used for destaining).

After the film has been developed, the autoradiogram should be digitized with a scanner. Band intensities should be quantified with the use of appropriate computer software.

References

Abe, H., Takio, K., Titani, K., and Furuya, M. (1989). Amino-terminal amino acid sequences of pea phytochrome II fragments obtained by limited proteolysis. *Plant Cell Physiol.* **30**, 1089–1097.

Berkelman, T. R., and Lagarias, J. C. (1986). Visualization of bilin-linked peptides and proteins in polyacrylamide gels. *Anal. Biochem.* **156**, 194–201.

Bhoo, S. H., Davis, S. J., Walker, J., Karniol, B., and Vierstra, R. D. (2001). Bacteriophytochromes are photochromic histidine kinases using a biliverdin chromophore. *Nature* **414**, 776–779.

Blumenstein, A., Vienken, K., Tasler, R., Purschwitz, J., Veith, D., Frankenberg-Dinkel, N., and Fischer, R. (2005). The *Aspergillus nidulans* phytochrome FphA represses sexual development in red light. *Curr. Biol.* **15**, 1833–1835.

Boylan, M. T., and Quail, P. H. (1996). Are the phytochromes protein kinases? *Protoplasma* **195**, 12–17.

Esteban, B., Carrascal, M., Abian, J., and Lamparter, T. (2005). Light-induced conformational changes of cyanobacterial phytochrome Cph1 probed by limited proteolysis and autophosphorylation. *Biochemistry* **44**, 450–461.

Fiedler, B., Broc, D., Schubert, H., Rediger, A., Börner, T., and Wilde, A. (2004). Involvement of cyanobacterial phytochromes in growth under different light qualities and quantities. *Photochem. Photobiol.* **79**, 551–555.

Fischer, A. J., and Lagarias, J. C. (2004). Harnessing phytochrome's glowing potential. *Proc. Natl. Acad. Sci. USA* **101**, 17334–17339.

Gambetta, G. A., and Lagarias, J. C. (2001). Genetic engineering of phytochrome biosynthesis in bacteria. *Proc. Natl. Acad. Sci. USA* **98**, 10566–10571.

Giraud, E., Vuillet, L., Hannibal, L., Fardoux, J., Zappa, S., Adriano, J. M., Berthomieu, C., Bouyer, P., Pignol, D., and Verméglio, A. (2005). A new type of bacteriophytochrome acts in tandem with a classical bacteriophytochrome to control the antennae synthesis in Rhodopseudomonas palustris. *J. Biol. Chem.* **280**, 32389–32397.

Giraud, E., Fardoux, J., Fourrier, N., Hannibal, L., Genty, B., Bouyer, P., Sreyfus, B., and Vermeglio, A. (2002). Bacteriophytochrome controls photosystem synthesis in anoxygenic bacteria. *Nature* **417**, 202–205.

Grimm, R., Eckerskorn, C., Lottspeich, F., Zenger, C., and Rüdiger, W. (1988). Sequence analysis of proteolytic fragments of 124-kilodalton phytochrome from etiolated *Avena sativa* L.: Conclusions on the conformation of the native protein. *Planta* **174**, 396–401.

Hübschmann, T., Jorissen, H. J., Börner, T., Gärtner, W., and Tandeau de Marsac, N. (2001). Phosphorylation of proteins in the light-dependent signaling pathway of a filamentous cyanobacterium. *Eur. J. Biochem.* **268**, 3383–3389.

Hübschmann, T., Yamamoto, H., Gieler, T., Murata, N., and Börner, T. (2005). Red and far-red light alter the transcript profile in the *cyanobacterium Synechocystis* sp. PCC 6803: Impact of cyanobacterial phytochromes. *FEBS Lett.* **579**, 1613–1618.

Hanzawa, H., Shinomura, T., Inomata, K., Kakiuchi, T., Kinoshita, H., Wada, K., and Furuya, M. (2002). Structural requirement of bilin chromophore for the photosensory specificity of phytochromes A and B. *Proc. Natl. Acad. Sci. USA* **99**, 4725–4729.

Hughes, J., Lamparter, T., Mittmann, F., Hartmann, E., Gärtner, W., Wilde, A., and Börner, T. (1997). A prokaryotic phytochrome. *Nature* **386**, 663.

Inomata, K., Hammam, M. A. S., Kinoshita, H., Murata, Y., Khawn, H., Noack, S., Michael, N., and Lamparter, T. (2005). Sterically locked synthetic bilin derivatives and phytochrome Agp1 from *Agrobacterium tumefaciens* form photoinsensitive Pr- and Pfr-like adducts. *J. Biol. Chem.* **280**, 24491–24497.

Jones, A. M., and Quail, P. H. (1986). Quaternary structure of 124-kilodalton phytochrome from *Avena sativa* L. *Biochemistry* **25**, 2987–2995.

Karniol, B., and Vierstra, R. D. (2003). The pair of bacteriophytochromes from *Agrobacterium tumefaciens* are histidine kinases with opposing photobiological properties. *Proc. Natl. Acad. Sci. USA* **100**, 2807–2812.

Kidd, D. G., and Lagarias, J. C. (1990). Phytochrome from the green alga *Mesotaenium caldariorum*. Purification and preliminary characterization. *J. Biol. Chem.* **265**, 7029–7035.

Krall, L., and Reed, J. W. (2000). The histidine kinase-related domain participates in phytochrome B function but is dispensable. *Proc. Natl. Acad. Sci. USA.* **5**, 8169–8174.

Laemmli, U. K. (1970). Cleavage of structural proteins during the assembly of the head of bacteriophage T4. *Nature* **227**, 680–685.

Lamparter, T. (2004). Evolution of cyanobacterial and plant phytochromes. *FEBS Lett.* **573**, 1–5.

Lamparter, T., Esteban, B., and Hughes, J. (2001). Phytochrome Cph1 from the cyanobacterium *Synechocystis* PCC6803: Purification, assembly, and quaternary structure. *Eur. J. Biochem.* **268**, 4720–4730.

Lamparter, T., Michael, N., Mittmann, F., and Esteban, B. (2002). Phytochrome from *Agrobacterium tumefaciens* has unusual spectral properties and reveals an N-terminal chromophore attachment site. *Proc. Natl. Acad. Sci. USA* **99**, 11628–11633.

Lamparter, T., Mittmann, F., Gärtner, W., Börner, T., Hartmann, E., and Hughes, J. (1997). Characterization of recombinant phytochrome from the cyanobacterium *Synechocystis*. *Proc. Natl. Acad. Sci. USA* **94**, 11792–11797.

Lagarias, J. C., and Mercurio, F. M. (1985). Structure function studies on phytochrome. Identification of light-induced conformational changes in 124-kDa Avena phytochrome *in vitro. J. Biol. Chem.* **260**, 2415–2423.

Lagarias, J. C., Kelly, J. M., Cyr, K. L., and Smith, W. O. (1987). Comparative photochemical analysis of highly purified 124 kilodalton oat and rye phytochromes *in vitro. Photochem. Photobiol.* **46**, 5–13.

Landgraf, F. T., Forreiter, C., Hurtado, P. A., Lamparter, T., and Hughes, J. (2001). Recombinant holophytochrome in *Escherichia coli. FEBS Lett.* **508**, 459–462.

Levskaya, A., Chevalier, A. A., Tabor, J. J., Simpson, Z. B., Lavery, L. A., Levy, M., Davidson, E. A., Scouras, A., Ellington, A. D., Marcotte, E. M., and Voigt, C. A. (2005). Synthetic biology: Engineering *Escherichia coli* to see light. *Nature* **438**, 441–442.

Mancinelli, A. (1994). The physiology of phytochrome action. *In* "Photomorphogenesis in Plants" (R. E. Kendrick and G. H. Kronenberg, eds.), 2nd ed., pp. 211–269. Kluwer Academic Publishers, Dordrecht, Netherlands.

Matsushita, T., Mochizuki, N., and Nagatani, A. (2003). Dimers of the N-terminal domain of phytochrome B are functional in the nucleus. *Nature* **424**, 571–574.

Noack, S., Michael, N., Rosen, R., and Lamparter, T. (2006). Protein conformational changes of Agrobacterium phytochrome Agp1 during chromophore assembly and photoconversion. *Biochemistry* **46**, 4164–4176.

Otto, H., Lamparter, T., Borucki, B., Hughes, J., and Heyn, M. P. (2003). Dimerization and inter-chromophore distance of Cph1 phytochrome from Synechocystis, as monitored by fluorescence homo and hetero energy transfer. *Biochemistry* **42**, 5885–5895.

Park, C. M., Shim, J. Y., Yang, S. S., Kang, J. G., Kim, J. I., Luka, Z., and Song, P. S. (2000). Chromophore–apoprotein interactions in *Synechocystis* sp. PCC6803 phytochrome Cph1. *Biochemistry* **30**, 6349–6356.

Pearson, K. M., Pannell, L. K., and Fales, H. M. (2002). Intramolecular cross-linking experiments on cytochrome c and ribonuclease A using an isotope multiplet method. *Rapid Commun. Mass Spectr.* **16**, 149–159.

Ryu, J. S., Kim, J. I., Kunkel, T., Kim, B. C., Cho, D. S., Hong, S. H., Kim, S. H., Fernandez, A. P., Kim, Y., Alonso, J. M., Ecker, J. R., Nagy, F., *et al.* (2005). Phytochrome-specific type 5 phosphatase controls light signal flux by enhancing phytochrome stability and affinity for a signal transducer. *Cell* **120**, 395–406.

Sage, L. C. (1992). "Pigment of the Imagination—A History of Phytochrome Research." Academic Press.

Scheerer, P., Michael, N., Park, J. H., Noack, S., Förster, C., Hammam, M. A. S., Inomata, K., Choe, H. W., Lamparter, T., and Krauß, N. (2006). Crystallization and preliminary X-ray crystallographic analysis of the N-terminal photosensory module of phytochrome Agp1, a biliverdin-binding photoreceptor from Agrobacterium tumefaciens. *J. Struct. Biol.* **153**, 97–102.

Schendel, R., Tong, Z., and Rüdiger, W. (1989). Partial proteolysis of rice phytochrome: Comparison with oat phytochrome. *Z. Naturforsch. [C]*. **44**, 757–764.

Schneider-Poetsch, H. A., Braun, B., Marx, S., and Schaumburg, A. (1991). Phytochromes and bacterial sensor proteins are related by structural and functional homologies. Hypothesis on phytochrome-mediated signal transduction. *FEBS Lett.* **281**, 245–249.

Strauss, H. M., Schmieder, P., and Hughes, J. (2005). Light-dependent dimerization in the N-terminal sensory module of cyanobacterial phytochrome 1. *FEBS Lett.* **579**, 3970–3974.

Tasler, R., Moises, T., and Frankenberg-Dinkel, N. (2005). Biochemical and spectroscopic characterization of the bacterial phytochrome of *Pseudomonas aeruginosa*. *FEBS J.* **272**, 1927–1936.

Wagner, J. R., Brunzelle, J. S., Forest, K. T., and Vierstra, R. D. (2005). A light-sensing knot revealed by the structure of the chromophore-binding domain of phytochrome. *Nature* **438**, 325–331.

Wu, S. H., and Lagarias, J. C. (2000). Defining the bilin lyase domain: Lessons from the extended phytochrome superfamily. *Biochemistry* **39**, 13487–13495.

Yeh, K. C., Wu, S. H., Murphy, J. T., and Lagarias, J. C. (1997). A cyanobacterial phytochrome two-component light sensory system. *Science* **277**, 1505–1508.

Yeh, K. C., and Lagarias, J. C. (1998). Eukaryotic phytochromes: Light-regulated serine/threonine protein kinases with histidine kinase ancestry. *Proc. Natl. Acad. Sci. USA* **95**, 13976–13981.

[10] A Temperature-Sensing Histidine Kinase—Function, Genetics, and Membrane Topology

By Yvonne Braun, Angela V. Smirnova, Helge Weingart, Alexander Schenk, and Matthias S. Ullrich

Abstract

Two-component systems provide a means for bacteria to sense and adapt to environmental signals in order to survive in a continuously changing environment. Understanding of the mechanism by which these systems function is important in combating bacterial infections because many bacterial two-component systems are associated with virulence. The plant pathogenic bacterium *Pseudomonas syringae* pv. glycinea PG4180 synthesizes high levels of the phytotoxin coronatine at the virulence-promoting temperature of 18°, but not at 28°, the optimal growth temperature. Temperature-dependent coronatine biosynthesis is regulated by a modified two-component system, consisting of the response regulator, CorR, the histidine protein kinase CorS, and a third component, CorP. To elucidate the mechanism by which CorRSP functions, genetic, transcriptional, and biochemical analyses were applied, including *in vitro* and *in planta* reporter gene analysis, mRNA quantification, protein expression, mutagenesis, and membrane topology analysis. A combination of these techniques helped to elucidate, to a considerable extent, the temperature-sensing activity of CorS, which seems to act as a membrane-bound molecular thermometer.

Introduction

Bacteria use two-component regulatory systems (TCSs) to adapt cellular functions to changes of diverse environmental parameters. TCSs usually consist of a membrane-bound sensor histidine protein kinase (HPK) that perceives environmental stimuli and a response regulator (RR) that affects gene expression (Calva and Oropeza, 2006). Environmental signals, in this context, can be osmolarity, pH, light, temperature, CO_2, ammonia, oxygen, metal ions, nutrients, or any host-borne factors (Beier and Gross, 2006; Stock *et al.*, 2000). Although there are many well-characterized bacterial TCSs, very few temperature-sensing TCSs have been studied thus far in detail. Temperature sensing plays a major role in many human pathogenic bacteria where the constant body temperature of the warm-blooded host activates expression of virulence factors and several modes of actions have

METHODS IN ENZYMOLOGY, VOL. 423   0076-6879/07 $35.00

been described (Hurme and Rhen, 1998). However, the situation is fundamentally different in environmental bacteria, in which low temperature signals often induce regulatory responses (Smirnova *et al.*, 2001).

Besides the herein described CorRSP system from the plant pathogen, *Pseudomonas syringae*, there are two other well-characterized TCSs for which low temperature seems to play a major role as an environmental stimulus: DesKR of *Bacillus subtilis* (Hunger *et al.*, 2004) and Hik33/Hik19/ Rer1 of *Synechocystis* sp. PCC 6803 (Suzuki *et al.*, 2001).

The DesKR system of *Bacillus subtilis*, in which the membrane-bound HPK, DesK, interacts with its cognate RR, DesR, regulates expression of *des* coding for $\Delta 5$ acyl lipid desaturase upon cold shock (Aguilar *et al.*, 2001; Albanesi *et al.*, 2004; Beckering *et al.*, 2002; Mansilla *et al.*, 2005). In contrast to heat shock proteins, which include chaperones and proteases and which help to cope with the detrimental effects of high temperature on proteins, cold-induced proteins are often involved in keeping up cellular functions, such as transcription, translation, and recombination, as well as in maintaining metabolism or the integrity of the cellular membrane (Phadtare *et al.*, 2000). Besides *de novo* synthesis of unsaturated fatty acids, *in situ* desaturation of membrane lipids serves *B. subtilis* cells to maintain the fluidity of the membrane upon a temperature downshift. DesK features four N-terminal transmembrane domains (TMDs) and a long cytosolic C-terminal domain, which contains a conserved His residue and the kinase domain (Albanesi *et al.*, 2004). Hunger *et al.* (2004) provided evidence that the membrane domain of DesK is the temperature-sensing part of the enzyme. The cytoplasmic kinase domain was able to activate transcription of *des* independent of temperature in a mutant lacking the N-terminus of DesK. How DesK senses the temperature signal has been proposed as follows: Decreased membrane fluidity caused by temperature downshift gives preference to a kinase-dominant state of DesK. DesK-mediated phosphorylation of DesR enables dimer formation and annealing to the promoter of *des* (Cybulski *et al.*, 2004; Mansilla and de Mendoza, 2005). The subsequent increase in membrane fluidity favors a phosphatase-dominant state of DesK which, in turn, results in dephosphorylation of DesR and decrease of *des* transcription.

The cyanobacterium *Synechocystis* sp. PCC 6803 adapts its membrane fluidity to low temperature by desaturation of membrane fatty acids and possesses four genes encoding acyl-lipid desaturases, designated *desA*, *desB*, *desC*, and *desD* (Los *et al.*, 1997). Gene *desC* is expressed constitutively, whereas transcription of *desA*, *desB*, and *desD* is induced after temperature downshift. Two HPKs controling desaturase gene expression, Hik33 and Hik19, were identified in the course of a genomewide mutagenesis approach of all HPKs of PCC 6803 (Suzuki *et al.*, 2000). Subsequent analysis revealed that induction of *desB* and *desD* but not of *desA* after temperature downshift was reduced in ΔHik19 and ΔHik33 mutants (Suzuki *et al.*, 2001). It was further demonstrated that Hik33 responds to osmotic stress and regulates a number of osmotic stress-induced genes (Mikami *et al.*, 2002). Hik33 comprises a typical HPK structure with two N-terminal TMDs and a classical C-terminal transmitter domain whereas Hik19 might be a cytosolic protein possessing a conserved histidine kinase domain and two signal receiver domains (Suzuki *et al.*, 2000). Consequently, Hik19 was considered a hybrid-type histidine kinase that might accept a phosphate group from and act downstream of Hik33. A hypothetical pathway for perception and transduction of low-temperature signals involves Hik33 as a membrane-embedded "cellular thermometer." Upon temperature-downshift and subsequent decrease of membrane fluidity, Hik33 autophos-phorylates and transfers its phosphate group to Hik19, and finally to the cognate RR, Rer1, which, in turn, regulate transcription of *desBD* genes (Suzuki *et al.*, 2000).

The plant pathogenic bacterium *Pseudomonas syringae* pv. glycinea PG4180 is the causal agent of bacterial blight in soybean plants. Due to its global occurrence and high losses in soybean cultivation, it causes an economically important disease (Wrather *et al.*, 1997). Phytopathogenic bacteria have evolved a number of mechanisms to colonize their host-plants and to evade detection through the plant defense system. Such mechanisms include, for example, production of exopolysaccharides, effec-tor proteins, cell-wall-degrading enzymes, and phytotoxins. Several patho-vars of *P. syringae* synthesize the nonhost specific phytotoxin coronatine (COR) to enhance virulence (Budde and Ullrich, 2000; Mittal and Davis, 1995). According to structural analyses and plant response studies, COR is believed to mimic plant signaling molecules such as methyl jasmonate (Palmer and Bender, 1995). Moreover, COR induces chlorosis, hypertro-phy, shrinkage of chloroplasts, and the synthesis of ethylene and proteinase inhibitors (Bender *et al.*, 1999). Structurally, COR consists of a polyketide component, coronafacic acid (CFA), which is coupled via amide bond formation to coronamic acid (CMA), an ethylcyclopropyl amino acid de-rivative. PG4180 synthesizes coronatine in a temperature-dependent man-ner, with maximum yields at 18° (Ullrich *et al.*, 1995). At 28°, the optimal growth temperature of *P. syringae*, COR biosynthesis is negligible. The COR biosynthetic and regulatory genes were shown to be encoded on a 90-kb indigenous plasmid in PG4180 (Bender *et al.*, 1993). Two biosynthetic regions, corresponding to CFA and CMA biosynthesis, are separated by a 3.4-kb regulatory region encoding a modified TCS. This TCS, consisting of the RR, CorR, the sensor kinase CorS, and a third component, CorP, was shown to regulate COR biosynthesis (Ullrich *et al.*, 1995). CorR was demonstrated to bind to the promoter regions of the *cma* operon in a temperature-dependent manner (Wang *et al.*, 1999). Despite its high degree of similarity to CorR, CorP lacks a typical DNA-binding motif, suggesting that it might modulate the function of CorR rather than bind to target DNA. Rangaswamy and Bender (2000) demonstrated that CorS transfers its phosphoryl group to CorR but not to CorP, which correlates to the presence of a receiver aspartate residue in the former but not the latter protein. These previous investigations had demonstrated that CorRSP is necessary to regulate COR gene expression in dependence of temperature. However, the mechanism by which CorS senses temperature changes and responds accordingly remained to be elucidated.

Genetic Approaches to Characterize CorRSP

Identification and Mapping of the COR Gene Cluster

Using transposon mutagenesis and Southern hybridization, Bender *et al.* (1991, 1993) demonstrated that COR biosynthetic genes are located in a 32-kb region on a 90-kb indigenous plasmid, designated p4180A in *P. syringae* pv. glycinea PG4180 (Fig. 1). Mobilization of p4180A into two COR-negative strains of *P. syringae* pv. syringae enabled these strains to synthesize COR. This experiment demonstrated that p4180A contains all genes necessary to synthesize COR. Subsequently, mapping of operons necessary for synthesis of the COR precursors, CFA and CMA, was conducted by saturated Tn5 mutagenesis and complementation analysis of mutants blocked in distinct biosynthetic steps, using external feeding of either CMA or CFA (Bender *et al.*, 1993; Ullrich and Bender, 1994;

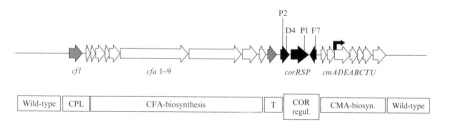

FIG. 1. The COR biosynthetic gene cluster of *P. syringae* pv. glycinea PG4180. Genes encoding the regulatory two-component system are indicated by black arrows, and white arrows indicate the two biosynthetic gene clusters, respectively. Two additional genes, cfl coding for coronafacate ligase and a transposase (T), are indicated in grey. The *cmaA* promoter is marked with a black arrow above the genetic map. Vertical bars indicate the location of antibiotics resistance cassette insertions in selected mutants.

Ullrich *et al.*, 1994; Young *et al.*, 1992). For this, fragments of p4180A obtained by restriction digests were cloned into the broad-host range vector pRK415, which is able to replicate in *P. syringae* (Keen *et al.*, 1988). The resulting plasmids were mutagenized with λ::Tn5 in E. *coli*, as described by de Bruijn and Lupski (1984). Derivatives of pRK415 containing single Tn5 insertions were mobilized into PG4180 by conjugation as described by Figurski and Helinski (1979) and recombined into plasmid p4180A. The presence of a mutagenized p4180A was subsequently verified using plasmid isolation, electrophoreses, and Southern blots. To characterize the mutants' phenotypes, biosynthesis of CFA, CMA, and COR was quantified using two different Reverse Phase-HPLC (RP-HPLC) approaches (Palmer and Bender, 1993; Ullrich *et al.*, 1994). The majority of the CMA biosynthetic region was found to be transcribed as a single operon, designated *cmaABT* (Ullrich and Bender, 1994).

The second biosynthetic region in the COR gene cluster comprises genes required for the production of CFA and the *cfl* gene, encoding coronafacate ligase (Fig. 1). Along with the genetic characterization of the CMA operon, the results for the CFA region allowed experimenters to set up a model of how COR is synthesized (Penfold *et al.*, 1996; Rangaswamy *et al.*, 1997). It was demonstrated that *cfl* and the CFA biosynthetic gene cluster are transcribed into one polycistronic mRNA and this transcription is directed by the *cfl* promoter (Liyanage *et al.*, 1995a,b).

Mutational Analysis of Genes Encoding corRSP

A series of PG4180 mutants, designated PG4180.D4 (Bender *et al.*, 1993), PG4180.P1 (Ullrich *et al.*, 1995), and PG4180.P2 (Peñaloza-Vázquez *et al.*, 1996), were found to lack production of CFA and CMA and could only be complemented for COR production by feeding both substances. Mutations in these three mutants were located between the gene clusters for CFA and CMA biosynthesis, respectively, suggesting that regulatory genes were affected. Introduction of plasmid pRGMU1 containing the biosynthetic *cmaABT* promoter fused to a promoterless β-glucuronidase (*uidA*) gene resulted in negligible reporter gene activity at 18 and 28°, demonstrating that the mutants were defective in transcriptional activation of *cmaABT*. The same plasmid introduced to PG4180 wild type resulted in a temperature-dependent reporter gene expression with ten-fold higher levels at 18 as compared to 28°. Several DNA fragments of the COR cluster were shown to complement the mutants' phenotype, suggesting that a 3.4-kb region was sufficient to restore wild type-level COR synthesis in these mutants. Nucleotide sequencing of the fragment revealed three open reading frames (ORF) coding for a modified two-component system, which consists of a sensor kinase (*corS*) and two putative response regulators (*corR* and *corP*).

In contrast to PG4180, *P. syringae* pv. tomato strain DC3000 synthesizes COR in a temperature-independent manner *in vitro* (Weingart *et al.*, 2004). A *corS*-negative mutant was constructed in DC3000 applying marker-exchange mutagenesis. Analysis of this mutant, DC3000.M1, revealed that the mutant was not able to produce COR. Reporter gene analysis using reporter plasmids in which β-glucuronidase was either fused to the *cma* promoter region from PG4180 or DC3000 showed no expression in DC3000.M1. In contrast to PG4180 where the COR biosynthetic gene cluster is plasmid-encoded, the DC3000 COR gene cluster is encoded on two separate entities in the chromosome (Brooks *et al.*, 2004). Despite very high nucleotide sequence similarities to *corRSP* of PG4180, the regulatory region comprising *corRSP* of DC3000 is flanking genes required for CMA biosynthesis but is separated by a 26-kb genomic region from the CFA biosynthetic gene cluster.

Transcriptional Analysis

To investigate the expression of COR biosynthetic genes in dependence of temperature *in vitro* and *in planta*, reporter gene analyses were performed making use of β-glucuronidase and EGFP as reporter enzymes.

Quantitative Analysis of Reporter Gene Activity: GUS-Assay

The promoter region of the *cmaABT* operon of PG4180 was identified by fusion of a 3.1-kb β-*PstI*-fragment to a promoterless glucuronidase gene (Ullrich and Bender, 1994). The resulting plasmid, designated pRGMU1, was introduced to PG4180 and its mutants via tri-parental matings. The *cfl* promoter region, driving transcription of the *cfl*/CFA operon, was localized to a 0.37-kb region by fusion to the promoterless β-glucuronidase gene as well (Liyanage *et al.*, 1995a). The reporter enzyme, β-glucuronidase, hydrolyzes the nonfluorescent substrate 4-methylumbelliferyl-β-D-glucuronid (MUG) to release the fluorescent 4-methylumbelliferone along with glucuronic acid. The quantification of the released fluorophore is used to determine promoter activity of the respectively fused gene. Herein, a high-throughput assay is described, which does not require cell lyses and which was optimized on the basis of assays developed by Jefferson *et al.* (1986) and Vidal-Aroca *et al.* (2006). Bacteria were cultured in Hointink-Sinden medium optimized for COR synthesis (HSC) (Palmer and Bender, 1993) at 18 and 28° until an OD_{600} of 1.5 to 2.0 was reached. One and a half ml of the cultures were pelleted and cells were resuspended in 500 μl of

GUS extraction buffer (50 mM Na$_2$HPO$_4$, pH 7.0; 10 mM EDTA; 0.1% N-Lauroylsarcosyl-sodiumsalt; 0.1% Triton; 0.07% β-mercaptoethanol). For additional dilution necessary to prevent saturation of the enzymatic reaction, 20 μl of the bacterial suspension were transferred into wells of a 96-well-microtiter plate containing 80 μl of GUS-extraction buffer. Subsequently, 25 μl of MUG substrate (1 mg/mL in GUS-extraction buffer) were added to the wells and mixed. After incubation at 37° for 10 min, the reaction was stopped by addition of 30 μl 1 M Na$_2$CO$_3$. Subsequently, the fluorescence generated by β-glucuronidase-dependent MUG hydrolysis was quantified in a Fluorolite-100 microplate reader (Dynatech Laboratories, Chantilly, VA) set to an excitation wavelength of 390 nm and an emission wavelength at 405 nm. For background determination, the assay was applied to a cellfree culture medium sample. The determined fluorescence corresponded to the amount of β-glucuronidase present in the assay. The enzymatic activity in units [U] was calculated as shown in Eq. (1) where $F_{390/405}$, t, and A_{600} represent sample fluorescence at the end of the reaction, time of reaction in minutes, and absorbance of the cell suspension.

$$\frac{F390 \div 405}{t \times A600} \tag{1}$$

When introduced to PG4180, the reporter gene fusion was used to show that the *cmaABT* promoter exhibits a temperature-dependent transcriptional activity (Budde *et al.*, 1998; Ullrich *et al.*, 1994). The plasmid was further introduced to *P. syringae* strains of different geographic origin to determine whether the *cmaABT* promoter is temperature-dependently expressed in these bacteria (Rohde *et al.*, 1998). Interestingly, that study revealed that there was no correlation between the degree of temperature responsiveness and the geographic origin of the isolated bacterial strains. The *cflCFA* promoter of PG4180 fused to *uidA* also showed a temperature-responsive activation and exhibited maximal expression at 18°. The enzyme, β-glucuronidase, is a stable, easy-to-detect reporter enzyme. This attribute, together with the described high-throughput assay saving elaborative cell lysis steps, make the GUS assay a convenient tool for the analysis of gene expression.

In Vitro *and* In Planta *Expression of COR Genes Using EGFP*

In vitro analysis using β-glucuronidase as reporter gene had clearly demonstrated that expression of COR biosynthetic genes is strictly temperature dependent in PG4180. However, the related strain, DC3000, exhibits a temperature-independent phenotype in this regard. Since both organisms are plant pathogenic, the next goal was to investigate the influence of temperature on the expression of COR biosynthetic genes *in planta*. A well-established reporter gene in this respect is *gfp* encoding for green fluorescent protein (GFP) (Chalfie *et al.*, 1994). Enhanced green fluorescent protein, EGFP, a red-shifted variant of GFP showing only minor differences in folding efficiency compared to GFP (Patterson *et al.*, 1997) was chosen as a reporter for the transcriptional analysis of COR gene expression *in vitro* and *in planta*. For this purpose, plasmid pHW01 containing a transcriptional fusion of the *cmaABT* promoter region of PG4180 to a promoterless *egfp* gene in the broad-host range vector pBBR1MCS was constructed (Weingart *et al.*, 2004). The reporter plasmid was introduced to PG4180 and DC3000 by tri-parental matings. The transcriptional activity of the *cmaABT* promoter was determined by measuring EGFP fluorescence of bacteria harvested at different optical densities (*in vitro*) and after re-isolation from infected leaf material (*in planta*), using either a fluorimeter or confocal laser scanning microscopy (CLSM). The analyses revealed that transcriptional activity of the *cmaABT* promoter in PG4180 is temperature dependent *in vitro* as well as *in planta*. A significantly stronger fluorescence was observed at 18 compared to 28° in both experiments. Maximal fluorescence was reached in the early stationary phase and remained constant until it declined in the late stationary phase. Fluorescence at 28° remained low regardless of the growth stage of the bacteria. When introduced to DC3000, the reporter plasmid revealed that transcription of COR biosynthetic genes *in vitro* is low at 18 as well as 28°. However, DC3000 (pHW01) showed an intensity of fluorescence which was three- to fourfold higher *in planta* than *in vitro*, indicating a clear plant-inducibility of COR biosynthetic genes in the latter strain. These results showed that COR biosynthesis is differently regulated in these two pathogens, possibly reflecting their individual adaptations to the respective host plants, that is, soybean and tomato, respectively.

Two factors contribute to the yield of fluorescence of EGFP: (1) its folding efficiency and (2) the fact that a certain threshold number of EGFP molecules per cell is required to detect its fluorescence by CLSM. These factors may limit the detection of fluorescence in cells harboring *egfp* transcriptional fusion. However, EGFP is very stable in bacteria and can be detected after several days (Ishii *et al.*, 2002). This fact should be considered when interpreting results obtained using EGFP as reporter protein, as fluorescent signals may be still detected once the analyzed promoter has been inactivated. Provided that re-isolation of bacterial cells from infected plant material is successful, EGFP can be used as a reliable reporter enzyme for analysis of bacterial gene expression *in planta*.

Determination of the Expression Levels of COR Biosynthetic Genes

Reporter gene analyses provide a rather indirect approach to determine gene expression levels since the measured promoter activity triggers synthesis and, consequently, enzymatic activity of the reporter protein. Furthermore, reporter gene constructs are often located on plasmids and thus do not necessarily represent the actual cellular situation, that is, the copy number and *in trans* effects. Aside from this, the antibiotics resistance markers encoded by the plasmid may impact bacterial growth. Thus, in order to investigate the thermoresponsive expression of COR biosynthetic genes directly, the respective promoter activity was studied at the transcriptional level. One option for doing so is the Northern blot technique, in which radioactively or fluorescently labeled DNA or RNA probes of the gene of interest are hybridized to total RNA. This method was applied by Budde *et al.* (1998) to demonstrate that transcription of the *cmaABT* operon is temperature-dependent in PG4180, with high levels of expression at 18° and 75% less at 28°. Furthermore, a significant decrease of *cmaABT* mRNA levels during stationary phase was demonstrated. When analyzing a bacterial operon of genes such as *cmaABT*, annealing of the probe to partially degraded mRNA molecules often results in a smear of the obtained signal, complicating the conclusions drawn.

Weingart *et al.* (2004) aimed to investigate the expression level of COR biosynthetic genes *in vitro* and after re-isolation of the bacterial RNA from infected leaf tissue, thus reflecting *in planta* expression. Re-isolation of *Pseudomonas* mRNA from infected leaf tissue using common techniques, such as commercially available RNA extraction kits, was unsuccessful. Therefore, a modified acid phenol/chloroform extraction was developed by Schenk *et al.* (submitted to *Molecular Plant Pathology*). Re-isolation of bacterial RNA from infected plant tissue using this novel extraction method was successful with high quality, as determined spectrophotometrically and by analyzing total RNA using a RNA 6000 Nano LabChip® Kit on the Agilent 2100 Bioanalyzer (Agilent Technologies, Palo Alto, CA). Samples contained a mixture of RNAs from the host plant and the inoculated bacterial cells. As expected, the bacterial RNA composed only 1 to 5% of total RNA isolated. However, this amount was still sufficient to conduct Northern blot experiments successfully. These RNA samples could also be successfully analyzed using a Spot blot technique, allowing for high throughput sampling. The method was used to confirm that *in vitro* expression of COR biosynthetic genes is highest in the early exponential phase and declines toward the stationary phase in PG4180 as well as in DC3000 (Weingart *et al.*, 2004). Moreover, it was shown for the first time that expression of *cmaABT* in PG4180 is also strongly thermoresponsive inside the infected plant tissue. Interestingly, and in line with our previous observations, expression of *cmaABT* in strain DC3000 did not show temperature dependence but was highly induced *in planta*. This substantiated that regulation of COR biosynthetic genes differs significantly between the two tested strains representing two distinct pathovars of *P. syringae*. The Spot blot technique was also used to analyze expression of the histidine protein kinase gene, *corS*, with regard to temperature, revealing that this gene is constitutively expressed at low levels. This result indicated that a certain threshold level of CorS may always be present in cells of PG4180. Once the temperature signal triggers activation of CorS, it may induce activity of the cognate DNA-binding protein, CorR, and this might consequently lead to expression of COR biosynthetic genes. One general disadvantage of the Spot blot technique is that the specificity of the signal needs to be verified by at least one alternative method, such as Northern blots. In the latter case, RNA molecules are separated by agarose gel electrophoresis and specific binding of the probe to the mRNA of interest can be visualized.

Biochemical Characterization of CorRSP

Overexpression and Enzymatic Analysis of the Response Regulator CorR

By means of mutagenesis, it had been demonstrated that the two putative RRs, CorR and CorP, of PG4180 were essential for the function of the TCS, CorRSP (Bender *et al.*, 1993; Peñaloza-Vázquez and Bender, 1998; Ullrich *et al.*, 1995). Translational fusions of CorR and CorP, respectively, to maltose-binding protein (MBP) were generated and overexpressed in *E. coli* and the native host, PG4180, to investigate their ability to complement *corR* and *corP* mutants of PG4180 and to study their DNA binding (Peñaloza-Vázquez *et al.*, 1996; Wang *et al.*, 1999).

For overexpression of CorR, plasmid pMal-c2 containing the *malE* gene encoding MBP under control of the *tac* promoter (P_{tac}) was first fused to the broad-host range plasmid pRK415. The obtained chimeric plasmid, pRK-Mal-c2, was used to express MBP in *E. coli* and *P. syringae* strains (Peñaloza-Vázquez *et al.*, 1996). A 0.63-kb DNA fragment containing the *corR* gene was then cloned into pRK-Mal-c2 and used for expression of MBP-CorR. For expression of CorP, the *corP* gene was PCR-amplified and cloned into pMal-c2. A 2.1-kb fragment containing the *malE::corP* fusion under P_{tac} control was subcloned in the broad-host range vector pBBR1MCS. Use of two different vectors was meant to eventually coexpress CorR-MPB and CorP-MPB in the same host cells, since any interaction or concerted function of CorR and CorP was to be investigated. Both chimeric plasmids were mobilized into PG4180 wild type, PG4180.P2 (*corR*), PG4180.F7 (*corP*), and PG4180.D4 (*corS*) by tri-parental mating (Wang *et al.*, 1999). Trans-conjugants of PG4180.P2 and PG4180.F7 showed complementation of COR synthesis while the respective trans-conjugant of PG4180.D4 did not, suggesting that MBP-CorR and MBP-CorP were functional.

Fusion protein overproduction in *E. coli* (at 18, 28, and 37°) and in *P. syringae* (at 18 and 28°) was induced using isopropyl-b-D-thiogalactopyranoside (IPTG) at an OD_{600} of 0.45 to 0.5. Bacteria were further incubated for additional 4 to 24 h or 24 to 48 h, respectively. Following centrifugation, cell pellets were resuspended in protein extraction buffer (50 mM TRIS-HCl, pH 7.4, 200 mM NaCl, 1 mM EDTA, 5 mM DTT, 1 mM magnesium acetate, 50 μg/ml lysozyme, and 2 μg/ml DNase), and incubated on ice for 30 min. Cells lysates were obtained by four passages through a French pressure cell. MBP and MBP fusions were purified from cell lysates by affinity chromatography on amylose columns, as recommended by the manufacturer (New England Biolabs, Schwalbach, Germany). Fusion of an RR protein to MBP for overexpression had the advantage that purification became simple and highly efficient. Additionally, the fusion protein might have been more stable than the native protein. As outlined later, expression of MBP-CorR in the native organism, PG4180, was necessary since MBP-CorR from *E. coli* did not show the expected DNA binding activity. One possible explanation for this might be misfolding of the RR in foreign host cells.

In order to test the function of CorR and CorP as potential DNA binding proteins, gel retardation experiments were conducted (Wang *et al.*, 1999). For this study, the 3′-ends of several DNA fragments composing the potential promoter region of *cmaABT* were labeled with digoxigenin-11-ddUTP (DIG) (Boehringer-Mannheim, Mannheim, Germany). The aim of using several DNA fragments was to identify the actual binding sites for CorR and CorP. The two fusion proteins were expressed in *E. coli* and PG4180, respectively. DIG-labeled DNA fragments were incubated with poly (dI-dC) and purified MBP-CorR, MBP-CorP, or a mixture of both fusion proteins at 0, 18, 28, or 37° for 30 min. After separation on polyacrylamide gels, samples were electroblotted onto positively charged nylon membranes (Boehringer-Mannheim), and fixed by ultraviolet (UV) cross-linking. Finally, DNA fragments were visualized using anti-DIG/alkaline phosphatase antibody conjugates and chemo-luminescent detection, as recommended by the manufacturer (Boehringer-Mannheim). MBP-CorR and MBP-CorP expressed in *E. coli* did not show any DNA binding activity regardless of the DNA target used, suggesting that both fusion proteins were inactive. However, MBP-CoR expressed in PG4180 at 18° specifically bound to a 218-bp fragment located upstream of the transcriptional start site of *cmaABT*, indicating a DNA-binding activity of CorR. Surprisingly, no DNA binding was observed when MBP-CorR was expressed at 28°. The temperature at which the actual DNA binding assay was conducted had no influence on binding efficiency of CorR. In contrast, MBP-CorP did not show any DNA binding activity under any of the tested conditions. Interestingly, MBP-CorR lost its DNA binding capability at 18° if it was expressed in the *corS* mutant, PG4180.D4 (Wang *et al.*, 1999). All attempts to co-incubate MBP-CorR and MBP-CorP resulted in the same DNA binding efficiency, suggesting that a direct interaction of CorR with CorP during target DNA binding is unlikely. MBP alone did not bind to any DNA fragment. Very similar results could be obtained for MBP-CorR and its interaction with the *cfl/CFA* promoter region (Peñaloza-Vázquez and Bender, 1998). In summary, MBP-CorR interacted with its target DNA sequences when synthesized at 18° and in presence of functional CorS, substantiating the temperature-dependent function of the CorRSP system. The actual role of CorP remained to be elucidated.

Expression of the Histidine Protein Kinase CorS

First of all, polyclonal antibodies specific to CorS had to be generated to allow for immunological detection of this enzyme using Western blot technique. Due to its membrane association (see the following text) and despite many different attempts, it was not possible to produce sufficient amounts of purified CorS for this purpose. Consequently, CorS peptide-specific polyclonal antibodies were generated. To choose the appropriate peptide sequences for commercial *in vitro* synthesis (Eurogentec, Hestal, Belgium), the following aspects had to be considered: Peptide sequences should neither be derived from the membrane-spanning domains (low accessibility) nor should they represent sequences conserved in other HPKs (low specificity). Additionally, they should exhibit high antigenicity and sufficient surface probability as determined by the program PROTEAN (Lasergene, Madison, WI). Two respective peptide sequences, NH$_2$-CRATNSQRARQ LAI-COOH and NH$_2$-CLQQLDRRARKTTED-COOH, within a linker region connecting the last of six TMDs and the conserved Histidine (H) box were chosen for synthesis and subsequent generation of polyclonal rabbit anti-CorS antibodies. As expected, due to its regulatory function, CorS abundance was found to be very low in *P. syringae* as determined by Western blot analysis. Consequently, CorS was expressed in *E. coli* using a StrepTag overexpression system with a tightly controlled *tetA*-promoter and translational fusion of CorS to a Strep tag II peptide (Degenkolb *et al.*, 1991; Skerra, 1994). The subsequent purification procedure was based on specific binding of the *Strep*-tag II peptide to immobilized StrepTactin (Schmidt *et al.*, 1996; Skerra and Schmidt, 2000). Specific competition with desthiobiotin enables elution of the recombinant protein from StrepTactin purification columns. To generate the construct, *corS* was amplified by PCR using oligonucleotides fcsB (5′-GAC*GGAT-CC*GTGACTCATTCTTACGAACTC-3′) and rcsB (5′-GAC*GGATCCG*-CTCTCACCGGCCTGACCAGG-3′) (*Bam*HI sites underlined). The amplified DNA fragment was cloned into *Bam*HI-treated vectors pASK-IBA3 and pASK-IBA7 (IBA, Göttingen, Germany). Cloning in pASK-IBA3 generated a translational fusion protein with CorS carrying a C-terminal *Strep*-tag II fusion. Respectively cloning in pASK-IBA7 generated a translational fusion protein with CorS carrying an N-terminal *Strep*-tag II fusion. Cells of *E. coli* were transformed with the resulting plasmids. Synthesis of the fusion protein was induced by addition of 3 mg of anhydrotetracycline per liter at an OD_{600} of 0.5 and 28°.

To isolate membrane proteins, cells were harvested by centrifugation at 7000 rpm at 4°. Pellets were resuspended in 10 ml of extraction buffer (50 mM Tris/HCl [pH 8.0], 10% sucrose, 0.2 mM dithiothreitol (DTT), 10 mg/ml of lysozyme, 0.2 mg/ml RNase A, and 0.2 mg/ml DNase) and incubated on ice for 1 h to generate spheroplasts. Spheroplasts were pelleted at 7000 rpm at 4° and subsequently resuspended in 50 mM Tris/HCl (pH 8.0) containing 0.2 mM DTT and 0.1 mM of the protease inhibitor AEBSF. Spheroplasts were then broken up using a French press apparatus. Intact spheroplasts were removed by an additional centrifugation step for 10 min at 8000 rpm at 4°. The membrane fraction was harvested by centrifugation for 2 h at 35,000 rpm at 4°. The pellet containing the membrane fraction was resuspended in 1 ml of 50 mM Tris/ HCl (pH 8.0) containing 0.1 mM AEBSF and stored at −80°. To solubilize CorS$_{Strep\text{-}tag}$ from the membrane, the pellet was resuspended in 800 μl solubilization buffer (62.5 mM Tris/HCl (pH 8.0), 12.5% glycerol, 1.25 mM EDTA, 2.5 mM β-mercaptoethanol). While stirring at 4°, 200 μl of TritonX-100 were added stepwise (10 × 20 μl). After stirring on ice for 1 h, the solubilizate was centrifuged for 30 min at 30,000 rpm. The supernatant was subsequently transferred to a fresh tube. The pellet was again resuspended in 800 μl of solubilization buffer and the solubilization step was repeated. Supernatants containing the solubilized CorS$_{Strep\text{-}tag}$ were combined and the fusion protein was further purified by StrepTactin affinity chromatography, according to the manufacturer's recommendation (IBA). The level of expression of CorS$_{Strep\text{-}tag}$ could be determined by Western blot analysis with streptavidin alkaline phosphatase antibody conjugates (Amersham-Pharmacia Biotech, Freiburg, Germany).

CorS$_{Strep\text{-}tag}$ could be detected in the membrane fraction in low amounts demonstrating membrane incorporation of the recombinant protein in *E. coli*. Possibly due to binding of biotin to the applied streptavidin antibody conjugates, some biotin-rich endogenous proteins of *E. coli* were detected along with CorS$_{Strep-tag}$. The low abundance of CorS might have been either due to toxic effects of this membrane protein in *E. coli* cells or due to losses during the solubilization process. We used TritonX-100 as it was success-fully applied to solubilize overexpressed membrane proteins in previous studies (Rübenhagen *et al.*, 2000; Tamai *et al.*, 1997). In future experi-ments, alternative detergents should be tested to improve the yield of the recombinant CorS protein.

Since expression of CorS was successfully achieved in *E. coli*, we next aimed to express this membrane protein in its native organism, *P. syringae*, in order to analyze its membrane topology at 18 and 28°. Recombinant proteins often do not undergo correct folding in foreign cellular back-grounds. To introduce the CorS$_{Strep-tag}$ construct to PG4180, the broad-host range vector pBBR1MCS was fused with pASK-IBA7-corS by ligation, resulting in the 9,9-kb plasmid pBBR-IBA7-corS, which could be mobilized into PG4180 cells. For expression of CorS$_{Strep-tag}$, transconjugant bacteria were next cultured in HSC medium at 18 or 28° until an OD$_{600}$ of 0.5 was reached. Subsequently, 3 mg/l anhydrotetracycline was added to induce expression of the recombinant protein and the cultures were further incu-bated at the respective temperatures. Proteins from isolated membrane fractions were solubilized using either n-octyl-β-D-glucopyranoside or Triton-X 100, as has been described. CorS$_{Strep-tag}$ was then further purified with the help of pre-packed StrepTactin columns, as recommended by the manufacturer (IBA).

Detection of CorS$_{Strep-tag}$ expressed in and purified from PG4180 was possible with either streptavidin antibody conjugates or the polyclonal anti-CorS peptide antibodies. In contrast, detection signals for native CorS from membrane fractions of PG4180 wild type using peptide-specific anti-bodies were much less pronounced. Either there were significantly more copies of expressed CorS$_{Strep-tag}$ in the cells as compared to native CorS or epitopes recognized by the peptide-specific antibodies were more accessible in CorS$_{Strep-tag}$ than in native CorS due to different folding properties of both protein variants.

Topological Analysis of the HPK CorS

Predictions of CorS Membrane Topology

Classic HPKs consist of a receiver domain, which senses the environ-mental stimulus, and a transmitter domain, in which autophosphorylation of a conserved histidine residue occurs followed by phospho-transfer to the cognate RR. The receiver domains of many HPK are membrane-embedded with several TMDs. In order to test whether this is true for CorS as well, different topology prediction programs were employed (Smirnova and Ullrich, 2004). The summarized results obtained from analyses with the programs TMHMM (Sonnhammer *et al.*, 1998), HMMTOP (Tusnady and Simon, 1998), DAS (Cserzo *et al.*, 1997), MEMSAT (Jones *et al.*, 1994), SOSUI (Hirokawa *et al.*, 1998), and Vector NTI (Informax Inc., USA) preferentially supported a topological model for CorS with six TMDs (Fig. 2). According to these predictions both the N- and the C-terminus of CorS are located in the cytoplasm. In contrast, a TopPred II (Claros and von Heijne, 1994) prediction indicated a possible 7th TMD located downstream of the conserved histidine residue (H-box). According to the later prediction, CorS topology organization would be rather atypical for HPKs owing to the fact that the H-box would then be located in the periplasmic space. Consequently, we considered this later prediction as highly unlikely. The biochemical analyses described in the following text aimed at elucidating the *in situ* CorS topology and thus aimed at proving the theoretical topology predictions and the importance of different domains for the function of CorS.

Deletion Analysis of the Membrane-Spanning Region of CorS

To analyze whether the membrane-spanning regions are in themselves important for CorS's ability to detect temperature changes, the gene region encoding either all six or the last four TMDs were removed by in-frame deletion (Smirnova and Ullrich, 2004). The resulting constructs, pASH31 (all TMDs deleted) and pASH29 (TMDs 4–6 deleted), were introduced to the *corS* mutant PG4180.D4, which additionally contained the GUS reporter plasmid pRGMU1 for complementation analysis. The truncated *corS* gene in pASH29 encodes a protein with two TMDs whereas in pASH31, it presumably encodes a soluble protein.

Complementation analysis revealed that neither of the two plasmids was able to restore wild type levels with respect to *cmaABT* expression in PG4180. D4 (pRGMU1). GUS activity in PG4180.D4 (pASH29; pRGMU1) was completely abolished. Introduction of pASH31 resulted in a basal level of GUS activity at 18°, comprising only 8% of the wild type activity at this temperature. At 28°, no promoter activity was detectable. A control plasmid containing intact *corS* was able to fully complement the mutant by restoring the typical temperature-dependent GUS activity. The deletion analysis demonstrated that CorS missing the hydrophobic N-terminal part is a non-functional protein. Hunger *et al.* (2004) expressed the cytoplasmic domain of the HPK DesK *in trans* and determined that kinase function is still present but temperature-independent. They concluded that the temperature-sensing part of the enzyme is the membrane-embedded domain. This may hold true for

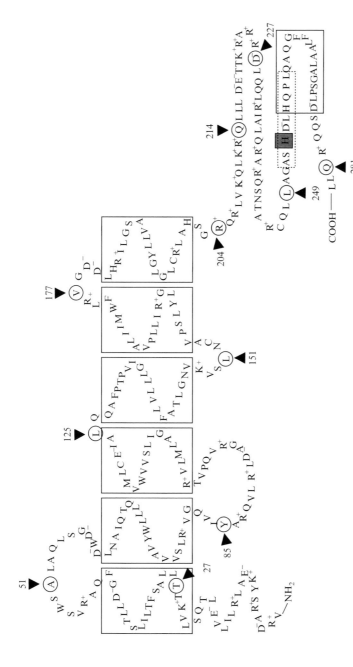

Fig. 2. Schematic representation of the membrane topology of CorS, derived from Smirnova and Ullrich (2004). Positions of PhoA and LacZ fusions are indicated by black arrows and amino acid sequence number. Amino acid residues are given in one-letter code. The conserved H-box is indicated by a dotted rectangle. TMDs are displayed by open rectangles.

CorS as well. However, whether or not kinase activity is still present in the truncated CorS versions remains to be elucidated in future experiments.

Validation of Membrane Topology Predictions Using Translational Fusions of CorS to PhoA and LacZ

To validate the predicted membrane topology for CorS, translational fusions to either alkaline phosphatase (PhoA) or β-galactosidase (LacZ) were constructed. This approach was based on differential enzymatic activities of these proteins in respective cellular compartments, that is, in the periplasm and the cytoplasm. When fused to periplasmic loops of a membrane protein, PhoA becomes active whereas LacZ is enzymatically active only when fused to cytoplasmic regions of a protein. Consequently, respective fusions provide direct evidence for the cellular location of a given protein region (Alexeyev and Winkler, 1999; Manoil et al., 1990). Use of PhoA and LacZ fusions had previously been used for topology analyses of several membrane proteins, including DctA, GlpT, and SecY (Akiyama and Ito, 1987; Gott and Boos, 1988; Jording and Puhler, 1993). In our study, PhoA and LacZ fusions were designed for every loop of CorS facing either the periplasm or the cytoplasm (Smirnova and Ullrich, 2004). Five additional PhoA fusions were generated upstream and downstream of the conserved H-box and the putative 7th TMD (Fig. 2).

For construction of the translational fusions, DNA fragments were PCR-amplified, those which contained the promoter region of corS and its ribosome binding site and terminated at the position of codons for amino acid residues located either in putative periplasmic or cytoplasmic loops of CorS. PCR fragments were cloned in pBluescript II SK (Stratagene, Heidelberg, Germany). Subsequently, a 2.6-kb DNA fragment from plasmid pPHO7 (Guttierrez and Devedjian, 1989) containing a signal peptide-free phoA gene was fused in-frame to all PCR fragments. This resulted in constructs bearing translational corS::phoA fusions under the control of the corS promoter. DNA sequence analysis of the junction regions verified precise in-frame fusions. Subsequently, these fragments were cloned into pBBR1MCS and transferred to cells of PG4180. To generate respective corS::lacZ translational fusions, the 2.6-kb fragment containing phoA was substituted by a 3.0-kb fragment encoding LacZ from plasmid pMC-1871 (Amersham-Pharmacia Biotech, Freiburg, Germany) in all corS::phoA fusions.

Qualitative assays were carried out on mannitol glutamate (MG) medium (Keane et al., 1970) agar plates containing the substrates for PhoA (X-Phos) and LacZ (X-Gal), respectively. For this study, 5-μl aliquots of cell suspensions were placed on the plates and incubated at 18 and 28° for

2 to 6 days. In addition, specific alkaline phosphatase and β-galactosidase activities were quantified using fluorometric assays, as described by Smirnova and Ullrich (2004). On agar plates, translational fusions predicted to be of periplasmic location showed PhoA activity at both temperatures. Similarly, fusions with assumed cytoplasmic location exhibited LacZ activity on agar plates. These results could be confirmed in quantitative enzymatic assays and indicated that CorS is embedded in the cytoplasmic membrane with six TMDs, as predicted by most of the applied membrane topology programs.

Interestingly, three additional PhoA fusions located downstream of the 6th TMD and upstream of the conserved H-box exhibited temperature-dependent enzymatic activities on MG agar plates, showing a PhoA-positive phenotype at 28° but no PhoA activity at 18°. Unfortunately, these results could not be confirmed in the quantitative assay. Since *P. syringae* does not grow on MG liquid medium, HSC and other liquid minimal media as well as different growth phases were tested. Under none of the tested conditions could the results obtained with agar plates be confirmed in liquid media. We lack a plausible explanation for this inconsistency. However, the qualitative results suggested an interesting temperature-dependent conformational change of this part of CorS for which a respective model is proposed in Fig. 3. When re-analyzing the above-mentioned membrane topology predictions, one realizes that the 6th TMD of CorS showed the lowest hydrophobicity of all TMDs (Smirnova and Ullrich, 2004). Likewise, the predicted but questionable 7th TMD is located downstream of the catalytic H-box of CorS. At 18° where CorS is active, the conserved H-box is accessible to autophosphorylation since the 6th TMD is embedded in the membrane. However, at 28°, the predicted 7th TMD might be located inside the membrane and the 6th TMD, together with the H-box, might form a periplasmic loop and thus render CorS inactive. Relocation of the 6th and 7th TMDs might be due to their relative low hydrophobicity and due to changes in the membrane's fluidity when temperature changes. In consequence, CorS might act as a fine-tuned and dynamic molecular thermometer of *P. syringae*.

Mutational Analysis of CorS

To elucidate how CorS functions in temperature signal perception and which particular amino acid sequence alterations of the TMDs make it insensitive to the temperature signal, a random *in vitro* mutagenesis approach was applied. Using a modified mini-*Mu* transposon, *corS* was mutagenized by random in-frame insertion of a pentapeptide sequence, according to the manufacturer's recommendations (Finnzymes, Espoo, Finland).

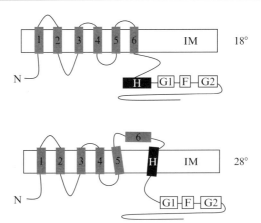

FIG. 3. Schematic representation of the proposed temperature-dependent conformational change of the sensor kinase CorS, during which the catalytic H-box becomes a TMD at 28°. Grey rectangles represent TMDs, and a black rectangle represents the conserved H-box. Conserved cytoplasmic HPK sequence motifs G1, F, and G2 are indicated.

After selection of proper insertion loci by endonuclease treatment, a total of 28 mutagenized plasmids harboring individual pentapeptide insertions in the N-terminal part of CorS and one wild type *corS* plasmid were mobilized into PG4180.D4 (pRGMU1). Resulting transconjugants were screened for GUS expression at 18 and 28° on X-Gluc-containing MG agar plates. Simultaneously, GUS activities were quantified using the fluorescence assay. The mutagenized plasmids neither restored temperature-dependent GUS expression nor did they cause a temperature-insensitive phenotype to PG4180.D4 (pRGMU1). In contrast and as expected, the wild type *corS* gene restored thermoresponsive GUS expression. Results of this mutational approach suggested that individual insertions of pentapeptides are not suitable for analysis of membrane proteins, for which proper folding might be essential for alterations of enzymatic function. In future studies, we plan to mutagenize *corS* using a site-directed approach. Suitable regions for such a mutational analysis might be the TMDs, in particular the 6th and the predicted 7th TMDs.

Generation of CorS Hybrid Proteins

Aside from the detailed analysis of CorS from strain PG4180, COR biosynthesis and CorS were also studied for various strains of other pathovars of *P. syringae* in dependence of temperature (Rohde *et al.*, 1998). From this analysis, three strains turned out to be of special interest since they showed remarkably different patterns of COR production. While PG4180 synthesizes COR temperature-dependently with a maximum at 18°, *P. syringae* pv. tomato DC3000 showed a temperature-independent COR phenotype and *P. syringae* pv. atropurpurea MAFF301309 surprisingly produced the phytotoxin with maximal rate at 28°. These results suggested that despite similar genetic backgrounds, a pathovar-specific regulation of COR biosynthesis exists. Just as in PG4180, the genome of DC3000 carries the regulatory gene triad, *corRSP* (Brooks *et al.*, 2004; GenBank accession number NC_004578). Sato *et al.* (1983) detected a 88-kb plasmid, designated pCOR1 in *P. syringae* pv. atropurpurea. Introduction of pCOR1 provided the ability to synthesize COR to a COR-negative *P. syringae* pv. atropurpurea strain. A 32-kb region containing COR biosynthetic and regulatory genes from p4180A of PG4180 was shown to hybridize to pCOR1 (Bender *et al.*, 1991).

Based on these findings, an approach is currently being undertaken to find out whether an exchange of *corS* alleles between different pathovars changes their COR phenotypes and transcriptional activation of *cmaABT* (Fig. 4). Mutants defective in *corS* in all three pathovar strains serve as host cells for this approach. The cloned *corS* gene of one strain is introduced *in trans* into the genetic loci of the modified two-component regulatory systems of the other two strains and vice versa, thereby replacing the native alleles. Subsequently, exchange of only the N- or C-termini of CorS will be used to fine-analyze the regulatory systems. The first steps of this analysis are described in the following text.

The genes *corR*, *corS*, and *corP* of PG4180 and DC3000 with their upstream sequences were separately amplified by PCR, thereby introducing

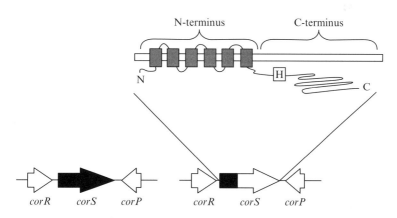

FIG. 4. Schematic representation of the *corRSP* hybrid system exemplified for two *P. syringae* strains. White arrows indicate sequences derived from PG4180, and gene sequences of DC3000 are represented in black. The proposed topology of CorS is depicted above.

unique restriction sites meant for later combinatorial cloning. PCR products were cloned into vector pGEM-T Easy (Promega, Mannheim, Germany) and their sequences were verified by nucleotide sequencing. Ligation of the three genes and cloning of the respective gene triad into the broad-host range vector pRK415 resulted in a system where each of the three genes can be easily exchanged with alleles from other strains. The recombinant *corRSP* systems were mobilized into the *corS* mutants of PG4180 and DC3000. CorS hybrids consisting of N- and C-termini from PG4180 and DC3000 were constructed by fusion-PCR. For this procedure, a 0.6-kb fragment encoding the six TMDs and a 1.2-kb fragment encoding the C-terminal part of CorS of both strains were separately amplified. In a subsequent PCR step, the outer oligonucleotide primers and the two PCR products were used to amplify a 1.8-kb fusion product. These fragments were then introduced to the recombinant *corRSP* constructs. In the near future, respective constructs will also be generated for strain MAFF301309.

Next, COR quantification, analyses of transcriptional activation using the reporter plasmid pRGMU1, and Spot blot transcriptional analysis of *cmaABT* at 18 and 28° will be conducted to characterize the effects of the hybrid *corRSP* systems in the genomic background of all three *corS* mutants. An important aspect to keep in mind is the copy number of COR regulatory genes in this exchange system. The genes *corR* and *corP* will be present in additional copies as well as the genomic copy after introduction of the plasmids. This might lead to elevated transcriptional levels of COR biosynthetic genes. Should this problem emerge, complete deletion mutants for *corRSP* will have to be constructed in all three strains.

Previous studies showed that COR production in DC3000 is induced by plant-borne signals (Boch *et al.*, 2002; Weingart *et al.*, 2004). Similarly, Sreedharan *et al.* (2006) provided evidence that CorR in DC3000 is regulated *in planta* by *hrpL*, an alternative sigma factor required for the expression of components of the type III *hrp* secretion system. These findings suggested that *corRSP* in DC3000 may be activated in the host plant despite its relatively low level of induction *in vitro*. Consequently, the hybrid *corRSP* systems will also be tested after inoculation of cells bearing them to host plants. Tools for this are now available and our future experiments will help to determine the differential regulation of COR synthesis and/or function(s) of CorS in the three pathovars of *P. syringae*.

Structural Comparison of Low Temperature Sensing HPKs

Low temperature is known to trigger incorporation of unsaturated fatty acids to bacterial membranes (Hazel, 1995; Murata and Los, 1997). This is necessary to maintain fluidity of the membrane which, in turn, is essential for the integrity of the bacterial cell and functionality of membrane-embedded enzymes. Conformational changes altering the activity of membrane-bound HPKs based on changes in membrane fluidity have been proposed for the temperature-affected DesK protein in *Bacillus subtilis* (Albanesi *et al.*, 2004) and for Hik33 from *Synecchocystis* sp. PCC 6803 (Suzuki *et al.*, 2000). Whether or not membrane fatty acid composition affects CorS conformational changes remains to be elucidated in future studies. The primary structures of CorS, Hik33, and DesK are only distantly related. As expected and as found for most HPKs, the C-terminal catalytic domains of the three HPKs show a moderately low similarity to each other. In contrast, no alignment was possible for the N-terminal receiver regions (http://www.ncbi.nlm.nih.gov/blast/bl2seq). Furthermore, Hik33 has two hydrophobic helices that may span the membrane and a long periplasmic loop, which is typical for HPKs that bind extracellular signal substrates. A HAMP region (Williams and Stewart, 1999), a leucin zipper, and a PAS domain (Taylor and Zhulin, 1999), which play a role in signal transduction, were identified in Hik33 (Mikami *et al.*, 2002). In contrast, CorS has six TMDs with very short periplasmic loops and does not possess a typical PAS domain or a HAMP region. DesK is more similar to CorS with respect to its N-terminal organization since it contains four TMDs (Albanesi *et al.*, 2004). However, DesK lacks classical PAS and leucine zipper domains in its cytosolic part (Murata and Los, 2006). Finally, comparative hydrophobicity profiling (Fig. 5) further substantiates that the structures of CorS, Hik33, and DesK differ considerably. Consequently, each of the three HPKs may have its very specific mode of signal perception. The functional roles in temperature adaptation are diverse: DesK and Hik33 sense temperature-mediated changes in the membrane fluidity and, consequently, activate transcription of desaturase genes to alter the membranes' composition. In contrast, CorS senses temperature changes to control biosynthesis of a secondary metabolite important for *P. syringae* infection of the host plant.

Concluding Remarks

This chapter summarized the work being done on a modified TCS that senses temperature changes and is induced by a temperature decrease. During the course of this work, it became clear that the HPK, CorS, is sensing low temperature by modulating its conformation inside the bacterial membrane, thus acting as a molecular thermometer. In this respect, CorS's function resembles that of two other HPKs, DesK from *B. subtilis* and Hik33 from *Synecchocystis* sp. However, the structures of all three enzymes are remarkably different from each other, suggesting that different *modi operandi* might exist. Important aspects of the mode of action of CorS

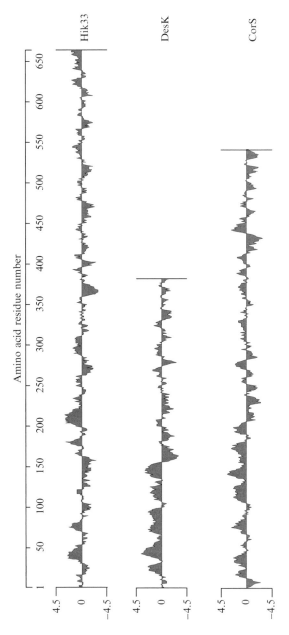

FIG. 5. Hydrophobicity profiles of histidine kinases, DesK, Hik33, and CorS, according to the algorithm of Kyte and Doolittle (1982) and generated with the program PROTEAN (DNA Star).

remain to be studied: (i) Which cellular and extracellular factors trigger the conformational changes of CorS, (ii) which particular domains are important for temperature sensing, and (iii) how fast and how durably is CorS responding to temperature alterations? Our future studies will be directed toward shedding light on these important questions.

Acknowledgments

The authors thank Ling Wang, Bettina Rohde, Ina Budde, Alejandro Penaloza-Vazquez, Vidhya Rangaswamy, Barbara Kunkel, Jens Boch, and Carol L. Bender for their continued support, technical help, and many fruitful discussions. This work was supported by the Deutsche Forschungsgemeinschaft and the Max Planck Society.

References

Aguilar, P. S., Hernandez-Arriaga, A. M., Cybulski, L. E., Erazo, A. C., and de Mendoza, D. (2001). Molecular basis of thermosensing: A two-component signal transduction thermometer in *Bacillus subtilis. EMBO J.* **20,** 1681–1691.

Akiyama, Y., and Ito, K. (1987). Topology analysis of the SecY protein, an integral membrane protein involved in protein export in *Escherichia coli. EMBO J.* **6,** 3465–3470.

Albanesi, D., Mansilla, M. C., and de Mendoza, D. (2004). The membrane fluidity sensor DesK of *Bacillus subtilis* controls the signal decay of its cognate response regulator. *J. Bacteriol.* **186,** 2655–2663.

Alexeyev, M. F., and Winkler, H. H. (1999). Membrane topology of the *Rickettsia prowazekii* ATP/ADP translocase revealed by novel dual pho-lac reporters. *J. Mol. Biol.* **285,** 1503–1513.

Beckering, C. L., Steil, L., Weber, M. H., Volker, U., and Marahiel, M. A. (2002). Genomewide transcriptional analysis of the cold shock response in *Bacillus subtilis. J. Bacteriol.* **184,** 6395–6402.

Beier, D., and Gross, R. (2006). Regulation of bacterial virulence by two-component systems. *Curr. Opin. Microbiol.* **9,** 143–152.

Bender, C. L., Alarcon-Chaidez, F., and Gross, D. C. (1999). *Pseudomonas syringae* phytotoxins: Mode of action, regulation, and biosynthesis by peptide and polyketide synthetases. *Microbiol. Mol. Biol. Rev.* **63,** 266–292.

Bender, C. L., Liyanage, H., Palmer, D., Ullrich, M., Young, S., and Mitchell, R. (1993). Characterization of the genes controlling the biosynthesis of the polyketide phytotoxin coronatine including conjugation between coronafacic and coronamic acid. *Gene* **133,** 31–38.

Bender, C. L., Young, S. A., and Mitchell, R. E. (1991). Conservation of plasmid DNA sequences in coronatine-producing pathovars of *Pseudomonas syringae. Appl. Environ. Microbiol.* **57,** 993–999.

Boch, J., Joardar, V., Gao, L., Robertson, T. L., Lim, M., and Kunkel, B. N. (2002). Identification of *Pseudomonas syringae* pv. tomato genes induced during infection of *Arabidopsis thaliana. Mol. Microbiol.* **44,** 73–88.

Brooks, D. M., Hernandez-Guzman, G., Kloek, A. P., Alarcon-Chaidez, F., Sreedharan, A., Rangaswamy, V., Penaloza-Vazquez, A., Bender, C. L., and Kunkel, B. N. (2004). Identification and characterization of a well-defined series of coronatine biosynthetic mutants of *Pseudomonas syringae* pv. tomato DC3000. *Mol. Plant Microbe Interact.* **17,** 162–174.

Budde, I. P., Rohde, B. H., Bender, C. L., and Ullrich, M. S. (1998). Growth phase and temperature influence promoter activity, transcript abundance, and protein stability during biosynthesis of the *Pseudomonas syringae* phytotoxin coronatine. *J. Bacteriol.* **180,** 1360–1367.

Budde, I. P., and Ullrich, M. S. (2000). Interactions of *Pseudomonas syringae* pv. glycinea with host and nonhost plants in relation to temperature and phytotoxin synthesis. *Mol. Plant. Microbe Interact.* **13,** 951–961.

Calva, E., and Oropeza, R. (2006). Two-component signal transduction systems, environmental signals, and virulence. *Microb. Ecol.* **51,** 166–176.

Chalfie, M., Tu, Y., Euskirchen, G., Ward, W. W., and Prasher, D. C. (1994). Green fluorescent protein as a marker for gene expression. *Science* **263,** 802–805.

Claros, M. G., and von Heijne, G. (1994). TopPred II: An improved software for membrane protein structure predictions. *Comput. Appl. Biosci.* **10,** 685–686.

Cserzo, M., Wallin, E., Simon, I., von Heijne, G., and Elofsson, A. (1997). Prediction of transmembrane alpha-helices in prokaryotic membrane proteins: The dense alignment surface method. *Protein Eng.* **10,** 673–676.

Cybulski, L. E., del Solar, G., Craig, P. O., Espinosa, M., and de Mendoza, D. (2004). *Bacillus subtilis* DesR functions as a phosphorylation-activated switch to control membrane lipid fluidity. *J. Biol. Chem.* **279,** 39340–39347.

de Bruijn, F. J., and Lupski, J. R. (1984). The use of transposon Tn5 mutagenesis in the rapid generation of correlated physical and genetic maps of DNA segments cloned into multicopy plasmids—A review. *Gene* **27,** 131–149.

Degenkolb, J., Takahashi, M., Ellestad, G. A., and Hillen, W. (1991). Structural require-ments of tetracycline-Tet repressor interaction: Determination of equilibrium binding constants for tetracycline analogs with the Tet repressor. *Antimicrob. Agents Chemother.* **35,** 1591–1595.

Figurski, D. H., and Helinski, D. R. (1979). Replication of an origin-containing derivative of plasmid RK2 dependent on a plasmid function provided in trans. *Proc. Natl. Acad. Sci. USA* **76,** 1648–1652.

Gott, P., and Boos, W. (1988). The transmembrane topology of the sn-glycerol-3-phosphate permease of *Escherichia coli* analyzed by *phoA* and *lacZ* protein fusions. *Mol. Microbiol.* **2,** 655–663.

Guttierrez, C., and Devedjian, J. C. (1989). A plasmid facilitating *in vitro* construction of *phoA* gene fusions in *Escherichia coli*. *Nucleic Acids Res.* **17,** 3999.

Hirokawa, T., Boon-Chieng, S., and Mitaku, S. (1998). SOSUI: Classification and secondary structure prediction system for membrane proteins. *Bioinformatics* **14,** 378–379.

Hunger, K., Beckering, C. L., and Marahiel, M. A. (2004). Genetic evidence for the temperature-sensing ability of the membrane domain of the *Bacillus subtilis* histidine kinase DesK. *FEMS Microbiol. Lett.* **230,** 41–46.

Hurme, R., and Rhen, M. (1998). Temperature sensing in bacterial gene regulation—What it all boils down to. *Mol. Microbiol.* **30,** 1–6.

Ishii, N., Takeda, H., Doi, M., Fuma, S., Miyamoto, K., Yanagisawa, K., and Kawabata, Z. (2002). A new method using enhanced green fluorescent protein (EGFP) to determine grazing rate on live bacterial cells by protists. *Limnology* **3,** 47–50.

Jefferson, R. A., Burgess, S. M., and Hirsh, D. (1986). beta-Glucuronidase from *Escherichia coli* as a gene-fusion marker. *Proc. Natl. Acad. Sci. USA* **83,** 8447–8451.

Jones, D. T., Taylor, W. R., and Thornton, J. M. (1994). A model recognition approach to the prediction of all-helical membrane protein structure and topology. *Biochemistry* **33,** 3038–3049.

Jording, D., and Puhler, A. (1993). The membrane topology of the *Rhizobium meliloti* C4-dicarboxylate permease (DctA) as derived from protein fusions with *Escherichia coli* K12 alkaline phosphatase (PhoA) and beta-galactosidase (LacZ). *Mol. Gen. Genet.* **241,** 106–114.

Keane, P. J., Kerr, A., and New, P. B. (1970). Grown gall of stone fruit. II. Identification and nomenclature of *Agrobacterium* isolates. *Aust. J. Biol. Sci.* **23,** 585–595.

Keen, N. T., Tamaki, S., Kobayashi, D., and Trollinger, D. (1988). Improved broad-host-range plasmids for DNA cloning in gram-negative bacteria. *Gene* **70,** 191–197.

Kyte, J., and Doolittle, R. F. (1982). A simple method for displaying the hydropathic character of a protein. *J. Mol. Biol.* **157,** 105–132.

Liyanage, H., Penfold, C., Turner, J., and Bender, C. L. (1995b). Sequence, expression and transcriptional analysis of the coronafacate ligase-encoding gene required for coronatine biosynthesis by *Pseudomonas syringae*. *Gene* **153,** 17–23.

Los, D. A., Ray, M. K., and Murata, N. (1997). Differences in the control of the temperature-dependent expression of four genes for desaturases in *Synechocystis* sp. PCC 6803. *Mol. Microbiol.* **25,** 1167–1175.

Manoil, C., Mekalanos, J. J., and Beckwith, J. (1990). Alkaline phosphatase fusions: Sensors of subcellular location. *J. Bacteriol.* **172,** 515–518.

Mansilla, M. C., Albanesi, D., Cybulski, L. E., and de Mendoza, D. (2005). Molecular mechanisms of low temperature sensing bacteria. *Ann. Hepatol.* **4,** 216–217.

Mansilla, M. C., and de Mendoza, D. (2005). The *Bacillus subtilis* desaturase: A model to understand phospholipid modification and temperature sensing. *Arch. Microbiol.* **183,** 229–235.

Mikami, K., Kanesaki, Y., Suzuki, I., and Murata, N. (2002). The histidine kinase Hik33 perceives osmotic stress and cold stress in *Synechocystis* sp PCC 6803. *Mol. Microbiol.* **46,** 905–915.

Mittal, S., and Davis, K. R. (1995). Role of the phytotoxin coronatine in the infection of *Arabidopsis thaliana* by *Pseudomonas syringae* pv. tomato. *Mol. Plant Microbe Interact.* **8,** 165–171.

Murata, N., and Los, D. A. (1997). Membrane fluidity and temperature perception. *Plant Physiol.* **115,** 875–879.

Murata, N., and Los, D. A. (2006). Histidine kinase Hik33 is an important participant in cold-signal transduction in cyanobacteria. *Physiologia Plantarum* **126,** 17–27.

Palmer, D. A., and Bender, C. L. (1995). Ultrastructure of tomato leaf tissue treated with the pseudomonad phytotoxin coronatine and comparison with methyl jasmonate. *Mol. Plant Microbe Interact.* **8,** 683–692.

Palmer, D. A., and Bender, C. L. (1993). Effects of environmental and nutritional factors on production of the polyketide phytotoxin coronatine by *Pseudomonas syringae* pv. Glycinea. *Appl. Environ. Microbiol.* **59,** 1619–1626.

Patterson, G. H., Knobel, S. M., Sharif, W. D., Kain, S. R., and Piston, D. W. (1997). Use of the green fluorescent protein and its mutants in quantitative fluorescence microscopy. *Biophys. J.* **73,** 2782–2790.

Peñaloza-Vázquez, A., and Bender, C. L. (1998). Characterization of CorR, a transcriptional activator which is required for biosynthesis of the phytotoxin coronatine. *J. Bacteriol.* **180,** 6252–6259.

Peñaloza-Vázquez, A., Rangaswamy, V., Ullrich, M., Bailey, A. M., and Bender, C. L. (1996). Use of translational fusions to the maltose-binding protein to produce and purify proteins in *Pseudomonas syringae* and assess their activity *in vivo*. *Mol. Plant Microbe Interact.* **9,** 637–641.

Penfold, C. N., Bender, C. L., and Turner, J. G. (1996). Characterization of genes involved in biosynthesis of coronafacic acid, the polyketide component of the phytotoxin coronatine. *Gene* **183,** 167–173.

Phadtare, S., Yamanaka, K., and Inouye, M. (2000). The cold shock response. *In* "Bacterial Stress Responses" (G. Stortz and R. Hengge-Aronis, eds.), pp. 33–47. Washington, DC.

Rangaswamy, V., and Bender, C. L. (2000). Phosphorylation of CorS and CorR, regulatory proteins that modulate production of the phytotoxin coronatine in *Pseudomonas syringae*. *FEMS Microbiol. Lett.* **193**, 13–18.

Rangaswamy, V., Ullrich, M., Jones, W., Mitchell, R., Parry, R., Reynolds, P., and Bender, C. L. (1997). Expression and analysis of coronafacate ligase, a thermoregulated gene required for production of the phytotoxin coronatine in *Pseudomonas syringae*. *FEMS Microbiol. Lett.* **154**, 65–72.

Rohde, B. H., Pohlack, B., and Ullrich, M. S. (1998). Occurrence of thermoregulation of genes involved in coronatine biosynthesis among various *Pseudomonas syringae* strains. *J. Basic Microbiol.* **38**, 41–50.

Rübenhagen, R., Rönsch, H., Jung, H., Krämer, R., and Morbach, S (2000). Osmosensor and osmoregulator properties of the betaine carrier BetP from *Corynebacterium glutamicum* in proteoliposomes. *J. Biol. Chem.* **275**, 735–741.

Sato, M., Nishiyama, K., and Shirata, A. (1983). Involvement of plasmid DNA in the productivity of coronatine by *Pseudomonas syringae* pv. atropurpurea. *Ann. Phytopath. Soc. Japan* **49**, 522–528.

Schenk, A., Weingart, H., and Ullrich, M. S. (2006). Extraction of high-quality bacterial and plant total RNA from infected leaf tissue for bacterial *in planta* gene expression analysis by multiplexed fluorescent Northern hybridization. Submitted to *Molecular Plant Pathology*.

Schmidt, T. G., Koepke, J., Frank, R., and Skerra, A. (1996). Molecular interaction between the Strep-tag affinity peptide and its cognate target, streptavidin. *J. Mol. Biol.* **255**, 753–766.

Skerra, A. (1994). Use of the tetracycline promoter for the tightly regulated production of a murine antibody fragment in *Escherichia coli*. *Gene* **151**, 131–135.

Skerra, A., and Schmidt, T. G. (2000). Use of the Strep-Tag and streptavidin for detection and purification of recombinant proteins. *Methods Enzymol.* **326**, 271–304.

Smirnova, A. V., and Ullrich, M. S. (2004). Topological and deletion analysis of CorS, a *Pseudomonas syringae* sensor kinase. *Microbiology* **150**, 2715–2726.

Smirnova, A., Li, H., Weingart, H., Aufhammer, S., Burse, A., Finis, K., Schenk, A., and Ullrich, M. S. (2001). Thermoregulated expression of virulence factors in plant-associated bacteria. *Arch. Microbiol.* **176**, 393–399.

Sonnhammer, E. L., von Heijne, G., and Krogh, A. (1998). A hidden Markov model for predicting transmembrane helices in protein sequences. *Proc. Int. Conf. Intell. Syst. Mol. Biol.* **6**, 175–182.

Sreedharan, A., Penaloza-Vazquez, A., Kunkel, B. N., and Bender, C. L. (2006). CorR regulates multiple components of virulence in *Pseudomonas syringae* pv. tomato DC3000. *Mol. Plant. Microbe Interact.* **19**, 768–779.

Stock, A. M., Robinson, V. L., and Goudreau, P. N. (2000). Two-component signal transduction. *Annu. Rev. Biochem.* **69**, 183–215.

Suzuki, I., Kanesaki, Y., Mikami, K., Kanehisa, M., and Murata, N. (2001). Cold-regulated genes under control of the cold sensor Hik33 in *Synechocystis*. *Mol. Microbiol.* **40**, 235–244.

Suzuki, I., Los, D. A., Kanesaki, Y., Mikami, K., and Murata, N. (2000). The pathway for perception and transduction of low-temperature signals in *Synechocystis*. *EMBO J.* **19**, 1327–1334.

Tamai, E., Fann, M. C., Tsuchiya, T., and Maloney, P. C. (1997). Purification of UhpT, the sugar phosphate transporter of *Escherichia coli*. *Protein Expr. Purif.* **10**, 275–282.

Taylor, B. L., and Zhulin, I. B. (1999). PAS domains: Internal sensors of oxygen, redox potential, and light. *Microbiol. Mol. Biol. Rev.* **63**, 479–506.

Tusnady, G. E., and Simon, I. (1998). Principles governing amino acid composition of integral membrane proteins: Application to topology prediction. *J. Mol. Biol.* **283,** 489–506.

Ullrich, M., and Bender, C. L. (1994). The biosynthetic gene cluster for coronamic acid, an ethylcyclopropyl amino acid, contains genes homologous to amino acid-activating enzymes and thioesterases. *J. Bacteriol.* **176,** 7574–7586.

Ullrich, M., Guenzi, A. C., Mitchell, R. E., and Bender, C. L. (1994). Cloning and expression of genes required for coronamic acid (2-ethyl-1-aminocyclopropane 1-carboxylic acid), an intermediate in the biosynthesis of the phytotoxin coronatine. *Appl. Environ. Microbiol.* **60,** 2890–2897.

Ullrich, M., Penaloza-Vazquez, A., Bailey, A. M., and Bender, C. L. (1995). A modified two-component regulatory system is involved in temperature-dependent biosynthesis of the *Pseudomonas syringae* phytotoxin coronatine. *J. Bacteriol.* **177,** 6160–6169.

Vidal-Aroca, F., Giannattasio, M., Brunelli, E., Vezzoli, A., Plevani, P., Muzi-Falconi, M., and Bertoni, G. (2006). One-step high-throughput assay for quantitative detection of beta-galactosidase activity in intact gram-negative bacteria, yeast, and mammalian cells. *Biotechniques* **40,** 433–434, 436, 438 passim.

Wang, L., Bender, C. L., and Ullrich, M. S. (1999). The transcriptional activator CorR is involved in biosynthesis of the phytotoxin coronatine and binds to the *cmaABT* promoter region in a temperature-dependent manner. *Mol. Gen. Genet.* **262,** 250–260.

Weingart, H., Stubner, S., Schenk, A., and Ullrich, M. S. (2004). Impact of temperature on *in planta* expression of genes involved in synthesis of the *Pseudomonas syringae* phytotoxin coronatine. *Mol. Plant Microbe Interact.* **17,** 1095–1102.

Williams, S. B., and Stewart, V. (1999). Functional similarities among two-component sensors and methyl-accepting chemotaxis proteins suggest a role for linker region amphipathic helices in transmembrane signal transduction. *Mol. Microbiol.* **33,** 1093–1102.

Wrather, J. A., Anderson, T. R., Arsyad, D. M., Gai, J., Ploper, L. D., Porta-Puglia, A., Ram, H. A., and Yorinori, J. T. (1997). Soybean disease loss estimates for the top 10 soybean producing countries in 1994. *Plant Disease* **81,** 107–110.

Young, S. A., Park, S. K., Rodgers, C., Mitchell, R. E., and Bender, C. L. (1992). Physical and functional characterization of the gene cluster encoding the polyketide phytotoxin coronatine in *Pseudomonas syringae* pv. glycinea. *J. Bacteriol.* **174,** 1837–1843.

[11] The Regulation of Histidine Sensor Kinase Complexes by Quorum Sensing Signal Molecules

By MATTHEW B. NEIDITCH and FREDERICK M. HUGHSON

Abstract

Two-component sensor kinase signaling systems are widespread in bacteria, but gaining mechanistic insight into how kinase activity is controlled by ligand binding has proved challenging. Here, we discuss this problem in the context of our structural and functional studies of bacterial quorum sensing receptors. Specifically, this chapter focuses on the transmembrane sensor kinase complex LuxPQ, which serves as the receptor for the "universal" quorum sensing signal molecule autoinducer-2 (AI-2). Methods are presented for the overproduction, purification, crystallization, and functional characterization of LuxPQ's ligand-binding (periplasmic) domain.

Introduction

A principal mechanism by which bacteria detect and respond to environmental signals makes use of two-component signaling systems (Inouye and Dutta, 2003; Stock *et al.*, 2000). At their simplest, such systems (as the name implies) have just two parts: an integral membrane histidine sensor kinase and a cytoplasmic protein, typically a transcription factor, called a response regulator. Environmental signals activate or inhibit the histidine sensor kinase. When active, the kinase phosphorylates itself on a conserved histidine residue. This same phosphoryl group is subsequently transferred to an aspartate residue within the response regulator protein, where it serves to modulate the activity of the response regulator. When the kinase is inactive, both the histidine and aspartate residues become dephosphorylated (through intrinsic or extrinsic phosphatase activities), and the response regulator reverts to its basal activity state.

In seeking to understand two-component signaling more fully, a key question is how the activity of the kinase is modulated by sensory input. The operational principles are not very well understood, in part because, for many histidine sensor kinases, the relevant ligands are not known. Moreover, even where ligands have been identified, the structural consequences of binding have not, in general, been fully elucidated. Here, we discuss our efforts to deduce the structural changes that accompany the binding of a quorum sensing signal, AI-2, to a histidine sensor kinase complex, LuxPQ

METHODS IN ENZYMOLOGY, VOL. 423
0076-6879/07 $35.00
DOI: 10.1016/S0076-6879(07)23011-3

(Neiditch *et al.*, 2005, 2006). Our results suggest that ligand binding induces quaternary structural changes that shut off the histidine kinase. The extent to which this conclusion is general remains to be established.

Bacterial Quorum Sensing

Populations of bacteria coordinate gene expression by means of a cell–cell communication process called quorum sensing (Waters and Bassler, 2005). Synchronizing gene expression can enhance the effectiveness of collective behaviors including light production, biofilm formation, and virulence factor production. Quorum-sensing responses are regulated by secreted signaling molecules, called autoinducers, that accumulate in proportion to cell density. Autoinducers affect gene expression either directly, by binding to transcription factors, or indirectly, through two-component signaling pathways. Typical autoinducers include acylated homoserine lactones, used by many Gram-negative bacteria, and modified oligopeptides, used by many Gram-positive bacteria.

Our work has been focused on an unusual autoinducer, called AI-2, that displays three unique properties. First, unlike other known autoinducers, AI-2 is both produced and detected by many different species of bacteria (Xavier and Bassler, 2003). As a result, it has the potential to mediate interspecies communication. Second, AI-2 is not a single compound, but a family of spontaneously interconvertable compounds (Miller *et al.*, 2004). Remarkably, different species of bacteria recognize different members of this family of compounds. Third, the marine bacteria *Vibrio harveyi*—the bacteria in which AI-2 signaling was discovered and in which it has been most intensively studied—recognizes AI-2 as a furanosyl borate diester (Chen *et al.*, 2002). AI-2 signaling in *V. harveyi* therefore utilizes one of the very few boron-containing compounds known to have a well-defined role in biological systems.

The *V. harveyi* AI-2 Signal Transduction Pathway

Quorum sensing controls light production in the bioluminescent bacterium *V. harveyi* (Bassler *et al.*, 1994; Cao and Meighen, 1989). At low cell density (that is, when autoinducer concentrations are low), individual cells are dark. At high cell density (when autoinducer concentrations are high), cells turn on light production by upregulating the production of luciferase enzymes. Although the ecological rationale for bioluminescence in *V. harveyi* is not entirely clear, the usefulness of light as a readily measured output has made this one of the best characterized model systems for studying quorum sensing.

To monitor the concentration of the autoinducer AI-2, *V. harveyi* use a histidine sensor kinase complex, LuxPQ (Bassler *et al.*, 1994; Neiditch *et al.*, 2005, 2006). Four polypeptide chains—two LuxP subunits and two LuxQ subunits—compose the heterotetrameric LuxPQ complex. There is some division of labor between the two types of subunit: LuxP subunits bind AI-2, while the LuxQ subunits contain the histidine kinase domains. LuxP is a so-called periplasmic binding protein; other members of this protein family include the maltose, ribose, and sulfate binding proteins. Each of the periplasmic binding proteins, including LuxP, binds its ligand by clamping it between two domains, in a manner reminiscent of a Venus flytrap capturing its prey (Chen *et al.*, 2002; Dwyer and Hellinga, 2004; Felder *et al.*, 1999).

To transduce AI-2 information across the bacterial inner membrane, the LuxP subunits collaborate with the two integral membrane LuxQ subunits. The external (periplasmic) portion of each LuxQ binds LuxP, whereas the internal (cytoplasmic) portion contains several domains, including the histidine kinase domain. A second cytoplasmic domain of LuxQ is called a receiver domain, because it contains a conserved aspartate residue that serves as the recipient for the phosphoryl group transferred from the histidine kinase domain. The phosphoryl group is subsequently transferred from the receiver domain aspartate residue of LuxQ to a histidine residue on the phosphotransfer protein LuxU, and is thence relayed to the response regulator (and transcription factor) LuxO (Freeman and Bassler, 1999a,b).

Like that of other histidine sensor kinases, the mechanism of LuxPQ kinase activity seems to entail the transphosphorylation of one LuxQ monomer by the other (Neiditch *et al.*, 2006). At low cell density, in the absence of AI-2, the histidine kinase is active. Under these conditions, an ATP bound by one LuxQ monomer is used to phosphorylate a conserved histidine residue on the other LuxQ monomer, and (potentially) vice versa. As has been mentioned, this phosphoryl group is then transferred sequentially to the receiver domain of LuxQ, then to LuxU, and finally to LuxO. Because phospho-LuxO represses luciferase expression, *V. harveyi* cells at low cell density are dark. At high cell density, by contrast, AI-2 binds to the LuxP subunits, and shuts off the LuxQ histidine kinase. As a consequence, LuxO is not phosphorylated, and luciferase expression is derepressed. Thus, at high density, *V. harveyi* cells are bright.

The external portion of the LuxPQ complex is entirely dispensable for LuxQ kinase activity (Neiditch *et al.*, 2005, 2006). Indeed, when either LuxP, or the external portion of LuxQ, is deleted, the remainder of LuxQ displays full, constitutive kinase activity. Thus, the LuxQ subunits possess intrinsic kinase activity, with the role of the external LuxPQ domains being to place this activity under the negative control of AI-2. This is an important point from the perspective of protein evolution, because a variety of mechanisms might be used to downregulate an intrinsically active receptor. We turned to X-ray crystallographic studies in order to begin learning how, at the level of molecular mechanism, AI-2 binding inhibits histidine kinase activity.

Regulation of the LuxPQ Receptor Complex by AI-2

To investigate how AI-2 binding to LuxP turns off LuxQ kinase activity, we determined X-ray crystal structures of the periplasmic portion of the LuxPQ complex with and without AI-2 (Neiditch *et al.*, 2005, 2006). Briefly, the structures reveal that the binding of AI-2 to LuxP causes a large movement of one domain versus the other, corresponding to the closing of the Venus flytrap. The consequence of this domain movement was completely unexpected: it allows a single LuxP monomer to bind the periplasmic portions of both of the LuxQ monomers within the $(LuxPQ)_2$ heterotetramer (Fig. 1). By contrast, each LuxQ is bound to only one LuxP in the X-ray structure of the AI-2-free protein. The ability of AI-2-bound LuxP to simultaneously bind to two LuxQ monomers pulls the monomers into an asymmetric conformation. We have hypothesized that it is this asymmetry in the external domains of the complex that shuts off the transphosphorylation activity of the internal domains. It is appealing to hypothesize that shutting

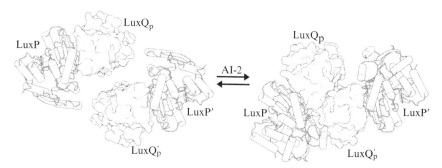

FIG. 1. AI-2 binding inhibits LuxPQ kinase activity by inducing an asymmetric interaction between the periplasmic domains. The LuxPQ complex is viewed from "above," looking toward the bacterial inner membrane. The LuxP subunits are shown using a cartoon representation (cylinders denote α-helices; arrows denote β-strands), while the LuxQ subunits are shown using a space-filling representation. The heterotetrameric complex contains two copies of each subunit, held together by the cytoplasmic domains of LuxQ (which are not shown but would lie behind the plane of the figure). AI-2 binding induces the periplasmic domains to associate asymmetrically (right panel). Our evidence suggests that, in the absence of AI-2, the periplasmic domains interact weakly, if at all (left panel). Ligand-induced asymmetry, if transduced across the membrane to the cytoplasmic domains of LuxQ, would provide a mechanism by which AI-2 binding could down-regulate the phosphorylation of each LuxQ monomer by the other (Neiditch *et al.*, 2006). Figure produced using PyMOL (DeLano, 2002).

off kinase activity requires breaking the symmetric relationship between one histidine kinase domain and the other (Neiditch *et al.*, 2006), but this remains to be established directly.

The LuxPQ structures described previously represent some of the first crystal structures of a bacterial periplasmic protein in complex with an inner membrane protein (albeit not an intact one). During the course of this work, a number of interesting technical challenges arose. Here, we outline some of the methods used, with the goal of helping to guide future studies of this and other periplasmic receptor complexes.

Expression of Wild-Type and Mutant LuxPQ$_p$

Overview

LuxP and the periplasmic domain of LuxQ (denoted LuxQ$_p$) can be overexpressed as individual subunits in *Escherichia coli*. Expressing and purifying the individual subunits allows their biochemical and structural characterization; for example, crystal structures of both AI-2 bound LuxP and LuxQ$_p$ have been determined (Chen *et al.*, 2002; Neiditch *et al.*, 2006). The LuxPQ$_p$ complex assembles spontaneously upon mixing the purified LuxP and LuxQ$_p$ subunits. Alternatively, it can be generated by coexpression of LuxP and LuxQ$_p$. Coexpression is technically convenient and proved necessary in specialized but important cases, as will be described.

Manipulating the N-terminus of the LuxP subunit was crucial to the success of our structural studies. The LuxP subunit is normally targeted for secretion into the periplasm by an N-terminal signal sequence that is removed by a periplasmic signal peptidase. As it turns out, the newly generated N-terminus of the mature LuxP subunit is functionally important. In the absence of AI-2, it nestles in a deep groove between the two tandem PAS (Per/Arnt/Sim) domains that compose LuxQ$_p$ (Neiditch *et al.*, 2005). In the presence of AI-2, on the other hand, the N-terminus disengages (Neiditch *et al.*, 2006). This disengagement, which we refer to as "unclasping," allows conformational flexibility critical for attainment of the kinase-inactive quaternary structure. As a result, crystallization of ligand-free LuxPQ$_p$ required generating complexes with the proper mature N-terminus of the LuxP subunit, while crystallization of ligand-bound LuxPQ$_p$ required generating constitutively unclasped complexes with the N-terminus deleted.

We typically overexpress LuxP as a glutathione S-transferase- (GST-) fusion protein in the cytoplasm of *E. coli*. The N-terminal signal sequence (residues 1–21) is simply excluded from the expression plasmid and, to generate the authentic N-terminal sequence of mature LuxP, we use a GST-LuxP fusion protein that can be cleaved by Factor X$_a$. Factor X$_a$ has the advantage that its recognition site lies entirely N-terminal to the cleavage site. As a result, no heterologous residues remain at the N-terminus of LuxP after cleavage.

Unfortunately, Factor X_a is ineffective in cleaving GST-LuxP$_{\Delta22-26}$, a variant lacking the N-terminal clasp sequence. Presumably, the clasp acts as a flexible linker in GST-LuxP, allowing FX$_a$ to access its recognition site. Therefore, we use a different strategy for overproducing LuxP$_{\Delta22-26}$ that entails coexpressing the truncated but untagged LuxP protein together with GST-LuxQ$_p$. This approach requires the addition of a heterologous Met residue to the N-terminus of LuxP$_{\Delta22-26}$ but—because residue 27 (glycine) is small—this Met is efficiently removed by endogenous bacterial aminopeptidases (Hirel *et al.*, 1989).

Expression of Wild-Type LuxP and LuxQ$_p$

One-liter cultures of *E. coli* yield approximately 15 to 20 mg of purified recombinant LuxP or LuxQ$_p$. Expression plasmids are based on pGEX-4T1 and encode LuxP residues 22 to 365 or LuxQ residues 39 to 278 (the periplasmic domain, denoted LuxQ$_p$) fused via a Factor X_a or thrombin cleavage site, respectively, to the C-terminus of GST. *E. coli* strain BL21 is transformed with one of these expression plasmids and plated on selective media containing 100 μg/ml carbenicillin. After overnight growth at 37°, single colonies are used to inoculate 72-ml starter cultures (LB media plus 100 μg/ml ampicillin), which are grown at 37° with shaking until stationary phase is reached. At this point, six 1-liter cultures are inoculated by adding 10 ml of the starter culture into 1 liter LB media (containing 100 μg/ml ampicillin) in each of six 2-liter baffled flasks. The cultures are incubated at 37° with shaking until their optical density at 600 nm (OD$_{600}$) reaches 0.3, when the incubator heater is turned off, allowing the cultures to cool down nearly to room temperature. When OD$_{600}$ reaches 0.9, protein expression is induced by the addition of 1/10,000 volume 1 M IPTG. Cells are harvested by centrifugation after a further 6 hours incubation at 23° with shaking.

Coexpression of LuxP and LuxQ$_p$

One liter of bacterial culture yields approximately 20 mg of purified recombinant complex. As has been explained, the development of a coexpression system was driven by our need to produce LuxPQ$_p$ complexes containing LuxP$_{\Delta22-26}$. In this protocol, untagged LuxP is coexpressed with GST-LuxQ$_p$. The fusion protein contains a thrombin cleavage site. Expression plasmids are based on pBB75 (LuxP) and pGEX-4T1 (LuxQ$_p$). The procedure is identical to that previously described, with three modifications. First, selective media contain 30 μg/ml kanamycin, in addition to carbenicillin or ampicillin, in order to maintain the pBB75-based expression plasmid. Second, *E. coli* BL21(DE3) is used instead of BL21, in order to provide the inducible T7 polymerase required for LuxP overexpression. Third, the cultures are grown for 12 h after the addition of IPTG before harvesting.

Purification of LuxP, LuxQ$_p$, and LuxPQ$_p$

Overview

In Protocol A, LuxP and LuxQ$_p$ are purified individually by making use of glutathione affinity, anion exchange, and size exclusion chromatographies. In Protocol B, the LuxPQ$_p$ complex is assembled from the separate subunits purified according to Protocol A, and then purified by anion exchange and size exclusion chromatographies. Finally, in Protocol C, the LuxPQ$_p$ complex, assembled *in vivo* during coexpression, is purified by glutathione affinity purification using a GST tag fused to the N-terminus of LuxQ$_p$, followed by anion exchange and size exclusion chromatographies.

Purification of LuxP and LuxQ$_p$

1. Resuspend cell pellets in 120 ml Lysis Buffer (20 mM Tris-HCl, pH 7.5, 150 mM NaCl, 8.3 mM MgCl$_2$, 2.5 μg/ml aprotinin, 2.5 μg/ml DNase I, 2 mM PMSF). It is straightforward to purify LuxP and LuxQ$_p$ in parallel; this protocol is equally effective for both. This and all subsequent steps are carried out at 0 to 4°.

2. Lyse cells. We use one passage through an Avestin EmulsiFlex-C5 cell disruptor operating at a pressure of approximately 12,000 psi, but other methods (e.g., sonication or French press) would likely be equally effective.

3. Clarify the lysate by centrifugation for 45 min at 14,000 rpm using a Sorvall SS-34 rotor. Discard the pellets.

4. Pour three columns, each containing 3 ml (packed volume) glutathione agarose resin. We use disposable Econo-Pac columns (BioRad #732-1010). Equilibrate each column with 30 ml Buffer A (20 mM Tris-HCl, pH 7.5, 150 mM NaCl).

5. Arrange the columns in series, so that the output of one column drips directly down the side of the top reservoir of the next column. Load the clarified lysate.

6. Wash each column separately with 30 ml Buffer A. Cap the bottom of each column, add 12 ml per column of Buffer A (for thrombin cleavage) or 20 mM Tris-HCl, pH 6.5, 50 mM NaCl, 1 mM CaCl$_2$ (for Factor X$_a$ cleavage), and resuspend the resin by capping the top of each column and inverting several times. Pool (resin plus buffer, approximately 45 ml total volume) in a 50 ml conical tube.

7. For cleavage of the immobilized GST fusion protein, the appropriate amount of thrombin or Factor X_a must be empirically determined using small-scale pilot experiments. Choose the lowest protease concentration that yields >90% cleavage during overnight digestion at 4°, as judged by polyacrylamide gel electrophoresis.

8. Carry out large-scale thrombin or FX_a cleavage on the remaining immobilized fusion protein using the conditions identified in the preceding step. The released LuxP or $LuxQ_p$ is conveniently recovered by pouring the resin into a clean, empty column and collecting the flowthrough.

9. To the collected protein, add an equal volume of 20 mM Tris-HCl, pH 7.5, in order to lower the salt concentration in preparation for ion exchange chromatography. Equilibrate a SourceQ 10/10 column attached to an ÄKTA FPLC system (GE Healthcare) with 20 mM Tris-HCl, pH 7.5, 75 mM NaCl. Fractionate the protein using 20 mM Tris-HCl, pH 7.5, with an 120-ml salt gradient from 75 to 500 mM NaCl. Identify LuxP- or $LuxQ_p$-containing fractions by SDS-PAGE and pool.

10. Concentrate the pooled protein by ultrafiltration to a final protein concentration of approximately 10 mg/ml. Either Amicon pressure cells with 10 kDa cutoff membranes or Amicon Ultra centrifugal filter devices can be used for concentrating the protein. Protein concentration is estimated using ultraviolet (UV) absorbance in conjunction with the calculated extinction coefficients.

11. Equilibrate an S200 16/60 column, attached to an ÄKTA FPLC system, with Buffer A. Load as much as 5 ml of the concentrated protein and run the column at 1 ml/min. Pool the fractions that contain suitably pure LuxP or $LuxQ_p$ as judged by SDS-PAGE.

12. Concentrate the pooled protein to a final protein concentration of approximately 200 μM, as in Step 10. Flash freeze 100 μl aliquots in liquid nitrogen and store at −80°.

Generation and Purification of LuxPQ_p from Pure Subunits

1. Mix equimolar quantities of purified LuxP and $LuxQ_p$. For optimal results, the final concentration of each subunit should be at least 30 μM. Incubate 1 h at 4°.

2. Purify using SourceQ 10/10 and S200 16/60 columns as described in steps 9 through 12 of Protocol A.

Purification of LuxPQ_p from Coexpressing Cells

The $LuxPQ_p$ complex is purified by glutathione affinity purification using a GST tag fused to the N-terminus of $LuxQ_p$. After affinity purification, the immobilized protein is digested with thrombin, after which the eluted complex is purified by anion exchange (SourceQ) and gel filtration (S200) chromatography, exactly as described in Protocol A.

Crystallization of LuxPQ$_p$ Complexes

Standard vapor diffusion screening techniques were used to identify conditions for growing diffraction-quality crystals of the LuxP subunit (with bound AI-2) (Chen et al., 2002), the LuxQ$_p$ subunit (Neiditch et al., 2006), and the LuxPQ$_p$ complex (unliganded) (Neiditch et al., 2005). Producing useful crystals of liganded LuxPQ$_p$, on the other hand, proved much more challenging (Neiditch et al., 2006). The first concern was the ligand. This issue had not arisen in earlier work, since crystals of the LuxP subunit retained the AI-2 that had bound during protein overproduction in E. coli BL21. In that case, the bound ligand was shown to be AI-2 by denaturing the protein and demonstrating that the liberated ligand is AI-2, as judged by a V. harveyi bioassay (Chen et al., 2002). It seems likely that the LuxP subunit, as purified from E. coli BL21, is actually a mixture of ligand-bound and ligand-free protein molecules. However, since the ligand-free polypeptide is not readily crystallized (X. Chen, M. B. N., and F. M. H., unpublished results), this recombinant LuxP gives rise to crystals containing only ligand-bound protein. LuxPQ$_p$, on the other hand, behaves differently. The ligand-free protein is the more readily crystallized, and all early attempts yielded exclusively ligand-free crystals. Again, it seems likely that the LuxPQ$_p$ complex purified from E. coli BL21 contains a mixture of ligand-bound and ligand-free complexes; indeed, because the subunit stoichiometry is LuxP$_2$LuxQ$_{p2}$ (Neiditch et al., 2006), there could, in principle, be species with zero, one, or two ligand molecules bound. We concluded that generating ligand-bound LuxPQ$_p$ crystals would be facilitated by the presence of excess AI-2.

AI-2 Production

In general, of course, the availability of the ligand is crucial not only for structural studies of regulated histidine sensor kinases, such as LuxPQ$_p$, but also for functional studies. Three means are available for producing useful quantities of pure AI-2. The first, originally developed in order to establish the metabolic pathway for AI-2 production in bacteria, is to recapitulate this pathway in vitro (Schauder et al., 2001). Thus, the recombinant enzymes Pfs and LuxS are employed to convert S-adenosylhomocysteine, a commercially available molecule, to dihydroxypentanedione (DPD). DPD converts spontaneously to AI-2 in the presence of boric acid (Meijler et al., 2004; Schauder et al., 2001). The second method for generating AI-2 is simply to release it from recombinant LuxP subunits that have bound AI-2 during overproduction in *E. coli* BL21 (Chen *et al.*, 2002). A concern with this method is the high temperature (70°) required to denature LuxP. Finally, the most elegant method is chemical synthesis. At least three different methods for synthesizing DPD (and therefore AI-2) have been developed (De Keersmaecker *et al.*, 2005; Meijler *et al.*, 2004; Semmelhack *et al.*, 2005). In practice, we used AI-2 produced by enzymatic synthesis in our efforts to crystallize ligand-bound LuxPQ$_p$.

Crystallization Strategies

To optimize our chances for identifying crystallization conditions, we took advantage of the high-throughput screening facility operated by the Hauptman-Woodward Medical Research Institute (Luft *et al.*, 2003) to screen over 1500 crystallization conditions using the microbatch-under-oil technique at room temperature. Two additional refinements were introduced into these experiments. First, we carried out parallel screens plus and minus added ligand. This allowed us to identify any conditions that yielded crystals only in the presence of excess ligand. Second, we used a variant of the LuxP subunit lacking the five N-terminal residues normally present in the full-length mature polypeptide. These residues, the "clasp" described earlier, stabilize the ligand-free form relative to the ligand-bound state (Neiditch *et al.*, 2005, 2006). Their deletion *in vivo* has the effect, therefore, of *increasing* the AI-2 sensitivity of *V. harveyi* cells. This result, initially surprising, is readily understood as a shift in equilibrium toward the ligand-bound state of the receptor. Taken together, these three tactics—high-throughput screening, parallel testing of conditions with and without excess ligand, and stabilization of the ligand-bound state by deleting the clasp sequence—yielded just one set of ligand-dependent crystallization conditions. The crystals were poorly formed and very small; furthermore, they grew out of a protein precipitate and did not appear to diffract X-rays, even when exposed to synchrotron radiation.

One additional step was required in order to obtain diffraction-quality LuxPQ$_p$ crystals containing AI-2. Our previous structure of the ligand-free complex displayed disorder (lack of interpretable electron density) at the N- and C-termini of LuxQ$_p$. This disorder was not entirely surprising, since each of the termini would normally be attached to a transmembrane helix. The production of useful crystals of AI-2-bound LuxPQ$_p$ depended crucially on the use of a new version of the complex in which the LuxQ$_p$ termini, in addition to the LuxP clasp, were deleted. This version formed high-quality crystals under the crystallization conditions identified by high-throughput screening. Since the crystals diffracted X-rays to 2.3 Å resolution, it was relatively straightforward (as these things go) to determine the structure.

Functional Analysis

The availability of structures for both ligand-free and ligand-bound LuxPQ$_p$ immediately suggested that ligand-induced quaternary structural changes regulate the activity of the histidine kinase (Fig. 1). To test this model, it was critical to be able to introduce structure-based modifications into the receptor *in vivo* and to test the impact of these modifications on AI-2 signaling in a well-defined functional assay. Together with our collaborators, we have further developed the bioluminescence-based bioassay in *V. harveyi* so that it is convenient and reproducible for these experiments. The foundation for the assay is the *V. harveyi* strain FED119, which lacks (1) LuxPQ, the AI-2 receptor; (2) LuxS, the AI-2 synthase; and (3) LuxN, the AI-1 receptor (Neiditch *et al.*, 2006). The last of these requires a word of explanation: AI-1 is a species-specific autoinducer produced and detected by *V. harveyi* (Cao and Meighen, 1989). Since the AI-1 signal transduction pathway feeds into the same downstream circuitry as the AI-2 pathway (Mok *et al.*, 2003), its operation can confound the AI-2 assay. FED119 cells, because they lack both AI-1 and AI-2 receptors, do not modulate their light production in response to cell density, nor do they respond to pure AI-2. Reintroducing plasmid-borne *luxPQ* into this strain restores AI-2 responsiveness. Furthermore, because the cells lack the AI-2 synthase, the precise AI-2 concentration can be set by the investigator.

Thus, the basis for the *V. harveyi* functional assay is the introduction of modified plasmid-encoded LuxPQ receptors into the strain FED119 and the subsequent measurement of bioluminescence in response to a range of AI-2 concentrations. AI-2 is created *in situ* by adding chemically synthesized DPD in the presence of 0.5 mM boric acid (Semmelhack *et al.*, 2005). By determining the DPD concentration required to induce 50% maximal bioluminescence (EC$_{50}$) for wild-type and mutant LuxPQ proteins, the impact of modifications on AI-2 signal transduction can be evaluated.

Bioluminescence Assay

1. These experiments use *V. harveyi* stain FED119 (*luxN$^-$*, *luxPQ$^-$*, *luxS$^-$*) transformed with a plasmid containing the *luxPQ* operon (either wild-type or modified). The plasmid, pFED368, also carries a chloramphenicol resistance gene (Neiditch *et al.*, 2006). From a freezer stock, grow cells in 5 ml AB medium (Greenberg *et al.*, 1979) containing 10 μg/ml chloramphenicol for about 24 h at 30° with shaking (200 rpm).

2. Dilute the overnight culture 1:10,000 into fresh medium supplemented with 10 μg/ml chloramphenicol and 0.5 mM boric acid.

3. Using the same medium, generate a five-fold dilution series containing fresh DPD in concentrations ranging from 100 μM down to 50 pM (a total of 10 different concentrations).

4. Dispense 90 μl of diluted bacteria into each well of a black-walled 96-well assay plate (Costar #3904). Aliquot 10 μl diluted AI-2 into each well; it is convenient to arrange the plate with AI-2 concentrations running across each row from high to low. The final concentration of AI-2 ranges from 10 μM to 5 pM.

5. Seal the plates with an adhesive plate cover and incubate in a shaking incubator at 30° for 16 to 20 h.

6. Measure the luminescence of each culture using a multi-well luminometer (Wallac Microbeta 1450). Determine the optical density of each culture (for example, by measuring the absorbance at 600 nm using a Wallac Victor 1420). Calculate normalized light units by dividing the luminescence by the optical density. For the greatest reproducibility, the optical density at which normalized light units are determined should be similar for all the cultures in an experiment.

7. Each dilution series should be performed in triplicate. Estimate the EC_{50} for each series using a curve-fitting program such as GraphPad Software's Prism. From the three independent estimates, calculate the mean EC_{50} and its standard deviation.

This assay was used to test models arising from the crystal structures of AI-2-free and AI-2-bound $LuxPQ_p$. The latter structure of AI-2-bound $LuxPQ_p$ displayed a set of novel subunit interfaces (Neiditch et al., 2006). To test whether these subunit interfaces were functionally significant, or simply crystallographic artifacts, over 40 single alanine substitutions were introduced into the interfaces and evaluated using the functional assay. About 40% of the alanine substitutions displayed a measurable change in the EC_{50}, with some causing desensitization by more than 100-fold. Thus, the results of the functional assay were in strong agreement with the inferences drawn from the crystal structure.

Conclusions

The integration of genetic/functional, biochemical, structural, and chemical approaches has proved useful in investigating AI-2-mediated quorum sensing at a molecular level. It appears likely that the application of a similar constellation of approaches will yield insight into other two-component systems, and will ultimately provide a more mechanistic understanding of the way—or ways—in which ligand binding is coupled to kinase regulation and gene expression in two-component systems.

Acknowledgments

We are immensely grateful to the Princeton colleagues who have participated in our studies of quorum sensing, including Bonnie Bassler, Mike Federle, Danielle Swem, Shawn Campagna, Xin Chen, Phil Jeffrey, Bob Kelly, Stephen Miller, Audra Pompeani, Stephan Schauder, Martin Semmelhack, Michi Taga, and Karina Xavier. Funding for our work in this area has been provided by the National Institutes of Health (AI-054442 and a postdoctoral fellowship to M. B. N.).

References

Bassler, B. L., Wright, M., and Silverman, M. R. (1994). Multiple signalling systems controlling expression of luminescence in *Vibrio harveyi*: Sequence and function of genes encoding a second sensory pathway. *Mol. Microbiol.* **13,** 273–286.

Cao, J. G., and Meighen, E. A. (1989). Purification and structural identification of an auto-inducer for the luminescence system of *Vibrio harveyi*. *J. Biol. Chem.* **264,** 21670–21676.

Chen, X., Schauder, S., Potier, N., Van Dorsselaer, A., Pelczer, I., Bassler, B. L., and Hughson, F. M. (2002). Structural identification of a bacterial quorum-sensing signal containing boron. *Nature* **415,** 545–549.

De Keersmaecker, S. C., Varszegi, C., van Boxel, N., Habel, L. W., Metzger, K., Daniels, R., Marchal, K., De Vos, D., and Vanderleyden, J. (2005). Chemical synthesis of (S)-4,5-dihydroxy-2,3-pentanedione, a bacterial signal molecule precursor, and validation of its activity in *Salmonella typhimurium*. *J. Biol. Chem.* **280,** 19563–19568.

DeLano, W. L. (2002). *The PyMOL Molecular Graphics System.* DeLano Scientific, San Carlos, CA, USA.

Dwyer, M. A., and Hellinga, H. W. (2004). Periplasmic binding proteins: A versatile superfamily for protein engineering. *Curr. Opin. Struct. Biol.* **14,** 495–504.

Felder, C. B., Graul, R. C., Lee, A. Y., Merkle, H. P., and Sadee, W. (1999). The Venus flytrap of periplasmic binding proteins: An ancient protein module present in multiple drug receptors. *AAPS PharmSci.* **1,** E2.

Freeman, J. A., and Bassler, B. L. (1999a). A genetic analysis of the function of LuxO, a two-component response regulator involved in quorum sensing in *Vibrio harveyi*. *Mol. Microbiol.* **31,** 665–677.

Freeman, J. A., and Bassler, B. L. (1999b). Sequence and function of LuxU: A two-component phosphorelay protein that regulates quorum sensing in *Vibrio harveyi*. *J. Bacteriol.* **181,** 899–906.

Greenberg, E. P., Hastings, J. W., and Ulitzer, S. (1979). Induction of luciferase synthesis in *Beneckea harveyi* by other marine bacteria. *Arch. Microbiol.* **120,** 87–91.

Hirel, P. H., Schmitter, M. J., Dessen, P., Fayat, G., and Blanquet, S. (1989). Extent of N-terminal methionine excision from *Escherichia coli* proteins is governed by the side-chain length of the penultimate amino acid. *Proc. Natl. Acad. Sci. USA* **86,** 8247–8251.

Inouye, M., and Dutta, R. (2003). "Histidine Kinases in Signal Transduction." Academic Press, San Diego.

Luft, J. R., Collins, R. J., Fehrman, N. A., Lauricella, A. M., Veatch, C. K., and DeTitta, G. T. (2003). A deliberate approach to screening for initial crystallization conditions of biological macromolecules. *J. Struct. Biol.* **142,** 170–179.

Meijler, M. M., Hom, L. G., Kaufmann, G. F., McKenzie, K. M., Sun, C., Moss, J. A., Matsushita, M., and Janda, K. D. (2004). Synthesis and biological validation of a ubiquitous quorum-sensing molecule. *Angew. Chem. Int. Ed. Engl.* **43,** 2106–2108.

Miller, S. T., Xavier, K. B., Campagna, S. R., Taga, M. E., Semmelhack, M. F., Bassler, B. L., and Hughson, F. M. (2004). *Salmonella typhimurium* recognizes a chemically distinct form of the bacterial quorum-sensing signal AI-2. *Mol. Cell* **15,** 677–687.

Mok, K. C., Wingreen, N. S., and Bassler, B. L. (2003). *Vibrio harveyi* quorum sensing: A coincidence detector for two autoinducers controls gene expression. *EMBO J.* **22,** 870–881.

Neiditch, M. B., Federle, M. J., Miller, S. T., Bassler, B. L., and Hughson, F. M. (2005). Regulation of LuxPQ receptor activity by the quorum-sensing signal autoinducer-2. *Mol. Cell* **18,** 507–518.

Neiditch, M. B., Federle, M. J., Pompeani, A. J., Kelly, R. C., Swem, D. L., Jeffrey, P. D., Bassler, B. L., and Hughson, F. M. (2006). Ligand-induced asymmetry in histidine sensor kinase complex regulates quorum sensing. *Cell* **126,** 1095–1108.

Schauder, S., Shokat, K., Surette, M. G., and Bassler, B. L. (2001). The LuxS family of bacterial autoinducers: Biosynthesis of a novel quorum sensing signal molecule. *Mol. Microbiol.* **41,** 463–476.

Semmelhack, M. F., Campagna, S. R., Federle, M. J., and Bassler, B. L. (2005). An expeditious synthesis of DPD and boron binding studies. *Org. Lett.* **7,** 569–572.

Stock, A. M., Robinson, V. L., and Goudreau, P. N. (2000). Two-component signal transduction. *Annu. Rev. Biochem.* **69,** 183–215.

Waters, C. M., and Bassler, B. L. (2005). Quorum sensing: Cell-to-cell communication in bacteria. *Annu. Rev. Cell Dev. Biol.* **21,** 319–346.

Xavier, K. B., and Bassler, B. L. (2003). LuxS quorum sensing: More than just a numbers game. *Curr. Opin. Microbiol.* **6,** 191–197.

Section II

Reconstitution of Heterogeneous Systems

[12]   Liposome-Mediated Assembly of Receptor Signaling Complexes

By DAVID J. MONTEFUSCO, ABDALIN E. ASINAS, and ROBERT M. WEIS

Abstract

The reconstitution of membrane-associated protein complexes poses significant experimental challenges. The core signaling complex in the bacterial chemotaxis system is an illustrative example: The soluble cytoplasmic signaling proteins CheW and CheA bind to heterogeneous clusters of transmembrane receptor proteins, resulting in an assembly that exhibits cooperative kinase regulation. An understanding of the basis for the cooperativity inherent in the receptor/CheW/CheA interaction, as well as other membrane phenomena, can benefit from functional studies under defined conditions. To meet this need, a simple method was developed to assemble functional complexes on lipid membranes. The method employs a receptor cytoplasmic domain fragment (CF) with a histidine tag and liposomes that contain a Ni^{2+}-chelating lipid. Assemblies of CF, CheW, and CheA form spontaneously in the presence of these liposomes, which exhibit the salient biochemical functions of kinase stimulation, cooperative regulation, and CheR-mediated receptor methylation. Although ligand binding phenomena cannot be studied directly with this approach, other factors that influence kinase stimulation and receptor methylation can be explored systematically, including receptor density and competition among stimulating and inhibiting receptor domains. The template-directed assembly of proteins leads to relatively well-defined samples that are amenable to analysis by a number of methods, including light scattering, electron microscopy, and fluorescence resonance energy transfer. The approach promises to be applicable to many systems involving membrane-associated proteins.

Introduction

This chapter presents the use of liposomes (vesicles) as templates for the assembly of membrane-associated signaling proteins. It focuses principally on applications to the *Escherichia coli* chemotaxis pathway, although it should be apparent that the approach is broadly applicable to many membrane systems, including other chemotaxis-like signaling pathways, two-component pathways, and eukaryotic signaling pathways involving single-pass transmembrane receptors.

METHODS IN ENZYMOLOGY, VOL. 423   0076-6879/07 $35.00
          DOI: 10.1016/S0076-6879(07)23012-5

The character actors in this method have been in existence for some time; they have simply assumed somewhat modified roles here. Specifically, lipids for tethering proteins to the membrane surfaces first found use in structural biology—to crystallize proteins in two dimensions (Uzgiris and Kornberg, 1983), which has the advantage of simplicity and requires small amounts of protein. The synthesis of lipids with a nickel nitrilotriacetic acid (Ni^{2+}-NTA) headgroup was significant, because it allowed the method to be extended to histidine-tagged proteins (Kubalek *et al.*, 1994; Schmitt *et al.*, 1994). As simply a tool for structure determination, the proteins studied with this approach were often not normally associated with membranes as a part of their biochemical function. On the other hand, engineered fragments of transmembrane receptors and membrane-associated proteins removed from the membrane are often devoid of biochemical function. Thus, the use of the Ni^{2+}-NTA lipid to regenerate the membrane association seemed propitious as an approach for restoring biochemical function to receptor fragments and for hierarchically assembling complexes of membrane-associated proteins. This chapter hopes to convince the reader that this expectation has been met, and of the general utility of the approach.

In this chapter, biochemical studies of kinase stimulation and methylation are described to highlight the utility of the method. Following this and/or coincident with it, the analysis of template-assembled samples with other methods—dynamic light scattering, fluorescence energy transfer, and electron microscopy—are described (in their most basic forms) to verify surface-assembly and to probe the properties of surface-assembled proteins in greater detail. These descriptions are really just vignettes, since each of these approaches merits a chapter of its own. The *Methods* section outlines the basics of template-directed assembly and provides leading references to the extensive literature on liposomes. The chapter concludes with a summary and forward-looking statements.

Transmembrane Receptors and Reconstitution

Receptor proteins are located in the two-dimensional environment of the cell membrane, an environment that promotes essential functional interactions among receptors, with membrane-associated signaling proteins. The removal of the membrane results in the loss of these functional interactions. Such features are evident in the chemotaxis signaling systems of *E. coli* and *Salmonella enterica* serovar Typhimurium, well-studied examples of two-component signaling systems that utilize transmembrane receptor proteins known as methyl-accepting chemotaxis proteins (MCPs) (Antommattei and Weis, 2006). In the membrane, MCPs form extended arrays that recruit the adaptor protein, CheW, and the kinase, CheA (Lybarger and Maddock, 2001; Maddock and Shapiro, 1993); the resulting complex of proteins exhibits cooperative regulation of the kinase activity (Bornhorst and Falke, 2000; Levit *et al.*, 2002; Li and Weis, 2000; Sourjik and Berg, 2004). The conserved cytoplasmic signaling domain is organized as a coiled-coil, which dimerizes to form a long ($\sim$200 Å) four-helix bundle (Kim *et al.*, 1999; Park *et al.*, 2006); this domain is the defining feature of MCPs (Le Moual and Koshland, 1996). Removed from the membrane, the cytoplasmic fragment (CF) often exhibits none of the biochemical function associated with the cytoplasmic domain in the intact receptor (Long and Weis, 1992; Seeley *et al.*, 1996). Assemblies of CF, CheW, and CheA that stimulate kinase activity can be formed either by protein engineering (Cochran and Kim, 1996; Francis *et al.*, 2004; Surette and Stock, 1996) or through spontaneous assembly (Ames and Parkinson, 1994; Ames *et al.*, 1996; Wolanin *et al.*, 2006), but the two-dimensional organization of the membrane, and the relevant interactions that go with it, are not restored in these complexes.

The membrane environment is essential for detailed studies of function. Unfortunately, the expression, purification, and reconstitution of membrane proteins adds significant complexities that are not encountered in studies of soluble proteins (Racker, 1985; Rigaud and Levy, 2003; Seddon *et al.*, 2004). Challenges that must be faced during reconstitution include (a) the determination of protein orientation in the membrane, (b) the presence of hard-to-remove impurities in membrane preparations, (c) achieving control over the protein-to-lipid ratio, and (d) the incompatibility of detergents with assembly and function. Some early success was achieved in reconstituting the basic methylation and kinase-stimulating functions of MCPs in vesicles (Ninfa *et al.*, 1991) and a soluble lipid–detergent system (Bogonez and Koshland, 1985), but these approaches have given way to the use of inner membrane preparations, in which MCPs are overexpressed (Bornhorst and Falke, 2000; Gegner *et al.*, 1992; Li and Weis, 2000; Lin *et al.*, 1994; Russo and Koshland, 1983). The electron microscopic analyses of overexpressed Tsr in cells and inner membrane preparations reveal extensive interactions among the cytoplasmic signaling domains of receptors and different modes of interaction, each possibly with a role to play in signaling (Lefman *et al.*, 2004; Weis *et al.*, 2003; Zhang *et al.*, 2004). Reconstitution methods that facilitate the systematic assembly of signaling components can help to identify the relevant interactions.

Template-Directed Assembly of Receptor Fragments and
  Signaling Complexes

Assembling MCP cytoplasmic domain fragments on the surface of a liposome recreates, in a controlled manner, the 2-dimensional organization that these domains possess in the intact receptor in the cell

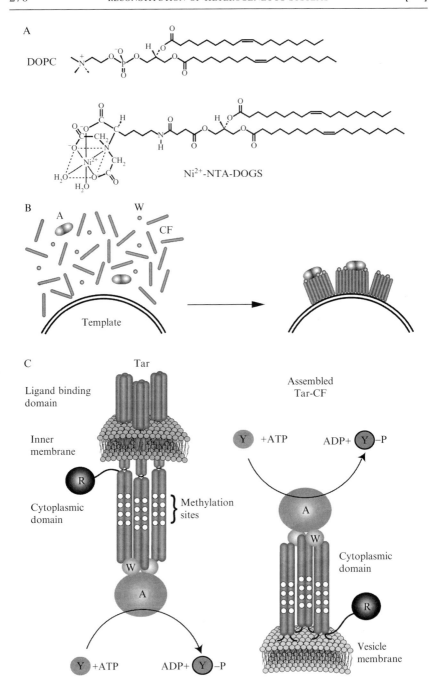

(Asinas and Weis, 2006; Montefusco *et al.*, 2007; Shrout *et al.*, 2003). Figure 1 depicts the concept, the components, and the process. The vesicle (liposome) membrane contains a chelating lipid, such as the nickel salt of 1,2-Dioleoyl-*sn*-Glycero-3-[[N(5-Amino-1-Carboxy-pentyl)iminodiAcetic Acid]Succinyl] (Ni^{2+}-NTA-DOGS) (Kubalek *et al.*, 1994; Schmitt *et al.*, 1994). Typically, the liposomes also contain another lipid, such as 1,2-Dioleoyl-*sn*-Glycero-3-Phosphocholine (DOPC), to stabilize the bilayer structure and to generate different surface concentrations of protein. The liposomes function as the templates (Fig. 1B) that bind histidine-tagged forms of MCP cytoplasmic domain fragments (CFs), which—by localizing and orienting the CFs on the membrane surface—promote interaction. In comparison to CFs in solution, surface-associated CFs are able to participate in efficient methylation, bind CheW and CheA, and stimulate kinase activity (Shrout *et al.*, 2003). Template-directed assembly offers several advantages over other methods used to reconstitute two component signaling systems: (1) the active system is assembled from purified *soluble* proteins, providing a high degree of control over sample composition; (2) there are no inherent limitations on treatment of reagents prior to mixing, and proteins may be labeled or pooled into mixtures prior to binding to vesicles; and (3) the method does not rely on detergents, but it is compatible with mild detergents.

As Fig. 1C depicts, the regulation of CheA activity occurs via clustered receptor dimers; a cluster of receptors in the membrane facilitates formation of the ternary MCP-CheW-CheA protein complex (Gegner *et al.*, 1992; Schuster *et al.*, 1993). Receptor dimers must also be in close proximity for receptor transmethylation to take place (Antommattei and Weis, 2006; Antommattei *et al.*, 2004; Le Moual *et al.*, 1997; Li *et al.*, 1997; Wu *et al.*, 1996). During methylation, the methyltransferase (CheR, R) binds to a conserved segment at the C-terminus of either Tsr or Tar (Antommattei and Weis, 2006; Wu *et al.*, 1996), where it is tethered close to the methyl-accepting glutamate residues on the cytoplasmic domains of all nearby MCP dimers (Tap, Tar, Trg, or Tsr) (Li and Hazelbauer, 2005; Windisch *et al.*, 2006; Wu *et al.*, 1996). Based on the published and unpublished data

---

FIG. 1. Template-directed assembly. (A) Structures of DOPC and the nickel-chelating lipid Ni^{2+}-NTA-DOGS. (B) Histidine-tagged chemoreceptor cytoplasmic domain fragments (CF) bind to vesicles (template) containing Ni^{2+}-NTA-DOGS. CheW (W) and CheA (A) then bind to surface-assembled CF. (C) Clusters of receptor dimers (left) regulate the kinase CheA, which phosphorylates CheY (Y) and CheB (not shown). Receptor methylation, which is catalyzed by CheR (R) and occurs on specific glutamate residues (depicted as open circles) in the cytoplasmic domain, also requires close proximity among receptors. Template-assembled CFs (right) have a similar organization, which facilitates transmethylation and kinase stimulation.

summarized in this chapter, it seems that template-assembled cytoplasmic domain fragments form, to a significant extent, the same interactions as the cytoplasmic domains in the intact receptor dimer (Fig. 1C, right).

Results—Biochemical Activity of Liposome-Assembled Receptor Fragments

Effects of Assembly on Kinase Activity and Methylation

Figure 2 shows a typical effect on kinase activity when the signaling proteins (30 μM Tar CF_{QEQE}, 5 μM CheW, 1.2 μM CheA) are incubated with vesicles—in this case, sonicated unilamellar vesicles (SUVs). Specific kinase activities were measured with the coupled ATPase assay (Shrout *et al.*, 2003). The two samples differed only by the presence of SUVs (580 μM total lipid, 1:1 Ni^{2+}-NTA-DOGS:DOPC). The specific activity of CheA in the presence of the SUVs (9.5 s^{-1}) is ~130-fold larger than the specific activity of dimeric CheA in solution (0.075 s^{-1}, 5.0 μM in CheA monomer). In some samples, 400-fold increases in kinase activity can be generated. Several factors influence the stimulation of kinase activity by vesicles, including the total lipid and protein concentrations, lipid composition, the method of liposome preparation, and the CF protein itself. Some of these influences are easily understood, for example, mass-action phenomena, while others are leading to a better understanding of the requirements for kinase stimulation and regulation.

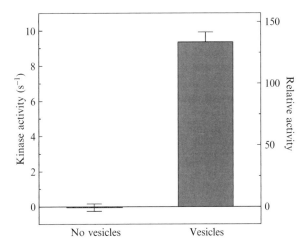

FIG. 2. Stimulation of CheA activity by vesicle-mediated assembly. Protein concentrations: 30 μM CF_{QEQE}, 1.2 μM CheA, and 5 μM CheW. The vesicle-containing sample contains SUVs (580 μM total lipid, 1:1 Ni^{2+}-NTA-DOGS:DOPC mole ratio).

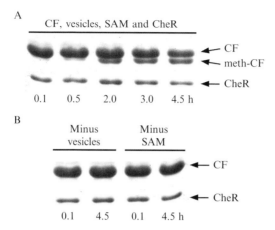

Fig. 3. Template-directed assembly facilitates methylation of CFs. (A) The time course of methylated CF (meth-CF) formation in a sample with 30 μM CF$_{EEEE}$, 6 μM CheR, 10 μM s-adeonsyl-L-methionine (SAM), and SUVs (580 μM lipid, 1:1 Ni^{2+}-NTA-DOGS:DOPC). (B) Control samples with the composition as the sample in A, but without either SUVs or SAM. Reprinted from Shrout *et al.* (2003) with the permission of the American Chemical Society.

SUVs containing the metal-chelating lipid also restore the substrate activity of CF$_{EEEE}$ in the methylation reaction. The Coomassie-stained protein bands in the SDS polyacrylamide gels of Fig. 3 show that Tar CF$_{EEEE}$ is methylated when SUVs, the methyltransferase (CheR), and s-adenosyl-L-methionine (SAM) are all present, but not when either SUVs or SAM are omitted (controls to show, respectively, the effect of template and eliminate proteolysis as the cause of the mobility shift).

Liposome Preparation, Protein Assembly, and Activity

Liposomal membranes are adaptable platforms for assembling proteins. The five volumes of *Methods in Enzymology* edited by Düzgünes treat numerous aspects of a large literature on preparing liposomes, their properties, and biomedical applications (Düzgünes, 2003a,b,c, 2004, 2005). Although the procedures for making liposomes are many, only two—bath sonication (Asinas and Weis, 2006; Shrout *et al.*, 2003) and extrusion through filter membranes of defined porosity (Montefusco *et al.*, 2007)—have been used to assemble chemoreceptor signaling complexes. Both methods reliably generate unilamellar vesicles, in which 50% of the lipid resides in the membrane outer leaflet, accessible to added protein. SUVs (for *small* as well as *sonicated*) are small by comparison to unilamellar vesicles prepared through the process of extrusion (LUVs).

The association of CF with vesicles approaches 100% at micromolar concentrations of histidine tagged protein and Ni^{2+}-NTA. The extent of association can be determined readily by sedimentation and gel-scanning to separate and measure the relative amounts of bound and unbound CF. Empirical determinations of the appearance of kinase activity are likewise simple to conduct, because of the relatively slow rate ($t_{1/2} \sim 30$ min) of the process. The activity appearance and disappearance data of Fig. 4 are representative time courses observed with SUV templates. An appearance-of-activity experiment (Fig. 4A) starts when liposomes, CF, CheW, and CheA are mixed together under prescribed assembly conditions: 580 μM total lipid (1:1 DOPC:Ni^{2+}-NTA-DOGS), 30 μM CF, 5 μM CheW, and 1.2 μM CheA (Montefusco et al., 2007). Aliquots are withdrawn at various times and assayed for ATPase activity. The relatively slow rate of appearance suggests that the molecular requirements for stimulating kinase activity are more involved than simple protein association, as in, for example, the formation of a multi-subunit protein complex.

The rates of activity disappearance are determined with pre-assembled samples, that is, samples that have been given sufficient time to reach a steady extent of assembly. Assembled samples are diluted, for example, 100-fold, incubated for an indicated time period (Fig. 4B), and assayed for kinase activity. The disappearance of activity provides evidence that assembly is

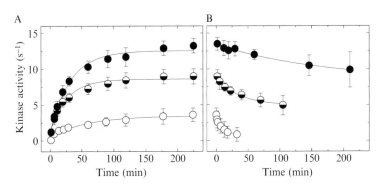

FIG. 4. Appearance and disappearance of stimulated CheA activity. Activities in (A) were measured as a function of time after mixing SUVs and proteins to give 30 μM CF, 1.2 μM CheA, 5 μM CheW, and 580 μM lipid (1:1 DOPC:Ni^{2+}-NTA-DOGS). Activities in B were measured as a function of time following a 100-fold dilution of samples assembled (as in A) for 4 h. Curves are least-squares fits to single-exponential associations (A), and single-exponential decays (B). The data points and uncertainties are averages and standard deviations, respectively, of six independent measurements. Symbol legend: CF_{EEEE} (○), CF_{QEQE} (◕), and CF_{QQQQ} (●). Reprinted from Montefusco et al. (2007) with permission of the American Chemical Society.

reversible and implies that the rate data can be used to infer qualitative trends in the stability of signaling complexes. With SUV templates, the activity disappearance rates decrease as the covalent modification on the CF increases. The $t_{1/2}$ (disappearance), determined from single exponential decays, are 6, 25, and 125 min for unmodified, partly modified, and fully modified CF, respectively. That covalent modification contributes to the stability of the signaling complex is a plausible explanation for the trend in activity disappearance ($t_{1/2}$) data, although additional experiments are required to discern the molecular basis of stability.

The acquisition of data is made more feasible with the simple procedure for making samples by liposome-mediated assembly and the comparatively modest requirements for purified protein and lipid ($\sim$2 μg CF, 0.2 μg CheA, 0.2 μg CheW and 1.0 μg lipid per measuring aliquot). Figure 5A and B are kinase activities and CheA binding, respectively—at equilibrium extents of

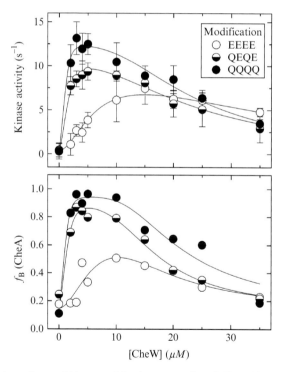

FIG. 5. The dependence of kinase activity (upper panel) and CheA binding (lower panel) on the CheW concentration. All samples contained 30 μM CF, 1.2 μM CheA, and SUVs made with a 1:1 mole ratio of Ni^{2+}-NTA-DOGS and DOPC (560 μM total lipid). Data are shown for CF$_{EEEE}$ ($\bigcirc$), CF$_{QEQE}$ ($\ominus$), and CF$_{QQQQ}$ ($\bullet$). Reprinted from Shrout $et\ al.$ (2003) with permission.

assembly—measured as a function of the CheW concentration and CF covalent modification (Shrout et al., 2003). These data are informative in several ways: (i) CheW is required to stimulate kinase activity and promotes CheA binding. The binding of CheA and CheW to the CF is synergistic, which is displayed as a rapid rise in activity and the fraction of CheA bound (f_B) at the low CheW concentrations (1 to 5 μM)—especially in complexes formed with CF in intermediate and high levels of glutamine substitution (CF_{QEQE} and CF_{QQQQ}). The synergy that CheA and CheW exhibit in binding to the CF, evident in the data of Fig. 5, has been observed in other experiments of template-assembled complexes (Asinas and Weis, 2006) and also in studies of signaling complexes with full-length Tsr (Levit et al., 2002). (ii) The data in Fig. 5 are also consistent with greater stability of complexes containing CF at higher modification levels. For CF_{EEEE}-containing complexes, the maximum fraction of bound CheA ($f_B \sim 0.5$) is significantly smaller than in complexes containing either CF_{QEQE} and CF_{QQQQ}, and the CheW concentration required to reach this maximum is larger. Both of these features are consistent with a CF_{EEEE}/CheW/CheA complex of lower stability. (iii) CheW competes with CheA for binding to the receptor at larger CheW concentrations. (iv) Complexes formed with CF in any of the three levels of covalent modification (EEEE, QEQE, or QQQQ) stimulate kinase activity.

Experiments comparing SUVs and LUVs have been conducted to determine their similarities and differences as templates for signaling complex assembly and restoring kinase activity (Montefusco et al., 2007). Figure 6 presents specific activity and f_B(CheA) data obtained with template-assembled complexes containing CF_{EEEE}, CF_{QEQE}, or CF_{QQQQ} (low, intermediate, and high levels of modification, which are represented by the open, gray, and black bars, respectively). In all samples, the kinase activities of surface-associated CheA (the specific activities of Fig. 6A divided by the f_B of Fig. 6B) are significantly greater than the kinase activity of dimeric CheA in solution (0.075 s^{-1}), even with the most weakly stimulating sample (CF_{EEEE} on SUVs). The salient differences between SUVs and LUVs are best observed as trends in CheA binding and stimulation as a function of CF modification. On SUVs, CheA binding and stimulation both increase with modification; the activity is significantly lower in signaling complexes assembled on SUVs with CF_{EEEE}. The disparity is reduced, but not eliminated, when the activities of surface-associated CheA are computed (as described previously and in Montefusco et al., 2007). It seems that SUVs decorated with CF_{EEEE} are less effective at recruiting and stimulating CheA, by comparison to SUVs decorated with either CF_{QEQE} or CF_{QQQQ}. The situation is different with LUV templates; the trend in

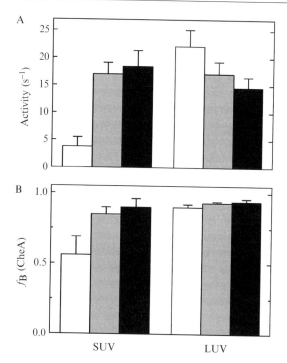

Fig. 6. Kinase activity (A) and the fraction of CheA bound (B) as a function of vesicle preparation method (sonicated unilamellar vesicles, SUVs, and large unilamellar vesicles, LUVs) and CF modification (CF_{EEEE}, CF_{QEQE}, and CF_{QQQQ}: open, gray and black bars, respectively). Vesicles were formed with a 1:1 Ni^{2+}-NTA-DOGS:DOPC (580 μM total lipid). Proteins were present at 30 μM CF, 1.2 μM CheA, and 5 μM CheW. The data points and uncertainties are averages and standard deviations, respectively, from at least four independent measurements. Reprinted from Montefusco et al. (2007) with permission.

activity is *reversed*, which decreases modestly between CF_{EEEE}- and CF_{QQQQ}-containing complexes ($\sim$35%). In addition, CheA binds effectively (=90%) to LUVs decorated with CFs at all three modification levels. In the results of Fig. 6, the principal difference between SUV and LUV templates is the lower recruitment and stimulation of CheA on SUVs into CF_{EEEE}-containing complexes.

The practical consequence of these and other observations is a preference for using LUVs as templates whenever reconstitution of function is the primary objective. For reconstitution of function, the low recruitment efficiency (f_B) and activity of surface-associated CheA are undesirable and, by contrast, the high levels observed with LUVs decorated with CF in any level of covalent modification are preferred. Apart from the observations in Montefusco et al. (2007), little has been done to explore the differences among SUV and LUV templates. In our studies, SUVs were prepared by bath sonication of multilamellar vesicles (MLVs) until the milky appearance of the MLV suspension was lost (Asinas and Weis, 2006; Montefusco et al., 2007; Shrout et al., 2003). The marked reduction in turbidity is a convenient indicator that the conversion to small unilamellar vesicles is essentially complete. When it is necessary to remove residual MLVs, a separation step should be added, for example, sedimentation or size-exclusion chromatography (Cornell et al., 1982; Szoka and Papahadjopoulos, 1980). Moreover, if the objective is to achieve the small size limit for vesicles, then (more powerful) probe sonication should be used (Cornell et al., 1982).

Extrusion of LUVs through filters is, like sonication, simple. It has been tested and found to be reliable on small and large (process) scales (Mayer et al., 1986; Mui et al., 2003; Olson et al., 1979). The extrusion method has the additional important advantage of control over vesicle diameter. Relatively narrow size distributions can be obtained with different mean vesicle diameters, which are determined by the pore size in the filters; unilamellar vesicles are more easily generated with smaller pore diameter filters (<200 nm). Vesicles produced by extrusion have been characterized thoroughly for size and lamellarity (as reviewed in Mui et al., 2003) through the use of various methods, such as electron microscopy, light scattering, and NMR. The average number of lamellae of a vesicle preparation containing NTA-DOGS can be determined by ITC from titrations of vesicles with Ni^{2+}, where 50% of the NTA sites in a preparation of unilamellar vesicles should be accessible for Ni^{2+} binding. The precision of the measurement is improved with a matched pair of samples, in which one sample is titrated without detergent present (~50% accessible) and the other is titrated after disruption by detergent, in which case all the sites are accessible (Montefusco et al., 2007). The hydrodynamic radii of unfractionated SUVs prepared by bath sonication, and of LUVs prepared by extrusion, are listed in Table I.

TABLE I
HYDRODYNAMIC RADII (nm) OF LIPOSOMES USED AS TEMPLATES[a]

SUVs	LUVs (Pore Diameter, nm)		
	50	100	1000
24 ± 18	39 ± 31	59 ± 34	320 ± 210

[a] Radii are maxima, and ranges are half-widths at half height, of distributions.

*Characterization of Protein Assembly with Light Scattering and
Electron Microscopy*

Dynamic light scattering (DLS) is a useful diagnostic for monitoring template-directed assembly of protein complexes. Figure 7A shows background-subtracted, normalized autocorrelation data of LUVs, LUVs in the presence of CF, and LUVs in the presence of CF, CheW, and CheA. It is evident that CF promotes vesicle aggregation, because the correlation time at half-maximum ($\tau_{1/2}$) is largest in these samples ($\tau_{1/2}$ grows in proportion to the size of the scatterers), and extensive aggregation is suppressed by the addition of CheW and CheA. When the autocorrelation data are deconvoluted, single mode distributions are obtained. From these distributions, the hydrodynamic radii (R_H) at the maxima of the distributions are computed to serve as the single parameter characteristic of the particle sizes in the sample. Measurements of the kinase activity and R_H on the same samples, over a range of CheW concentrations (Fig. 7A and B, respectively), have helped to elucidate the extent of vesicle aggregation in the system when the level of kinase stimulation is at its greatest. The activities and R_H are negatively correlated with each other; R_H trends toward a minimum and activity trends toward a maximum between 0 and 20 μM CheW. These data provide further evidence that pronounced vesicle aggregation is induced by the CF, but only when CheW and CheA are not present. The values of R_H in the presence of CF (30 μM), CheA (1.2 μM), and CheW ($\geq$10 μM) are larger than the R_H of vesicles alone by the length of the CF (20 nm). It is tempting to interpret this increase in R_H literally, but the assumptions made in the course of generating the estimates of R_H warrant additional investigation to establish the relationship between changes in R_H and the organization of protein on the vesicle surface. Nonetheless, these data do demonstrate that extensive vesicle aggregation—and thus extensive interactions between CFs in different vesicles—is not required to stimulate kinase activity.

The complexities observed in the assembly of CF/CheW/CheA complexes are a consequence of the numerous possible interactions among the different protein components: the tendency of cytoplasmic domain to self-associate, which is tunable by covalent medication, the hierarchic and polymorphic properties of cytoplasmic domain assembly, and the complexities of the interaction of cytoplasmic domains with CheW and CheA, which are synergistic under certain conditions and competitive under others. The observations of CF-induced vesicle aggregation, and the suppression of it by CheW, are plausible consequences of these interactions. The CF has a relatively weak propensity to self-associate in solution (but tunable through mutation). Attachment of the CF to the liposome surface promotes self

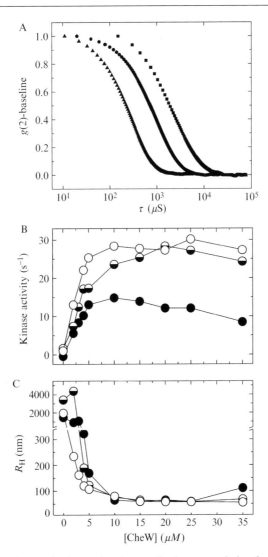

Fig. 7. (A) Background-subtracted and normalized autocorrelation functions of LUVs, (▲), LUVs with CF$_{QEQE}$, CheA and CheW (●), and LUVs with CF$_{QEQE}$ (■) determined by dynamic light scattering (DLS). The LUVs were formed by extrusion, using filters with 50 nm diameter pores. (B) Kinase activity and (C) hydrodynamic radii (R_H, nm) of CF/CheW/CheA complexes assembled with either CF$_{EEEE}$ (○), CF$_{QEQE}$ (◒), or CF$_{QQQQ}$ (●) on LUVs as a function of the CheW concentration. Reprinted from Montefusco *et al.* (2007) with permission.

association and vesicle aggregation, probably through the types of interactions observed in electron microscope images of the intact receptor in membranes and in crystal structures of the CF (Kim *et al.*, 1999; Lefman *et al.*, 2004; Park *et al.*, 2006; Weis *et al.*, 2003; Zhang *et al.*, 2004). Also, CheW and CheA can promote end-to-end arrangements of CFs by a bridging interaction (Francis *et al.*, 2004; Wolanin *et al.*, 2006), but these are disrupted when CheW and CheA are present at larger concentrations.

Intentionally, Fig. 1A does not illustrate all the complexities in the assembly of liposome-CF/CheW/CheA samples, since conditions can readily be found that correspond well to the scheme depicted in the figure. The micrographs in Fig. 8 show liposome-associated signaling complexes in samples applied to carbon-coated EM grids and stained with uranyl acetate. The liposomes in these samples were prepared by extrusion, through filters with 50 nm or 1 μm diameter pores (Fig. 8A and B, respectively). The shapes of vesicles in Fig. 8A (circular, sausage-like) are representative of the forms observed. These shapes may reflect the influence of osmotic effects—since similar shapes are observed in samples without protein present (Mui *et al.*, 2003)—but it is probable that CF binding (alone, and with CheW and CheA) contributes to shape changes (Montefusco *et al.*, 2007). In the sample prepared with the larger vesicles (Fig. 8B), the mottling produced by the stain is plausibly interpreted as protein clusters on the liposome surface (vesicles without protein do not show this effect; Montefusco *et al.*, 2007). Also, at the vesicle edge, staining patterns consistent with side-on views of signaling complexes, such as those shown in Fig. 1A, can be seen (noted by the arrows in Fig. 8B).

Signaling Complexes Assembled from Binary Mixtures of CFs

The ability to assemble mixtures of two (or more) different CFs onto the same membrane is a significant advantage of liposome-mediated assembly, because these samples can be used to probe interactions among cytoplasmic domains (Asinas and Weis, 2006). A large number of mutations in the cytoplasmic domain of the MCPs have been generated in various ways, such as through random chemical mutagenesis and selection, or structure-guided mutagenesis, and the effects on kinase activity have been characterized (Ames and Parkinson, 1988; Ames *et al.*, 2002; Mutoh *et al.*, 1986). Figure 9 indicates the kind of results that are obtained with binary mixtures of wildtype (kinase-stimulating) CF and point mutants of the CF, which fail to stimulate kinase activity (*ns*-CF). Each *ns*-CF, when it is the only CF present in the sample, is unable to stimulate kinase activity. Each *ns*-CF, in mixtures with the *wt*-CF, exhibits a different manner in which the kinase-stimulating activity is lost. Template-directed assembly in binary mixtures gives clues to the mechanism of kinase regulation that neither the mutant cell phenotype, the absence

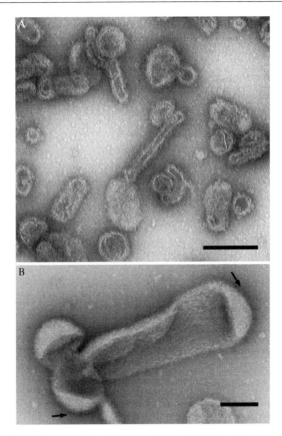

FIG. 8. Negative-stain transmission electron micrograph of signaling proteins assembled on LUVs. LUVs were prepared by extrusion through filters with pore diameters of either 50 nm (A) or 1 μM (B), and were incubated with either CF_{QQQQ} (A) or CF_{EEEE} (B), and CheW and CheA for 4 h. Samples contained LUVs (580 μM total lipid, 1:1 Ni^{2+}-NTA-DOGS:DOPC), 30 μM CF, 5 μM CheW, 1.2 μM CheA. Scale bars are 200 nm. Reprinted from Montefusco *et al.* (2007) with permission.

of kinase activity caused by the mutant protein by itself, nor a single wildtype/ mutant protein mixture, can provide. The properties displayed by the three mutants in Fig. 9 are representative of results obtained with a larger panel of *ns*-CFs (Asinas and Weis, 2006), which were sorted into three classes that: (1) bound CheW and CheA more weakly than *wt*-CF (○), (2) bound CheW and CheA more tightly (▲), and (3) inhibited kinase activity in a cooperative manner (●). The analysis of these and other mutants supports a model of CF/ CheW/CheA interaction in which CheW and CheA bind synergistically to the CF, but also compete for binding to the same or overlapping sites on the CF.

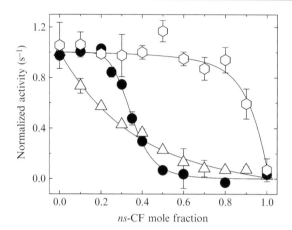

Fig. 9. Kinase activity as a function of the mole fraction of nonstimulating (point mutant) CF_{EEEE} (*ns*-CF) in mixtures with wildtype CF_{EEEE}. Inhibition curves are shown for three *ns*-CFs: S325L ($\triangle$), S461L ($\bullet$) and E383A ($\bigcirc$). Experiments were carried out with SUVs (580 μM total lipid, 1:1 Ni^{2+}-NTA-DOGS:DOPC), 30 μM total CF, 5 μM CheW and 1.2 μM CheA. Reprinted from Asinas and Weis (2006) with the permission of the ASBMB.

In addition, some mixtures exhibit cooperative change in CheA activity and binding. Importantly, the data procured with template-assembled CFs show a good correspondence with studies conducted on intact receptors.

Effects of Varying Surface Concentration

The liposome-mediated surface-assembly method is also very well suited for generating sets of samples in which the average area available to the CF is adjusted. To estimate the average surface area per CF, three features of the system must be known: (1) the percentage binding of the CF to liposomes, (2) the concentrations of CF and lipid, and (3) the fraction of the lipid in the vesicle preparation in the outermost leaflet of the membrane (and accessible for binding CF). This percentage is 50 for unilamellar vesicles prepared by extrusion, <50% for MLVs, and >50% for SUVs in the small size limit. The experiments described so far in this chapter have been conducted with one lipid-CF composition that corresponds to "assembly conditions": 30 μM CF, 580 μM lipid (1:1 Ni^{2+}-NTA-DOGS:DOPC). Under these conditions, essentially all the CF (>95%) is associated with the vesicle surface (which is estimated through the co-sedimentation of CF with vesicles), at a molar ratio of accessible lipid to CF close to 10 (290/30). The area of membrane surface per molecule of DOPC is reported to be 0.70 nm^2

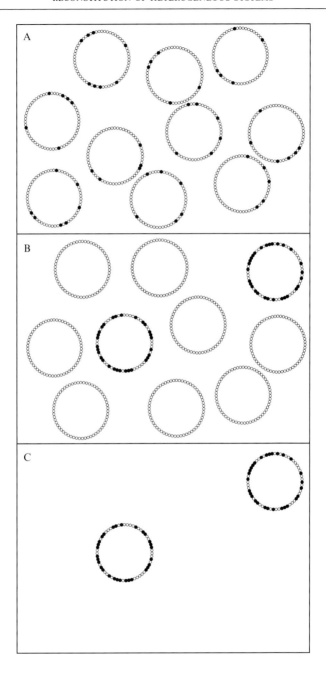

(Nagle and Tristram-Nagle, 2000) and, assuming this same area for Ni^{2+}-NTA-DOGS (both molecules have oleic acid acyl chains, Fig. 1A), the area per CF is 7 nm^2 (or 14 nm^2 per CF dimer).

Experiments that probe binding and activity of surface-associated proteins as a function of surface density are typically conducted by varying the mole fraction of Ni^{2+}-NTA-DOGS in the membrane, while holding the Ni^{2+}-NTA-DOGS and CF concentrations constant. In these experiments, it is preferable to have a consistent (high) fraction of bound protein through the range of densities studied. Under certain circumstances, such as when the membrane association of CheA is not complete ($<90\%$), relatively small changes in the experimental parameters can exert an influence on the extent of protein binding (and activity). In these situations, it may be desirable to keep the total lipid content in the sample constant, which can be achieved by keeping the vesicle concentration constant. To do this, liposomes of two different compositions are mixed: liposomes that contain Ni^{2+}-NTA-DOGS to bind CF at the specified average surface concentration, and liposomes that do not contain Ni^{2+}-NTA-DOGS, which serve to keep the total liposome concentration, and the excluded volume, constant. Figure 10 illustrates how the excluded volume may be kept constant with liposome mixtures. Panel A depicts a uniform sample of 1:10 Ni^{2+}-NTA-DOGS/DOPC LUVs. Panel B depicts a sample that is composed of the same amount of lipid, and thus the same number of liposomes (and excluded volume), but a different distribution of Ni^{2+}-NTA-DOGS within the sample, in this case, a 20 to 80% mixture of 1:1 Ni^{2+}-NTA-DOGS/DOPC LUVs and DOPC LUVs. For comparison, Fig. 10C shows a sample of 1:1 Ni^{2+}-NTA-DOGS/DOPC LUVs, which can generate the same surface density of template-assembled proteins, but has a lower excluded volume. Figure 10 exaggerates the changes in fraction of volume excluded for purposes of illustration. With the typical lipid concentration used in liposome-mediated assembly (600 μM), the entrapped volume is small ($\sim$0.2% for 100 nm diameter LUVs). Consequently, the change in the *accessible* volume will be small, for example, $\sim$1% for the situations that correspond to the differences in lipid composition between Fig. 10B and C. Excluded volume effects generally become significant only under highly crowded conditions (Chebotareva *et al.*, 2004).

FIG. 10. Excluded volume effects in liposome solutions. Liposomes are depicted by mixtures of DOPC (○) and Ni^{2+}-NTA-DOGS (●). In (A) and (B) the excluded volumes (number of liposomes) and the CF binding site concentration ([Ni^{2+}-NTA-DOGS]) are kept constant, but the binding site density is changed: 10% Ni^{2+}-NTA-DOGS (A) and 50% Ni^{2+}-NTA-DOGS (B) in the Ni^{2+}-NTA-DOGS-containing liposomes. In (B), liposomes composed of DOPC are added to keep the excluded volume constant. These are not included in the sample depicted in (C).

In experiments with the template-assembled signaling system, both receptor methylation and kinase stimulation have been found to depend significantly on the average surface concentration of CF, CheW, and CheA (Bessechetnova, Montefusco, Shrout, Antommattei, Asinas and Weis, in preparation). Representative results from kinase stimulation experiments conducted as a function of the surface concentration are plotted in Fig. 11A. LUVs prepared with different binary mixtures of Ni^{2+}-NTA-DOGS and DOPC, where the mole percentage of Ni^{2+}-NTA-DOGS ranged from 5 to 60%, were used to assemble CF/CheW/CheA complexes. The LUVs were present in different amounts, calculated to give the same overall Ni^{2+}-NTA-DOGS (290 μM) and protein concentrations (30 μM CF, 5 μM CheW, and 1.2 μM CheA). CF (and CheA) bound to all these LUVs at levels close to 100% so that the approximate membrane surface areas per CF were estimated to vary from 70 nm^2 (140 nm^2 per CF dimer) on LUVs containing 5% Ni^{2+}-NTA-DOGS to 6 nm^2 (12 nm^2 per CF dimer) on

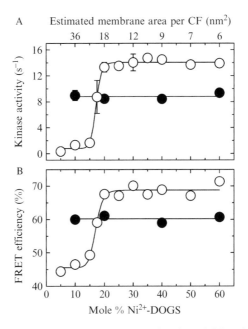

FIG. 11. Kinase activity and FRET efficiency as a function of CF surface density. Kinase activity (A) and FRET efficiency (B) were measured as a function of the surface concentrations of CF_{EEEE} (○) and CF_{QQQQ} (●). All samples contained vesicles prepared by extrusion through 50 nm pore diameter filters (580 μM lipid, 1:1 Ni^{2+}-NTA-DOGS:DOPC), 30 μM CF, 1.2 μM CheA, and 5 μM CheW. (Asinas & Weis, unpublished data).

LUVs containing 60% Ni^{2+}-NTA-DOGS. The kinase activity of signaling complexes assembled with CF_{EEEE} increased from a low to a high level over a narrow range of surface concentrations centered at 17% Ni^{2+}-NTA-DOGS (40 nm^2 per CF dimer); by contrast, the kinase activity of samples containing CF_{QQQQ} remained level over the same range. These data provide evidence that the kinase activity is regulated in a manner that depends on both the receptor density and covalent modification.

To determine the properties of density-dependent kinase regulation, a fraction of the CFs in these same samples were labeled with extrinsic fluorescent probes—a mixture of donor-labeled and acceptor-labeled CFs—to detect changes in proximity among the CFs through changes in the fluorescence resonant energy transfer (FRET) efficiency. The FRET efficiencies that are plotted in Fig. 11B show a clear increase with the CF_{EEEE}-containing samples, but not CF_{QQQQ}-containing samples. The activity and FRET data show a remarkable correspondence. The ability to probe enzyme activity and physical properties in the same samples, generated by liposome-mediated assembly, can help to establish complex structure–function relationships.

Methods

Purification of Chemotaxis Proteins

The histidine-tagged ctyoplasmic fragment (CF) of the aspartate receptor from *E. coli*, and also CheW, CheA, CheY, and CheR, are isolated according to published protocols (Kott *et al.*, 2004; Montefusco *et al.*, 2007; Shrout *et al.*, 2003; Wu *et al.*, 1996). Protein stock solutions are aliquoted and stored at $-70°$. Typical yields of purified protein, per liter of cell culture, are 50 to 175 mg CF (which is mutation specific), 10 mg $CheA_L$; 25 mg CheW; 15 mg CheY; and 15 mg CheR. Typical stock concentrations are 0.5 m*M* CF, 75 μM CheA, 250 μM CheW, and 1 m*M* CheY.

Preparation of Liposomes

Numerous methods are available to make liposomes, a result of the many fundamental studies of membranes and their technological applications (Düzgünes, 2003a; Szoka and Papahadjopoulos, 1980). Lipids are available from a number of reputable vendors; the nickel-chelating lipid, in both the metal free (NTA-DOGS) and nickel-chelated forms (Ni^{2+}-NTA-DOGS), can be purchased from Avanti Polar Lipids (www.avantilipids.com; Alabaster, AL). Abbreviated procedures are given for MLVs (steps 1 to 4), SUVs (steps 1 to 5) and LUVs (steps 1 to 4, 6):

1. Dissolve lipids (2–10 mg) in a test tube or a small round-bottom flask with an effective low-boiling solvent, such as chloroform, at the desired molar ratio.

2. Remove the solvent, either by flash rotary evaporation or a gas stream (N_2 or Ar), to generate a uniform film of lipid on the inner surface of the vessel. Remove residual solvent under high vacuum for at least 1 h.

3. Hydrate the lipid film at a temperature above the main (solid–fluid) transition temperature with enough buffer to produce a ~2 to 10 mg/ml lipid suspension (~2–10 mM). Thirty degrees Celsius can be used with Ni^{2+}-NTA-DOGS/DOPC mixtures. The buffer composition depends on the experiment.

4. Vortex the hydrated lipid film to generate an MLV suspension.

5. To generate SUVs, place the vessel with the MLV suspension in a bath sonicator, taking care to position it for "effective" sonication, for example, at an antinode of a standing wave, and sonicate the suspension above the main phase transition temperature for a period of 1.5 to 3 h, at least until the solution clarifies. A more uniform SUV preparation, at the small size limit, requires more vigorous sonication and a fractionation step (Cornell *et al.*, 1982; Huang and Thompson, 1974).

6. LUVs are prepared conveniently on small scales by an extrusion device, consisting of two 1 ml glass syringes (MacDonald *et al.*, 1991), which are used to push the liposome solution back and forth through polycarbonate filters. Commercial extruders are available from Avanti Polar Lipids Inc. (Alabaster, AL) and Avestin Inc. (Ottawa, Ontario, Canada). Nucleopore® polycarbonate track-etched membrane filters are available to fit these extruders in a range of pore diameters (0.03–1.0 μm). LUVs are made by passing the solution through the filter 12 to 15 times (Mui *et al.*, 2003), at a temperature above the main phase transition temperature.

With respect to bulk properties, such as size and the average number of lamellae, the reproducibility of liposome preparations has been established through decades of research, which supports the use of these materials in the basic experiments of biochemical activity that are described here without further characterization. Additional characterization of the vesicles can be carried out, as needed, with established methods (Szoka and Papahadjopoulos, 1980), which can also provide insight into the assembly of vesicle-associated proteins. The fraction of lipid in the outermost leaflet can, for example, be determined with NMR (Kumar *et al.*, 1989), or for vesicles that contain NTA-DOGS, with isothermal titration calorimetry (Montefusco *et al.*, 2007). Population-averaged hydrodynamic properties of vesicles can be characterized by dynamic light scattering

(Patty and Frisken, 2003, 2006). Although most analyses assume spherical shapes for the vesicles, this assumption is reasonable for SUVs and the smaller LUVs. Electron microscopy complements scattering methods very well, by providing microscopic detail of vesicle structure in a direct manner, as simple transmission electron microscope images of either stained or unstained samples (Frederik and Hubert, 2005; Huang, 1969) or with freeze-fracture techniques (Hope et al., 1989). Cryoelectron tomography can provide structural information in much greater detail (Lucic et al., 2005; Subramaniam, 2005).

Formation of CF/CheW/CheA Signaling Complexes

The formation of template-assembled protein complexes is initiated simply by mixing a solution of liposomes with a solution containing all the signaling proteins (CF, CheW, and CheA). In many of the experiments described here, the concentrations after mixing are 580 μM total lipid (1:1 Ni^{2+}-NTA-DOGS:DOPC), 30 μM CF, 5.0 μM CheW, and 1.2 μM CheA, but there are no limitations in this regard. Other concentrations of lipid, other ratios of Ni^{2+}-NTA-DOGS to DOPC, and other concentrations of protein can be used, depending on the objective of the experiment. The protein–lipid mixture is incubated for a sufficient period of time ($\sim$5 h at 25°), that is, until the system reaches a steady level of activity (see Fig. 4 and Montefusco et al., 2007). The requirement for a long incubation is specific to the chemotaxis system; other systems are more likely to require a significantly shorter time period to achieve equilibrium assembly and activity.

Assays of Biochemical Activity

Protein Binding. Liposome-associated proteins can be separated from unassociated protein by sedimentation in a tabletop ultracentrifuge (Beckman TLX, TLA 120.2 rotor, 125 000g, 15 min). In a typical procedure to determine the extents of CheW and CheA membrane association under "assembly conditions," an aliquot (10 μl) is removed from the sample (150 μl) prior to sedimentation (free plus bound protein), protein–vesicle complexes are sedimented, and an equal aliquot (free protein) is removed. Proteins are separated on SDS-PAGE (12.5 or 15% acrylamide), stained (Gelcode Blue, Pierce Biotechnology), and analyzed with scanning densitometry (Bio-Rad Molecular Imager GS-700). The bound fractions are computed as $(OD_{total} - OD_{free})/(OD_{total})$ (Shrout et al., 2003).

Kinase Activity. The slow rate of activity disappearance (Fig. 4) is advantageous in kinase activity measurements, because it is possible to dilute substantially the complexes that form under assembly conditions, such as by 100-fold. Established protocols, which measure initial rates or extents of ^{32}P-phosphate incorporation into either CheA or CheY should also be compatible with template-assembled complexes (Borkovich and Simon, 1991; Chervitz and Falke, 1995; Li and Weis, 2000); these methods can be conducted in small volumes on either diluted or undiluted samples.

An enzyme-coupled ATPase assay (Norby, 1988), which has been used to measure the steady-state activity of CheA in free-standing, self-assembled complexes (Liu et al., 1997), is also effective with template-assembled complexes (Asinas and Weis, 2006; Montefusco et al., 2007; Shrout et al., 2003). The procedure for measuring ATPase activity using complexes formed under standard assembly conditions is given here as an example.

In outline, a CF/CheW/CheA solution is prepared from protein stock solutions and ATPase buffer. The CF/CheW/CheA solution and vesicles (in ATPase buffer) are then mixed to initiate assembly. Before measuring activity, a 2 μl aliquot of the CF/CheW/CheA/vesicle solution is mixed with 180 μl ATPase buffer containing CheY and pyruvate kinase/lactate dehydrogenase (PK/LDH, Sigma P0294). Immediately prior to measuring activity, 20 μl of an ATP/PEP/β-NADH solution is added (with mixing) to give the final 200 μl volume. Activity is measured as the rate of change in the absorbance of β-NADH at 340 nm (ε_{340nm} = 6220 M^{-1}cm^{-1}) using quartz microcuvettes (200 μl capacity).

1. ATPase Buffer: 75 mM Tris, pH 8.0, 100 mM KCl, 2 mM Tris[2-carboxyethyl]phosphine, 5 mM MgCl$_2$, 5% v/v DMSO.

2. Solutions:
   a. ATP/PEP/β-NADH solution: Using ATPase buffer, prepare fresh, a solution that is 40 mM ATP (Sigma A2383), 25 mM PEP (phosphoenolpyruvate, Sigma P3637), and 2.5 mM β-NADH (Sigma N8129). Adjust the pH to 8.0.
   b. CheY/PK/LDH solution: 200 μl CheY (1 mM), 100 μl PK/LDH (Sigma, P0294), 3.30 ml ATPase buffer. (3.6 ml provides for 20 samples.)
   c. CF/CheA/CheW solution: Mix aliquots of the protein stock solutions and ATPase buffer to give the concentrations required for the experiment (see following text).
   d. Vesicles solution: Using ATPase buffer to hydrate and resuspend the lipid, prepare vesicles as has been described ($\sim$2 mg/ml total lipid).

3. Mix the CF/CheW/CheA and vesicle solutions to initiate assembly. As an example, mix 55 μl of a CF/CheW/CheA solution (41 μM CF, 6.8 μM CheW, and 1.6 μM CheA) with 20 μl of a vesicle solution (2.18 mM in lipid, 1:1 DOPC/Ni^{2+}-NTA-DOGS) to give 75 μl of protein/vesicle solution at the assembly condition: 30 μM CF, 5.0 μM CheW, 1.2 μM CheA, and 580 μM lipid. Incubate the sample at 25° in a water bath.

4. As needed, withdraw 2 μl aliquots of the protein/vesicle solution and mix with 180 μl aliquots of the CheY/PK/LDH solution.

5. Immediately prior to measurement, add and mix 20 μl of the ATP/PEP/β-NADH solution. Concentrations at measurement: 100-fold dilution of CF/CheW/CheA and vesicles relative to assembly conditions, 50 μM CheY, 4.0 mM ATP, 2.5 mM PEP, 0.25 mM β-NADH, ~25 units LDH, and ~18 units PK.

Notes: (a) To estimate the stimulation factors (kinase activity in template-assembled complexes relative to CheA dimer in solution), measure the CheA activity in solution at concentrations where it will be dimeric (= 5.0 μM). (b) Nonspecific ATPase activity should be determined and subtracted from the CheA-specific reactions. The CheY solution often accounts for the majority of the background. (c) That CheY is not limiting the reaction rate can be determined by doubling or halving the aliquot of the CF/CheW/CheA/vesicle mixture, which should either double or halve, respectively, the rate obtained with the standard (2 μl) aliquot.

Methylation. Methylation of template-assembled CF_{EEEE} can be monitored by simple variations of established methods, such as by using either the increase in electrophoretic mobility that accompanies receptor methylation (Borkovich *et al.*, 1992), or the time course of tritiated methyl group incorporation (Antommattei *et al.*, 2004; Chalah and Weis, 2005). The gel shift assay, illustrated in Fig. 3, using the conditions described in the legend to Fig. 3, is initiated by the addition of CheR and *s*-adenosyl-L-methionine (SAM). Aliquots are removed and quenched by the addition of gel-loading buffer and analyzed by SDS-PAGE.

Methylation, like kinase stimulation, can be interrogated as a function of sample composition, such as with different CF surface densities and in the presence versus absence of CheA and CheW. Typically, experiments use samples assembled in 100 μl ATPase buffer containing 25 μM CF_{EEEE}, 200 μM Ni^{2+}-NTA-DOGS (as a binary mixture with DOPC, in amounts necessary to yield mole percentages of Ni^{2+}-NTA-DOGS between 10 and 60%), 5.0 μM CheW, 1.2 μM CheA, and 10 μM [^{3}H-methyl]-SAM (Perkin Elmer, NET115H). The reaction is initiated by the addition of CheR, to give a final concentration of 4 μM. Fourteen microliter aliquots are withdrawn at 20, 40, 60, 90, and 120 seconds and the aliquots are quenched by the addition of 2× SDS buffer. Samples are resolved on 12.5% SDS-acrylamide gels, stained briefly (10 min) with Gel-code Blue (Pierce Biotechnology, Inc.), and rinsed in deionized water. The CF-containing gel bands are excised with a razor blade and placed in 1.5 ml plastic microcentrifuge tubes that contain 1 ml 1 M NaOH. These tubes are placed within scintillation vials containing 2 ml scintillation fluid (Fisher Scientific, Scintiverse II). The capped scintillation vials are incubated overnight at room temperature, during which time methyl groups released through base-catalyzed ester hydrolysis transfer through the vapor phase to the scintillation fluid (Chelsky et al., 1984). Initial rates of methyl group incorporation are determined as the slope of calibrated DPMs versus time data.

FRET Measurements as a Function of Surface Density

To conduct the energy transfer experiments, the expression plasmids for E. coli aspartate receptor cytoplasmic domain fragments CF_{EEEE} and CF_{QQQQ} were engineered, using standard mutagenesis approaches, to replace six residues within the CheR tethering region of the CF with a tetra-cysteine motif, which binds bis-arsenical derivatives of fluorescein (FlAsH) and resorufin (ReAsH) with high affinity (Adams et al., 2002; Giepmans et al., 2006). The CF residues within the region ^{513}FRLAASPLTNKP QTP^{527} were altered to ^{513}FRLAA**CCPGCC**PQTP^{527}, where the segment in bold is the tetra-cysteine motif (and the numbers denote residue positions in full-length receptor). The spectral overlap in FlAsH emission and ReAsH excitation are compatible with the requirements of a FRET pair (Adams et al., 2002).

FlAsH- and ReAsH-labeled CFs are prepared in ATPase buffer containing 10 mM TCEP. FlAsH or ReAsH and *tetra*Cys CF are mixed in a 1:1 ratio and incubated in the dark for 1 h with gentle rocking. Labeled CF is prepared less than 24 h before use and is stored at 4°.

Each FRET measurement is conducted with a matched pair of samples. One is designed to measure donor fluorescence in the absence of the acceptor chromophore; the other provides for a measurement of donor fluorescence in the presence of acceptor. The CF in donor-only samples is a 0.85/0.15 unlabeled/donor-labeled (FlAsH-labeled) mixture. In samples containing donor- and acceptor-labeled (ReAsH-labeled) CF, the unlabeled/ acceptor-labeled/donor-labeled composition is 0.50/0.35/0.15. Otherwise, the vesicle–protein mixtures are assembled according to the protocol described for the ATPase assay. The CF/CheW/CheA mixture is prepared and assembly is initiated by mixing vesicles with protein. Typical protein concentrations under assembly conditions are 30 μM CF, 5 μM CheW, and 1.2 μM CheA. The lipid concentration, composed of vesicles that are 360 μM in Ni^{2+}-NTA-DOGS, with DOPC in amounts to generate Ni^{2+}-NTA-DOGS mole percentages over a range from 5 to 60%. For FRET, emission spectra (510 to 590 nm) are collected using an excitation wavelength of 495 nm, and the FRET efficiency, E, is calculated as $1 - F'_D/F_D$, where F'_D and F_D are donor fluorescence intensities with and without the acceptor present, respectively (Sabanayagam et al., 2005).

Conclusion

Reconstituting the enzymatic activity of a transmembrane signaling system using liposomes as templates is an innovation in a lineage of technological advances involving engineered membrane surfaces. Much of the previous research effort has focused on problems in structural biology and biophysics, which illuminate aspects of function. The intentional emphasis of the method described here is geared toward the restoration of biochemical function, from components that lack full function on their own, specifically, the cytoplasmic domain fragment of the chemoreceptors. The wider applicability of this approach to other systems rests on the similarity of membrane protein organization and the contribution made by the membrane in promoting functional interactions among the component parts. In addition to the successful application of template-directed assembly of the *E. coli* chemotaxis system (Asinas and Weis, 2006; Montefusco *et al.*, 2007; Shrout *et al.*, 2003), surface templating has also proven to be effective in enhancing activity and function of the epidermal growth factor receptor tyrosine kinase domain (Zhang *et al.*, 2006). Vesicle-mediated assembly is also proving to be an effective method for restoring biochemical function to membrane-associated protein assemblies involving extracellular receptor domains. For example, tissue factor (TF) extracellular domain – assembled on nickel-lipid-containing vesicles via a histidine tag introduced at the C-terminus – facilitated the formation of complexes with factor VIIa, and restored the activity of TF/factor VIIa complexes to the levels observed with reconstituted full-lengh TF (Water and Morrissey, 2006). In an investigation of membrane pore formation by the protective antigen (PA) of anthrax tocxin, Sun *et al.* demonstrated that vesicle-assembled anthrax toxin receptor domain (histidine-tagged von Willebrand factor A) was effective in recruiting PA, and that the membrane-associated complexes could be readily triggered to form pores at pH 5 (Sun *et al.*, 2007). Moreover, template-directed assembly has been used to study the initial events that accompany polio virus–target cell interaction (Tuthill *et al.*, 2006). Vesicle-mediated assembly is also proving to be an effective method for restoring biochemical function to membrane-associated protien assemblies involving extracellular receptor domains. For example, tissue factor (TF) extracellular domain - assembled on nickel-lipid-containing vesicles via a histidine tag introduced at the C-terminus- facilitated the formation of complexes with factor VIIa, and restored the activity of TF/factor VIIa complexes to the levels observed with reconstituted full-lenth TF (Waters and Morrissey, 2006). In an investigation of membrane pore formation by the protective antigen (PA) of anthrax toxin, Sun *et al.* demonstrated that vesicle-assembled anthrax toxin receptor domain (hisitidine-tagged von Willebrand factor A) was effective in recruiting PA, and that the membrane-associated complexes could be readily triggered to from pores at pH 5 (Sun *et al.,* 2007). In this case, liposomes decorated with the extracellular domain of the polio virus receptor were used to generate a simple, but effective, target cell membrane surrogate in order to study the early structural transitions that accompany virus binding to the membrane surface (Bubeck *et al.,* 2005). From studies like these, it can be expected that liposome-mediated assembly of functional complexes—generated via engineered interactions between proteins and membranes—will find wider application.

What are some likely future developments? First, the chemistry of Ni^{2+}-NTA-histidine tag interaction has improved through the synthesis of new lipid molecules that are bivalent (or trivalent) in the Ni^{2+}-NTA moiety, which has generated increased affinity and improved stoichiometric properties of the Ni^{2+}-NTA-histidine tag interaction (Gavutis *et al.,* 2005; Lata *et al.,* 2006). Investigations of growth factor binding kinetics and energetics to the membrane-associated binding domains of the cognate receptors are improved as a result (Gavutis *et al.,* 2006). These reagents should generate similar benefits to biochemical studies of proteins, or portions of proteins, that normally reside on the membrane leaflet. Second, a substantial body of work describes the development of methods to generate planar-supported monolayer and bilayer membranes as a platform to probe membrane-related phenomena (McConnell *et al.,* 1986; Sackmann and Tanaka, 2000; Tanaka and Sackmann, 2005). In parallel to these efforts, dramatic improvements have been made in the number and sophistication of the methods that probe the structure and dynamics of surface-associated assemblies. Again, these methods have been applied usually to proteins that reside on the *outer* membrane leaflet to probe protein binding and surface dynamics. The application of these same approaches to membrane-associated protein complexes—which are also functionally (enzymatically) active—should pay dividends in understanding the biology of the inner *and* outer membrane leaflets.

Acknowledgment

RMW thanks Lubna Al-Challah, Frances M. Antommattei, Tatiana Y. Besschetnova, Anas Chalah, Anthony L. Shrout, Hoa Tran, and Li Zhi for helpful discussions and contributions. This work was supported with grants from the NIH (5R01GM53210) and NSF (SGER CHE0346419).

References

Adams, S. R., Campbell, R. E., Gross, L. A., Martin, B. R., Walkup, G. K., Yao, Y., Llopis, J., and Tsien, R. Y. (2002). New biarsenical ligands and tetracysteine motifs for protein labeling *in vitro* and *in vivo*: Synthesis and biological applications. *J. Am. Chem. Soc.* **124,** 6063–6076.
Ames, P., and Parkinson, J. S. (1988). Transmembrane signaling by bacterial chemoreceptors: *E. coli* transducers with locked signal output. *Cell* **55,** 817–826.

Ames, P., and Parkinson, J. S. (1994). Constitutively signaling fragments of Tsr, the *Escherichia coli* serine chemoreceptor. *J. Bacteriol.* **176**, 6340–6348.

Ames, P., Studdert, C. A., Reiser, R. H., and Parkinson, J. S. (2002). Collaborative signaling by mixed chemoreceptor teams in *Escherichia coli*. *Proc. Natl. Acad. Sci. USA* **99**, 7060–7065.

Ames, P., Yu, Y. A., and Parkinson, J. S. (1996). Methylation segments are not required for chemotactic signalling by cytoplasmic fragments of Tsr, the methyl-accepting serine chemoreceptor of *Escherichia coli*. *Mol. Microbiol.* **19**, 737–746.

Antommattei, F. M., Munzner, J. B., and Weis, R. M. (2004). Ligand-specific activation of *Escherichia coli* chemoreceptor transmethylation. *J. Bacteriol.* **186**, 7556–7563.

Antommattei, F. M., and Weis, R. M. (2006). Reversible methylation of glutamate residues in the receptor proteins of bacterial sensory systems. *In* "Protein Methytransferases" (S. G. Clarke and F. Tamanoi, eds.), Vol. 24, pp. 325–382. Elsevier/Academic Press, Amsterdam.

Asinas, A. E., and Weis, R. M. (2006). Competitive and cooperative interactions in receptor signaling complexes. *J. Biol. Chem.* **281**, 30512–30523.

Bogonez, E., and Koshland, D. E., Jr. (1985). Solubilization of a vectorial transmembrane receptor in functional form: Aspartate receptor of chemotaxis. *Proc. Natl. Acad. Sci. USA* **82**, 4891–4895.

Borkovich, K. A., Alex, L. A., and Simon, M. I. (1992). Attenuation of sensory receptor signaling by covalent modification. *Proc. Natl. Acad. Sci. USA* **89**, 6756–6760.

Borkovich, K. A., and Simon, M. I. (1991). Coupling of receptor function to phosphate-transfer reactions in bacterial chemotaxis. *Methods Enzymol.* **200**, 205–214.

Bornhorst, J. A., and Falke, J. J. (2000). Attractant regulation of the aspartate receptor-kinase complex: Limited cooperative interactions between receptors and effects of the receptor modification state. *Biochemistry* **39**, 9486–9493.

Bubeck, D., Filman, D. J., and Hogle, J. M. (2005). Cryo-electron microscopy reconstruction of a poliovirus-receptor-membrane complex. *Nat. Struct. Mol. Biol.* **12**, 615–618.

Chalah, A., and Weis, R. M. (2005). Site-specific and synergistic stimulation of methylation on the bacterial chemotaxis receptor Tsr by serine and CheW. *BMC Microbiol.* **5**, 12.

Chebotareva, N. A., Kurganov, B. I., and Livanova, N. B. (2004). Biochemical effects of molecular crowding. *Biochemistry (Mosc.)* **69**, 1239–1251.

Chelsky, D., Gutterson, N. I., and Koshland, D. E., Jr. (1984). A diffusion assay for detection and quantitation of methyl-esterified proteins on polyacrylamide gels. *Anal Biochem.* **141**, 143–148.

Chervitz, S. A., and Falke, J. J. (1995). Lock on/off disulfides identify the transmembrane signaling helix of the aspartate receptor. *J. Biol. Chem.* **270**, 24043–24053.

Cochran, A. G., and Kim, P. S. (1996). Imitation of *Escherichia coli* aspartate receptor signaling in engineered dimers of the cytoplasmic domain. *Science.* **271**, 1113–1116.

Cornell, B. A., Fletcher, G. C., Middlehurst, J., and Separovic, F. (1982). The lower limit to the size of small sonicated phospholipid vesicles. *Biochim. Biophys. Acta* **690**, 15–19.

Düzgünes, N. (ed.) (2003a). Liposomes, Part A Elsevier Academic Press, San Diego.

Düzgünes, N. (ed.) (2003b). Liposomes, Part B Elsevier Academic Press, San Diego.

Düzgünes, N. (ed.) (2003c). Liposomes, Part C Elsevier Academic Press, San Diego.

Düzgünes, N. (ed.) (2004). Liposomes, Part D Elsevier Academic Press, San Diego.

Düzgünes, N. (ed.) (2005). Liposomes, Part E Elsevier Academic Press, San Diego.

Francis, N. R., Wolanin, P. M., Stock, J. B., Derosier, D. J., and Thomas, D. R. (2004). Three-dimensional structure and organization of a receptor/signaling complex. *Proc. Natl. Acad. Sci. USA* **101**, 17480–17485.

Frederik, P. M., and Hubert, D. H. (2005). Cryoelectron microscopy of liposomes. *Methods Enzymol.* **391**, 431–448.

Gavutis, M., Jaks, E., Lamken, P., and Piehler, J. (2006). Determination of the two-dimensional interaction rate constants of a cytokine receptor complex. *Biophys. J.* **90**, 3345–3355.

Gavutis, M., Lata, S., Lamken, P., Muller, P., and Piehler, J. (2005). Lateral ligand-receptor interactions on membranes probed by simultaneous fluorescence-interference detection. *Biophys. J.* **88,** 4289–4302.

Gegner, J. A., Graham, D. R., Roth, A. F., and Dahlquist, F. W. (1992). Assembly of an MCP receptor, CheW, and kinase CheA complex in the bacterial chemotaxis signal transduction pathway. *Cell* **70,** 975–982.

Giepmans, B. N., Adams, S. R., Ellisman, M. H., and Tsien, R. Y. (2006). The fluorescent toolbox for assessing protein location and function. *Science* **312,** 217–224.

Hope, M. J., Wong, K. F., and Cullis, P. R. (1989). Freeze-fracture of lipids and model membrane systems. *J. Electron Microsc. Tech.* **13,** 277–287.

Huang, C. (1969). Studies on phosphatidylcholine vesicles. Formation and physical characteristics. *Biochemistry* **8,** 344–352.

Huang, C., and Thompson, T. E. (1974). Preparation of homogeneous, single-walled phosphatidylcholine vesicles. *Methods Enzymol.* **32,** 485–489.

Kim, K. K., Yokota, H., and Kim, S. H. (1999). Four-helical-bundle structure of the cytoplasmic domain of a serine chemotaxis receptor. *Nature* **400,** 787–792.

Kott, L., Braswell, E. H., Shrout, A. L., and Weis, R. M. (2004). Distributed subunit interactions in CheA contribute to dimer stability: A sedimentation equilibrium study. *Biochim. Biophys. Acta* **1696,** 131–140.

Kubalek, E. W., Le Grice, S. F., and Brown, P. O. (1994). Two-dimensional crystallization of histidine-tagged, HIV-1 reverse transcriptase promoted by a novel nickel-chelating lipid. *J. Struct. Biol.* **113,** 117–123.

Kumar, V. V., Malewicz, B., and Baumann, W. J. (1989). Lysophosphatidylcholine stabilizes small unilamellar phosphatidylcholine vesicles. Phosphorus-31 NMR evidence for the "wedge" effect. *Biophys. J.* **55,** 789–792.

Lata, S., Gavutis, M., Tampe, R., and Piehler, J. (2006). Specific and stable fluorescence labeling of histidine-tagged proteins for dissecting multi-protein complex formation. *J. Am. Chem. Soc.* **128,** 2365–2372.

Le Moual, H., and Koshland, D. E., Jr. (1996). Molecular evolution of the C-terminal cytoplasmic domain of a superfamily of bacterial receptors involved in taxis. *J. Mol. Biol.* **261,** 568–585.

Le Moual, H., Quang, T., and Koshland, D. E., Jr. (1997). Methylation of the *Escherichia coli* chemotaxis receptors: Intra- and interdimer mechanisms. *Biochemistry* **36,** 13441–13448.

Lefman, J., Zhang, P., Hirai, T., Weis, R. M., Juliani, J., Bliss, D., Kessel, M., Bos, E., Peters, P. J., and Subramaniam, S. (2004). Three-dimensional electron microscopic imaging of membrane invaginations in *Escherichia coli* overproducing the chemotaxis receptor Tsr. *J. Bacteriol.* **186,** 5052–5061.

Levit, M. N., Grebe, T. W., and Stock, J. B. (2002). Organization of the receptor-kinase signaling array that regulates *Escherichia coli* chemotaxis. *J. Biol. Chem.* **277,** 36748–36754.

Li, G., and Weis, R. M. (2000). Covalent modification regulates ligand binding to receptor complexes in the chemosensory system of *Escherichia coli*. *Cell* **100,** 357–365.

Li, J., Li, G., and Weis, R. M. (1997). The serine chemoreceptor from *Escherichia coli* is methylated through an inter-dimer process. *Biochemistry* **36,** 11851–11857.

Li, M., and Hazelbauer, G. L. (2005). Adaptational assistance in clusters of bacterial chemoreceptors. *Mol. Microbiol.* **56,** 1617–1626.

Lin, L. N., Li, J., Brandts, J. F., and Weis, R. M. (1994). The serine receptor of bacterial chemotaxis exhibits half-site saturation for serine binding. *Biochemistry* **33,** 6564–6570.

Liu, Y., Levit, M., Lurz, R., Surette, M. G., and Stock, J. B. (1997). Receptor-mediated protein kinase activation and the mechanism of transmembrane signaling in bacterial chemotaxis. *EMBO J.* **16,** 7231–7240.

Long, D. G., and Weis, R. M. (1992). Oligomerization of the cytoplasmic fragment from the aspartate receptor of *Escherichia coli*. *Biochemistry* **31,** 9904–9911.

Lucic, V., Forster, F., and Baumeister, W. (2005). Structural studies by electron tomography: From cells to molecules. *Annu Rev. Biochem.* **74,** 833–865.

Lybarger, S. R., and Maddock, J. R. (2001). Polarity in action: Asymmetric protein localization in bacteria. *J. Bacteriol.* **183,** 3261–3267.

MacDonald, R. C., MacDonald, R. I., Menco, B. P., Takeshita, K., Subbarao, N. K., and Hu, L. R. (1991). Small-volume extrusion apparatus for preparation of large, unilamellar vesicles. *Biochim. Biophys. Acta* **1061,** 297–303.

Maddock, J. R., and Shapiro, L. (1993). Polar location of the chemoreceptor complex in the *Escherichia coli* cell. *Science* **259,** 1717–1723.

Mayer, L. D., Hope, M. J., and Cullis, P. R. (1986). Vesicles of variable sizes produced by a rapid extrusion procedure. *Biochim. Biophys. Acta* **858,** 161–168.

McConnell, H. M., Watts, T. H., Weis, R. M., and Brian, A. A. (1986). Supported planar membranes in studies of cell-cell recognition in the immune system. *Biochim. Biophys. Acta* **864,** 95–106.

Montefusco, D. J., Shrout, A. L., Besschetnova, T. Y., and Weis, R. M. (2007). Formation and activity of template-assembled receptor signaling complexes. *Langmuir* **23,** 3280–3289.

Mui, B., Chow, L., and Hope, M. J. (2003). Extrusion Technique to Generate Liposomes of Defined Size. *In* "Methods in Enzymology" (D. Nejat, ed.), Vol. 367, p. 3. Academic Press.

Mutoh, N., Oosawa, K., and Simon, M. I. (1986). Characterization of *Escherichia coli* chemotaxis receptor mutants with null phenotypes. *J. Bacteriol.* **167,** 992–998.

Nagle, J. F., and Tristram-Nagle, S. (2000). Structure of lipid bilayers. *Biochim. Biophys. Acta* **1469,** 159–195.

Ninfa, E. G., Stock, A., Mowbray, S., and Stock, J. (1991). Reconstitution of the bacterial chemotaxis signal transduction system from purified components. *J. Biol. Chem.* **266,** 9764–9770.

Norby, J. G. (1988). Coupled assay of Na^+,K^+-ATPase activity. *Methods Enzymol.* **156,** 116–119.

Olson, F., Hunt, C. A., Szoka, F. C., Vail, W. J., and Papahadjopoulos, D. (1979). Preparation of liposomes of defined size distribution by extrusion through polycarbonate membranes. *Biochim. Biophys. Acta* **557,** 9–23.

Park, S. Y., Borbat, P. P., Gonzalez-Bonet, G., Bhatnagar, J., Pollard, A. M., Freed, J. H., Bilwes, A. M., and Crane, B. R. (2006). Reconstruction of the chemotaxis receptor-kinase assembly. *Nat. Struct. Mol. Biol.* **13,** 400–407.

Patty, P. J., and Frisken, B. J. (2003). The pressure-dependence of the size of extruded vesicles. *Biophys. J.* **85,** 996–1004.

Patty, P. J., and Frisken, B. J. (2006). Direct determination of the number-weighted mean radius and polydispersity from dynamic light-scattering data. *Appl. Opt.* **45,** 2209–2216.

Racker, E. (1985). *Reconstitutions of Transporters, Receptors, and Pathological States.* Academic Press, Orlando.

Rigaud, J. L., and Levy, D. (2003). Reconstitution of membrane proteins into liposomes. *Methods Enzymol.* **372,** 65–86.

Russo, A. F., and Koshland, D. E., Jr. (1983). Separation of signal transduction and adaptation functions of the aspartate receptor in bacterial sensing. *Science* **220,** 1016–1020.

Sabanayagam, C. R., Eid, J. S., and Meller, A. (2005). Using fluorescence resonance energy transfer to measure distances along individual DNA molecules: Corrections due to noni-deal transfer. *The Journal of Chemical Physics* **122,** 061103.

Sackmann, E., and Tanaka, M. (2000). Supported membranes on soft polymer cushions: Fabrication, characterization and applications. *Trends Biotechnol.* **18,** 58–64.

Schmitt, L., Dietrich, C., and Tampé, R. (1994). Synthesis and Characterization of Chelator-Lipids for Reversible Immobilization of Engineered Proteins at Self-Assembled Lipid Interfaces. *Journal of the American Chemical Society* **116,** 8485–8491.

Schuster, S. C., Swanson, R. V., Alex, L. A., Bourret, R. B., and Simon, M. I. (1993). Assembly and function of a quaternary signal transduction complex monitored by surface plasmon resonance. *Nature* **365,** 343–347.

Seddon, A. M., Curnow, P., and Booth, P. J. (2004). Membrane proteins, lipids and detergents: Not just a soap opera. *Biochim. Biophys. Acta* **1666,** 105–117.

Seeley, S. K., Weis, R. M., and Thompson, L. K. (1996). The cytoplasmic fragment of the aspartate receptor displays globally dynamic behavior. *Biochemistry* **35,** 5199–5206.

Shrout, A. L., Montefusco, D. J., and Weis, R. M. (2003). Template-directed assembly of receptor signaling complexes. *Biochemistry* **42,** 13379–13385.

Sourjik, V., and Berg, H. C. (2004). Functional interactions between receptors in bacterial chemotaxis. *Nature* **428,** 437–441.

Subramaniam, S. (2005). Bridging the imaging gap: Visualizing subcellular architecture with electron tomography. *Curr. Opin. Microbiol.* **8,** 316–322.

Sun, J., Vernier, G., Wigelsworth, D. J., and Collier, R. J. (2007). Insertion of anthrax protective antigen into liposomal membranes: Effects of a receptor. *J. Biol. Chem.* **282,** 1059–1065.

Surette, M. G., and Stock, J. B. (1996). Role of alpha-helical coiled-coil interactions in receptor dimerization, signaling, and adaptation during bacterial chemotaxis. *J. Biol. Chem.* **271,** 17966–17973.

Szoka, F., Jr., and Papahadjopoulos, D. (1980). Comparative properties and methods of preparation of lipid vesicles (liposomes). *Annu. Rev. Biophys. Bioeng.* **9,** 467–508.

Tanaka, M., and Sackmann, E. (2005). Polymer-supported membranes as models of the cell surface. *Nature* **437,** 656–663.

Tuthill, T. J., Bubeck, D., Rowlands, D. J., and Hogle, J. M. (2006). Characterization of early steps in the poliovirus infection process: Receptor-decorated liposomes induce conversion of the virus to membrane-anchored entry-intermediate particles. *J. Virol.* **80,** 172–180.

Uzgiris, E. E., and Kornberg, R. D. (1983). Two-dimensional crystallization technique for imaging macromolecules, with application to antigen–antibody–complement complexes. *Nature* **301,** 125–129.

Waters, E. K., and Morrissey, J. H. (2006). Restoring full biological activity to the isolated ectodomain of an integral membrane protein. *Biochemistry* **45,** 3769–3774.

Weis, R. M., Hirai, T., Chalah, A., Kessel, M., Peters, P. J., and Subramaniam, S. (2003). Electron microscopic analysis of membrane assemblies formed by the bacterial chemotaxis receptor Tsr. *J. Bacteriol.* **185,** 3636–3643.

Windisch, B., Bray, D., and Duke, T. (2006). Balls and chains–a mesoscopic approach to tethered protein domains. *Biophys. J.* **91,** 2383–2392.

Wolanin, P. M., Baker, M. D., Francis, N. R., Thomas, D. R., DeRosier, D. J., and Stock, J. B. (2006). Self-assembly of receptor/signaling complexes in bacterial chemotaxis. *Proc. Natl. Acad. Sci. USA* **103,** 14313–14318.

Wu, J., Li, J., Li, G., Long, D. G., and Weis, R. M. (1996). The receptor binding site for the methyltransferase of bacterial chemotaxis is distinct from the sites of methylation. *Biochemistry* **35,** 4984–4993.

Zhang, P., Bos, E., Heymann, J., Gnaegi, H., Kessel, M., Peters, P. J., and Subramaniam, S. (2004). Direct visualization of receptor arrays in frozen-hydrated sections and plunge-frozen specimens of *E. coli* engineered to overproduce the chemotaxis receptor Tsr. *J. Microsc.* **216,** 76–83.

Zhang, X., Gureasko, J., Shen, K., Cole, P. A., and Kuriyan, J. (2006). An allosteric mechanism for activation of the kinase domain of epidermal growth factor receptor. *Cell* **125,** 1137–1149.

[13]   Analyzing Transmembrane Chemoreceptors Using *In Vivo* Disulfide Formation Between Introduced Cysteines

By Wing-Cheung Lai and Gerald L. Hazelbauer

Abstract

The sulfhydryl chemistry possible at the thiol group of cysteine provides a very useful tool for probing protein structure and function. The power of site-specific mutagenesis makes it possible to use this tool at essentially any position in a polypeptide sequence. The reactivity of introduced cysteines is often assessed *in vitro*, using purified proteins or cell extracts. However, it can be particularly informative to probe the protein of interest *in vivo*, in its native cellular environment. Our laboratory has used *in vivo* approaches extensively in studies of bacterial transmembrane chemoreceptors, particularly by utilizing disulfide formation between pairs of introduced cysteines to learn about structural organization and mechanisms of function. We have concentrated on experimental conditions in which the cellular system of interest remained functional and thus the protein we were characterizing maintained not only its native structure but also its natural interactions. For this reason, our studies of bacterial transmembrane chemoreceptors using disulfide formation *in vivo* have focused in large part on cysteines separated from the reducing environment of the cell interior, in transmembrane or periplasmic domains. In this chapter, we discuss the applications and limitation of these approaches as well as the details of experimental manipulations and data analysis.

Introduction

The thiol group of cysteine provides a very useful tool for probing protein structure and function using approaches that exploit sulfhydryl chemistry. The power of site-specific mutagenesis makes it possible to use this tool at essentially any position in a polypeptide sequence. The approach is to replace native cysteines, often few in number, by a residue lacking a sulfhydryl, such as serine, and introduce cysteine at one or two positions of choice. Fortunately, many proteins appear tolerant of replacing native cysteines and introducing a cysteine at almost any other position. Each introduced cysteine can be tested for its propensity for reactivity with sulfhydryl reagents or for disulfide formation with a second cysteine.

METHODS IN ENZYMOLOGY, VOL. 423   0076-6879/07 $35.00
DOI: 10.1016/S0076-6879(07)23013-7

In addition, the modified protein can be tested for function. The results are generally most informative if obtained for a series of sequential residues. This approach, cysteine and disulfide scanning, has been widely used for analysis of protein structure and function. The reactivity of introduced cysteines is often assessed *in vitro*, using purified proteins or cell extracts. However, it can be particularly informative to probe the protein of interest in its native cellular environment. Our laboratory has used this approach extensively in studies of bacterial chemoreceptors, utilizing disulfide formation between pairs of cysteines upon exposure to the natural oxidizing environment of the periplasm, to the oxidizing compound molecular iodine, or to the oxidation catalyst $Cu(II)$-$(o$-phenanthroline$)_3$ (hereafter, Cu-phenanthroline) and using migration in SDS polyacrylamide gel electrophoresis and immunoblotting to assess the extent of disulfide formation. In this chapter, we describe the background for and salient features of these experimental approaches. Our experience has been in characterization of transmembrane chemoreceptors in *Escherichia coli*, for the most part, chemoreceptor Trg. However, we expect that many aspects of the strategies and protocols we have used are applicable to other bacterial membrane proteins and to membrane proteins in other cells.

Disulfide Formation *In Vivo*: Applications and Limitations

Disulfide bonds are rarely found in cytoplasmic proteins (Schulz and Schirmer, 1979; Thornton, 1981). This is because the cytoplasm contains systems that maintain an effective reducing environment (Bader and Bardwell, 2001). If an externally applied oxidation treatment induced disulfide formation in the cytoplasm, the bonds would be rapidly reduced unless the cytoplasmic reducing environment were compromised or the disulfide were sequestered from its influence. Thus, the time course and extent of disulfide formation in the cytoplasm will reflect competing influences of an externally applied oxidation treatment, the mechanisms for maintaining a reducing environment in the cytoplasm, and protein-specific features that might protect the bond from reduction. Even with these competing factors, extents of disulfide formation could reflect the relative placement of cysteine pairs, particularly if proximity of sulfhydryls on neighboring structural units not only increased the probability of disulfide formation but also sequestered the bond from the general reducing environment. If the protein of interest contained a redox center, there could be additional protection from the general reducing environment. This may be the case for the chemoreceptor Aer, which includes a cytoplasmic FAD-binding domain (Amin *et al.*, 2006; Ma *et al.*, 2004; Watts *et al.*, 2006). Alternatively, the reducing environment of the cytoplasm can be diminished and thus disulfide formation made more likely by mutations that cripple components important for redox homeostasis (Bessette *et al.*, 1999; Derman *et al.*, 1993; Prinz *et al.*, 1997; Rietsch and Beckwith, 1998; Tiebel *et al.*, 2000) or by treatment of whole cells with diamide (Kosower and Kosower, 1995; Studdert and Parkinson, 2004). However, because of the potential complexities of interpreting disulfide formation in the cytoplasm, particularly in terms of rate of formation, we have not pursued studies of oxidative coupling between cysteine pairs in that environment. Instead, we aimed to limit experimental perturbations so that that the cellular system of interest remained functional and thus the protein we were characterizing maintained its native structure and interactions. With this goal in mind, our studies of chemoreceptors using disulfide formation *in vivo* have focused in large part on cysteines separated from the reducing environment of the cell interior, in the transmembrane or periplasmic domains.

The mildest condition that could induce disulfide formation between introduced cysteines in a transmembrane or periplasmic domain of a bacterial membrane protein is simply the oxidizing environment of the periplasm (Nakamoto and Bardwell, 2004). In our studies of chemoreceptor Trg, we found that simple exposure to this environment rarely resulted in significant disulfide formation between introduced cysteines. Among >100 cysteine pairs we have introduced into the Trg transmembrane helixes and their immediate extensions into the periplasm and cytoplasm, in the absence of a specific oxidation treatment only one pair exhibited as much as 50% disulfide formation, another just over 5%, and most exhibited none. This low frequency could reflect the fact that none of these positions participated in a disulfide bond in the native protein and that the cysteines were introduced as diagnostic probes for proximity and orientation, not with the goal of effective positioning for disulfide formation. Results were quite different when we introduced four pairs of cysteines in the periplasmic domain of Trg with the intent that they would be well-positioned for effective disulfide formation according to a model for the four-helix bundle of that domain (Lai *et al.*, 2006b). Three of four pairs exhibited >95% disulfide formation in the native oxidizing environment of the periplasm, without an additional oxidation treatment. Except for special situations like these, the use of *in vivo* disulfide formation to probe structure and function is likely to require a specific oxidation treatment. As outlined in the following sections, such treatments can be functionally benign, allowing reaction between favorably placed cysteines to proceed essentially to completion in a few minutes (Beel and Hazelbauer, 2001; Hughson and Hazelbauer, 1996; Lee *et al.*, 1995).

Before describing these oxidation treatments, we consider the issue of disulfide formation in the membrane. In the hydrocarbon environment of

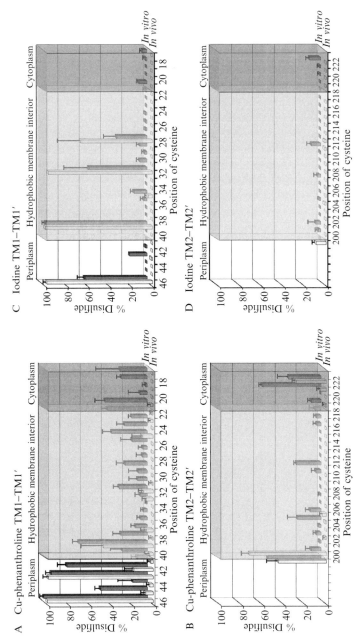

FIG. 1. Disulfide formation between homologously placed cysteines in the two subunits of the chemoreceptor Trg homodimer—the transmembrane domain and adjacent segments. Trg is a homodimer with four transmembrane segments, TM1 and TM2 in one subunit and TM1' and TM2' in the other. Trg with a single cysteine per polypeptide chain, thus two per dimer, and contained in intact chemotactically functional cells (*in vivo*) or in isolated membrane vesicles (*in vitro*) was treated with Cu-phenanthroline (panels A and B) or molecular iodine (panels C and D). The extent of disulfide formation between the homologously placed cysteines in the two subunits was determined after 10 min or 15 s, respectively. The light shading indicates segments embedded in the hydrophobic environment of the cytoplasmic membrane (Boldog and Hazelbauer, 2004) and the darker shading indicates segments in the cytoplasm. Data are from Hughson *et al.* (1997) and Boldog and Hazelbauer (2004). The figure is based on Fig. 3 in Hughson *et al.* (1997).

the membrane, the low dielectric constant would be expected to reduce the likelihood of formation of the reactive thiolate ion and thus the likelihood of disulfide formation. There are indications of such an effect in the pattern of disulfide formation induced by Cu-phenanthroline for cysteine pairs in Trg within and near the membrane (Fig. 1A,B). However, the hydrophobic membrane interior does not prohibit extensive disulfide formation. As illustrated in Fig. 1C, Trg cysteines at positions 32 and 32', embedded in middle of the cytoplasmic membrane at homologous positions on helices TM1 and TM1' of the two receptor subunits (Boldog and Hazelbauer, 2004), form disulfides almost quantitatively when intact cells are submitted to a functionally benign oxidation treatment with molecular iodine (Hughson and Hazelbauer, 1996; Lee *et al.*, 1995). In the same conditions, extensive disulfide formation also occurs for the nearby 28–28' pair (Fig. 1; Hughson *et al.*, 1997). The reducing environment of the cytoplasm might influence efficiency of disulfide formation *in vivo*. The data for the Trg transmembrane domain suggests this could occur, although the depth of the influence appears to be a function of the particular interface and oxidation treatment. (Fig. 1; Hughson *et al.*, 1997).

Oxidation Reagents

As has been discussed, utilization of most cysteine pairs to assess protein structure and function *in vivo* requires an external oxidizing treatment. We have used the oxidant molecular iodine or the oxidation catalyst Cu-phenanthroline. The two treatments have different features (Hughson *et al.*, 1997). Iodine-generated disulfide formation occurs rapidly after reagent addition, in a period shorter than the few seconds required to sample by hand, and the extent of disulfide formation does not change with longer incubation (Hughson *et al.*, 1997; Pakula and Simon, 1992). Reducing the concentration of iodine does not extend the time course of disulfide formation into a regime amenable to hand sampling; instead, it reduces the extent. These features probably reflect consumption of the reagent by oxidation of the many components in the experimental sample. In contrast, the time course of disulfide formation catalyzed by Cu-phenanthroline is a function of reagent concentration. Lowering its concentration lowers the rate of disulfide formation. Thus, at appropriate concentrations, Cu-phenanthroline-catalyzed disulfide formation occurs over a time course that can be monitored by hand sampling. These features are consistent with the catalytic nature of Cu-phenanthroline, that is, each chelated copper can catalyze multiple rounds of disulfide formation, using dissolved molecular oxygen as the chemical oxidant (Kobashi, 1968). At lower rates, disulfide formation reaches a plateau level that can be substantially less than 100%

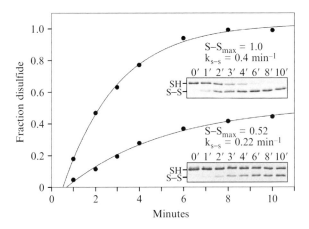

Fɪɢ. 2. Time courses of disulfide formation *in vivo*. The figure shows examples of disulfide formation between one cysteine in TM1 of Trg and the other in TM2 of the same polypeptide chain, specifically residues 38 and 203 (upper curve) and 38 and 202 (lower curve). Relevant portions of immunoblots are shown in the inserts. The chain constrained by a disulfide (S–S) migrates faster in SDS-PAGE than the polypeptide with unlinked cysteines (SH). Each band was quantified by densitometry, the fraction disulfide calculated for each sample, and the values plotted as a function of time. Values for the rate constant ($k_{S–S}$) and maximum extent ($S–S_{max}$) of disulfide formation were obtained by fitting the data to an equation with a term for disulfide formation and another for other oxidation reactions of the cysteine sulfhydryl (see text).

and is roughly proportional to the rate (Fig. 2). This pattern is thought to reflect competing oxidation reactions that convert cysteine sulfhydryls to sulphinic or sulphonic acids, thus irreversibly blocking the reacted residue for disulfide formation (Careaga and Falke, 1992). If the rate of disulfide formation is high relative to the rates of these alternative oxidations, then a large proportion of the cysteine pairs form disulfides. If the disulfide formation rate is relatively low, then a significant portion of the cysteines are individually oxidized and the proportion that forms disulfides is lower. The competing reactions should be taken into account when determining rate constants of disulfide formation by analyzing time courses (see the Data Analysis section).

Rates and extents, both relative and absolute, of disulfide formation by a cysteine pair can differ, depending on the oxidation treatment (Fig. 1; Hughson *et al.*, 1997; Lee *et al.*, 1995). Cu-phenanthroline or molecular iodine can be the more effective for a particular cysteine pair. In our studies of cysteine pairs in the transmembrane segments of Trg, differential effects of the two treatments were consistent with the notion that disulfide

formation by rapidly acting iodine would favor cysteines with high prob-
abilities of favorable distance and position whereas disulfide formation
catalyzed over a more extended time course by Cu-phenanthroline could
favor pairs that come into reactive proximity as the result of dynamic
fluctuations (Hughson *et al.*, 1997).

Oxidation Treatments That Preserve *In Vivo* Function

Characterization of disulfide formation *in vivo* has the advantage that a
membrane protein of interest can be probed in its natural environment,
embedded in its native membrane, and associated with its natural partners
both known and unknown. Ideally, the physiological process in which the
protein participates should be functional after the oxidation treatment, thus
providing an important indication that the protein structure and its interactions were essentially unperturbed by the treatment. For our studies, this
meant developing a procedure for disulfide formation in chemoreceptors
in vivo that preserved motility and its control by the chemosensory system. We had previously characterized disulfide formation *in vitro* between
cysteines of Trg contained in isolated membrane vesicles (Hughson *et al.*,
1997; Lee *et al.*, 1995), using exposure to 3 mM Cu-phenanthroline for
up to 10 min at 37° or 0.5 mM molecular iodine for 15 s at 25°. These
treatments allowed us to distinguish cysteine pairs with high propensities
for cross-linking from those with low or no propensity and thus deduce
relative positions of the transmembrane helices of the chemoreceptor.
However, exposing intact cells to either treatment greatly reduced motility.
Thus, we lowered reagent concentrations and found that the motility system, assessed by monitoring flagellar rotation, was minimally perturbed
when the concentration of Cu-phenanthroline was reduced tenfold to
0.3 mM or the concentration of iodine was reduced 50% to 0.25 mM.
Furthermore, these conditions allowed control of flagellar rotation by the
chemotaxis system (Lee *et al.*, 1995). Fortunately, these milder conditions
generated disulfide formation between many of the cysteine pairs that
exhibited oxidative cross-linking at the higher concentrations when treated
in isolated membrane vesicles (Fig. 1; Hughson *et al.*, 1997; Lee *et al.*, 1995).
Importantly, the patterns of disulfide formation *in vivo* suggested the
same orientation and relative proximity of the four transmembrane helices
of the Trg homodimer as suggested by the patterns of *in vitro* cross-linking.
In addition, several cysteine pairs could be cross-linked to ~100% by these
functionally benign treatments, thus allowing investigation of the functional
effects of restricting movement between neighboring helices (Lee *et al.*, 1995).
However, the oxidation treatments we identified as benign for chemotaxis
and motility were not without adverse effects. For instance, exposure to

0.25 mM iodine for 15 s at 25° reduced cell viability ∼60%. The same treatment at 37° left essentially no viable cells (iodine is an antiseptic). Thus, for studies of other proteins, it would be important to determine whether oxidation treatments that do not perturb motility or the chemotaxis signaling system are equally benign for the system of interest and, as necessary, optimize conditions to preserve relevant function.

Experimental Designs

We have used *in vivo* disulfide formation to address several issues of chemoreceptor structure and function: (1) definition of the three-dimensional organization of a transmembrane domain, (2) stringent testing of a three-dimensional homology model of a periplasmic domain, (3) identification of helical interfaces participating in the conformational change of transmembrane signaling, and (4) deduction of the conformational change in a transmembrane domain upon ligand-induced signaling and compensatory adaptational modification. An important feature of all of these experiments is that the sensitivity of immunoblotting with a high avidity antiserum makes it possible to assay the protein of interest at a level of expression equivalent to production from the native chromosomal gene. Our experimental designs could be applied to other transmembrane proteins.

Three-Dimensional Organization of a Transmembrane Domain

The three-dimensional organization of a transmembrane domain can be deduced by determining the relative propensities for disulfide formation between many pairs of cysteines placed at different positions in that domain. We did this for chemoreceptor Trg using isolated membrane vesicles (Lee *et al.*, 1994) and then replicated much of the analysis *in vivo* (Hughson *et al.*, 1997). It is important to keep in mind that patterns of disulfide formation, whether *in vivo* or *in vitro*, are empirical and thus their validity depends on internal checks and critical assessment of the data. Disulfide formation is a trap that can capture even low probability positioning of otherwise distant residues. For instance, cysteines on the opposite sides of adjacent helices by other criteria form disulfides upon sufficiently long exposure to sufficiently high concentrations of Cu-phenanthroline. Thus, we recommend two important features of a dependable strategy for deducing three-dimensional organization from propensities for disulfide formation: (1) Use an oxidation treatment (reagent concentration and time) for which the rate/extent of disulfide formation varies from extensive to zero among the cysteine pairs being tested, ideally determining relative time courses. In this way, conclusions will be based on observations in a regime in which reaction reflects relative positioning. (2) Assay as many cysteine pairs as feasible in a contiguous sequence; the most reliable conclusions are obtained by interpreting patterns.

Many transmembrane segments are α-helices. The periodicity of propensities for disulfide formation between cysteine pairs positioned with one on each of the two homologous segments of the Trg homodimer suggested that this was the case for the chemoreceptor (Hughson *et al.*, 1997; Lee *et al.*, 1994). Such periodicity can be assessed by Fourier analysis (Lee and Hazelbauer, 1995). The orientation of the two segments can be deduced by aligning the local maxima for disulfide formation (Lee *et al.*, 1994). A more inclusive analysis utilizes the quantitative values for disulfide formation at every position in the helix by using them to calculate a helical "cross-linking moment" (Lee and Hazelbauer, 1995), a vector sum inspired by the notion of the helical hydrophobic moment developed by Eisenberg *et al.* (1982). Vectors representing cross-linking moments can be used to construct a model for orientation of multiple transmembrane segments. The details of determining and utilizing cross-linking moments are described in Lee and Hazelbauer (1995).

Testing Structural Models

Propensity for disulfide formation *in vivo* can test models for the three-dimensional organization of a protein in its natural environment. Diamide treatment made it possible to test *in vivo* for close proximity of chemoreceptor cytoplasmic domains using disulfide formation (Studdert and Parkinson, 2004). For domains in the periplasm, diagnostic disulfide formation does not require perturbing the reducing environment of the cytoplasm and thus the cell can remain functional. For instance, introduced cysteine pairs provided stringent tests of a four-helix bundle homology model for the periplasmic domain of Trg (Lai *et al.*, 2006b). The tests were designed to probe the details of helical packing. Sets of two related diagnostic cysteine pairs were placed on helices predicted by the model to be adjacent in the helical bundle. For each set of cysteine pairs, one pair was predicted to be very favorably positioned to form disulfides and the second pair differed from the first only by the shift of one cysteine by a single position in the sequence. This shift would make the second pair less favorably positioned for disulfide formation. Four such sets of diagnostic cysteine pairs were tested for disulfide formation *in vivo* (Lai *et al.*, 2006b). The model was strongly validated. For the three sets of diagnostic pairs that the model predicted to be on firmly packed neighboring helices, the pair predicted to be favorably placed exhibited essentially 100% disulfide formation without addition of an oxidation reagent and the less favorably placed pair exhibited almost no disulfide formation in the same conditions. Note that disulfides did form between the less favorably placed pairs upon exposure to Cu-phenanthroline. The fourth set of diagnostic pairs spanned a helical interface known to be mobile (Falke and Hazelbauer, 2001) and thus both pairs formed disulfides to a significant extent. However, the pair predicted to be more favorably oriented exhibited a greater extent of reaction. Thus, the strategy of determining disulfide formation between two sets of two cysteine pairs was very useful in assessing the detailed validity of a three-dimensional structural model of a chemoreceptor periplasmic domain. The same strategy could be applied easily to models for other periplasmic domains.

Identification of Signaling Interfaces

Disulfides that restrict movement between structural elements can identify the interfaces that need to shift for the protein to perform its function. The approach is particularly informative if disulfides can be formed *in vivo* between introduced cysteines in conditions in which the system of interest remains functional. For instance, for a transmembrane receptor, the effect of a constraining disulfide can be assayed by testing for the native physiological coupling of stimulus to response. The strategy is to introduce cysteine pairs that could form disulfides across an interface between structural elements, such as adjacent helices in a helical bundle, and identify those pairs for which a functionally benign oxidation treatment results in essentially quantitative disulfide formation for the cellular population of the subject protein. Such a disulfide would effectively restrict relative movement of the two helices for the entire cellular receptor population. If the conformational change of signaling involved a movement greater than allowed by the restrictive disulfide, then stimulation would not result in response. Otherwise, stimulus would be coupled to response. We used this approach in intact cells of *E. coli* to investigate the effect of restricting movement between TM1 and TM1′ of a chemoreceptor across the interface between the two subunits of the receptor homodimer and between TM1 and TM2 within one subunit (Lee *et al.*, 1995). By identifying functionally benign oxidation treatments, we were able to assess the consequences of these disulfide restrictions by assaying the behavioral response of intact, chemotactically functional cells to temporal gradients of attractant. The results showed that effective transmembrane signaling required an unrestricted helical interface within a subunit but signaling was not perturbed by constraining movement at the interface between subunits (Lee *et al.*, 1995).

Deduction of the Nature of Conformational Changes

Disulfides that connect structural units of a protein can be used not only to identify interfaces that shift but also to deduce the nature of that change. Experiments on native protein *in vivo* allow assessment at natural dosages and physiological conditions. The logic of the experimental design is that a conformational shift can change the positioning of cysteine pairs and thus the likelihood of disulfide formation. In the extreme, a cysteine pair with high propensity for oxidative cross-linking could be sufficiently separated to prohibit reaction or a pair that did not form disulfides could become efficient reactants. Such effects would require large-magnitude conformational movements. For more modest shifts, effects can be detected as altered rates of disulfide formation. As in determination of static structure, it is important to test multiple cysteine pairs and make conclusions based on consistent patterns.

We utilized this approach to identify the ligand-induced conformational change in chemoreceptor Trg, using eight cysteine pairs. The four that bridged the TM1–TM1′ interface showed no significant change in rates of disulfide formation between the ligand-free and ligand-saturated state but the four that bridged the TM1–TM2 interface all exhibited changed rates (Beel and Hazelbauer, 2001; Hughson and Hazelbauer, 1996; Lai *et al.*, 2006a). Importantly, ligand-induced changes were in both directions: two rates decreased and two increased. This made it unlikely that the changes reflected nonspecific effects of experimental manipulations, since these would be expected to affect all rates in the same direction. Moreover, a combination of increases and decreases allowed us to assess which possible movement between two helices could account for the pattern. Only axial sliding (a "piston" movement) did so and thus we concluded the conformational change of transmembrane signaling was a helical piston movement, a conclusion consistent with a large body of parallel and independent observations (Falke and Hazelbauer, 2001). We also used measurement of rates of disulfide formation between diagnostic cysteine pairs to deduce the conformational change generated in the transmembrane domain by adaptational modification (methylation) of the cytoplasmic domain (Lai *et al.*, 2006a). Using the same TM1–TM2 cysteine pairs as in studies of ligand-induced signaling, we found that all four rates were altered by adaptational modification. For each pair, the modification-induced change was opposite the ligand-induced change, that is, an increase versus a decrease or vice versa, indicating that adaptational modification reversed the axial sliding caused by ligand binding.

Procedures

To take full advantage of the *in vivo* procedure, which provides the opportunity to probe a transmembrane protein involved in its native interactions and at native copy number in an intact and functional cell, it is important to ensure that the cells being assayed are grown and manipulated in a way that minimizes perturbations to the system of interest. For our studies of the sensory system that controls motility, we followed procedures developed to maintain maximal cell motility. In our experience, the proportion of *E. coli* cells in a culture or suspension that are vigorously motile is very sensitive to many physiological perturbations. Thus, vigorous and uniform motility is a good indication of physiologically unperturbed cells, not only for studies of the chemotaxis system but also for studies of other cellular systems. Of particular importance in maintaining motility are (1) inoculation procedures into liquid growth media that provide sufficient cell density to avoid the shock of drastic dilution of cells into fresh medium and (2) growth regimens that maintain cells in the logarithmic phase of growth and avoid the transition into stationary. We used strains with a chromosomal deletion of the gene for the chemoreceptor to be studied but harboring a plasmid carrying that gene under tight control of an inducible promoter and altered to code for the desired cysteine-containing receptor. Stocks of such strains were produced by growing to late logarithmic phase a vigorously motile culture in Luria broth (1% bactotryptone, 0.5% yeast extract, 0.5% NaCl) plus appropriate plasmid-maintaining antibiotic(s), adding glycerol to 20%, freezing rapidly in liquid nitrogen, and storing at $-70°$.

To initiate testing of a particular cysteine pair, a portion of a frozen stock of the appropriate strain was scraped off by an inoculation loop and used to inoculate to a visibly discernible turbidity a few tenths ml of tryptone broth (1% bactotryptone, 0.5% NaCl) containing appropriate antibiotic(s) in a 20-ml, 16-mm (OD) test tube. To ensure efficient aeration, the tube was tipped far on its side in a reciprocating shaker (model R-2, New Brunswick Scientific, Edison, NJ) and incubated with vigorous shaking (200 rpm) at $35°$. Once turbidity was visibly increasing, indicating a significant proportion of the cellular population was dividing, the culture was diluted to 4 ml and incubation continued until late log phase ($\sim 5 \times 10^8$ cells/ml). The tryptone culture was used immediately to inoculate 4 ml of minimal salts medium (we use H1, which is 100 mM potassium phosphate, pH 7.0; 15 mM (NH$_4$)$_2$SO$_4$, 1 mM MgSO$_4$; 1.1 μM Fe$_2$(SO$_4$)$_3$ adjusted to pH 7.0 and containing required amino acids at 0.5 mM, an appropriate carbon and energy source (we commonly use 20 mM sodium succinate or 13 mM ribose, each of which exerts little catabolite repression) and appropriate antibiotic(s), which were usually at half the concentration used in the richer medium. Alternatively, growth was stopped by rapid cooling with agitation in an ice slurry and the cooled culture was stored at 4° for use as an inoculum the following day. Cells were inoculated to ~2.5 × 10⁷ cells/ml and grown as above to ~2.5 × 10⁸ cells/ml. These logarithmic-phase cells were used, either immediately or after rapid cooling and storage as has been described, to inoculate the same minimal medium (volume dependent on needs of the experiment, often 10 ml in a 100 ml flask) to 2.5 × 10⁷ cells/ml. This medium was supplemented with sufficient inducer to produce the cysteine-containing protein at the desired cellular dosage, often the level equivalent to that produced by the native chromosomal gene. For experiments with Trg, this was achieved by addition of IPTG to 20 μM. The culture was grown at 35° with shaking to provide generous aeration to a density of ~2.5 × 10⁸ cells/ml and cooled rapidly in an ice slurry for use in experiments that day.

For experiments using Cu-phenanthroline, 0.5 ml of the cooled culture was placed into a pre-warmed 10 ml beaker, warmed for 5 min by incubation at 35° with vigorous agitation to provide generous aeration, and mixed with an equal volume of pre-warmed Cu-phenanthroline solution at twice the desired final concentration. Some experiments included controls in which cells were mixed with a volume lacking Cu-phenanthroline. Other experiments involved adding a chemoattractant 5 min prior to addition of the oxidation catalyst. The Cu-phenanthroline solution added to cells should be prepared on the day of the experiment by diluting into water a 60 mM solution of Cu-phenanthroline stored at 4°. The 60 mM stock was made by adding 1,10-phenanthroline dissolved in methanol to NaH_2PO_4 in H_2O and then adding $CuSO_4$ to achieve final concentrations of 180, 60, and 50 mM, respectively. This order of addition is necessary to avoid formation of a precipitate. For our experiments, we used a rotary water bath at 200 rpm (New Brunswick Scientific, Edison, NJ, model Gyrotary® water bath shaker G76) equipped with clamps that hold the 10 ml beakers. The large surface-to-volume ratio for the 0.5 ml sample in the beaker and the vigorous agitation are important to ensure a sufficient supply of oxygen for reproducible time courses of sulfhydryl oxidation in a suspension of respiring cells. The same volume in a small test tube can produce inconsistent results. For preparation of cells to be subsequently tested for function, disulfide formation was stopped by addition of EDTA to 10 mM. For monitoring disulfide formation by immunoblotting, samples of 40 μl were removed just before and at various times after addition of the catalyst, immediately mixed with 10 μl of stop solution, which is our non-reducing sample buffer for SDS polyacrylamide gel electrophoresis supplemented with EDTA and N-ethylmaleimide (20 mM Tris-HCl, pH 7.8, 8 mM NaH_2PO_4, 12.5 mM EDTA, 5 mg/ml N-ethylmaleimide, 1.25% SDS,

25% w/v sucrose, and 2.5 mg/ml bromphenol blue), and placed on ice. EDTA chelates Cu^{2+}, thus removing it from further action on sulfhydryls; N-ethylmaleimide reacts with remaining cysteine sulfhydryls, thus avoiding disulfide formation that would otherwise likely occur during sample processing; and sucrose substitutes for glycerol as a density agent in the electrophoresis sample that does not accumulate the reducing power that is created when glycerol is exposed to water. For determining time courses and rate constants of disulfide formation, we routinely took samples each minute for 4 min, then every 2 min to 10 min. For a particular diagnostic cysteine pair, preliminary experiments were performed to identify a concentration of catalyst that provided rates of disulfide formation that generated a time course that extended 5 to 10 min.

Experiments using molecular iodine were performed in essentially the same way as those using Cu-phenanthroline except that 2.5 ml of culture was placed in a 10 ml beaker, warmed by incubation on a shaker at 25° for 5 min, and 2.5 ml of 0.5 mM iodine solution at 25° was added. After 15 s, cells were tested for function and/or analyzed for extent of disulfide formation, as described for the Cu-phenanthroline treatment. Our iodine stock was a 100 mM solution in ethanol, which takes ~0.5 h of stirring to go into solution. On the day of an experiment, a 1 mM is prepared by adding the appropriate volume of the ethanol solution to water. The limit of iodine solubility in water is 1.18 mM at 25°.

Samples for analysis of disulfide formation by either reagent were boiled 4 min as soon as an experiment was completed and analyzed by nonreducing SDS-PAGE and immunoblotting immediately thereafter or stored at −20° for later analysis. Electrophoresis should be in conditions known to resolve the unreacted and disulfide-containing protein of interest. The position of disulfide cross-linked proteins on SDS-PAGE reflects the positions that the disulfide joins (Lee *et al.*, 1994). If the bond links two different polypeptide chains, the cross-linked product migrates in SDS-PAGE more slowly than either individual polypeptide. If the disulfide links two positions on the same chain, the constrained polypeptide migrates more rapidly than the unreacted chain. The apparent molecular mass of the cross-linked species in SDS-PAGE varies as a function of the particular position of the disulfide. Thus, with appropriate standards, positions can be used to make deductions about the nature of the cross-linked species. For instance, in our work with the ~60 kDa chemoreceptor Trg (Lee *et al.*, 1994), a disulfide that joined cysteines in the same polypeptide chain, near residues 40 and 200, respectively, resulted in migration at an apparent molecular mass of ~55 kDa. A disulfide between two monomers at cysteines near residue 40, close to the amino terminus, resulted in a species that migrated at an apparent molecular mass of ~120 kDa, twice the monomer molecular mass. In contrast, a disulfide between the two subunits at a cysteine near residue

200 of the 535-residue polypeptide migrated significantly more slowly, with an apparent molecular mass $\sim$180 kDa, three times the monomer molecular mass, even though the species was a cross-linked dimer of two Trg chains, not a trimer. Finally, two polypeptides joined by a disulfide between the position near residue 40 on one chain and near 200 on the other chain migrated with an apparent molecular mass of $\sim$150 kDa, between 120 and 180 kDa.

Analysis

The intensities of the disulfide-linked and unreacted protein are determined by densitometric quantification of the respective bands in an immunoblot lane (Fig. 2). The extent of disulfide formation is calculated as the intensity of the band of cross-linked species divided by the sum of the intensities in the two bands. For time courses, the fraction of protein that has formed the disulfide is plotted as a function of time after exposure to the oxidation treatment (Fig. 2). These time courses reflect the two oxidation reactions undergone by cysteine sulfhydryls: to disulfide or to sulfinic or sulfonic acid. The latter reactions block cysteine from participation in disulfide formation. As explained by Careaga and Falke (1992), the net result can be expressed as $F_{SS}(t) = (k_{SS}/k_{SS} + k_2) (1-\exp[-(k_{SS} + k_2)t])$, in which $F_{SS}(t)$ is the fraction of protein in the disulfide form at time t and k_{SS} and k_2 are the rate constants for formation of the disulfide and the other oxidation products, respectively. In our studies of *in vivo* disulfide formation in chemoreceptors, the data implied a modest time lag between addition of Cu-phenanthroline to a cell suspension and initiation of sulfhydryl oxidation (Fig. 2). Correspondingly, mathematical fits to the data were most satisfactory if a lag time, t_0, were introduced into the fitting equation. Operationally, we use a commercial program to fit our data to $y = a(1-\exp[-b(x-x_0)])$ in which $a = (k_{SS}/k_{SS} + k_2)$, $b = (k_{SS} + k_2)$, and x_0 = time lag. Using the values for a and b determined by the fit, $(a \times b) = k_{SS}$, the rate constant for disulfide formation, and $b = S-S_{max}$, maximal disulfide formation.

Values for fraction disulfide formation must be critically evaluated for two potentially misleading factors that can distort the apparent extent of reaction: (1) band intensities outside the range of a linear relationship between intensity and protein amount and (2) selective loss of a protein species. Quantification of band intensities is challenging because there is a limited range over which intensities on immunoblots or stained gels are directly proportional to the amount of protein. In our experience with many different proteins and antisera, this range hardly exceeds one order of magnitude; it is usually 15- to 20-fold. It is even less for immunoblots developed using chemiluminescence. If the extent of disulfide formation is small and thus the band of cross-linked protein is faint, it is tempting to load more material on a gel, use higher antiserum concentration, or extend the duration of the chemiluminescence reaction so that the band becomes darker. The danger in this procedure is that the intensity of the corresponding band of unreacted protein becomes so intense that it is "overloaded," that is, its intensity underrepresents amount of material relative to the lighter disulfide-linked species. In such a case, the extent of disulfide formation is exaggerated, either as viewed by eye or quantified by densitometry. In the extreme, this effect can lead to conclusions based on reactions that involve only a few percent or a few tenths of a percent of the protein of interest and thus may not represent the behavior of most of the protein. An easy empirical check to determine whether intensities are within the range of proportionality to protein amounts is to run three two-fold serial dilutions of the same sample on an immunoblot. The intensities of the band of a given protein species should differ by the dilution factor and the proportion of the protein that is disulfide-linked material should be the same for each dilution. Systematic disparities are indications of intensities outside the regime of proportionality. The solution is to analyze samples with intensities within the proportional range.

It is possible that, during exposure to an oxidation treatment, a particular protein species is operationally lost. The loss could reflect proteolysis or creation of a heterodisperse population of oxidation products with many positions on an immunoblot of a gel and thus not visible as a distinct band. A heterodisperse population could occur by coupling of the protein with many different reactive partners or by multiple oxidations of side chains at a variable number of positions. Species-specific losses could distort apparent time courses of disulfide formation. For instance, steady proteolysis of protein that had not formed a disulfide and little proteolysis of the disulfide-linked product would result in an increase in the proportion of disulfide product greater than the increase of disulfide formation. The opposite is also possible. An easy check for loss of material is to plot the sum of intensities of disulfide-containing and unreacted species as a function of time after initiation of the oxidation treatment. In the absence of losses, this sum should be constant over time. If there is a systematic loss of total intensity, it is crucial to assess the impact of this loss on interpretation of the data.

Closing Comments

Disulfide formation *in vivo* between cysteines introduced into the transmembrane receptor Trg has proven very useful in deducing structural organization and functionally important conformational changes. We hope that the experimental strategies and details will be useful for the investigation of other chemoreceptors, other bacterial transmembrane proteins, and membrane proteins in general.

Acknowledgments

Geoff Lee and Michael Lebert initiated exploration of *in vivo* disulfide formation in our laboratory and Lee did many experiments defining and characterizing the procedures that he and others used. Andy Hughson and Bryan Beel utilized and refined the procedures. The work was supported, in part, by grant GM29963 from the National Institute of General Medical Sciences.

References

Amin, D. N., Taylor, B. L., and Johnson, M. S. (2006). Topology and boundaries of the aerotaxis receptor Aer in the membrane of *Escherichia coli*. *J. Bacteriol.* **188,** 894–901.

Bader, M. W., and Bardwell, J. C. A. (2001). Catalysis of disulfide bond formation and isomerization in *Escherichia coli*. *Adv. Protein Chem.* **59,** 283–301.

Beel, B. D., and Hazelbauer, G. L. (2001). Signaling substitutions in the periplasmic domain of chemoreceptor Trg induce or reduce helical sliding in the transmembrane domain. *Mol. Microbiol.* **40,** 824–834.

Bessette, P. H., Aslund, F., Beckwith, J., and Georgiou, G. (1999). Efficient folding of proteins with multiple disulfide bonds in the *Escherichia coli* cytoplasm. *Proc. Natl. Acad. Sci. USA* **96,** 13703–13708.

Boldog, T., and Hazelbauer, G. L. (2004). Accessibility of introduced cysteines in chemoreceptor transmembrane helices reveals boundaries interior to bracketing charged residues. *Protein Sci.* **13,** 1466–1475.

Careaga, C. L., and Falke, J. J. (1992). Thermal motions of surface alpha-helices in the D-galactose chemosensory receptor. Detection by disulfide trapping. *J. Mol. Biol.* **226,** 1219–1235.

Derman, A. I., Prinz, W. A., Belin, D., and Beckwith, J. (1993). Mutations that allow disulfide bond formation in the cytoplasm of *Escherichia coli*. *Science* **262,** 1744–1747.

Eisenberg, D., Weiss, R. M., and Terwilliger, T. C. (1982). The helical hydrophobic moment: A measure of the amphiphilicity of a helix. *Nature* **299,** 371–374.

Falke, J. J., and Hazelbauer, G. L. (2001). Transmembrane signaling in bacterial chemoreceptors. *Trends Biochem. Sci.* **26,** 257–265.

Hughson, A. G., and Hazelbauer, G. L. (1996). Detecting the conformational change of transmembrane signaling in a bacterial chemoreceptor by measuring effects on disulfide cross-linking *in vivo*. *Proc. Natl. Acad. Sci. USA* **93,** 11546–11551.

Hughson, A. G., Lee, G. F., and Hazelbauer, G. L. (1997). Analysis of protein structure in intact cells: Crosslinking *in vivo* between introduced cysteines in the transmembrane domain of a bacterial chemoreceptor. *Protein Sci.* **6,** 315–322.

Kobashi, K. (1968). Catalytic oxidation of sulfhydryl groups by o-phenanthroline copper complex. *Biochimica et Biophysica Acta (BBA)—General Subjects* **158,** 239–245.

Kosower, N. S., and Kosower, E. M. (1995). Diamide: An oxidant probe for thiols. *Methods Enzymol.* **251,** 123–133.

Lai, W.-C., Peach, M. L., Lybrand, T. P., and Hazelbauer, G. L. (2006b). Diagnostic cross-linking of paired cysteine pairs demonstrates homologous structures for two chemoreceptor domains with low sequence identity. *Protein Sci.* **15,** 94–101.

Lee, G. F., Burrows, G. G., Lebert, M. R., Dutton, D. P., and Hazelbauer, G. L. (1994). Deducing the organization of a transmembrane domain by disulfide cross-linking. The bacterial chemoreceptor Trg. *J. Biol. Chem.* **269,** 29920–29927.

Lee, G. F., and Hazelbauer, G. L. (1995). Quantitative approaches to utilizing mutational analysis and disulfide crosslinking for modeling a transmembrane domain. *Protein Sci.* **4,** 1100–1107.

Lee, G. F., Lebert, M. R., Lilly, A. A., and Hazelbauer, G. L. (1995). Transmembrane signaling characterized in bacterial chemoreceptors by using sulfhydryl cross-linking *in vivo. Proc. Natl. Acad. Sci. USA* **92,** 3391–3395.

Ma, Q., Roy, F., Herrmann, S., Taylor, B. L., and Johnson, M. S. (2004). The Aer protein of *Escherichia coli* forms a homodimer independent of the signaling domain and flavin adenine dinucleotide binding. *J. Bacteriol.* **186,** 7456–7459.

Nakamoto, H., and Bardwell, J. C. A. (2004). Catalysis of disulfide bond formation and isomerization in the *Escherichia coli* periplasm. *Biochimica et Biophysica Acta* **1694,** 111–119.

Pakula, A. A., and Simon, M. I. (1992). Determination of transmembrane protein structure by disulfide cross-linking: The *Escherichia coli* Tar receptor. *Proc. Natl. Acad. Sci. USA* **89,** 4144–4148.

Prinz, W. A., Aslund, F., Holmgren, A., and Beckwith, J. (1997). The role of the thioredoxin and glutaredoxin pathways in reducing protein disulfide bonds in the *Escherichia coli* cytoplasm. *J. Biol. Chem.* **272,** 15661–15667.

Rietsch, A., and Beckwith, J. (1998). The genetics of disulfide bond metabolism. *Annu. Rev. Genet.* **32,** 163–184.

Schulz, G. E., and Schirmer, R. H. (1979). "Principles of Protein Structure." Springer-Verlag, New York.

Studdert, C. A., and Parkinson, J. S. (2004). Crosslinking snapshots of bacterial chemoreceptor squads. *Proc. Natl. Acad. Sci. USA* **101,** 2117–2122.

Thornton, J. M. (1981). Disulphide bridges in globular proteins. *J. Mol. Biol.* **151,** 261–287.

Tiebel, B., Garke, K., and Hillen, W. (2000). Observing conformational and activity changes of Tet repressor *in vivo.* **7,** 479–481.

Watts, K. J., Sommer, K., Fry, S. L., Johnson, M. S., and Taylor, B. L. (2006). Function of the N-terminal cap of the PAS domain in signaling by the aerotaxis receptor Aer. *J. Bacteriol.* **188,** 2154–2162.

[14]   Using Nanodiscs to Create Water-Soluble Transmembrane Chemoreceptors Inserted in Lipid Bilayers

By THOMAS BOLDOG, MINGSHAN LI, and GERALD L. HAZELBAUER

Abstract

In this chapter we describe application of the emerging technology of Nanodiscs® to chemoreceptors, a class of transmembrane proteins that presents many challenges to the investigator. Nanodiscs are soluble, nano-scale (~10 nm diameter) particles of lipid bilayer surrounded by an annulus of amphipathic protein, the membrane scaffold protein. A transmembrane protein inserted in a Nanodisc is surrounded by a lipid bilayer much as it is prior to detergent solublization. Thus, the Nanodisc-inserted protein is in an environment that approximates its native state. Yet, that membrane protein is also water-soluble and segregated from other membrane proteins because the bilayer into which it is inserted is of very limited size and, with appropriate preparation, contains only a single protein. In a Nanodisc, the water-soluble, bilayer-inserted membrane protein can be purified by conventional techniques and analyzed for activities and interactions as a pure entity. Thus, Nanodisc technology has great promise for improving isolation, purification, and characterization of the many membrane proteins that are difficult to handle, become unstable, or lose native activity when surrounded by detergent instead of lipid bilayer. The technology has proven useful for the investigation of chemoreceptor activity as a function of oligomeric state.

Introduction

Biochemical and structural characterization of a transmembrane protein requires isolation of the purified molecule independent of the many protein neighbors that are inserted in the same lipid bilayer. This can be accomplished using a detergent to dissociate the membrane and solubilize its individual components, then purifying the protein of interest using standard techniques performed in the presence of the detergent. However, the structure and activity of many membrane proteins is disrupted in the detergent-solubilized state. A commonly used approach to address this problem is to reinsert the protein into a lipid bilayer by reconstitution procedures. In such procedures, solubilized lipid and purified protein are mixed and detergent removed, resulting in spontaneous formation of vesicles of

METHODS IN ENZYMOLOGY, VOL. 423
0076-6879/07 $35.00
DOI: 10.1016/S0076-6879(07)23014-9

lipid bilayer into which the protein has been incorporated. Such reconstituted membrane protein can exhibit improved activity and more native structure. Optimized reconstitution procedures can be quite effective but they create two important limitations: (1) the purified protein is now part of a much larger entity, a proteoliposome, which is not soluble and thus cannot be manipulated or analyzed by many of the techniques applied to purified proteins, and (2) the protein is embedded in an extended two-dimensional solution of lipid bilayer in which multiple protein molecules incorporated into the same bilayer can readily interact in associations that are difficult to define or control. In most cases, it has not been possible to avoid these limitations if detergent disrupts the structure or function of the membrane protein. The emerging technology of Nanodiscs®, developed in the laboratory of Stephen Sligar, provides a promising alternative (Nath et al., 2007). In this chapter, we describe the application of Nanodiscs to manipulation of chemoreceptors, a class of transmembrane proteins that presents many challenges to the investigator.

Nanodiscs (Fig. 1) are soluble, nanoscale ($\sim$10 nm diameter) particles of lipid bilayer surrounded by an annulus of amphipathic protein (Bayburt et al., 2002; Denisov et al., 2004). The particles have a defined, homogeneous size and composition (Denisov et al., 2004). Their key component is a fragment of human apolipoprotein A-1 engineered in the Sligar laboratory.

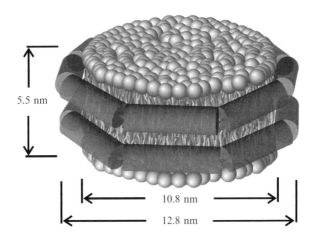

FIG. 1. Cartoon of a Nanodisc showing two copies of a membrane scaffold protein (MSP) surrounding the hydrocarbon side chains of a lipid bilayer. Phospholipid head groups are represented by spheres and the series of amphipathic helices of MSP are represented as cylinders. Dimensions are shown for a disc formed with MSP1E3 (Table I). Based on Fig. 1B in Boldog et al. (2006).

Native apolipoprotein A-1 creates water-soluble particles of lipid that are transported in the bloodstream (Segrest *et al.*, 2000). The particles can take the form of a simple disc of bilayer, surrounded by a portion of the protein. This portion is a series of amphipathic helices punctuated by prolines, providing a water-protective coating around the hydrophobic core of the bilayer. Sligar's laboratory manipulated the coding sequence for apolipoprotein A-1 and introduced it into a vector that, when introduced into *E. coli*, produces large amounts of the amphipathic, bilayer-surrounding segment of apolipoprotein A-1. The product is termed "membrane scaffold protein," MSP. There are several sizes of membrane scaffold protein, with different numbers of amphipathic helices and thus enclosing different diameters of lipid bilayer (Table I). Scaffold proteins are water-soluble, but have the ability to create an annulus around a plug of lipid bilayer. Nanodiscs form spontaneously when detergent is removed from a mixture of detergent-solubilized phospholipid and scaffold protein. Two amphipathic MSP molecules form a helical belt that encircles a plug of bilayer, shielding the hydrophobic lipid acyl chains (Fig. 1) and making the particle water soluble. If the initial mixture of MSP and detergent-solubilized lipid contains detergent-solubilized membrane protein, the protein can be incorporated into Nanodiscs that are formed upon detergent removal (Fig. 2) (Bayburt and Sligar, 2003; Duan *et al.*, 2004). A protein inserted in a Nanodisc is surrounded by a lipid bilayer, much as it is prior to solubilization. Thus, the Nanodisc-inserted protein is in an environment that approximates its native state. Yet, that membrane protein is also water soluble and segregated from other membrane proteins because the bilayer into which it is inserted is of very limited size and, with appropriate preparation, contains only a single protein. In a Nanodisc, the water-soluble, bilayer-inserted membrane protein can be purified by conventional techniques and analyzed for activities and interactions as a pure entity. Thus, Nanodisc technology has great promise for improving isolation, purification, and characterization of the many membrane proteins that are difficult to handle, become unstable, or lose native activity when surrounded by detergent instead of lipid bilayer.

Developing a Protocol for Producing Nanodisc-Embedded Protein

Formation of Nanodiscs is an empirical process that requires optimum ratios of the constituents in the mixture from which detergent is removed. Presumably, this is because detergent removal creates a molecular scramble in which hydrophobic surfaces suddenly exposed to an aqueous environment are driven to cluster and, in this scramble, only certain ratios result in a high proportion of the molecules forming homogeneous Nanodiscs,

TABLE I

PARAMETERS FOR NANODISCS MADE WITH DIPALMITOYL PHOSPHATIDYL CHOLINE[a]

MSP[b]	"Belt" residues	DPPC/disc	Disc dimensions		Bilayer dimensions	
			Diameter (nm)	Thickness (nm)	Diameter (nm)	Area (sq nm)
MSP1	178	160	9.8	5.75	7.7	47
MSP1 E1	200	212	10.6	5.6	8.5	57
MSP1 E2	222	268	11.9	5.5	9.8	75
MSP1 E3	244	334	12.9	5.5	10.8	92

[a] Based on data from Tables 1 and 2, Denisov et al. (2004).

[b] The MSPs listed are the variants that make different-sized Nanodiscs. E1, E2, and E3 signify extended proteins in which 1, 2, or 3 22-amino acid repeats are inserted between residues Q55 and P56. The extensions are duplications of residues 56–77, 56–99, and 56–121, respectively. MSPs carry a 6- or 7-histidine amino terminal tag connected to the body of the protein by a short linker, plus a protease cleavage site recognized by Factor X or TEV protease (see Denisov et al., 2004, for details). MSP in which protease has removed the tag are designated (−). In addition, each MSP can be deleted for 11 (D1) or 22 (D2) amino-terminal residues which do not participate in enclosing lipid in the disc and thus are not considered "belt" residues.

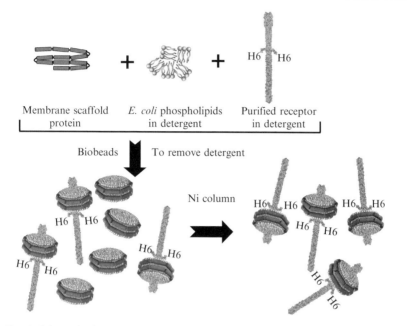

FIG. 2. Schematic diagram of preparation of Nanodisc-embedded chemoreceptor Tar-6H. See text for details.

instead of lipid vesicles or aggregates of scaffold protein and lipid in shapes other than defined-sized discs. The need for optimized ratios is important for Nanodiscs formed from only two constituents, a membrane scaffold protein and a single kind of lipid, and becomes even more important when the goal is to incorporate a transmembrane protein. Thus, the first step in developing a Nanodisc protocol is to screen ratios of MSP and lipid, assessing for production of particles of homogeneous size by size-exclusion chromatography. We screened a range of MSP:lipid ratios guided by previous optimizations for different-sized scaffold proteins and different lipids (Denisov *et al.*, 2004) and then tested a narrower range based on the initial results. This process identified an optimized molar ratio of 1:120 *E. coli* lipids/MSP1D1E3(–) (see legend to Table I for explanation of nomenclature) (Boldog *et al.*, 2006). As might be expected, this ratio is different from optimia for other combinations of MSPs and lipids. MSP1D1E3(–) forms a larger-diameter Nanodisc than does MSP1D1 (Table I), but we found that the optimized ratio of MSP to *E. coli* lipid determined for the larger protein could be translated into an effective ratio for the smaller protein by adjusting the number of lipid molecules by the ratio of areas enclosed by the two sizes of scaffold protein.

Optimized incorporation of a transmembrane protein into Nanodiscs can require adjustment of the preparation ratio of MSP to lipid if a significant area of the lipid bilayer will be occupied by the protein. The appropriate reduction in number of lipid molecules per scaffold protein can be calculated if the transmembrane area of the protein to be incorporated is available from structural characterization or modeling. The MSP/lipid ratio can be adjusted as required. In any case, an experimental check of the appropriateness of the ratio used is to assess the Nanodisc preparation for homogeneous size-by-size-exclusion chromatography and SDS-polyacrylamide gel electrophoresis of eluted fractions. In optimized preparations, the transmembrane and scaffold proteins will co-elute in a symmetrical peak ahead of empty Nanodiscs, consistent with a relatively homogeneous population larger than empty discs because of inserted transmembrane protein (Fig. 3). As necessary, further optimizations are performed to identify ratios of the constituents that result in preparations with the greatest proportion of particles eluting as a population of homogeneous particles.

Components and Affinity Tags

Nanodisc technology gives the investigator the option of using whatever lipid or lipids best serve the particular experimental purpose. Most studies from the Sligar laboratory have used a single lipid type, usually a phosphatidyl choline with a specific fatty acid composition (e.g., dipalmitoyl, dimyristoyl, or palmitoyl-oleoyl forms of phosphatidyl choline). Such lipids are commonly utilized in many membrane reconstitution procedures, particularly for transmembrane proteins from eukaryotic cells. In our initial attempts to incorporate a chemoreceptor from *E. coli* into Nanodiscs, we used palmitoyl-oleoyl phosphatidyl choline, but receptor activity was low. We reasoned that activity might be enhanced if the lipids in the Nanodisc were the same as those of the *E. coli* cells from which the protein had been extracted, that is, a mixture primarily of phosphatidyl ethanolamine, phosphatidyl glycerol, and diphosphatidyl glycerol (Cronan and Rock, 1996). Thus, we turned to a commercially available extract of lipids from *E. coli* (Avanti, *E. coli* polar lipid extract order # 101601). In combination with several other changes, the resulting Nanodisc-embedded chemoreceptors exhibited substantial activity. We have not studied the effects of lipid type or composition for otherwise optimized preparations of chemoreceptor-containing Nanodiscs, so we do not yet know the extent to which native lipids are important for chemoreceptor activity.

The constructs that produce membrane scaffold proteins in *E. coli* are engineered so that the proteins carry a seven-histidine, amino-terminal tag. This allows easy purification using a Ni column. For some purposes, this tag

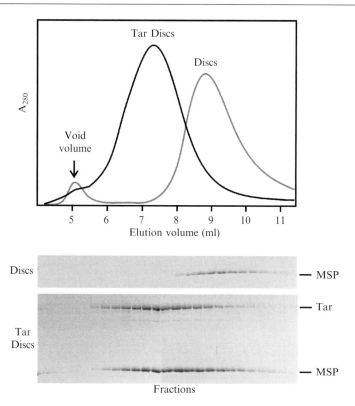

FIG. 3. Size-exclusion chromatography of Nanodiscs containing only *E. coli* lipids (Discs) or lipid plus chemoreceptor (Tar Discs). The upper panel shows continuous recording of absorbance and the lower panel, Coomassie Blue-stained SDS-polyacrylamide gels of samples from the fractions collected from a TSK G5000 PWXL column (7.8 mm inner diameter × 30 cm, Tosho Haas) in 50 m*M* Tris·HCl pH 7.5, 10% w/v glycerol, 100 m*M* NaCl, 0.5 m*M* EDTA 19° at 0.7 ml/min with 0.3 ml fractions. The figure is a relabeled version of Fig. 2B, Boldog *et al.* (2006).

provides a convenient means to isolate and manipulate Nanodiscs formed with the MSP. In other situations, the histidine tag would interfere with desired manipulations. In these cases, the tag can be removed by utilizing a specific protease cleavage site that has been placed between the amino-terminal histidine tag and the body of the recombinant scaffold protein. Cleavage at this site is effective and efficient, resulting in high yields of otherwise intact MSP (see protocol in the following text). In our manipulations of chemoreceptors in Nanodiscs, we used membrane scaffold proteins from which the affinity tag had been removed. This allowed us to use a histidine tag on the chemoreceptors for convenient purification of the detergent-solubilized protein from the other proteins in isolated bacterial membranes and also for purification of Nanodiscs that contained receptor from "empty" Nanodiscs, those lacking inserted receptor. The latter manipulation was particularly important for studies in which we varied the input ratio of MSP to receptor in order to make Nanodisc preparations with different numbers of receptor dimers/Nanodisc (see following section).

Manipulating the Number of Inserted Proteins per Nanodisc

Preparations of Nanodisc-embedded receptors with different average numbers of receptor dimers per disc can be obtained by varying the preparation ratio of MSP to 6-histidine-tagged receptor (Fig. 4). These preparations should contain few "empty" discs, those that have not incorporated a receptor. This is because discs lacking a his-tagged receptor should be not retained by the Ni column, and thus not present in the imidazole-eluted material. Because each disc has two membrane scaffold polypeptides (Bayburt and Sligar, 2003; Bayburt *et al.*, 2006; Denisov *et al.*, 2004; Duan *et al.*, 2004), the concentration of discs in this preparation can be determined by quantifying the amount of scaffold protein. We routinely quantify membrane scaffold protein by comparing on the same SDS-polyacrylamide gel the intensity of Coomassie Blue staining of a series of dilutions of a Nanodisc preparation and the intensity of bands in a standard curve constructed from a dilution series of a standard solution of that scaffold protein quantified by

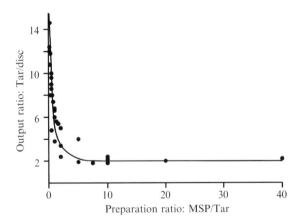

FIG. 4. Tar polypeptides/disc as a function of preparation ratio of MSP/receptor determined by quantification of Coomasie Blue-stained bands in SDS-polyacrylamide gels of the respective disc preparations. The figure is a relabelled version of Fig. 2C, Boldog *et al.* (2006).

amino acid analysis. The determinations include only those intensities that fall in the range of a linear relationship between intensity and amount of material loaded on the gel. Determination of chemoreceptor concentration in a Nanodisc preparation is performed in the same way, using a receptor stock quantified by quantitative amino acid analysis. In our experience, it is important to use quantified standards of the respective scaffold protein and transmembrane protein. In the case of MSP1D1E3(−) and 6-histidine-tagged chemoreceptor Tar, Tar-6H, the respective relationships of Coomassie Blue staining intensity to amount of protein varied by a factor of two!

With input ratios at eight-fold or greater excess of MSP over chemoreceptor, the number of receptor polypeptides per disc reached a lower limit of two (Fig. 4), indicating that the core structural unit incorporated in a disc was a receptor homodimer and implying that each disc contained a single dimer. Nanodiscs produced from preparation mixtures with less MSP/Tar contained on the average more than one receptor dimer in the same 10.8 nm diameter plug of bilayer (Denisov *et al.*, 2004). Using the data in Fig. 4 as a calibration curve, we could produce a Nanodisc preparation with the desired average content of receptor dimers per disc, within the limits indicated by the scatter of the data. There was no evident preference for integer values of dimers/disc (Fig. 4), arguing that dimers incorporated individually into discs and that the number of receptors per disc represented a mean among a population with different integer numbers of dimers/disc. For discs containing more than one receptor dimer, independent insertion implies that multiply inserted dimers could be parallel or antiparallel. Initial electron micrographs of negative stained Nanodisc preparations indicate that both orientations occur, but we have not yet determined the relative frequencies of parallel versus antiparallel insertion.

Preparation of Nanodisc-Embedded Chemoreceptor

In this section, we describe the detailed procedure for producing Nanodisc-embedded chemoreceptor Tar-6H, membrane scaffold protein MSP1D1E3(−) and *E. coli* lipid.

Materials

Membrane Scaffold Protein. MSP is available upon request from Dr. Stephen G. Sligar (Department of Biochemistry, University of Illinois Urbana-Champaign, 116 Morrill Hall, 505 S. Goodwin, Urbana, IL 61801; sligar@uiuc.edu). For our published experiments, we used the largest membrane scaffold protein, MSP1D1E3(−). This protein differs from MSP1 by a truncation, designated D1, and an insertion, designated E3 (Table I). The former deletes 11 amino-terminal residues not required for Nanodisc formation and the latter adds three 22-residue helices, bracketed by prolines, in the amphipathic belt (Denisov et al., 2004). Freeze-dried protein is solubilized by addition of room temperature H_2O to the original volume of the protein solution, thus reconstituting a solution of $\sim$4.5 mg/ml MSP in 20 mM Tris·HCl, pH 7.4, 100 mM NaCl, 0.5 mm EDTA, 0.01% NaN_3. The solution is passed through a 0.2 μm filter and stored at $-70°$. MSP concentration can be determined spectrophotometrically using the following extinction coefficients: MSP1D1E3($-$) $e_{280} = 0.868$ (mg/ml)$^{-1}$ cm^{-1} (26,024 M^{-1} cm^{-1}), MSP1D1($-$) $e_{280} = 0.825$ (mg/ml)$^{-1}$ cm^{-1} (18,200 M^{-1} cm^{-1}), MSP1D1 $e_{280} = 0.852$ (mg/ml)$^{-1}$ cm^{-1} (21,000 M^{-1} cm^{-1}).

Removal of the 6-histidine tag from MSP is accomplished by treatment with the appropriate protease. Initial versions carried a Factor X site between the 6-histidine tag and the MSP. Newer and more efficiently cleaved versions carry a TEV protease site. A protocol for cleavage of MSP at the TEV protease is as follows:

1. Add distilled H_2O to MSP that had been freeze-dried in 20 mM Tris-HCl, pH 7.5, 100 mM NaCl, 0.5 mm EDTA, 0.01% NaN_3 to the original volume of the protein solution and dilute with 50 mM Tris-HCl, pH 7.5, 0.5 mm EDTA to a 5 mg MSP/ml and add DTT to provide 1 mM final concentration. Reserve 5 μl for subsequent analysis by SDS-PAGE.

2. Add 1 mg/ml 6-histidine-tagged TEV-protease in 50 mM Tris/HCl pH 7.5, 1 mM EDTA, 5mM DTT, 50% glycerol, 0.1% Triton X-100 to 1 mg protease/200 mg MSP and incubate overnight ($\sim$22 h) at room temperature. Take 5 μl samples during the incubation (e.g., at 2, 4, 6, and 22 h).

3. Check progression of the proteolysis reaction by SDS-PAGE and staining with Coomassie Brilliant Blue of samples taken before protease addition and during incubation, loading the same amount of MSP ($\sim$1 μg) in each lane. Cleaved MSP is $\sim$2.6 kD smaller than intact protein and is resolved as a slightly more rapidly migrating band in SDS-PAGE.

4. Remove EDTA and DTT by dialysis against 3 × 15 volumes 50 mM Tris-HCl, pH 7.5, 100 mM NaCl in a cold box. Removal of these compounds is necessary in order to use Ni affinity chromatography.

5. Apply the dialyzed protein solution to a 1 cm diameter Ni-Agarose column (e.g., 8 ml bed, for a 27 ml sample) and wash with 4 bed volumes of 50 mM Tris-HCl pH 7.5, 100 mM NaCl, 25 mM imidazole. Pool fractions collected during sample application and wash. These fractions should contain cleaved MSP but not the 6-histidine-tagged protease or uncleaved MSP, which should be retained by the column.

6. Concentrate the pooled fractions to 20 ml using an Amicon Ultra 15 concentrator and dialyze against 3 × 350 ml 20 mM Tris-HCl pH7.5, 100 mM NaCl, 0.5 mM EDTA.

Phospholipid. Lipid used in formation of Nanodiscs is added to the preparation mixture as a detergent-solubilized solution. *E. coli* lipids are prepared by placing 1 ml portions of the manufacturer-supplied 25 mg/ml chloroform solution of *E. coli* polar lipids (Avanti #101601) in individual glass tubes, evaporating the solvent with a gentle stream of N_2 or argon while rotating the tube to create a thin lipid film, and drying the lipid overnight in a vacuum desiccator. To suspend the lipid as vesicles in an aqueous medium, 0.7 ml H_2O is added to each tube, the tubes sealed with a rubber stopper and placed at 35° on a reciprocating shaker at 200 rpm for 2 h. The result is cloudy suspensions of lipid vesicles. Lipids are solubilized by addition to each tube of 0.2 ml 500 mM sodium cholate in H_2O and periodic gentle agitation with a vortex mixer until the solutions clear. Each solution is then brought to 100 mM Na-cholate and 50 mM Tris·HCl pH 7.5 by addition of 50 μl of 1 M Tris·HCl, pH 7.5, adjustment with pH paper to pH 7.5 using 1 N HCl and adjustment of the volume to 1 ml. If the suspension is not clear, it can be sonicated 6 × 5 s with 25 s pauses using a Tekmar TM-250 sonic disruptor equipped with a microtip at output 2. The pooled solution is passed through a 0.2 μm filter attached to a syringe, distributed into 1.5 ml microcentrifuge tubes, flushed with N_2 or argon, frozen by immersion in liquid N_2, and stored at −70°. The concentration of lipid in the aqueous stock will be close to the concentration of the original chloroform stock. Its precise value is determined using a phosphorus assay.

Phosphorus Assay. We used the Fiske and Subbarow method (Fiske and Subbarow, 1925) as described by Chen *et al.* (1956). See also useful information at http://www.avantilipids.com. In this assay, phosphorus-containing compounds are hydrolyzed by H_2SO_4 and released phosphorus is oxidized by H_2O_2 to inorganic orthophosphate (PO_4^{3-}). Addition of molybdate results in formation of a heteropolyphosphomolybdate complex, which upon reduction with ascorbic acid has a distinct blue color that can be quantified spectrophotometrically.

Solutions

1. 8.9 N H_2SO_4: Slowly pour 99 ml 36 N H_2SO_4 (Fisher ACS plus A300) into ∼ 250 ml MilliQ H_2O. Bring to 400 ml with MilliQ H_2O. Store at room temperature.
2. 10% w/v ascorbic acid: Dissolve 5.0 g ascorbic acid (SigmaUltra A5960) in MilliQ H_2O. Bring to 50 ml with MilliQ H_2O. Store at 4° protected from light; use within 1 month.

3. 2.5% w/v ammonium molybdate tetrahydrate. Dissolve 1.25 g $(NH_4)_2MoO_4 \cdot H_2O$ (Baker Analyzed 0716–01) in MilliQ H_2O. Bring to 50 ml with MilliQ H_2O. Store at $4°$ protected from light; use within 1 month.

Procedure

1. Place a metal heating block with holes that fit 150×85 mm glass test tubes on a hot plate and allow the block to reach $>200°$ while the assay is being prepared.

2. Add to individual 150×85 mm glass test tubes: nothing = blank; 25 μl (16.25 nmol), 50 μl (32.5 nmol), 75 μl (48.75 nmol), 100 μl (65.0 nmol), 125 μl (81.25 nmol), and 150 μl (97.5 nmol) of a 0.65 mM phosphorus standard solution (Sigma P3869 stored at $4°$); and three different volumes of each experimental sample with different amounts designed to contain a quantity of phosphorus within the range of the standard curve. For samples in chloroform, evaporate the solvent with a gentle stream of N_2.

3. Place the uncovered tubes in a fume hood. Add 2 PFTE boiling stones (Chemware D1069103) and 225 μl 8.9 N H_2SO_4. Insert the tubes in the heating block and incubate 25 min at $200–215°$. It is crucial that the temperature is $>200°$. Remove tubes from block and allow to cool 5 min. Add 75 μl 30% H_2O_2 (Sigma 216763 stored at $4°$) to the bottom of each tube and reinsert in the heating block. After 30 min, remove tubes from block and check that all samples are colorless. If still brown, add 25 μl H_2O_2 and heat an additional 15 min. Once samples are colorless, remove the tubes from the heating block and allow to cool to room temperature.

4. During sample heating, equilibrate a water bath to $100°$ and turn on a spectrophotometer to allow it to stabilize. Add to each tube 1.95 ml MilliQ H_2O, 0.25 ml 2.5% w/v ammonium molybdate, and 0.25 ml 10% w/v ascorbic acid. Mix well by gentle vortexing. Cover each tube with a marble, incubate in a water bath 7 min, $100°$, and allow tubes to cool to room temperature. Using quartz cuvettes, zero the spectrophotometer at 820 nm and measure A_{820} of standards and samples. Determine phosphorus in samples using the linear portion of the standard curve.

Chemoreceptors. We used purified 6-histidine-tagged chemoreceptor to produce chemoreceptor-containing Nanodiscs. An alternative possibility is to use detergent-solubilized membranes that contain 6-his receptor and depend on affinity purification to separate Nanodiscs containing the tagged receptor from Nanodiscs containing other membrane proteins (Civjan *et al.*, 2003). Tar-6H is produced at a high level in an appropriate strain, membranes are prepared by osmotic lysis of spheroplasts, and receptor-containing membranes are isolated by sucrose gradient centrifugation (Osborn and Munson, 1974). Cell growth, gene expression, and membrane isolation are performed in ways we have found reliable for high-level production of well-behaved, functional chemoreceptors. Receptor-containing membranes are solubilized with β-octylglucoside and Tar-6H purified with a Ni column. The following sections provide the details for these procedures.

Preparation of Cytoplasmic Membranes with High Tar-6H Content

For production of high levels of Tar-6H, we use RP3098, a derivative of *E. coli* K-12 devoid of other chemoreceptors and chemotaxis proteins (Parkinson and Houts, 1982) that harbors plasmid pAL67 (Lai and Hazelbauer, 2005), which carries the gene for Tar-6H under the control of a modified *lac* promoter as well as the *lacI^q* gene. A sample of a frozen 20% glycerol stock of this strain is streaked on an agar plate containing Luria Broth $+$ 100 μg/ml ampicillin (LB-amp) and incubated at 35° overnight. The next day 6 good-sized colonies are used to inoculate 0.5 ml LB-amp in a culture tube to a visible turbidity and the tube agitated (reciprocal shaker at 200 rpm) at 35°. Once turbidity increases noticeably, LB-amp is added to 4 ml and incubation with agitation continued. Cell density is monitored with a densitometer. The aim is to ensure that the cultures remain well aerated and in the logarithmic phase of growth. As the culture approaches 7×10^8 cells/ml (OD_{560} 1 with our densitometer), it is used to inoculate to a cell density of $\sim 3.5 \times 10^7$ a larger volume of LB-amp in a flask that provides sufficient aeration (no more than 10–15% of the nominal flask volume). This process is continued until 120 ml of a $\sim 7 \times 10^8$ cells/ml culture is obtained. This culture is rapidly cooled in an ice-water bath and stored overnight in the cold for inoculation of the production culture. For this culture, 2.4 l LB-amp is inoculated to $\sim 3.5 \times 10^7$ cells/ml, the volume divided equally among three 6000 ml Erlenmeyer flasks and the flasks agitated on a rotary shaker at 35°. Cell density is monitored. At $\sim 2.8 \times 10^8$ cells/ml (our OD_{560} 0.4), IPTG is added to 0.5 mM and incubation continued for 3.5 h, at which point the OD_{560} will have increased ~ 10-fold. The cultures are fast cooled by swirling the flasks in an ice-water bath, kept cold, and centrifuged 10 min, 8000 rpm, 4° in a SCL-6000 rotor. Pelleted cells are suspended in a small volume of 30% sucrose, 10 mM Tris·HCl, pH 8.0, 5 mM EDTA by forcing the suspension in and out of a 10 ml glass pipette. Once homogeneously suspended, the cell suspension is brought to 100 ml, which should be at $OD_{560} \sim 100$.

This cell suspension is used immediately to prepare membranes without washing or freezing. All manipulations are performed on ice or in a cold room ($= 8°$) with pre-cooled solutions. Lysozyme from a freshly prepared stock (10 mg/ml) is added to 100 μg/ml and the suspension incubated 30 min with gentle stirring. Spheroplasts are prepared by gradually adding 200 ml 0.5 mM EDTA, 1 mM phenanthroline, 1 mM PMSF (add PMSF freshly from a 100 mM stock in ethanol) using a peristaltic pump at 5 ml/min. Spheroplasts are lysed by rapid addition of 1.2 l H_2O. The solution will become highly viscous due to release of DNA (it is hardly possible to pipette the solution). Add DNaseI to 10 μg/ml and $MgCl_2$ to 2mM. Monitor for lowered viscosity with a pipette. After $\sim$20 min, the suspension should flow in a pipette essentially as if it were only water. Centrifuge 2.5 h, 12,500 rpm, 4° in a GSA rotor. Use the centrifugation time to prepare density gradients. To make density gradients, layer sequentially in six thin-walled SW32 tubes sucrose solutions previously prepared in 10 mM Tris·HCl, pH 8.0, 5 mM EDTA: 5 ml 55%, 11 ml 50%, 5 ml 45%, 6 ml 40%, and 6 ml 35%. Add to each membrane pellet $\sim$0.5 ml 30% sucrose, 5 mM Tris·HCl, pH 8.0, 5 mM EDTA, and suspend, using first a glass rod and then a 5 ml syringe equipped first with a blunted 16G1 and then a blunted 22G11/2 needle. Pool the suspensions, bring the volume to 30 ml, layer 5 ml on each of the six sucrose gradients, and centrifuge 14 to 16 h, 32,000 rpm, 4°, with a SW32 rotor. Receptor-containing cytoplasmic membranes form one or two turbid, slightly brown bands which are the uppermost bands visible by eye and are at $\sim$46% sucrose (at higher amounts of material, the two bands fuse to a single band). These bands are removed by illuminating the tube from above and using an inserted Pasteur pipette to remove the turbid material. The harvested suspensions should have a refractive index of $\sim$1.41 as measured by a refractometer. Dilute the pooled membrane suspension with H_2O $\sim$4.6-fold to $\sim$10% sucrose. Centrifuge 1 h, 60,000 rpm, 4°, in a 60Ti rotor. Suspend pelleted membranes in a small volume 50 mM Tris·HCl, pH 7.5, 10% w/v glycerol using a glass rod, then a 5 ml syringe equipped with a blunted 22G11/2 needle, bring volume to $\sim$7 ml, distribute in $\sim$1 ml samples into 1.5 ml microcentrifuge tubes, freeze by immersing the sealed tube into liquid N_2, and store at -70°. Protein concentration is determined by an appropriate assay, such as BCA (Pierce, Inc., Rockford, IL 61105) and receptor concentration determined by quantitative immunoblotting using a standard of known concentration. Typically, our preparations of $\sim$8 ml have $\sim$20 mg/ml membrane protein and $\sim$5 mg/ml Tar-6H.

Receptor Purification

Tar-6H is purified in a two-step procedure, solubilization with octyl-β-D-glucopuranoside (octylglucoside) and affinity separation on a Ni-agarose column. Octylglucoside is particularly effective in receptor solubilization but because of its high cost and the wide use of sodium cholate for reconstitution procedures, we replaced octylglucoside with sodium cholate at the affinity column step. Detergent-solubilized, not-yet-purified receptor is prone to degradation. Thus, solubilization is performed on ice in the presence of protease inhibitors, subsequent manipulations are on ice or at 4° with pre-cooled solutions, glycerol is added to all buffers to stabilize the receptor, and the period between detergent addition and separation on the affinity column is as short as possible. Purified receptor is significantly more stable. Reducing agents are not added to avoid interference with the Ni-column. The detailed procedure is as follows:

1. One or more tubes of frozen membrane stored at $-70°$, such as ~200 mg of membrane protein, are thawed in an ice bath, the volume measured, 50 mM Tris·HCl, pH 7.5, 10% w/v glycerol added to yield $=10$ mg/ml membrane protein, 1 mM leupeptin in H_2O added to 2 μM, 1 mM pepstatin in DMSO added to 2 μM, 100 mM PMSF in ethanol added to 1 mM, and 500 μM N^{α}-p-tosyl-L-lysine chloromethyl ketone (TLCK) in H_2O added to 100 μM. TLCK is unstable and must be prepared freshly. Immediately, add a 20% stock solution of octylglucoside to achieve a 1.25:1 wt:wt ratio of detergent to membrane protein, such as 1.25% octylglucoside, for a suspension of 10 mg membrane protein/ml. Incubate 10 min on ice.

2. Centrifuge 15 min, 100,000 rpm, 4° in a TLA 100.4 rotor to pellet unsolubilized material.

3. Immediately apply the supernant to a Ni-NTA agarose (Qiagen) column (1 ml bed volume per 2.5 mg receptor) equilibrated in 50 mM Tris·HCl, pH 9.0 (high pH minimizes receptor aggregation), 10% w/v glycerol, 100 mM NaCl, 25 mM sodium cholate, and 15 mM imidazole. Apply 5 column volumes of equilibration buffer followed by 5 column volumes of that buffer containing 300 mM imidazole. Collect bulk fractions corresponding to application of the sample, 5 volumes equilibration buffer, and 5 volumes 300 mM imidazole.

4. The fraction collected during elution with 300 mM imidazole is concentrated 10-fold in Centriprep concentrators, 30,000 MWCO, dialyzed in Spectra/Por tubing, 12,000 to 14,000 MWCO, against 100 volumes 50 mM Tris·HCl, pH 7.5, 10% w/v glycerol, 25 mM cholate, 8 to 12 h for each of three buffer changes, and used directly for Nanodisc preparation or frozen in volumes of ~1 ml in 1.5 ml microcentrifuge tubes by immersion in liquid N_2 and stored at $-70°$. Tar-6H concentration is determined using a Tar-6H standard by quantitative Coomassie Blue staining or quantitative immunoblotting of bands on an SDS-polyacrylamide gel.

Preparation of Receptor-Containing Nanodiscs

Our Nanodisc protocol was developed for MSP1D1E3($-$) and *E. coli* lipids. For other MSPs or lipids the MSP:lipid employed should be tested and adjusted as necessary. Removal of detergent using Biobeads is rapid and convenient, but Nanodiscs can also be formed using dialysis to remove detergent (Bayburt et al., 2002; Denisov et al., 2004). The preparation mixture after detergent removal includes not only receptor-containing Nanodiscs but also "empty" discs with no receptor. Receptor-containing discs are purified from these and other components by affinity purification with a Ni resin. The detailed procedure is as follows:

1. Thaw on ice detergent-solubilized E. coli lipids in 50 mM Tris pH 7.5, 100 mM sodium cholate, and purified Tar-6H in 50 mM Tris·HCl, pH 7.5, 10% w/v glycerol, 25 mM sodium cholate.

2. Mix receptor and lipid. Add PMSF to 1 mM, leupeptin to 1 μM, and pepstatin to 1 μM. Cholate (CMC = 12 mM) is maintained =25 mM to insure solubility of receptor and lipid. Add MSP in 20 mM Tris·HCl, pH 7.4, 100 mM NaCl, 0.5 mM EDTA, and 0.01% NaN$_3$ to a molar ratio 1 MSP/120 lipid and incubate 1 h, room temperature, with gentle rocking to allow the mixture to equilibrate. The amount of materials used depends on the desired amount of receptor-containing Nanodiscs and the Tar-6H/MSP ratio. For instance, a 500 μl preparation mixture of 1:1 Tar-6H/MSP could represent additions of 19 μl 32 mM lipid, 95 μl 53 μM Tar-6H and 36 μl 140 μM MSP.

3. Add 2/3 volume of hydrated Biobeads SM-2 (BioRad) to 1 volume of preparation mixture (e.g., 330 μl Biobeads to 500 μl preparation mixture in a 1.5 ml microcentrifuge tube) and incubate 1 h, gently rocking at room temperature. Prior to use, Biobeads are washed with methanol and hydrated in water. Specifically, 2 volumes of methanol are added to 1 volume of beads, mixed, and the methanol decanted for three such washes. Water is added, the slurry poured on a filter placed on a filter holder inserted in a vacuum flask, and the beads washed extensively with water by alternately sucking off the water with vacuum, disconnecting the vacuum, and adding water. The slurry of washed and hydrated Biobeads is drawn into a plastic syringe with a wide opening created by cutting off its tip, excess H$_2$O extruded by exerting pressure with the syringe plunger while pressing tip against a clean wipe, and the Biobeads added to the assembly mixture from the syringe.

4. Pierce with a 26 G needle the bottom and cap of a microcentrifuge tube containing the Biobead/assembly mixture, place the pierced tube in a hole cut in the cap of a larger tube, such as a 14 ml, 17 × 100 mm polypropylene culture tube with cap, and centrifuge 1 min, 3700 rpm in a GH 3.8 rotor. The liquid sample but not the Biobeads passes essentially quantitatively through the hole to the larger tube, from which it is collected. The hole in the cap of the microcentrifuge tube allows liquid to flow freely. Alternatively, spin columns can be used; however, a wide mouth pipette is

required for transferring liquid and beads and it is difficult to pipette small volumes.

5. Apply the preparation mixture to a Ni-NTA agarose column (bed volume = volume preparation mixture) equilibrated in 50 mM Tris·HCl, pH 9.0, 10% w/v glycerol, 100 mM NaCl, and 15 mM imidazole. Apply 12 to 15 column volumes of equilibration buffer and elute receptor-containing Nanodiscs with 5 column volumes of 50 mM Tris·HCl, pH 9.0, 10% w/v glycerol, 100 mM NaCl, and 300 mM imidazole. Collect fractions throughout the procedure and analyze by SDS-PAGE. An example of purification of Tar-6H-containing Nanodiscs is shown in Fig. 5.

6. Concentrate the 300 mM imidazole-eluted sample in an Amicon Ultra 4 concentrator with 30,000 MWCO filter to ∼10% the volume of

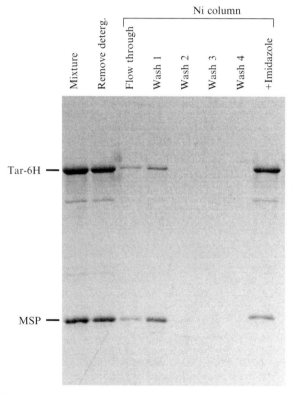

Fig. 5. Coomassie Blue-stained SDS-polyacrylamide gel of samples (equivalent proportions of the total material at each stage) from stages of Nanodisc preparation with chemoreceptor (Tar-6H, 60 kDa) and membrane scaffold protein (MSP, 30 kDa) indicated. See text for details. The figure is Fig. 2C of Boldog et al. (2006).

the original assembly mixture, place in Spectra/Por 12,000 to 14,000 MWCO, 6.4 mm inner-diameter tubing, dialyze against 3×100 volumes 50 mM Tris·HCl, pH 7.5, 10% w/v glycerol, 100 mM NaCl, and 0.5 mM EDTA, 8 to 12 h. Place the dialyzed sample in one or more 1.5 ml microcentrifuge tubes, flush with N_2 to minimize oxidation of lipid or protein, freeze by immersion in liquid N_2 to minimize disruption of Nanodiscs in the course of freezing, and store at $-70°$. Determine concentrations of receptor and MSP by SDS-PAGE using standards of known concentration.

Analysis of Receptor-Containing Nanodiscs

The number of receptor molecules per Nanodisc for a given preparation is determined by the intensity of staining of a dilution series on SDS-PAGE compared to dilution series of receptor and MSP standards. Initial criteria for homogeneity are symmetry and constancy of receptor:MSP ratio across the elution profile in size exclusion chromatography. Receptor activity is assessed by assays for covalent modification (usually methylation and deamidation), kinase activation, and effects of ligand occupancy on both modification and kinase activity (Boldog *et al.*, 2006).

Acknowledgments

Steve Sligar and Steve Grimme provided expertise and material that allowed us to develop a procedure for incorporating chemoreceptors into Nanodiscs. Throughout the development of this procedure, Linda L. Randall provided much-appreciated encouragement, advice, and enthusiasm. The work was supported in part by grant GM29963 to GLH from the National Institute of General Medical Sciences.

References

Bayburt, T. H., Grinkova, Y. V., and Sligar, S. G. (2002). Self-assembly of discoidal phospholipid bilayer nanoparticles with membrane scaffold proteins. *Nano Lett.* **2**, 853–856.

Bayburt, T. H., Grinkova, Y. V., and Sligar, S. G. (2006). Assembly of single bacteriorhodopsin trimers in bilayer nanodiscs. *Arch. Biochem. Biophys.* **450**, 215–222.

Bayburt, T. H., and Sligar, S. G. (2003). Self-assembly of single integral membrane proteins into soluble nanoscale phospholipid bilayers. *Protein Sci.* **12**, 2476–2481.

Boldog, T., Grimme, S., Li, M., Sligar, S. G., and Hazelbauer, G. L. (2006). Nanodiscs separate chemoreceptor oligomeric states and reveal their signaling properties. *Proc. Natl. Acad. Sci. USA* **103**, 11509–11514.

Chen, P. S., Jr., Toribara, T. Y., and Warner, H. (1956). Microdetermination of phosphorus. *Anal. Chem.* **28**, 1756–1758.

Civjan, N. R., Bayburt, T. H., Schuler, M., and Sligar, S. G. (2003). Direct solubilization of heterologously expressed membrane proteins by incorporation into nanoscale lipid bilayers. *Biotechniques* **35**, 556–563.

Cronan, J. E., Jr., and Rock, C. O. (1996). Biosynthesis of membrane lipids. In "Escherichia coli and Salmonella: Cellular and Molecular Biology" (F.C Neidhardt, ed.), Vol. 1, pp. 612–636. ASM Press, Washington, DC.

Denisov, I. G., Grinkova, Y. V., Lazarides, A. A., and Sligar, S. G. (2004). Directed self-assembly of monodisperse phospholipid bilayer nanodiscs with controlled size. J. Am. Chem. Soc. 126, 3477–3487.

Duan, H., Civjan, N. R., Sligar, S. G., and Schuler, M. A. (2004). Co-incorporation of heterologously expressed Arabidopsis cytochrome P450 and P450 reductase into soluble nanoscale lipid bilayers. Arch. Biochem. Biophys. 424, 141–153.

Fiske, C. H., and Subbarow, Y. (1925). The colorimetric determination of phosphorus. J. Biol. Chem. 66, 375–400.

Lai, W.-C., and Hazelbauer, G. L. (2005). Carboxyl-terminal extensions beyond the conserved pentapeptide reduce rates of chemoreceptor adaptational modification. J. Bacteriol. 187, 5115–5121.

Nath, A., Atkins, B., and Sligar, S. (2007). Applications of phospholipid bilayer Nanodiscs in the study of membranes and membrane proteins. Biochemistry 46, 2059–2069.

Osborn, M. J., and Munson, R. (1974). Separation of the inner (cytoplasmic) and outer membranes of Gram-negative bacteria. Methods Enzymol. 31, 642–653.

Parkinson, J. S., and Houts, S. E. (1982). Isolation and behavior of Escherichia coli deletion mutants lacking chemotaxis functions. J. Bacteriol. 151, 106–113.

Segrest, J. P., Li, L., Anantharamaiah, G. M., Harvey, S. C., Liadaki, K. N., and Zannis, V. (2000). Structure and function of apolipoprotein A-I and high-density lipoprotein. Curr. Opin. Lipidol. 11, 105–115.

[15] Assays for CheC, FliY, and CheX as Representatives of Response Regulator Phosphatases

By TRAVIS J. MUFF and GEORGE W. ORDAL

Abstract

Much study of two-component systems deals with the excitation of the histidine kinase, activation of the response regulator, and the ultimate target of the signal. Removal of the message is of great importance to these signaling systems. Many methods have evolved in two-component systems to this end. These include autodephosphorylation of the response regulator, hydrolysis of the phosphoryl group by the kinase, or a dedicated phosphatase protein. It has long been known that CheZ is the phosphatase in the chemotaxis system of *Escherichia coli* and related bacteria. Most bacteria and archaea, however, do not have a cheZ gene, but instead rely on the CheC, CheX, and FliY family of CheY-P phosphatases. Here, we describe assays to test these chemotactic phosphatases, applicable to many other response regulator phosphatases.

Introduction

Central to two-component signal transduction systems is the activation of the histidine kinase and the transfer of the phosphoryl group to the response regulator (Stock *et al.*, 1989). However, the rate of phosphoryl group hydrolysis from the response regulator, a reaction that is sometimes enzymatically facilitated, is often an important factor. Two methods are employed in various systems to control the dephosphorylation of the response regulator. First, response regulator proteins have an innate rate of dephosphorylation, which varies from system to system (Stock *et al.*, 1989). While the half life of Spo0A in the sporulation pathway is on the order of hours, CheY in the chemotaxis system has a half life of one to two minutes (Stock *et al.*, 1989). The lifetime of the phosphorylated response regulator is related to its function. A protein controlling gene transcription needs to be active long enough to activate or repress the genes it controls, whereas CheY must be activated and deactivated quickly to affect flagellar rotation on a much shorter time scale.

Facilitation of dephosphorylation by another protein may also help control the level of response regulator phosphorylation. This may be accomplished by the histidine kinase itself or by a dedicated phosphatase.

METHODS IN ENZYMOLOGY, VOL. 423   0076-6879/07 $35.00

In systems where the kinase is responsible for the phosphatase action, removal of the activating signal or a negative signal induces the histidine kinase to essentially operate in reverse by transferring the phosphoryl group back to the histidine residue and then to a water molecule (Stock *et al.*, 1989).

Dedicated aspartyl–phosphate phosphatases can be found in various two-component systems, most notably, sporulation and chemotaxis. The sporulation phosphorylation cascade involves two response regulators and at least six phosphatases (Perego, 1998, 2001; Piggot and Hilbert, 2004). Some of these phosphatases are located in operons with genes encoding corresponding inhibitors (Stephenson *et al.*, 2003). These multiple points of regulation allow cues from the state of the cell and environmental signals to be integrated at many steps in the cascade for tight regulation of the important decision to sporulate.

For chemotaxis in the proteobacteria such as *E. coli*, the predominant phosphatase for CheY-P is CheZ or a homolog (Simon *et al.*, 1989). CheZ binds to the receptor complex via an alternative form of the CheA kinase termed "CheA-short" (CheA$_S$) (Cantwell *et al.*, 2003), which is missing the first 97 amino acids of normal CheA (CheA$_L$) (O'Connor and Matsumura, 2004). Increase of CheY-P levels causes CheZ to oligomerize at the receptor and become much more active (Lipkow, 2006). Thus, CheY-P levels quickly return to normality (which produces a sharp signal) and uniformity throughout the cell (compared to what occurs when, as in certain mutants, CheZ is not located at the receptors) (Vaknin and Berg, 2004). Thus, each flagellum experiences similar concentrations of CheY-P as a function of time.

While CheZ is well known as the chemotaxis phosphatase, it is not found in the genomes of archaea or bacteria outside of the proteobacteria. It has been shown in *Bacillus subtilis* (Szurmant *et al.*, 2004), *Thermotoga maratima* (Park *et al.*, 2004), and *Borrelia burgdorferi* (Motaleb *et al.*, 2005) that a family of proteins distinct from CheZ are the CheY-P phosphatases. This family includes the proteins CheC, CheX, and FliY (see Fig. 1). These proteins share a highly conserved D/S-X$_3$-E-X$_2$-N-X$_{22}$-P motif, which forms the putative active site for dephosphorylation of CheY-P (Szurmant *et al.*, 2004). The structures for CheC, CheX, and the more distantly related FliM are available. These structures have a common secondary structural motif and tertiary $\alpha/\beta/(\alpha$ or $\beta)$ sandwich fold with the 2α helices containing the active sites on one side, 2α helices or β strands on the other, and a central 6-membered β sheet (Park *et al.*, 2004, 2006).

CheC is the best known member of this family and is widespread throughout bacteria and archaea (Szurmant and Ordal, 2004). In genomes, cheC is usually found preceding cheD and translationally coupled to it. CheD has been shown to enhance the phosphatase activity of CheC

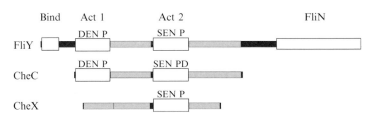

FIG. 1. Domain architecture of FliY, CheC, and CheX. White boxes represent regions of high sequence and structural conservation. Black areas represent important structural regions. Grey areas are thought to be linker regions of low conservation. Conserved putative active site residues are highlighted in their approximate locations, as is the conserved aspartate residue unique to CheC.

(Szurmant et al., 2004). CheD is also a receptor deamidase (Kristich and Ordal, 2002), an activity that is blocked by CheC. These regulations are due to CheD's active site binding to an α helix of CheC. This helix mimics CheD's target for deamidation (Chao et al., 2006), the chemoreceptors or methyl-accepting chemotaxis proteins (MCPs). The enhancement of CheC is important due to its weak phosphatase activity compared to FliY and CheX (Park et al., 2004; Szurmant et al., 2004). This interaction has also been shown to be critical in vivo for CheC's function (Chao et al., 2006). While most cheC's are found translationally coupled to cheD, there are those that have no cheD gene coupled to them or in the genome at all. Most of these cheC-like genes lack the conserved aspartate residue after the proline of the second active site (see Fig. 1). This aspartate has been shown to be important for the CheC–CheD interaction (Chao et al., 2006). Most of these cheC-like genes are found in che-like operons such as DifG of the *Myxococcus xanthus* dif pathway (Black and Yang, 2004). These proteins may represent a fourth member of the family, distinct from CheC and CheX.

CheX is smaller than CheC (18 versus 23 kDa) and has only one active site, which corresponds to the second active site on CheC and FliY, as seen in Fig. 1. Instead of α helices on both sides of the structure's central β sheet, CheX has a two-stranded β sheet on the side opposite the active site. This smaller β sheet associates with the central β sheet of another monomer to form the CheX dimer analogous to the CheC–CheD complex. In contrast to the important conserved aspartate residue on CheC, CheX has a glycine residue conserved through most CheX sequences (Park et al., 2004).

FliY was the first member of this family to be identified as a phosphatase. FliY is a multidomain protein with a short N-terminal α-helix that binds to CheY-P, the central CheC-homolgous domain with two active sites, and a C-terminal domain homologous to the *E. coli* FliN flagellar switch protein (see Fig. 1). FliY is necessary for the formation of flagella, as is FliN in *E. coli*, and is assumed to localize to the flagella basal bodies (Szurmant *et al.*, 2003).

Here, we describe three assays for the CheC phosphatase family, but these can also be applied to most other response regulator phosphatases. Two of the assays are used to measure the effectiveness of a phosphatase. The third is a pulldown assay for quickly determining protein–protein associations.

Assays

^{32}P-Response Regulator Dephosphorylation Assay

This is a relatively simple assay to monitor the phosphorylation/dephosphorylation of CheY (or another response regulator) and the effect of a potential phosphatase on this rate (Boesch *et al.*, 2000; Szurmant *et al.*, 2004). The assay can be used to observe multiple conditions relatively quickly and allows for side-by-side comparison. Briefly, CheY-^{32}P is generated in the reaction and immediately begins to dephosphorylate. Time-points are taken and resolved on SDS-PAGE. Results are visualized via a storage phosphor screen and phosphor imager or by X-ray film. In the example in Fig. 2, CheA pre-incubated with $[\gamma\text{-}^{32}P]$-ATP is used to generate a pool of CheA-^{32}P; this serves as the phospho-donor for CheY. A small molecule ^{32}P donor, such as acetyl phosphate, could also be used with modification of the protocol. The ^{32}P donor is added to a solution containing CheY and the potential phosphatase to be tested to begin the reaction.

The half-life of CheY-P is generally in the 1- to 2-min range (Stock *et al.*, 1989), so for the example, 10 s, 1 min, and 2 min time-points are used. For another response regulator with a much longer lifetime, such as Spo0A, the time-points would need to be adjusted accordingly.

Buffers, Reagents, and Equipment Used

TKMD (50 mM Tris, pH 8.0, 50 mM KCl, 5 mM MgCl2, 0.1 mM DTT, 10% glycerol)

Stop buffer (2× SDS-Loading Buffer, 100 mM EDTA)

100 mM ATP

$[\gamma\text{-}^{32}P]$-ATP (10 μCi/μl) (Amersham Biosciences)

Storage Phosphor Screen and Storm PhosphorImager (Molecular Dynamics)

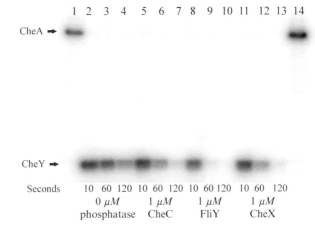

FIG. 2. CheY-^{32}P dephosphorylation assay. The SDS-PAGE image shows CheY-^{32}P hydrolysis at 10 s, 1-, and 2-min time-points under different conditions. Lanes 1 and 14 show CheA-^{32}P before and after the 4 reactions were carried out. All four reactions contained 10 μM CheA and 20 μM CheY with either no phosphatase (lanes 2–4), 1 μM CheC (lanes 5–7), 1 μM FliY (lanes 8–10), or 1 μM CheX (lanes 11–13).

^{32}P-Response Regulator Dephosphorylation Protocol

1. All proteins used were dialyzed into TKMD buffer. All reactions are performed in TKMD buffer.

2. Set up one reaction vial for each condition to be tested. In Fig. 2, the four conditions tested are 1 μM CheC, 1 μM FliY, 1 μM CheX, and the control with CheY only. Reactions were 30 μl with 10 μl CheA-^{32}P to be added later, for a total volume of 40 μl. Larger reaction volumes would be needed if more time-points were taken. Appropriate volumes of ATP, proteins, and TKMD were added for the reactions to contain 5 mM ATP, 20 μM CheY, and 1 μM phosphatase.

3. Set up one 1.5 ml tube with 10 μl stop buffer for each time-point for each reaction, plus two for CheA-^{32}P controls.

4. Phosphorylate CheA with [γ-^{32}P]-ATP. CheA was diluted to 40 μM in 50 μl with 5 μl [γ-^{32}P]-ATP (.05 μCi) for 10 min. This volume was enough for the 4 conditions tested; larger or smaller volumes would be needed for more or fewer conditions, respectively. The CheA or histidine kinase should be tested for optimal phosphorylation time if not previously determined.

5. Take 2 μl of CheA-^{32}P solution and add to 8 μl of water and 10 μl of stop buffer. This control will show that the CheA was phosphorylated.

6. To start a reaction, take 10 μl of the CheA-^{32}P solution, add to the reaction, and pipette up and down several times to mix. When the CheA

solution is added to the reaction, a timer or stopwatch should be started simultaneously.

7. Remove 10 μl of the reaction after 10 s and immediately transfer to 10 μl of stop buffer previously aliquoted in a 1.5 ml tube. Repeat at 1- and 2-min time-points.

8. Repeat steps 6 and 7 for the other conditions prepared (other phosphatases or concentrations).

9. Take another 2 μl aliquot of the CheA-^{32}P solution, as in step 3.

10. Apply 10 μl of each sample to SDS-PAGE. Care must be taken in this step not to run the CheY off the gel as it migrates close behind the dye front.

11. Wrap the gel in plastic wrap and expose to a Storage Phosphor Screen for at least 1 h. Alternatively, X-ray film may be used.

12. Scan the Storage Phosphor Screen with PhosphoImager to obtain the image.

Phosphate Release Assay

This assay is a steady state, quantitative method to measure the rate of dephosphorylation of a response regulator. This method employs the EnzChek Phosphate Assay Kit (E-6646) from Molecular Probes/Invitrogen to detect inorganic phosphate (P_i) as it is released from CheY by its innate dephosphorlyation (Boesch *et al.*, 2000). The kit detects P_i using the enzyme purine nucleoside phosphorylase (PNP). The PNP converts the included substrate 2-amino-6-mercapto-7-methylpurine riboside (MESG) and P_i to ribose 1-phosphate and 2-amino-6-mercapto-7-methylpurine, the latter of which is detectable by absorbance at 360nm. This kit is very sensitive to the presence of free P_i, so care must be taken to not contaminate the reaction with P_i. The absorbance at 360 nm is observed for 2 to 3 min in an ultraviolet (UV)/Vis spectrophotometer to find the rate of P_i release.

In order to remove possible effects of CheA interactions with CheY and phosphatases and to achieve steady state conditions with dephosphorylation of CheY as the rate-limiting step, this assay uses a small molecule phosphodonor to phosphorylate CheY. Acetyl phosphate is commonly used as a phosphodonor for CheY phosphorylation instead of CheA-P *in vitro* (Szurmant *et al.*, 2003), but this is unsuitable for the assay because acetyl phosphate cannot be obtained free of P_i. For this assay, monophosphorylimidizole (MPI) is used as the phosphodonor (Silversmith *et al.*, 1997) because it can be made as a soluble Ca^{2+} salt. The CaPO$_4$ resulting from spontaneous hydrolyzation of MPI is insoluble and can be removed by centrifugation once the MPI is put into solution.

This assay can be used to compare phosphate release rates from most response regulators. Since it is quantifiable, it is possible to directly compare the dephosphorylation rates of different response regulators, phosphatases, response regulator or phosphatase mutants, and different purified preparations of a response regulator to ensure consistency among preparations. In the example case, the assay is used to compare the effects of two different phosphatases on CheY-P.

To compare the activity of a set of phosphatases, a set concentration of the response regulator is observed with a range of phosphatase concentrations. In Fig. 3, the release of P_i from 5 uM CheY-P is observed over a large range of CheC and FliY concentrations. At each concentration, the value of A_{360} is followed as a function of time and converted to concentration of P_i using a standard curve of A_{360} and P_i. The experimental data is plotted as P_i released per min versus concentration of phosphatase (see Fig. 3). Rates in the linear range can then be compared for the different phosphatases. In this example, CheC has a rate of 3 $P_i/min/\mu M$ and FliY has a rate of 23 $P_i/min/\mu M$ in the linear portion of each plot. Thus, it is seen that FliY has over seven times the activity of CheC on CheY-P, or CheC has 13% of FliY's activity.

The phosphatase concentrations used will depend on each one's activity. In the example, CheC concentrations from .5 to 2 μM and FliY from .01 to .5 μM were used for determining the initial linear rate. FliY not only is a

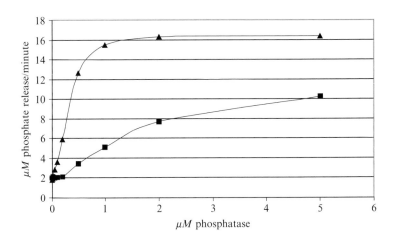

Fig. 3. P_i release assay plot. Rates of phosphate evolution from CheY were plotted versus the concentration of FliY (triangles) or CheC (squares) present. Linear ranges were used to find rate of P_i release per μM of phosphatase. Lines on graph are included only to guide the eye.

more active phosphatase, but it also has an N-terminal peptide that binds to CheY-P. For a phosphatase of unknown strength, start with a wide range (0.01 to 5 μM) and find the linear range. The concentration of response regulator must also be optimized since different response regulators will have different autodephosphorylation rates.

The kit procedure calls for a 1 ml reaction volume, but 250 μl works well and will conserve protein. The 20× buffer solution included in the kit, when diluted, is 50 mM Tris, pH 7.5, 1 mM MgCl$_2$. Addition of 10 μl of 250 mM MgCl$_2$ to the 250 μl reaction gives a final concentration of 11 μM MgCl$_2$. The additional MgCl$_2$ was added to ensure proper concentration of Mg^{2+}, which is a co-factor for CheY.

The MPI should be kept desiccated at $-20°$ when not in use. The MPI solution should always be made fresh and kept on ice. Vortexing of the MPI in water should cause most of the solid to go into solution, but there will be precipitate (CaPO$_4$) left, which can simply be removed by centrifugation. Synthesis of the MPI is fairly straightforward with minimal organic chemistry experience. However, use of fresh, dry POCl$_3$ is essential for the synthesis of phosphoramidate to be effective since that compound is very reactive with water.

Phosphate Release Protocol

> Sample reaction mix, 250 μl total volume
> x μl water
> 12.5 μl 20× buffer
> 10 μl 250 mM MgCl2 (10 mM)
> 50 μl 1 mM MESG (200 μM)
> y μl phosphatase
> 25 μl 50 mM MPI (5 mM)
> 2.5 μl PNP
> z μl CheY (5 μM)

1. Make 250 mM solution of MgCl$_2$.
2. Make 50 mM MPI solution in 1.5 ml tube. Once the MPI is dissolved, spin in microcentrifuge at maximum speed for 2 min to pellet insoluble CaPO$_4$. Keep supernatant.
3. Mix buffer, MgCl$_2$, MESG, phosphatase, and water for each reaction.
4. Add 25 μl of the MPI solution and 2.5 μl PNP. Vortex.
5. Incubate for 2 min to allow the PNP to scavenge contaminating P$_i$ in the reaction.
6. Add CheY to a final concentration of 5 μM and vortex.
7. Transfer to 250 μl cuvette, insert in spectrophotometer, and record data for 2 to 3 min. Record data as rate of dA$_{360}$/min.

8. Repeat steps 4 to 7 for each phosphatase concentration to be tested.
9. Repeat experiment for each different phosphatase to be tested and compared.
10. Convert dA_{360}/min data to P_i/min using P_i standard curve.
11. Plot P_i/min versus phosphatase concentration for each phosphatase.
12. Find linear range for each phosphatase and the slope of the line. The slope will be P_i released/min/concentration of phosphatase.

MPI Synthesis

Phosphoramidate, ammonium salt (Sheridan *et al.*, 1972)
Combine 200 ml H_2O, 100 ml NH_4OH, 2 l flask; stir on ice.
Add 18.3 ml $POCl_3$ over 5 min (must be fresh).
Add 800 ml acetone; collect bottom aqueous layer with separatory flask.
Bring pH to 6.0 with acetic acid; leave at $4°$ overnight.
Filter crystals, wash with ethanol, diethyl ether, dry (yield $\sim$7 g).
Phosphoramidate, potassium salt (Sheridan *et al.*, 1972)
Add 1.75 ml 50% KOH for each gram phosphoramidate.
Heat to $50°$ for 10 min.
Cool to $<10°$.
Bring pH to 6 with acetic acid.
Add 500 ml 95% ethanol.
Filter crystals, wash with ethanol, diethyl ether, dry (yield $\sim$8 g).
MPI, calcium salt (Rathlev and Rosenberg, 1956)
Combine 4 g phosphoramidate (K salt), 1.06g imidazole, 40 ml H_2O.
Bring pH to 7.2 with HCl.
Cover, incubate RT overnight.
Make $CaCl_2$ solution (4.56 g $CaCl_2 \cdot H_2O$ in 3–5 ml H_2O).
Add to stirring MPI solution.
Filter, keep filtrate.
Add 70 ml 95% ethanol, RT 15 min.
Filter crystals, wash 65% EtOH, 95% EtOH, ether.
Dry in hood (yield $\sim$1.2 g).

Pulldowns

The glutathione-S-transferase (GST) pulldown, similar to immunoprecipitation, is a simple qualitative assay to determine whether two proteins have a specific interaction. The primary protein is expressed with an affinity tag (GST), which can be immobilized on a surface linked to a binding partner (Glutathione-Sepharose). Other affinity tags, such as $6\times$ histidine, may also be used in place of GST. Once the primary protein is bound to the immobilized surface or bead, the secondary protein is added and allowed to bind. The beads are then washed to remove nonspecific interactions. An appropriate elution buffer is then used to elute the proteins, SDS-Loading Buffer is added, the sample is resolved on SDS-PAGE, and the gel is analyzed for the presence of the secondary protein by Coomassie stain, Western Blot, or radiological methods.

For the example in Fig. 4, small Handee Spin Cup columns (Pierce Scientific) were used. These retain the Glutathione-Sepharose beads during the centrifugation steps while allowing the liquid wash volume to simply flow out. Alternatively, normal 1.5 ml microcentrifuge tubes may be used with the supernatant removed via pipette. However, care must be taken to not remove the beads with the wash volume.

Proper controls must be performed in the experiment to rule out non-specific binding. In this case, the beads themselves may be incubated with the secondary protein or, more appropriately, GST can be used as the control primary protein.

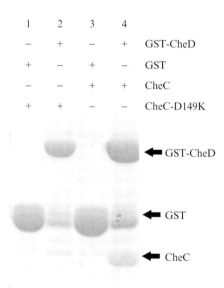

FIG. 4. GST-pulldown assay. The Coomassie stained SDS-PAGE shows a pulldown experiment demonstrating the interaction of CheC with GST-CheD (lane 4) and loss of the interaction due to a D149K mutation in CheC (lane 2). Lanes 1 and 3 are controls, with GST replacing GST-CheD to show the interaction is specific to CheD. Reprinted from (Chao *et al.*, 2006) with permission from Elsevier.

The primary and secondary proteins may be previously purified; alternatively, lysates of overexpressed proteins may be used. Use of the latter would require more washes to remove nonspecific binding by the additional proteins found in the lysate. Using the purified proteins also allows for greater reproducibility and is better for comparing the binding of mutants since equimolar amounts can be added.

Choice of which protein to tag can be important. In Fig. 4, CheD is tagged with GST since CheD is relatively unstable and the fusion with GST aids folding and stability. CheC is quite stable without a tag.

The example in Fig. 4 compares the binding of wild-type CheC to a defective CheC mutant to demonstrate the defect is due specifically to disruption of the CheC-CheD interaction (Chao et al., 2006). GST is included as a control to show that the CheC retention is due to the interaction with CheD and not to nonspecific interactions with GST or the beads.

This assay may also be used to demonstrate binding to a phosphorylated response regulator such as CheY-P. This method is similar to the described protocol, but in order to ensure the CheY is phosphorylated, acetyl phosphate is added to the incubation and wash steps. This technique has been used to show a phosphorylation-dependent interaction between GST-CheY and FliM and FliY (Szurmant et al., 2003). This experiment was carried out in duplicate, one of the experiments having 20 mM acetyl phosphate in the washes and incubations.

Buffers and Reagents

> PBS/Triton (150 mM NaCl, 20 mM Na_2HPO_4, pH 7.4, 1% Triton X-100)
> GEB (10 mM Glutathione, 50 mM Tris Base, pH 8.0)
> Glutathione-Sepharose beads (Amersham Biosciences)
> Handee Spin Cup columns (Pierce Scientific)

GST-Pulldown Protocol

1. Aliquot 50 μl Glutathione beads into spin columns for each condition.
2. Wash beads twice with 400 ul PBS/Triton, spinning in microcentrifuge at $1000 \times g$ for 30 s each time. Triton X-100 or another detergent will help remove nonspecific binding.
3. Add primary (tagged) protein to the beads. In the example, 100 μl GST-CheD or GST was added to a concentration of 75 μM. If the interaction is expected to be weak, adding more primary protein will increase the signal.
4. Incubate 10 min at room temperature with slow rocking or inversion.

5. Repeat wash as in step 2. Do 2 washes at a minimum.
6. Add secondary protein. In the example, CheC and CheC(D149K) were added at 100 µl of 100 µM. The concentration will depend on the strength of binding, but should be maximal for at least the first attempt.
7. Incubate 10 min at room temperature with slow rocking or inversion.
8. Repeat wash as in step 2. Do 4 to 6 wash steps.
9. Add 75 µl GEB for elution. Incubate 10 min.
10. Centrifuge at max speed 1 min, add 25 µl 4× SDS-Loading Buffer.
11. Load 10 µl of each sample on SDS-PAGE.
12. Visualize by Coomassie stain or Western blot.

Acknowledgments

We thank Vincent Cannistraro and George Glekas for critical reading and suggestions on this manuscript. Work in the laboratory of GWO was supported by NIH grant GM054365.

References

Black, W. P., and Yang, Z. (2004). Myxococcus xanthus chemotaxis homologs DifD and DifG negatively regulate fibril polysaccharide production. *J. Bacteriol.* **186,** 1001–1008.

Boesch, K. C., Silversmith, R. E., and Bourret, R. B. (2000). Isolation and characterization of nonchemotactic CheZ mutants of *Escherichia coli. J. Bacteriol.* **182,** 3544–3552.

Cantwell, B. J., Draheim, R. R., Weart, R. B., Nguyen, C., Stewart, R. C., and Manson, M. D. (2003). CheZ phosphatase localizes to chemoreceptor patches via CheA-short. *J. Bacteriol.* **185,** 2354–2361.

Chao, X., Muff, T. J., Park, S. Y., Zhang, S., Pollard, A. M., Ordal, G. W., Bilwes, A. M., and Crane, B. R. (2006). A receptor-modifying deamidase in complex with a signaling phosphatase reveals reciprocal regulation. *Cell* **124,** 561–571.

Kristich, C. J., and Ordal, G. W. (2002). *Bacillus subtilis* CheD is a chemoreceptor modification enzyme required for chemotaxis. *J. Biol. Chem.* **277,** 25356–25362.

Lipkow, K. (2006). Changing cellular location of CheZ predicted by molecular simulations. *PLoS Comput. Biol.* **2,** e39.

Motaleb, M. A., Miller, M. R., Li, C., Bakker, R. G., Goldstein, S. F., Silversmith, R. E., Bourret, R. B., and Charon, N. W. (2005). CheX is a phosphorylated CheY phosphatase essential for Borrelia burgdorferi chemotaxis. *J. Bacteriol.* **187,** 7963–7969.

O'Connor, C., and Matsumura, P. (2004). The accessibility of cys-120 in CheA(S) is important for the binding of CheZ and enhancement of CheZ phosphatase activity. *Biochemistry* **43,** 6909–6916.

Park, S. Y., Chao, X., Gonzalez-Bonet, G., Beel, B. D., Bilwes, A. M., and Crane, B. R. (2004). Structure and function of an unusual family of protein phosphatases: The bacterial chemotaxis proteins CheC and CheX. *Mol. Cell* **16,** 563–574.

Park, S. Y., Lowder, B., Bilwes, A. M., Blair, D. F., and Crane, B. R. (2006). Structure of FliM provides insight into assembly of the switch complex in the bacterial flagella motor. *Proc. Natl. Acad. Sci. USA* **103,** 11886–11891.

Perego, M. (1998). Kinase–phosphatase competition regulates Bacillus subtilis development. *Trends Microbiol.* **6,** 366–370.

Perego, M. (2001). A new family of aspartyl phosphate phosphatases targeting the sporulation transcription factor SpoOA of *Bacillus subtilis*. *Mol. Microbiol.* **42,** 133–143.

Piggot, P. J., and Hilbert, D. W. (2004). Sporulation of *Bacillus subtilis*. *Curr. Opin. Microbiol.* **7,** 579–586.

Rathlev, T., and Rosenberg, T. (1956). Non-enzymic formation and rupture of phosphorus to nitrogen linkages in phosphoramido derivatives. *Arch. Biochem. Biophys.* **65,** 319–339.

Sheridan, R. C., McCullough, J. F., and Wakefeild, Z. T. (1972). Phosphoramidic acid and its salts. *Inorg. Synth.* **13,** 23–26.

Silversmith, R. E., Appleby, J. L., and Bourret, R. B. (1997). Catalytic mechanism of phosphorylation and dephosphorylation of CheY: Kinetic characterization of imidazole phosphates as phosphodonors and the role of acid catalysis. *Biochemistry* **36,** 14965–14974.

Simon, M. I., Borkovich, K. A., Bourret, R. B., and Hess, J. F. (1989). Protein phosphorylation in the bacterial chemotaxis system. *Biochimie* **71,** 1013–1019.

Stephenson, S., Mueller, C., Jiang, M., and Perego, M. (2003). Molecular analysis of Phr peptide processing in *Bacillus subtilis*. *J. Bacteriol.* **185,** 4861–4871.

Stock, J. B., Ninfa, A. J., and Stock, A. M. (1989). Protein phosphorylation and regulation of adaptive responses in bacteria. *Microbiol. Rev.* **53,** 450–490.

Szurmant, H., Bunn, M. W., Cannistraro, V. J., and Ordal, G. W. (2003). *Bacillus subtilis* hydrolyzes CheY-P at the location of its action, the flagellar switch. *J. Biol. Chem.* **278,** 48611–48616.

Szurmant, H., Muff, T. J., and Ordal, G. W. (2004). *Bacillus subtilis* CheC and FliY are members of a novel class of CheY-P-hydrolyzing proteins in the chemotactic signal transduction cascade. *J. Biol. Chem.* **279,** 21787–21792.

Szurmant, H., and Ordal, G. W. (2004). Diversity in chemotaxis mechanisms among the bacteria and archaea. *Microbiol. Mol. Biol. Rev.* **68,** 301–319.

Vaknin, A., and Berg, H. C. (2004). Single-cell FRET imaging of phosphatase activity in the *Escherichia coli* chemotaxis system. *Proc. Natl. Acad. Sci. USA* **101,** 17072–17077.

[16]   Genetic Dissection of Signaling Through the Rcs Phosphorelay

By NADIM MAJDALANI and SUSAN GOTTESMAN

Abstract

The Rcs phosphorelay, consisting of a hybrid sensor kinase, a phospho-transferase, and a response regulator, regulates a large number of bacterial functions. These include capsule production, the target originally defined for these regulators, a small regulatory RNA, and a growing list of additional genes, many of unknown function. At the core of this phosphorelay is the response regulator RcsB that activates the expression of the target genes. In addition to RcsB, some but not all of these targets require a co-regulator. One such co-regulator is RcsA, which has not been described as working except with RcsB; RcsA is itself regulated at both the transcriptional and post-transcriptional levels. Signaling to the system is also complex, and numerous plasmids, mutations, and environmental conditions have been described as activating this system. Activation of the system on cell surfaces and the nature of some of the regulated functions suggest a role for this phosphorelay in biofilm formation. Here, we describe reporters and mutants that allow the genetic dissection of the system from two directions. In cases where a condition activates the system, for instance, causing an increase in capsule synthesis (a phenotype easily observed in colonies), specific tests can identify at what stage the signal feeds into the system. In cases where a target of the phosphorelay is identified, specific tests can define the genetic requirements for regulation of the target. Finally, in cases where overproduction of capsule interferes with other studies, mutants allow the study of cells in the absence of capsule formation.

Overview

The RcsC/RcsD/RcsB phosphorelay (Fig. 1) and its associated proteins regulate a variety of cellular functions, all of which may have in common a role during biofilm formation. RcsC is a hybrid sensor kinase, with two transmembrane domains and a reasonably large periplasmic domain. RcsD, previously called YojN, is a histidine phosphotransferase (HPt) protein. RcsB, the response regulator of the system, is absolutely required in all cases; the roles of the other components of the system may vary with conditions. This system has been reviewed in Huang et al. (2006) and Majdalani and Gottesman (2005).

METHODS IN ENZYMOLOGY, VOL. 423   0076-6879/07 $35.00
                                                          DOI: 10.1016/S0076-6879(07)23016-2

Positively controlled genes include the genes for the colanic acid capsular polysaccharide, the stress sigma factor RpoS via the small RNA RprA, and a number of other genes, some with roles in biofilm synthesis and some known to respond to osmolarity (Boulanger et al., 2005; Francez-Charlot et al., 2005; Majdalani et al., 2002; Stout and Gottesman, 1990; Sturny et al., 2003). In addition, flagellar genes are negatively regulated by the system (Francez-Charlot et al., 2003).

The complexity of the system is increased by the finding that RcsB can act alone on some promoters (for example, see Boulanger et al., 2005; Francez-Charlot et al., 2005; Majdalani et al., 2002), with the unstable regulatory protein RcsA on other promoters (Ebel and Trempy, 1999; Stout et al., 1991), and, in at least one case, with other regulatory proteins (Mouslim et al., 2003). Thus, signaling to the regulated promoters can proceed through RcsC, RcsD, to RcsB, or through the synthesis or activity of the accessory regulatory proteins, or possibly through direct signaling to RcsD or RcsB, bypassing RcsC (Fig. 1). Signaling to the sensor kinase RcsC is similarly complex (see, for instance, Majdalani et al., 2005). A set of reporter fusions as well as mutations in various components of the system make it possible to dissect the inputs to regulating a specific gene by this system, and to dissect the specific components of the output of the system. The strains and discussion here focus on E. coli; similar approaches can be used in related organisms, but additional components have been identified in some other organisms (for instance, see Minogue et al., 2005).

Also not addressed here are in vitro approaches to studying the Rcs phosphorelay. RcsB and RcsA binding and activity in vitro has been demonstrated in a number of systems (Davalos-Garcia et al., 2001; Pristovsek et al., 2003; Wehland and Bernhard, 2000; Wehland et al., 1999).

Flowchart of Testing: Signaling Inputs

Implication of the Rcs phosphorelay in a particular process usually comes about in one of two ways. In some cases, mutations or plasmids render colonies mucoid. In E. coli K12, this is always due to the overproduction of the colanic acid polysaccharide. Thus, any genetic or environmental condition leading to mucoidy is likely to be the result of activating some aspect of the signaling system for colanic acid, dependent on the Rcs phosphorelay, and it may be of interest to figure out why cells have become mucoid. In other experiments, tests to look at regulators of a gene of interest may turn up components of the Rcs phosphorelay. Demonstrating that this gene is, in fact, regulated by the Rcs system, and how it is regulated, will require dissecting the system.

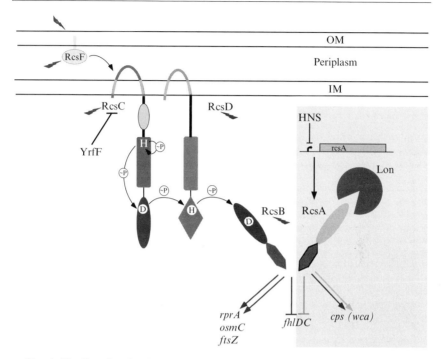

FIG. 1. The Rcs phosphorelay. The core protein components of the phosphorelay include the RcsC sensor kinase, the RcsD histidine phosphotransfer protein, and the RcsB response regulator. Some targets of RcsB also require the accessory factor, RcsA. RcsA is subject to Lon-dependent proteolysis and its synthesis is regulated, in part by Hns. Upstream signaling to RcsC is sometimes, but not always, via the outer membrane lipoprotein RcsF. The jagged line indicates possible signaling inputs. See text and Majdalani and Gottesman, 2005, for further details. (See color insert.)

Figure 1 shows the components of the phosphorelay and associated proteins. Figure 2 gives a schematic of decision trees for mutations that help to identify specific steps in the relay. Strains referred to in the text are listed in Table I.

Analysis of the Regulation of a Target Gene

We will first consider the analysis of the upstream regulators of a target gene of interest. Presumably, as a result of mutation, array analysis, or screening of multicopy libraries, some component of the Rcs system has been implicated in regulation.

Step 1: Is regulation at the level of transcription initiation? The most efficient way to determine this, and to carry out further steps of analysis,

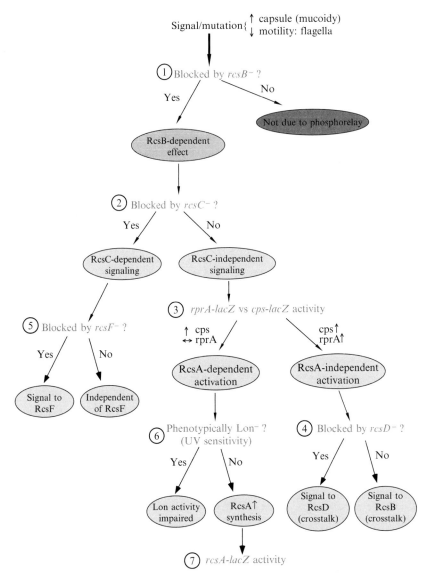

FIG. 2. Decision tree for analyzing Rcs phosphorelay. The flow chart is keyed to the Steps for testing various models, as outlined in the text. Conclusions based on results are shown in blue ovals. The questions answered at each step are: Step 1: Is it dependent upon the Rcs phosphorelay? Step 2: Is it dependent upon RcsC activation? Step 3: Is there increased RcsA? Step 4: Is signaling directly to RcsD? Step 5: Is signaling RcsF dependent? Step 6: Is Lon activity impaired? Step 7: Is RcsA synthesis increased? (See color insert.)

TABLE I
STRAINS AND PLASMIDS

Strain	Strain background and genotype	References or comments
SG12047	C600 lon-146::tet	Source of lon mutation (Gottesman, 1990)[a]
SG22812	MC4100 cpsB-lacZ	Reporter for RcsA, RcsB-dependent signaling. Derivative of SG20781 (Brill et al., 1988)[b]; lab strain
SG30074	MG1655 rprA1::kan	Allele described in (Majdalani et al., 2002)[c]; lab strain
NM20785	MC4100 cpsB-lacZ trp::Tn10 rcsF118::kan	Source for rcsF118::kan; allele described in (Majdalani et al., 2005)[d]
DDS1301	MC4100 rcsA::lacZ	Reporter for rcsA synthesis (Sledjeski and Gottesman, 1995)[e]
DH300	DJ480 rprA142p::lacZ	Reporter for RcsB-dependent, RcsA-independent signaling (Majdalani et al., 2002)[c]
DH310	DJ480 rprA142p::lacZ rcsA::kan	Source for rcsA::kan (Majdalani et al., 2002)[c]
DH311	DJ480 rprA142p::lacZ rcsB::kan	Source for rcsB::kan (Majdalani et al., 2002)[c]
DH312	DJ480 rprA142p::lacZ rcsC::kan	Source for rcsC::kan (Majdalani et al., 2002)[c]
DH335	DJ480 rprA142p::lacZ rscB::tet, omp::Tn5(kan)	Source for rcsB::tet; lab strain
DH366	DJ480 rprA142p::lacZ, rcsC137::cm	Source for rcsC137::cm (Majdalani et al., 2005)[d]
DH368	DJ480 rprA142p::lacZ ΔrcsD543::kan (low polarity)	Source for rcsD::kan (Majdalani et al., 2005)[d]
DH510	DJ480 rprA142p::lacZ rcsC::tet	Source for rcsC::tet; lab strain
Plasmid		
pATC400	rcsA[+]	Multicopy constitutive rcsA expression (Majdalani et al., 2002)[c]
pJB100	rcsB[+]	Multicopy constitutive rcsB expression (Brill et al., 1988)[b]
pMH300	pTrc99-rcsF[+]	IPTG-inducible rcsF expression (Majdalani et al., 2005)[d]
pPSG961	pBAD33-djlA[+]	Arabinose-inducible djlA expression (Clarke et al., 1997)[f]

[a] Gottesman, 1990. Minimizing proteolysis in *Escherichia coli*: Genetic solutions. *Methods Enzymol.* **185**, 119–129.

[b] Brill et al., 1988. Fine-structure mapping and identification of two regulators of capsule synthesis in *Escherichia coli* K-12. *J. Bacteriol.* **170**, 2599–2611.

[c] Majdalani et al., 2002. Regulation and mode of action of the second small RNA activator of RpoS translation, RprA. *Mol. Microbiol.* **46**, 813–826.

[d] Majdalani et al., 2005. Role of RcsF in signaling to the Rcs phosphorelay in *Escherichia coli*. *J. Bacteriol.* **187**, 6770–6778.

[e] Sledjeski and Gottesman, 1995. A small RNA acts as an antisilencer of the H-NS-silenced rcsA gene of *Escherichia coli*. *Proc. Natl. Acad. Sci. USA* **92**, 2003–2007.

[f] Clarke et al., 1997. Point mutations in the transmembrane domain of DjlA, a membrane-linked DnaJ-like protein, abolish its function in promoting colanic acid production via the Rcs signal transduction pathway. *Mol. Microbiol.* **25**, 933–944.

is to create a transcriptional fusion to the promoter of the gene of interest. Our usual tools for doing this are the plasmids and phage described by Simons and coworkers (Simons *et al.*, 1987) using a Δ*lac* derivative of MG1655, DJ480 (Cabrera and Jin, 2001), as the host, but any other reporter of gene transcription or Δ*lac* strain will do. If the +1 region of the transcript is known, the fusion can be made not far downstream of this. If the +1 of the transcript is not known, it may be useful to create both a transcriptional and a translational fusion, to help distinguish post-initiation regulation (see "complications" in the following text).

a. Introduce the *rcsC*137 allele by P1 transduction from DH366 into a Δ*lac* strain carrying the reporter fusion. This is a constitutively active allele of the sensor kinase and results in very high levels of expression of all RcsB activated genes. Does expression increase? If so, the reporter is likely regulated by RcsB. Note that repression by RcsB is also a possibility (Francez-Charlot *et al.*, 2003) and should demonstrate the opposite effects in all tests described here.

b. An alternative strategy for activating RcsB-dependent genes is to introduce a plasmid that activates the system. An *rcsB*⁺ plasmid such as pJB100 (Brill *et al.*, 1988) or an *rcsF*⁺ plasmid such as pMH300 (Majdalani *et al.*, 2005) or a plasmid expressing DjlA (pPSG961; Clarke *et al.*, 1997) all will activate target genes without further signaling. The latter two plasmids require *rcsC*⁺ for activation while pJB100 does not.

c. If gene expression is activated by any of the treatments already described, RcsB is likely a regulator of the gene. However, as noted in Fig. 1, some but not all RcsB-regulated genes are also regulated by RcsA. Is the expression of the gene in question RcsA dependent? Introduce *rcsA*:: kan by P1 transduction from DH310 into the *rcsC*137 derivative. Does expression decrease significantly? If it does, the gene is probably dependent upon both RcsB and RcsA. If it does not decrease, it is likely to be dependent upon RcsB alone or RcsB and another regulator.

Alternatively, the *rcsA*⁺ plasmid pATC400 can be introduced into the otherwise wild-type strain carrying the reporter; if expression increases significantly (more than five-fold), the promoter is likely to be RcsA dependent. Note that even RcsA-independent fusions may increase a bit in the presence of excess RcsA.

Details and Cautions

• If possible, use strains for the fusion that do not carry Cm, Tet, or Kan markers, to allow introduction of mutations to analyze the expression of the fusion. The experiments with mutants will be more definitive than those with plasmids.

• The binding sites for RcsB and RcsA can be significantly upstream of the promoter elements. Therefore, fusions should include 200 nt upstream of the likely +1, if at all possible. If only RcsB, and not RcsA, is involved in the regulation, the binding site may be immediately upstream of the −35, and a fusion with 60 nt upstream of the +1 will show the correct regulation (Majdalani *et al.*, 2002). Deletion analysis as well as sequence comparison to the somewhat defined consensus sequences can be used to further identify essential binding sites (Carballes *et al.*, 1999; Wehland and Bernhard, 2000). If repression by the RcsB system is suspected, reporter fusions should begin somewhat downstream of the +1.

• The Rcs phosphorelay regulates at least one small regulatory RNA, RprA (Majdalani *et al.*, 2002). RprA has both positive effects on targets (*rpoS*) and negative effects on other targets (Majdalani *et al.*, 2001); (J. Benhammou, N. Majdalani, and S. Gottesman, in preparation), and acts at the level of mRNA stability and translation by pairing with target mRNAs. Even a transcriptional fusion that carries a region that RprA can pair with may be regulated, via changes in the stability of the mRNA. Because mutations or plasmids that can activate RcsB-dependent targets will also increase synthesis of RprA, a reporter can, in theory, be regulated by RprA (or potentially another, undiscovered, RcsB-regulated RNA) at a post-transcriptional level. A fusion that does not carry any of the transcribed region of the reporter avoids this complication. If the +1 is not known, this possibility should be kept in mind; mutation of *rprA* (introduced from SG30074) in the *rcsC*137 host will eliminate Rcs regulation if regulation is via RprA.

More generally, it is possible that genes defined as regulated by the Rcs phosphorelay by the tests already described could be indirectly regulated. A search for the RcsB binding site will help to confirm direct regulation, and further genetic analysis in which specific RcsB-regulated genes other than *rprA* are mutated in strains activated for the phosphorelay can help eliminate possible indirect regulation, although this can be tedious.

Analysis of Signaling via the Rcs Phosphorelay

We next consider the analysis of why cells may demonstrate a mucoid phenotype. The assumption here will be that some component of the Rcs phosphorelay is activated, but what component it is may be of interest. Note that while these tests can be done using the phenotype of mucoidy, transcriptional reporter fusions for either mucoidy (*cps*-lacZ; SG22812) or another RcsB-dependent gene (for instance, the small RNA *rprA*-lacZ, DH300) may provide a more sensitive test. However, as will be noted, not every stimulatory signal will act similarly on these two reporters. The steps listed are keyed to numbers in Fig. 2.

Step 1: Introduce an *rcsB*::tet (DH335) or *rcsB*::kan (DH311) allele by P1 transduction into the mucoid strain in question. Does the phenotype disappear? The answer to this test is generally "yes," particularly for mucoidy, and doesn't provide much additional information except that RcsB is a required regulator. If the answer is "no," the phenotype is not due to this phosphorelay! We are not aware of any conditions where colanic acid over-production is not RcsB dependent; possibly the observed phenotype could be due to a different capsule or is being misinterpreted as mucoidy.

Assuming the answer is yes, the analysis proceeds to:

Step 2: Is the increased signaling via RcsC? Two major pathways of information transfer are known to affect capsule synthesis. One of them depends on activation of the phosphorelay, frequently via RcsC; the other depends upon increasing the synthesis or stability of the co-activator of capsule synthesis, RcsA (Fig. 1). If the situation that led to increased capsule is signaling via RcsC, such as that seen with mutations in the *rfa* or *mdo* genes (Ebel *et al.*, 1997; Parker *et al.*, 1992), a cell carrying an *rcsC* loss of function mutant will no longer activate the pathway, and the phenotype (of mucoidy, for instance) will disappear in the *rcsC* null mutant strain. Introduce a loss of function mutation in *rcsC* (*rcsC*::tet (DH510) or *rcsC*::kan (DH312)). If the mucoidy disappears, activation is via RcsC, and the next analysis that may be of interest is to dissect where upstream of RcsC it occurs. This is discussed in Step 5.

Note that, for a very sensitive reporter, some increase in reporter activity can be seen with an *rcsC* mutation, but this is not usually sufficient to give mucoidy (Majdalani *et al.*, 2002). Comparison of an otherwise wild-type *rcsC* mutant with the original "mucoid" strain should be done if possible.

If mucoidy does not disappear, proceed to:

Step 3: Is the increased signaling via RcsA? If an *rcsC* mutation fails to reverse the phenotype, signaling may be either downstream from RcsC in the phosophorelay, or via RcsA. An increase in RcsA synthesis or stability (for instance, from a *lon* mutation; Torres-Cabassa and Gottesman, 1987) will increase capsule synthesis and expression of a *cps*-lacZ fusion.

Mutations in *rcsA* will become nonmucoid in almost all cases, so using mucoidy as a measure does not help determine whether the levels of RcsA have changed. A better approach is to test the signaling condition that led to mucoidy (for instance, a plasmid or mutation) with the *rprA*-lacZ (DH300) and *cps*-lacZ (SG22812) fusions, in parallel. *rprA* is dependent upon RcsB, but not upon RcsA (Majdalani *et al.*, 2002); therefore, if the signal activates *rprA*-lacZ, it is likely to be via RcsB and not RcsA. The *cps*-lacZ fusion requires both RcsA and RcsB. Therefore, a situation or mutation that activates the *cps* fusion but not *rprA* is likely to be acting via an increase in RcsA synthesis or stability. Depending on the answer to this question, the next steps in analysis will vary. If the effect is via RcsA (the *cps* fusion but not the *rprA* fusion is activated), steps six and seven discuss how to determine how this is occurring. If the effect is independent of RcsA (also activates the *rprA* fusion) and is not blocked by an *rcsC* null mutation, signaling is downstream of RcsC in the phosphorelay, and is discussed in step four.

Signaling Downstream of RcsC in the Rcs Phosphorelay

Signals that activate both the *cps* and *rprA* fusions but are not blocked (or are increased) by an *rcsC* mutation are likely to act downstream of RcsC in the phosphorelay. Because RcsC can act as both a kinase, activating RcsB via RcsD, and as a phosphatase, reducing activity of RcsB activated in other ways, in some cases, mutations in *rcsC* may also increase expression (Majdalani *et al.*, 2002, 2005).

There are a few examples of signaling that are independent of both RcsC and RcsA. High expression of RcsB (for instance, from a multicopy plasmid) can activate RcsB-dependent promoters in a RcsC and RcsA-independent fashion (Brill *et al.*, 1988), but it is unclear whether there is any natural mechanism of increasing RcsB-dependent expression to this extent. Cross-talk from other sensor kinases or small molecules may activate RcsB or possibly RcsD (Fredericks *et al.*, 2006).

Signals that activate both the *cps* and *rprA* fusions but are not blocked (or are increased) by an *rcsC* mutation are likely to act downstream of RcsC in the phosphorelay, via direct effects on the phosphorylation levels of RcsD or RcsB (see Fig. 1). Because RcsC can act as both a kinase, activating RcsB via RcsD, or as a phosphatase, reducing activity of RcsB activated in other ways, in some cases, mutations in *rcsC* will increase expression (Fredericks *et al.*, 2006; Majdalani *et al.*, 2002, 2005).

Step 4: Signaling to RcsD? Introduction of a mutation in *rcsD* (DH368) may block mucoidy/reporter expression if signaling is via RcsD (or above RcsD); no examples have yet been presented where signaling is RcsC independent but RcsD dependent. Note that the *rcsD* insertion is partially polar on *rcsB*, so a partial decrease in phenotype may reflect a reduction in RcsB (Majdalani *et al.*, 2005). Alternatively, since RcsD appears to act to move phosphates both to RcsB (positive regulation) and away from RcsB (negative regulation), eliminating RcsD may enhance expression of a reporter if there is significant signaling directly to RcsB (Majdalani *et al.*, 2002, 2005).

Signaling directly to RcsB is likely to be the default explanation if all else has been ruled out.

RcsC-Dependent Signaling

If *rcsC* mutations block the phenotype, some additional dissection is possible. In particular, some but not all signals have been demonstrated to activate the phosphorelay via the outer membrane lipoprotein RcsF (reviewed in Majdalani and Gottesman, 2005).

Step 5: RcsF-dependence of signaling. Introduction of a mutation in *rcsF* (from NM20785, *rcsF*118::kan) may block activation/mucoidy, much as an *rcsC* null mutation does. If this is observed, signaling is through RcsF. Thus far, the signals that have acted via this system somehow perturb the cell surface, including growth of cells on a surface, mutations that lead to changes in LPS, and mutations that increase outer membrane permeability (Huang *et al.*, 2006; Lazzaroni *et al.*, 1999; Parker *et al.*, 1992). Further understanding of the signal transduction pathway upstream of RcsF is not yet available.

Other treatments and/or mutations (for instance, *dsbA* mutants and multicopy DjlA) are independent of RcsF, but still dependent upon RcsC (reviewed in Majdalani and Gottesman, 2005). How these entities act is not fully understood. DjlA may activate via interactions with the cytoplasmic domains of RcsC (Clarke *et al.*, 1997); *dsbA* mutations presumably change a protein or proteins other than RcsF that interact with RcsC, but the nature of this signaling is not really understood. One gene with a profound but thus far little understood effect on RcsC-dependent signaling is *igaA* (*Salmonella*) or *yrfF* (*E. coli*). Loss-of-function mutations in this gene lead to lethality unless RcsC or a downstream component of the phosphorelay is inactivated, and partial loss of function leads to high constitutive levels of Rcs activity (Cano *et al.*, 2002; Dominguez-Bernal *et al.*, 2004).

RcsA-Dependent Signaling: Increased RcsA Synthesis or Stability

RcsA is an unstable protein that is believed to act by interacting with RcsB on the DNA to activate transcription. The sites for RcsB/RcsA binding are far upstream of the −35 region in the best-characterized genes, *cpsB* and the *rcsA* gene itself; sites where RcsB appears to act alone are frequently just upstream of the −35 (Ebel and Trempy, 1999; Kelm *et al.*, 1997; Majdalani *et al.*, 2002; Wehland *et al.*, 1999). Mutations in the Lon protease (that degrades RcsA) activate the *cps* genes by increasing RcsA accumulation, even when no active signaling via RcsC is taking place (Torres-Cabassa and Gottesman, 1987) (see Fig. 1). Similarly, mutations or plasmids that increase RcsA synthesis will also activate capsule production. Therefore, if increased mucoidy is found and not abolished by an *rcsC* mutation, the possibility of increased RcsA synthesis or stability should be considered.

Step 6: Increased stability of RcsA usually means that the activity of the Lon protease is compromised. An easy test of this is whether another substrate of Lon, the SulA protein, also shows increased stability (Mizusawa and Gottesman, 1983). Because SulA is SOS induced and causes lethal cell filamentation if it is induced and cannot be degraded (Huisman and D'Ari, 1981; Huisman *et al.*, 1984), increased ultraviolet (UV) sensitivity is manifested. Increased UV sensitivity can be easily determined in a test on LB plates; alternatively, cells will be seen to filament upon growth at 37° in rich broth. Generally, less Lon activity is due to a mutation in the *lon* gene itself, but there are examples of conditions and/or mutations that lead to perturbation of Lon activity in ways that are still not fully understood (Snyder and Silhavy, 1992). Mapping studies and/or Western blots can be used to clarify the presence or absence of a *lon* mutation or change in Lon levels.

Determining Whether a Strain Carries a *lon* Mutation or is Phenotypically Lon⁻

If possible, include the wild-type parent and a known *lon* mutant (for instance, by P1 transduction of *lon*-146::tet into the parent from SG12047 (Gottesman, 1990)) as well as the strain in question. For each, resuspend cells from a colony or colonies in a drop of broth and spread a thin line across an LB plate. Covering the plate (without the cover) with a piece of paper or card, place it under a UV lamp, and slowly move the card sideways during the UV exposure, so that one side of the plate (and of each streak) receives a high dose and the other side receives a low dose. Cover the plate, wrap in foil, and incubate overnight. If the dose is correct, the *lon* mutant should grow poorly at all except the lowest doses, while the wild-type grows reasonably well. A dose close to that necessary to induce a prophage is usually sufficient.

Step 7: Alternatively, increased *rcsA* synthesis can be tested with an *rcsA*-lacZ fusion (DDS1301; (Sledjeski and Gottesman, 1995)) or considered as the most likely alternative if Lon activity appears to be normal. Conditions in which RcsA synthesis has been found to increase are the overexpression of the small regulatory RNA, DsrA, and in cells carrying mutations in *hns* (Sledjeski and Gottesman, 1995). In addition, LeuO has been found to regulate *rcsA* transcription (Klauck *et al.*, 1997;

Repoila and Gottesman, 2001). Other components of RcsA transcriptional and translational regulation remain to be defined, and thus a conclusion that this is the level at which a novel condition acts may be of special interest.

Conclusions

The Rcs phosphorelay is one of the most complex of the "two-component systems" in *E. coli*. Studies have demonstrated its role in regulation of multiple promoters, many of them with roles in biofilm formation. An understanding of the mechanisms of signaling to the components of this phosphorelay, as well as identification of the direct and indirect targets of the phosphorelay, will be important in understanding both bacterial development and the degree of plasticity in signaling systems of this sort.

Acknowledgments

This research was supported by the Intramural Research Program of the NIH, National Cancer Institute, Center for Cancer Research.

References

Boulanger, A., Francez-Charlot, A., Conter, A., Castanie-Cornet, M.-P., Cam, K., and Gutierrez, C. (2005). Multistress regulation in *Escherichia coli*: Expression of *osmB* involves two independent promoters responding either to σ^S or to the RcsCDB His-Asp phosphorelay. *J. Bacteriol.* **187,** 3282–3286.

Brill, J. A., Quinlan-Walshe, C., and Gottesman, S. (1988). Fine-structure mapping and identification of two regulators of capsule synthesis in *Escherichia coli* K-12. *J. Bacteriol.* **170,** 2599–2611.

Cabrera, J. E., and Jin, D. J. (2001). Growth phase and growth rate regulation of the *rapA* gene, encoding the RNA polymerase-associated protein RapA in *Escherichia coli. J. Bacteriol.* **183,** 6126–6134.

Cano, D. A., Dominguez-Bernal, G., Tierrez, A., Garcia-Del Portillo, F., and Casadesus, J. (2002). Regulation of capsule synthesis and cell motility in *Salmonella enterica* by the essential gene *igaA. Genetics* **162,** 1513–1523.

Carballes, F., Bertrand, C., Bouche, J. P., and Cam, K. (1999). Regulation of *Escherichia coli* cell division genes *ftsA* and *ftsZ* by the two-component system *rcsC-rcsB. Mol. Microbiol.* **34,** 442–450.

Clarke, D. J., Holland, I. B., and Jacq, A. (1997). Point mutations in the transmembrane domain of DjlA, a membrane-linked DnaJ-like protein, abolish its function in promoting colanic acid production via the Rcs signal transduction pathway. *Mol. Microbiol.* **25,** 933–944.

Davalos-Garcia, M., Conter, A., Toesca, I., Gutierrez, C., and Cam, K. (2001). Regulation of *osmC* gene expression by the two-component system *rcsB-rcsC* in *Escherichia coli. J. Bacteriol.* **183,** 5870–5876.

Dominguez-Bernal, G., Pucciarelli, M. G., Ramos-Morales, F., Garcia-Quintanilla, M., Cano, D. A., Casadesus, J., and Garcia-del Portillo, F. (2004). Repression of the RcsC-YojN-RcsB phosphorelay by the IgaA protein is a requisite for *Salmonella* virulence. *Mol. Microbiol.* **53,** 1437–1449.

Ebel, W., and Trempy, J. E. (1999). *Escherichia coli* RcsA, a positive activator of colanic acid capsular polysaccharide synthesis, functions to activate its own expression. *J. Bacteriol.* **181,** 577–584.

Ebel, W., Vaughn, G. J., Peters, H. K., 3rd, and Trempy, J. E. (1997). Inactivation of *mdoH* leads to increased expression of colanic acid capsular polysaccharide in *Escherichia coli.* *J. Bacteriol.* **179,** 6858–6861.

Francez-Charlot, A., Castanie-Cornet, M. P., Gutierrez, C., and Cam, K. (2005). Osmotic regulation of the *Escherichia coli bdm* (biofilm-dependent modulation) gene by the RcsCDB His-Asp phosphorelay. *J. Bacteriol.* **187,** 3873–3877.

Francez-Charlot, A., Laugel, B., Van Gemert, A., Dubarry, N., Wiorowski, F., Castanie-Cornet, M. P., Gutierrez, C., and Cam, K. (2003). RcsCDB His-Asp phosphorelay system negatively regulates the *flhDC* operon in *Escherichia coli.* *Mol. Microbiol.* **49,** 823–832.

Fredericks, C. E., Shibata, S., Aizawa, S., Reimann, S. A., and Wolfe, A. J. (2006). Acetyl phosphate-sensitive regulation of flagellar biogenesis and capsular biosynthesis depends on the Rcs phosphorelay. *Mol. Microbiol.* **61,** 734–747.

Gottesman, S. (1990). Minimizing proteolysis in *Escherichia coli*: Genetic solutions. *Methods Enzymol.* **185,** 119–129.

Huang, Y.-H., Ferrieres, L., and Clarke, D. J. (2006). The role of the Rcs phosphorelay in *Enterobacteriaceae.* *Res. Microbiol.* **157,** 206–212.

Huisman, O., and D'Ari, R. (1981). An inducible DNA-replication-cell division coupling mechanism in *E. coli.* *Nature* **290,** 797–799.

Huisman, O., D'Ari, R., and Gottesman, S. (1984). Cell division control in *Escherichia coli*: Specific induction of the SOS SfiA protein is sufficient to block septation. *Proc. Natl. Acad. Sci. USA* **81,** 4490–4494.

Kelm, O., Kiecker, C., Geider, K., and Bernhard, F. (1997). Interaction of the regulator proteins RcsA and RcsB with the promoter of the operon for amylovoran biosynthesis in *Erwinia amylovora.* *Mol. Gen. Genet.* **256,** 72–83.

Klauck, E., Bohringer, J., and Hengge-Aronis, R. (1997). The LysR-like regulator LeuO in *Escherichia coli* is involved in the translational regulation of *rpoS* by affecting the expression of the small regulatory DsrA-RNA. *Mol. Microbiol.* **25,** 559–569.

Lazzaroni, J. C., Germon, P., Ray, M.-C, and Vianney, A. (1999). The Tol proteins of *Escherichia coli* and their involvement in the uptake of biomolecules and outer membrane stability. *FEMS Microbiol. Lett.* **177,** 191–197.

Majdalani, N., Chen, S., Murrow, J., St. John, K., and Gottesman, S. (2001). Regulation of RpoS by a novel small RNA: The characterization of RprA. *Mol. Microbiol.* **39,** 1382–1394.

Majdalani, N., and Gottesman, S. (2005). The Rcs phosphorelay: A complex signal transduction system. *Annu. Rev. Microbiol.* **59,** 379–405.

Majdalani, N., Heck, M., Stout, V., and Gottesman, S. (2005). Role of RcsF in signaling to the Rcs phosphorelay in *Escherichia coli.* *J. Bacteriol.* **187,** 6770–6778.

Majdalani, N., Hernandez, D., and Gottesman, S. (2002). Regulation and mode of action of the second small RNA activator of RpoS translation, RprA. *Mol. Microbiol.* **46,** 813–826.

Minogue, T. D., Carlier, A. L., Koutsoudis, N. D., and Beck von Bodman, S. (2005). The cell density-dependent expression of stewartan exopolysaccharide in *Pantoea stewartii* ssp. *stewartii* is a function of EsaR-mediated repression of the *rcsA* gene. *Mol. Microbiol.* **56,** 189–203.

Mizusawa, S., and Gottesman, S. (1983). Protein degradation in *Escherichia coli*: The *lon* gene controls the stability of the SulA protein. *Proc. Natl. Acad. Sci. USA* **80,** 358–362.

Mouslim, C., Latifi, T., and Groisman, E. A. (2003). Signal-dependent requirement for the co-activator protein RcsA in transcription of the RcsB-regulated *ugd* gene. *J. Biol. Chem.* **278,** 50588–50595.

Parker, C. T., Kloser, A. W., Schnaitman, C. A., Stein, M. A., Gottesman, S., and Gibson, B. W. (1992). Role of the *rfaG* and *rfaP* genes in determining the lipopolysaccharide core structure and cell surface properties of *Escherichia coli* K-12. *J. Bacteriol.* **174,** 2525–2538.

Pristovsek, P., Sengupta, K., Lohr, F., Schafer, B., von Trebra, M. W., Ruterjans, H., and Bernhard, F. (2003). Structural analysis of the DNA-binding domain of the *Erwinia amylovora* RcsB protein and its interaction with the RcsAB box. *J. Biol. Chem.* **278,** 17752–17759.

Repoila, F., and Gottesman, S. (2001). Signal transduction cascade for regulation of RpoS: Temperature regulation of DsrA. *J. Bacteriol.* **183,** 4012–4023.

Simons, R. W., Houman, F., and Kleckner, N. (1987). Improved single and multicopy *lac*-based cloning vectors for protein and operon fusions. *Gene* **53,** 85–96.

Sledjeski, D., and Gottesman, S. (1995). A small RNA acts as an antisilencer of the H-NS-silenced *rcsA* gene of *Escherichia coli*. *Proc. Natl. Acad. Sci. USA* **92,** 2003–2007.

Snyder, W. B., and Silhavy, T. J. (1992). Enhanced export of β-galactosidase fusion proteins in *prlF* mutants is Lon dependent. *J. Bacteriol.* **174,** 5661–5668.

Stout, V., and Gottesman, S. (1990). RcsB and RcsC: A two-component regulator of capsule synthesis in *Escherichia coli*. *J. Bacteriol.* **172,** 659–669.

Stout, V., Torres-Cabassa, A., Maurizi, M. R., Gutnick, D., and Gottesman, S. (1991). RcsA, an unstable positive regulator of capsular polysaccharide synthesis. *J. Bacteriol.* **173,** 1738–1747.

Sturny, R., Cam, K., Gutierrez, C., and Conter, A. (2003). NhaR and RcsB independently regulate the *osmC*p1 promoter of *Escherichia coli* at overlapping regulatory sites. *J. Bacteriol.* **185,** 4298–4304.

Torres-Cabassa, A. S., and Gottesman, S. (1987). Capsule synthesis in *Escherichia coli* K-12 is regulated by proteolysis. *J. Bacteriol.* **169,** 981–989.

Wehland, M., and Bernhard, F. (2000). The RcsAB box. Characterization of a new operator essential for the regulation of exopolysaccharide biosynthesis in enteric bacteria. *J. Biol. Chem.* **275,** 7013–7020.

Wehland, M., Kiecker, C., Coplin, D. L., Kelm, O., Saenger, W., and Bernhard, F. (1999). Identification of an RcsA/RcsB recognition motif in the promoters of expopolysaccharide biosynthetic operons from *Erwinia amylovora* and *Pantoea stewartii* subspecies *stewartii*. *J. Biol. Chem.* **274,** 3300–3307.

Section III

Intracellular Methods and Assays

[17] *In Vivo* Measurement by FRET of Pathway Activity in Bacterial Chemotaxis

By Victor Sourjik, Ady Vaknin, Thomas S. Shimizu, and Howard C. Berg

Abstract

The two-component pathway in *Escherichia coli* chemotaxis has become a paradigm for bacterial signal processing. Genetics and biochemistry of the pathway as well as physiological responses have been studied in detail. Despite its relative simplicity, the chemotaxis pathway is renowned for its ability to amplify and integrate weak signals and for its robustness against various kinds of perturbations. All this information inspired multiple attempts at mathematical analysis and computer modeling, but a quantitative understanding of the pathway was hampered by our inability to follow the signal processing *in vivo*. To address this problem, we developed assays based on fluorescence resonance energy transfer (FRET) and bioluminescence resonance energy transfer (BRET) that enabled us to monitor activity-dependent protein interactions in real time directly in living cells. Here, we describe quantitative applications of these assays in cell populations and on a single-cell level to study the interaction of the phosphorylated response regulator CheY with its phosphatase CheZ. Since this interaction defines the rate of CheY dephosphorylation, which at steady state equals the rate of CheY phosphorylation, it can be used to characterize intracellular kinase activity and thus to analyze properties of the chemotaxis signaling network.

Introduction

E. coli *Chemotaxis*

Motile chemotactic bacteria are able to follow gradients of certain chemicals in the surrounding medium. They move toward higher concentrations of attractants, while avoiding higher concentrations of repellents. *E. coli* swimming consists of periods of smooth translation, or "runs," interrupted by periods of erratic reorientation, or "tumbles." These two modes correspond to counterclockwise (CCW) and clockwise (CW) directions of flagellar motor rotation, respectively. Swimming direction is chosen randomly, but swimming in a favorable direction results in longer runs,

METHODS IN ENZYMOLOGY, VOL. 423   0076-6879/07 $35.00
            DOI: 10.1016/S0076-6879(07)23017-4

while swimming in an unfavorable direction causes a cell to revert to its baseline behavior. Bacteria thus detect spatial gradients by sensing temporal changes in chemoeffector concentrations; such changes can be simulated in the lab simply by adding or removing chemoeffectors.

The signal transduction pathway in *E. coli* is well characterized; for reviews on chemotaxis, see Bourret and Stock, 2002, Bren and Eisenbach, 2000, Parkinson *et al.*, 2005, Sourjik, 2004, Vaknin and Berg, 2004, Wadhams and Armitage, 2004. As in many other signaling systems, signaling in chemotaxis relies on protein phosphorylation. The key enzyme in the pathway is a histidine kinase, CheA, that forms a sensory complex with transmembrane chemoreceptors (so-called methyl-accepting chemotaxis proteins, or MCPs) and an adaptor protein, CheW. Autophosphorylation activity of CheA is inhibited by binding to receptors of attractant and enhanced by binding of repellent. When autophosphorylated, CheA can transfer a phosphate group to the response regulator, CheY, that subsequently binds to the flagellar motors, increasing the probability that they spin CW and thus promoting tumbles. When a cell moves up an attractant gradient, the P-CheY level is reduced, so the cell tends to continue running in that direction. In addition to CheA, CheW, and CheY, the signal transduction pathway includes CheZ, a phosphatase of P-CheY, and a receptor methylation system that consists of a methyltransferase, CheR, and a methylesterase, CheB. The phosphatase ensures a rapid turnover of P-CheY, whereas methylation mediates adaptation to the ambient attractant concentration. Receptor methylation on four specific glutamate residues increases CheA activity and decreases sensitivity to attractants. CheB phosphorylation by CheA provides negative feedback from receptor activity to the adaptation system by promoting demethylation of receptors and lowering CheA activity.

E. coli has five types of receptors, Tsr, Tar, Tap, Trg, and Aer, that sense a variety of amino acids, sugars, and dipeptides, as well as pH, temperature, redox state, etc. Major receptors, those for aspartate (Tar) and serine (Tsr), are abundant, numbering several thousand copies per cell. These receptors can also sense other attractants and repellents. Minor receptors specific for dipeptides (Tap), ribose and galactose (Trg), and redox potential (Aer) are much less abundant, with only a few hundred copies per cell (Li and Hazelbauer, 2004). Receptors form clusters at the cell pole where different species of receptors are intermixed and interact to integrate and amplify stimuli (Gestwicki and Kiessling, 2002; Lai *et al.*, 2005; Maddock and Shapiro, 1993; Sourjik and Berg, 2002b, 2004; Studdert and Parkinson, 2004). All other cytoplasmic chemotaxis proteins localize to the clusters to create a signaling scaffold (Banno *et al.*, 2004; Cantwell *et al.*, 2003; Shiomi *et al.*, 2002; Sourjik and Berg, 2000).

FRET

FRET assays rely on nonradiative distance-dependent energy transfer from one fluorescent molecule (donor) to another (acceptor) and permit study of interactions of fluorescently labeled proteins in living cells (Wouters *et al.*, 2001). The major advantages of FRET are that the measurements of intracellular protein interactions are noninvasive, quantitative, and performed in real time. It is thus particularly useful for observing transient protein interactions, such as those involved in signal transduction. Changes in protein conformation can be monitored as well, if they alter the distances between fluorophores attached at different points of the protein of interest, a technique used to develop a number of FRET-based reporter assays (Miyawaki and Tsien, 2000; Ting *et al.*, 2001; Zaccolo *et al.*, 2001; Zhang *et al.*, 2001). Either fluorescent dyes or fluorescent proteins (FPs) can be used as labels (Miyawaki *et al.*, 2003), but specific *in-vivo* labeling with fluorescent dyes is difficult in bacteria, due to the low permeability of cell membranes and the small target for microinjection. Expressing fluorescent protein fusions is thus the approach of choice for bacteria. Several green fluorescent protein (GFP) mutants can be used as donor/acceptor pairs for FRET, with cyan and yellow fluorescent proteins, CFP and YFP (Miyawaki and Tsien, 2000) being the most commonly used pair. These are available commercially in forms enhanced for human codon usage, as ECFP and EYFP, respectively (Clontech, Mountain View, CA). FP-labeling has an advantage of absolute specificity, and in bacteria with inducible expression systems, gives control over levels of donor and acceptor proteins in the cell. A disadvantage of FPs is their relatively large size, which does not allow their fluorescent centers to come into very close proximity, and thus lowers FRET efficiency. Fusions to FPs can also perturb the function of targets, so functionality tests of fusion proteins need to be performed whenever possible.

Efficiency of energy transfer in a protein complex depends upon several parameters, such as spectral properties of the donor and acceptor fluorophores, but most importantly on the distance between the fluorophores, and can be expressed as

$$E_{FRET} = \frac{R_0^6}{R^6 + R_0^6} \tag{1}$$

where R is the distance between the centers of the fluorophores and R_0 is the Förster radius (Förster, 1948), the distance at which the energy is transferred from the donor to acceptor with 50% efficiency. R_0 can be calculated from the overlap of the emission spectrum of the donor with the emission spectrum of the acceptor, the optical properties of the surrounding medium, and the relative orientation of the donor and acceptor fluorophores (Selvin, 1995). Assuming free rotation of FPs in the complex, which is the case if they are fused to the target proteins with flexible peptide linkers, R_0 for the CFP/YFP pair in bacterial cytoplasm is 4.9 nm (Sourjik and Berg, 2002a; Tsien, 1998), about the same size as the diameter of the FPs. Steep dependence of the FRET efficiency upon distance means that very little energy transfer takes place if the fusion proteins are farther apart than a critical distance of about 10 nm. This makes chances of transfer through random encounters of non-interacting proteins minuscule, and essentially eliminates false-positive FRET signals. However, the steep dependence of the FRET signal upon spacing between FPs in a multiprotein complex also can lead to false negatives. Thus, whenever the relative positions of proteins in a complex are unknown, it is advisable to test all combinations of FP fusions to the N- and C-termini of both proteins.

There are several commonly used techniques for measuring FRET *in vivo*. Generally, one excites a donor fluorophore and monitors donor and/or acceptor fluorescence, or the lifetime of the excited donor (Wouters *et al.*, 2001). In the sections that follow, we describe two microscopy-based assays to determine stimulation-dependent interactions between the response regulator CheY and its phosphatase CheZ in a population bacteria or in single cells, respectively. The first assay provides high sensitivity and time resolution and can be used to measure intracellular kinase activity or response times, whereas the second allows one to measure spatial distributions of protein interactions in a cell. In addition, we describe a related assay based upon bioluminescence resonance energy transfer (BRET), where responses are measured in suspensions of swimming cells. In this case, *Renilla* luciferase (RLUC) is used rather than CFP, and one does not need to use light to excite the donor fluorophore. All three assays are based upon measuring fluorescence intensity in two channels that correspond to CFP (or RLUC) and YFP emission, where energy transfer is detected as a decrease (or quenching) in emission of CFP (or RLUC) and an increase in the emission of YFP.

FRET Measurement of the Interaction Between CheY-YFP and CheZ-CFP in a Population of Bacteria Fixed to a Microscope Cover Slip

Overview

The steady-state concentration of phosphorylated response regulator P-CheY is determined by a balance between CheY phosphorylation by the kinase CheA and P-CheY dephosphorylation by the phosphatase CheZ.

Assuming Michaelis-Menten kinetics, the rate of P-CheY dephosphorylation is determined by the concentration of the enzyme-substrate complex CheZ·P-CheY. This concentration, which can be determined by FRET, serves as a measure of the intracellular kinase activity in steady state (or close to steady state), and can be used to monitor signal processing by the cell. Addition of a large amount of attractant reduces this concentration close to zero, providing an essential FRET control; see the following text.

To find an optimal FRET pair, we tested functionality and FRET efficiency of all combinations of N- and C-terminal fusions. Proteins were fused by a short flexible linker of three to five glycines that produced products that were stable against proteolysis in the cell (Sourjik and Berg, 2000). All fusions complemented chemotaxis-driven spreading of the corresponding knock-out mutants on soft agar plates (Sourjik and Berg, 2000), but only two combinations, YFP-CheY/CheZ-CFP and CheY-YFP/CheZ-CFP, showed a FRET signal. Maximal FRET efficiency was observed for the pair of two C-terminal fusions, as could be understood from the subsequently published structure of the complex (Zhao *et al.*, 2002), where the C-terminus of CheZ is close to P-CheY, but the N-terminus of CheZ is not.

In our early FRET assays, the CheY-YFP and CheZ-CFP fusion proteins were expressed from two different plasmids, pVS18 and pVS54 (Sourjik and Berg, 2002b). The former has *cheY-eyfp* under control of an isopropyl β-D-thiogalactoside (IPTG)-inducible pTrc promoter, a pBR322 replication origin, and confers ampicillin resistance. The latter has *cheZ-ecfp* under control of an arabinose-inducible pBAD promoter, a pACYC184 replication origin, and confers chloramphenicol resistance. This allowed us to independently optimize the expression level of each fusion protein. We found that, under our standard growth conditions, the FRET signal was maximal when cells were induced with 50 μM IPTG and 0.01% arabinose, resulting in average intracellular expression levels of CheY-YFP and CheZ-CFP of approximately 18 and 8 μM, respectively (Sourjik and Berg, 2002a; Sourjik, unpublished). In later experiments, similar expression levels and FRET signal were obtained when both fusions were expressed at 50 μM IPTG induction from a pVS88 plasmid, which encodes *cheY-eyfp* and *cheZ-ecfp* as parts of one artificial operon under control of a pTrc promoter (Sourjik and Berg, 2004). An advantage of the latter approach is that an additional (chemotaxis) protein, for example, receptor, can be expressed at varying levels from a second plasmid.

Experimental Setup for a Population-Based FRET Measurement

The experiment can be performed with an upright or inverted fluorescence microscope equipped with a second fluorescence cube and two photodetectors, as shown in Fig. 1. Bacteria expressing CheZ-CFP and CheY-YFP

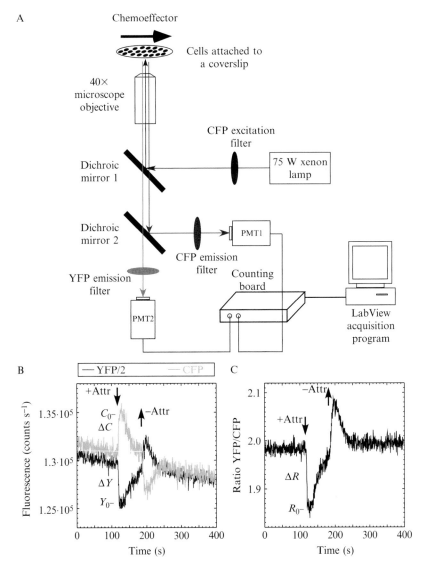

FIG. 1. Population-based FRET measurement of activity-dependent protein interactions. (A) Experimental setup, shown for an inverted microscope. (B,C) FRET response of LL4 [$\Delta(cheY\ cheZ)$ $\Delta flgM$] cells that express CheY-YFP and CheZ-CFP from pVS88 at 50 μM IPTG induction, plotted in individual fluorescence channels (B) and as a change in the ratio of YFP to CFP fluorescence (C). Cells were stimulated by addition and subsequent removal of attractant, 10 μM α-methyl-DL-aspartate (MeAsp), at time points indicated by arrows. A gradual decrease in the yellow and cyan fluorescence due to photobleaching (B) has little effect on the ratio (C). A 520 nm long-pass emission filter was used for the YFP channel in this experiment; note that the YFP fluorescence signal was rescaled for convenience. Amplitudes of changes in signals and their values upon saturating stimulation are indicated; see text.

are attached as a dense monolayer to a round 12-mm glass cover slip that is then mounted in a flow chamber (Sourjik and Berg, 2002a). The chamber is kept under a constant flow (0.3–1 ml min^{-1}) of buffer by a syringe pump (Harvard Apparatus, Holliston, MA), which provides aeration and allows one to add and remove attractants and repellents. At a flow rate of 1 ml min^{-1}, 99% of the solution in the flow chamber is replaced in about 6 s (Sourjik and Berg, 2002b). Fluorescence is excited through the objective (epifluorescence) at CFP-specific wavelengths. In the setup at Heidelberg, we use a 436/20 nm excitation filter and a 75 W super quiet xenon lamp (Hamamatsu, Bridgewater, NJ) that is attenuated by a factor of 550 with neutral density filters; in the work at Harvard, a 442 nm 18 mW He–Cd laser (Melles Griot, Carlsbad, CA), attenuated by a factor of 400, was used. The emission from a field of cells (300–500) is separated from the excitation light by a 455 nm dichroic mirror and split in two spectral parts by a second 515 nm dichroic mirror. Signals are collected in two spectral channels, using a 480/40 nm band-pass emission filter for CFP fluorescence and a 520 nm long-pass or a 535/30 nm band-pass emission filter for YFP fluorescence. For detection, we use Peltier-cooled photon-counting photomultipliers (PMTs, H7421-40, Hamamatsu) with high quantum efficiency and low noise. This allows one to work at relatively low light levels, minimizing bleaching and cell damage; *E. coli* is sensitive to blue light (Wright *et al.*, 2006). Data are acquired as photon counts using a PCI-6034E counting board and acquisition program written in LabView 7 (both from National Instruments, Austin, TX). Depending upon the desired time resolution, counts are integrated for 0.2 to 1 s, with the longer integration times reducing measurement noise. Alternatively, the counter outputs can be converted to analogue signals by ratemeters (RIS-375, Rowland Institute, Cambridge, MA), filtered by 8-pole low-pass Bessel filters (Krohn-Hite 3384, Krohn-Hite, Avon, MA) at the desired cutoff frequency (e.g., 2 Hz) and sampled at a corresponding frequency (e.g., 5 Hz), using the same PCI-6034E board and a different LabView 7 acquisition program. Performance of either acquisition setup is similar, although measurement noise is somewhat lower in the latter. Data are saved as text files and analyzed using KaleidaGraph (Synergy Software, Reading, PA), MS Excel, or any other graphing software.

To achieve higher time resolution for measurements of response kinetics, chemoeffectors can be released from an inactive, "caged" state on a ms time scale by a short (~1 ms) flash of ~350 nm ultraviolet (UV) light using a xenon flash lamp (e.g., model 35S, driven by model 238 power supply, 200 Ws, 50 μs, Chadwick-Helmuth, El Monte, CA). In our experiments (Sourjik and Berg, 2002a), the flash lamp was mounted in a housing (66055, Oriel, Stamford, CT) equipped with a fused silica condensing lens (60076) and a UG-5 filter

(330 nm peak, Schott, Curyea, PA). For photolysis on an upright microscope, the light was fed to the microscope with a liquid light guide (Oriel 77554) and focused onto the flow cell with a 25-mm fused silica lens (Oriel 41220). The light was reflected by a UV-enhanced aluminum mirror (Edmund Scientific, Barrington, NJ) and entered the flow cell from below, through a fused quartz bottom window. The focused spot was adjusted to be about 4 mm wide, much larger than the field of view. A UV-blocking filter (Chroma Technology, Brattleboro, VT) should be inserted into the light path above the lower dichroic mirror to prevent UV light from reaching the photomultipliers. For these experiments, data were acquired with a time resolution of 5 ms (200 Hz cutoff frequency and 500 Hz sampling frequency).

FRET Measurement of Receptor Kinase Activity Using Flow Stimulation

1. Grow the overnight culture at 30 to 34° in tryptone broth (TB; 1% tryptone, 0.5% NaCl, pH 7.0) in the presence of 100 μg ml^{-1} ampicillin and 35 μg ml^{-1} chloramphenicol. Dilute the overnight culture 1:100 in 10 ml prewarmed TB in a 125-ml flask containing antibiotics, 50 μM IPTG and 0.01 % arabinose, and grow the culture at 33 to 34° in a rotary shaker at 275 rpm for about 4 h to mid-exponential phase (OD600 = 0.45–0.5). Note that LB (Luria-Bertani) medium and temperatures higher than 37° repress expression of chemotaxis genes and are thus not suited for growing chemotactic E. coli cultures. Our experience is that keeping constant growth conditions is extremely important in reducing day-to-day variations in results.

2. Collect cells by centrifugation (3500 rpm, 10 min), wash them with 5 ml tethering buffer (10 mM potassium phosphate, 0.1 mM EDTA, 1 μM L-methionine, 10 mM sodium lactate, pH 7), and resuspend them in 10 ml tethering buffer. One can immediately proceed with the measurements, but we routinely pre-incubate the cells for 1 h at 4 to 10° in the buffer to stop cell growth and protein expression. Cells can be kept at this temperature for up to 5 h without loss of FRET response.

3. Place 20 μl of a 0.1 (w/v) solution of poly-L-lysine (Sigma, St. Louis, MO) on an ethanol-cleaned cover slip for 15 to 30 min, then aspirate the rest and allow the cover slip to dry. Wash the cover slip three times with ~100 μl of water to remove the excess polylysine and allow the cover slip to dry again. The thickness of polylysine-coating is an important factor, since it determines the efficiency of cell attachment. Even under seemingly identical conditions, it might vary depending upon the batches of polylysine and cover slips, so the optimal coating and subsequent attachment conditions have to be determined in every specific case.

4. Collect cells from 1.5 ml by centrifugation at 8000 rpm in an Eppendorf microfuge, discard the supernatant, and resuspend the pellet in the

remaining liquid (~100 μl). Place the cells on the polylysine-coated cover slip and incubate for 15 to 30 min.

5. Fill the flow chamber with tethering buffer, coat the edges of the flow chamber and/or the cover slip with Apiezon L grease (Apiezon, London), and seat the cover slip. Then, turn on the flow: apply suction to the output port rather than pressure to the input port; otherwise, the cover slip will pop off. Unstuck cells are rapidly washed away, and the efficiency of attachment can be examined by phase-contrast or bright-filed illumination. Optimally, cells should be attached as a dense monolayer. If no such field of view can be found, repeat steps (3) and (4) with varying coating and attachment times to optimize the attachment efficiency.

6. Start the measurement by switching to the fluorescence illumination, turning on the PMTs and starting the acquisition program. To ensure linearity of the H7421-40 PMTs, fluorescence signals must not exceed 10^6 counts per second. If the signals are higher, the illumination intensity can be reduced by inserting additional neutral density filters. If the signals are lower than 10^5 counts per second, the illumination intensity can be increased.

7. Perform the measurement by replacing the buffer in the flow chamber with a chemoeffector solution of interest. Recall that it takes several seconds for the solution to be replaced. The same field of cells can be stimulated multiple times by adding and removing different solutions. Allow enough time (up to several minutes, depending upon the preceding stimulus strength) for the cells to re-equilibrate in buffer between stimulations. At room temperature (20 to 23°), cells remain fully chemotactic for up to 1 h, after which time the responses might start to deteriorate.

Data Analysis

Results of a typical FRET experiment using flow stimulation are shown in Fig. 1B,C. LL4 [Δ(*cheY cheZ*) $\Delta flgM$] cells expressing CheZ-CFP and CheY-YFP were stimulated by step-addition of a saturating amount of attractant, 10 μM MeAsp, indicated by an arrow, which inhibits kinase activity and lowers the concentration of P-CheY. Since only phosphorylated CheY binds CheZ at the protein expression levels that we use, interaction between CheY-YFP and CheZ-CFP and energy transfer from CFP to YFP decrease upon stimulation. As a result, CFP fluorescence goes up and YFP fluorescence goes down. Routinely, we monitor the ratio of YFP to CFP fluorescence (R, black line) because it is less sensitive to perturbations that affect both channels simultaneously, including fluctuations in the intensity of the excitation light, changes in the number of cells observed, or movement of the cover slip. R is proportional to the number of interacting P-CheY-YFP/CheZ-CFP FRET pairs and thus to kinase activity. In CheR$^+$ CheB$^+$ cells, the initial fast response to a step in attractant concentration is followed by a slower CheR-dependent adaptation. Subsequent removal of attractant elicits a repellent response, followed by a CheB-dependent adaptation.

In our subsequent analysis, we define the value of FRET as the fractional change in CFP fluorescence, $\Delta C/C_0$, where ΔC is a decrease in the CFP signal due to energy transfer and C_0 is the signal in the absence of FRET (Sourjik and Berg, 2002a,b, 2004). Assuming that only one type of complex is formed by P-CheY-YFP and CheZ-CFP, FRET as so defined is related to the concentration of the complex as

$$\frac{\Delta C}{C_0} = E_{FRET} \frac{[P\text{-}CheY\text{-}YFP \cdot CheZ\text{-}CFP]}{[CheZ\text{-}CFP]_T} \qquad (2)$$

where E_{FRET} is the efficiency of energy transfer in the complex as defined in Eq. (1), $[P\text{-}CheY\text{-}YFP \cdot CheZ\text{-}CFP]$ is the intracellular concentration of the complex, and $[CheZ\text{-}CFP]_T$ is the total concentration of CheZ-CFP. Since E_{FRET} is constant for a given FRET pair and $[CheZ\text{-}CFP]_T$ can be fixed by using the same growth and induction conditions, FRET is proportional to $[P\text{-}CheY\text{-}YFP \cdot CheZ\text{-}CFP]$. Normalizing to $[CheZ\text{-}CFP]_T$ makes the value of FRET independent of the total number of cells in a field of view, as long as the background signal from the optics is much smaller than fluorescence from the cells. Such normalization is advantageous because the exact number of cells varies from experiment to experiment. It can be used when expression level of donor (CheZ-CFP) is kept constant, which is true in most of our measurements of kinase activity; in other cases, when the expression of donor varies, FRET should be defined instead as an absolute change in CFP fluorescence. When needed, the background from the optics can be determined separately and subtracted from the measured values.

In principle, FRET as defined in Eq. (2) can be determined solely from changes of fluorescence in the CFP channel upon stimulation; however, as has been mentioned, we found a measurement of the ratio of YFP to CFP signals, R, to be less noisy. We showed earlier (Sourjik and Berg, 2002a,b) that FRET can be recalculated from changes in this ratio as

$$\frac{\Delta C}{C_0} = \frac{\Delta R}{\alpha + R_0 + \Delta R} \qquad (3)$$

where ΔR is the change in ratio due to energy transfer, $R_0 = Y_0/C_0$ is the ratio in the absence of FRET, and $\alpha = |\Delta Y/\Delta C|$ is the absolute value of the ratio of changes in the fluorescence signal in YFP and CFP channels due to energy transfer. α depends upon quantum yields of YFP and CFP, on properties of the channels (emission filters and dichroic mirrors) used for detection of YFP and CFP fluorescence, and upon the sensitivity of the detectors. Values of ΔR and of R_0 are measured in each experiment; the value of α is constant for any given setup and has to be determined only once.

Typically, we use flow experiments to measure dose-response curves of cellular responses to attractants. In this case, we start with the cells that are equilibrated in the buffer and have a certain level of FRET and then stimulate them by adding and subsequently removing increasing amounts of attractant until the kinase activity (and FRET) is completely inhibited by saturating levels of stimulation (Fig. 2A). It is thus more convenient to measure ΔR_S, a change between the ratio in buffer (R_B) and the ratio at a defined level of stimulation (R_A), and then recalculate ΔR from these values as

$$\Delta R = R_B - R_0 - \Delta R_S \tag{4}$$

Since there is usually a slow drift in the YFP/CFP ratio, caused by a faster bleaching of CFP compared to YFP (Fig. 1B), it is important to acquire the signal in the buffer prior to and after the measurement for a sufficient time to be able to accurately interpolate the base line (dashed line in Fig. 2A) and measure the responses (ΔR_S).

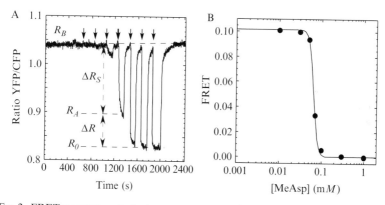

FIG. 2. FRET measurement of a dose–response curve for an adaptation-deficient strain that expresses only an aspartate receptor Tar. (A) FRET response, plotted as a change in a ratio of YFP to CFP fluorescence, to a sequential addition (indicated by arrows) and subsequent removal of MeAsp steps of increasing size: 10 μM, 30 μM, 50 μM, 70 μM, 100 μM, 300 μM, and 1 mM. Ratio values used to calculate FRET for a 70-μM step of MeAsp (thicker arrow) using Eqs. (3) and (4) are shown. A 535/30 nm band-pass filter was used for YFP emission in this experiment; hence, the difference in ratio values from Fig. 1C. (B) Dose–response curve calculated from the data in (A) using $\alpha = 1.05$. The solid line is a fit using Eq. (5) with best-fit values $K_{1/2} = 65\ \mu M$ and $H = 9$. In this experiment, receptorless strain VH1 [Δtsr-7028 $\Delta(tar$-$tap)5201$ Δtrg-100 Δaer-1 $\Delta(cheR\ cheB\ cheY\ cheZ)$] was used as a background, CheY-YFP and CheZ-CFP were expressed as described in Fig. 1, and aspartate receptor, Tar, was expressed at 2 μM salicylate induction from a pVS123 plasmid, constructed as previously described (Sourjik and Berg, 2004). This strain is less sensitive to MeAsp than the strain used for Fig. 1.

To calculate the value of FRET correctly, it is important to make sure that the value of R at saturating levels of a given attractant indeed corresponds to zero kinase activity and can be used as R_0. This can be done either by stimulating cells with attractants that are known to completely inhibit kinase activity in a given strain (e.g., a combination of high concentrations of serine and aspartate) or by determining the absolute value of FRET by bleaching the YFP fluorescence and measuring the resulting increase in the CFP fluorescence. The latter approach requires additional equipment and has been used to confirm that essentially all of CheY-YFP binding to CheZ-CFP at the induction levels that we use is phosphorylation-dependent (Sourjik and Berg, 2002b).

As has been discussed, at steady state or close to steady state, FRET can be used as a measure of the intracellular kinase activity. The steady-state assumption is obvious for adaptation-deficient *cheR cheB* cells (Fig. 2A), but it works well even in wild-type cells, because adaptation is much slower than the initial response (Fig. 1C). Measured FRET values at a range of attractant concentrations can thus be directly plotted as the dose-response curve of a receptor-kinase complex (Fig. 2B). These experiments allow a relatively quick analysis of several key parameters of the signal processing by receptor-kinase complexes in the cell—steady-state kinase activity before stimulation and sensitivity and apparent cooperativity of the response—in a variety of mutants with altered signaling properties (e.g., receptor mutants). We routinely fit the dose-response curves using the Hill equation in the form

$$FRET([L]) = \frac{[L]^H}{[L]^H + K_{1/2}^H} \qquad (5)$$

where $[L]$ is a ligand concentration, $K_{1/2}$ is the midpoint of the response curve, and H is a Hill coefficient. Using the best-fit values of Eq. (5), the response sensitivity can be described as $K_{1/2}^{-1}$ or $\log(K_{1/2}^{-1})$, whereas H characterizes the apparent cooperativity of the response. The obtained dose-response data can be used to analyze signal amplification and integration by the sensory complexes (Sourjik and Berg, 2002b, 2004), and also for a detailed computer modeling of the complex behavior (Keymer *et al.*, 2006; Mello and Tu, 2003, 2005; Shimizu *et al.*, 2003).

FRET Measurement of Response Times

In flow experiments, the exchange of a solution in the flow chamber is too slow to enable a measurement of response times. The necessary time resolution can be achieved when the cells are stimulated by release of a caged chemoeffector. Cells are prepared and the measurement is done as for flow stimulation before with the following modifications:

8. NPE-caged L-aspartate {*N*-[1-(2-nitrophenyl)ethyloxycarbonyl] aspartic acid, Calbiochem, San Diego, CA}, an attractant, or NPE-caged proton (2-hydroxyphenyl-1-(2-nitrophenyl)ethyl phosphate, Molecular Probes, Eugene, OR), a repellent, are drawn into the flow chamber and the flow is stopped.

9. Pre-stimulus CFP and YFP signals are acquired for 2 s, then the chemoeffector is released by flash, and the signals are acquired for a further 5 s. The efficiency of flash release can be calibrated by comparing the amplitudes of response at defined concentrations of caged aspartate to the dose-response curve measured by flow stimulation. (See Sourjik and Berg, 2002a, for more details.)

10. Fresh solution of caged chemoeffector is drawn into the chamber and step (9) is repeated. Up to 10 flash experiments can be performed on the same field of cells.

Data are analyzed the same way as in the flow experiments. However, as a result of short integration times (5 ms), these measurements are much noisier and data from multiple experiments have to be averaged. Observed response kinetics can then be combined with computer models of the pathway to derive values of the intracellular rate constants (Sourjik and Berg, 2002a).

FRET Measurement of the Interaction Between CheY-YFP and CheZ-CFP in Single Bacteria Fixed to a Microscope Cover Slip

Setup

In the experiments described in Vaknin and Berg (2004), we used an upright microscope (Nikon Optiphot) equipped with a Plan Fluor 100×/1.30 numerical aperture Oil Ph4 DLL objective, as shown in Fig. 3. Cells expressing CheZ-CFP and CheY-YFP were attached to a glass coverslip mounted in a flow chamber subjected to constant flow (0.4 ml/min), as has been described. The fluorescence was excited by a 75 W xenon lamp using an HQ440/20 excitation filter and a 455 DCLP dichroic mirror, and the fluorescent light was split into cyan and yellow beams with a 515 DCXR dichroic mirror followed by HQ485/40 and HQ535/30 emission filters, essentially as before. These optics (Chroma Technology) were mounted in two 25 mm Nikon fluorescence cubes. Half of each image was blocked, and the two identical halves were recombined by using a 515 DCSP mirror (reflecting the yellow light and passing the cyan) and projected side-by-side onto the face of a sensitive charge-coupled device camera (DV887ECS-BV, Andor Technology, South Windsor, CT). The path lengths of the cyan and

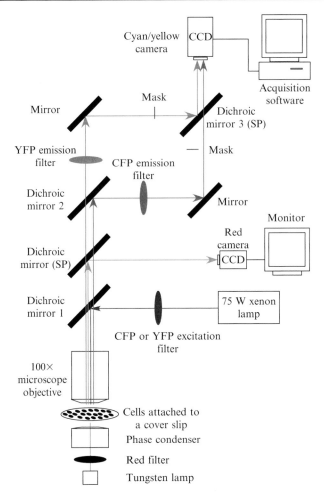

FIG. 3. Setup for FRET imaging. A short-pass (SP) dichroic mirror extracts red light for a phase-contrast recording. Dichroic mirrors 1 and 2 and the emission filters are the same as in Fig. 1A. The PMTs have been replaced by two front-surface mirrors, dichroic mirror 3, which is a short-pass filter that allows one to display the cyan and yellow images side by side, and a CCD camera.

yellow beams were matched and contained the same projection lenses (Nikon CF PL2x), so that the magnification of the cyan and yellow images was the same.

It is essential in these experiments that cell images remain in focus. To monitor the focus, we added a third, red channel that could be viewed in parallel with the cyan and yellow channels. A 620/40 bandpass filter was placed between the tungsten light source and the phase-contrast condenser, and the transmitted light was extracted with a 580 DCSP dichroic mirror and directed to a second video camera (BP550; Panasonic, Secaucus, NJ). The red extractor was placed in one port of a Zeiss Optovar mounted between dichroic mirrors 1 and 2 (Figs. 1A, 3), so that it could be switched out of the beam path when not required. The red light was not detectable in either the cyan or yellow channels, but it gave vivid images that could be used to optimize the focus or detect cell movement. Finally, a counterweight was added to the microscope stage to reduce its drift.

An additional factor that is important in FRET imaging is redistribution of proteins. While the total fluorescence might not be sensitive to such changes, the spatial distribution of the fluorescence clearly is. We measured the effect of the addition of attractants on the distribution of CheZ-CFP (excited at 440 nm) and CheY-YFP (excited at 515 nm) when expressed separately in cells. No change was observed with CheZ-CFP, but a substantial shift was detected with CheY-YFP, consistent with the fact that CheY binds to CheA, while P-CheY binds to CheZ, which, in turn, binds to the short form of CheA (Cantwell *et al.*, 2003). Redistribution of CheY-YFP was measured in each FRET experiment (when both fusions were present) by changing excitation cubes: to one containing an HQ515/5 excitation filter, a Q525LP dichroic mirror, and an HQ530LP emission filter.

The optical parameters of the system can be checked by using a purified CFP-YFP fusion protein with a trypsin-sensitive site in the middle that enables one to separate the two fluorophores enzymatically. This experiment was done as follows. About 0.6 ml of PBS containing the purified CFP-YFP protein was drawn into the chamber from a 1.5-ml test tube, and a few images were taken. Then, trypsin was added to the test tube; the mixture was allowed to incubate for ~2 min and then drawn into the chamber, and additional images were taken, as shown in Fig. 4A. In our system with excitation at 440 nm, trypsin digestion increased the cyan signal by 32% and decreased the yellow signal by 70% (after subtracting the leakage of CFP emission; see following sections). However, with excitation at 515 nm, the YFP emission was not affected by digestion, affirming the use of the 515 nm excitation as a measure of YFP distribution even in the presence of CFP. After digestion, both fluorophores were present at identical concentrations, and the cyan signal from CFP with excitation at 440 nm was four times stronger than the yellow signal from YFP with excitation at 515 nm. In addition, we found that when cells expressing only CFP were excited at 440 nm, the intensity in the yellow channel was 30% of that in the cyan channel. With cells excited at 515 nm, the leakage of YFP emission into

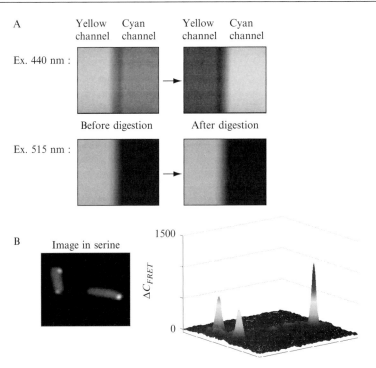

Fig. 4. (A) Demonstration of FRET changes with a CFP-YFP fusion protein. Images were taken during a single experiment before (left) and after (right) digestion of the fusion protein at a trypsin-sensitive site, located between YFP and CFP, and with excitation at 440 nm (top images) and 515 nm (bottom images). Note that while substantial changes in fluorescence occurred upon digestion with excitation at 440 nm, no change in fluorescence was observed with excitation at 515 nm. The detected light intensity was similar to that detected from receptor clusters (a few thousand photons per pixel per second). The vertical band in the middle of each image is a region of overlap between the projections of the two channels. (B) Changes in the cyan channel due to FRET. The image (left), shows two cells, one with two clusters and another with a single cluster. Three-dimensional plots (right) showing the spatial distribution of changes in intensity, ΔC_{FRET}, observed upon addition of serine. Adapted from Fig. 2 of Vaknin and Berg (2004).

the cyan channel was negligible. The background signals from cells that expressed neither CFP nor YFP was relatively small, 100 photons/s per pixel.

Experimental Procedure

The experimental procedure is the same as that described in the previous section (for cell population measurements), with the following exceptions:

Steps 3–4. For imaging, a dense layer of cells is not desired; sparse attachment is preferred. Also, it is more crucial here to get cells that are firmly stuck with their long axes parallel to the cover slip. While such attachment can be achieved with polylysine-coated glass, we immobilized cells using anti-FliC antibody. Mix antibody in 50 μl of buffer with 20 μl of cell suspension and place on a 12-mm round cover slip (pre-cleaned with ethanol). After 15 to 25 min, seat the cover slip on the flow chamber, as has been described. The exact amounts of antibody and cells can vary, depending upon the strain, antibody purification, etc., but the concentration of the antibody should be relatively high and the number of cells relatively low.

Step 7. Acquire a series of images under continuous flow, alternating between buffer and attractant; see following text.

Data Analysis

In a homogeneous mixture, the change in the concentration of interacting YFP and CFP molecules upon stimulation, Δn, is proportional to the changes in the intensity of the cyan and yellow channels due to changes in FRET: ΔC_{FRET} and ΔY_{FRET}, respectively. In our case, Δn is equal to the change in the concentration of P-CheY-YFP/CheZ-CFP pairs, $[P\text{-}CheY\text{-}YFP \cdot CheZ\text{-}CFP]$; see Eq. (2). If the distribution of CFP does not change (wild-type CheZ-CFP, in our case), and the leakage of YFP fluorescence into the cyan channel is negligible, then ΔC_{FRET} is equal to the observed intensity change in the cyan channel, ΔC, from which Δn can be extracted, as shown in Fig. 4B. If redistribution of proteins is apparent, it is essential to measure these changes independently of the FRET measurement on the same cells. In our case, this could be done by exciting CheY-YFP directly (at 515 nm, as has been explained) in the presence and absence of attractant; the difference between these fluorescence distributions is defined as ΔY_{515}. Given this information, one can infer Δn by calculating ΔY_{FRET} as shown in Eq. (6) and explained below.

With excitation at 440 nm, the cyan signal represents only the CFP emission, but the yellow signal represents three contributions:

i. Emission from YFP due to FRET.
ii. Emission from YFP due to direct excitation of the YFP fluorophore.
iii. Emission from CFP due to spectral overlap (in our case, 30% of the cyan emission intensity).

Changes in FRET affect contributions (i) and (iii) by an amount defined as ΔY_i and ΔY_{iii}, respectively. Even in the absence of FRET, redistribution of YFP during a response would yield a change in the yellow emission, due to contribution (ii), by an amount defined as ΔY_{ii}. Given the properties of the setup (defined previously), ΔY_{FRET} is given by:

$$\Delta Y_{FRET} = \Delta Y_{tot} - \Delta Y_{iii} - \Delta Y_{ii} = \Delta Y_{tot} - 0.3 \cdot \Delta C - \beta \cdot \Delta Y_{515} \qquad (6)$$

where ΔY_{tot} is the total change in the yellow channel, ΔC is the change in the cyan channel, and ΔY_{515} is the change observed when only YFP is excited at 515 nm. Both ΔY_{ii} and ΔY_{515} represent the redistribution of YFP during a response and thus ought to be proportional. The proportionality factor, β, is given by the relative intensity between the yellow emissions due to the excitation at 440 nm and that at 515nm, defined as Y_{440} and Y_{515} respectively; thus, $\beta = Y_{440}/Y_{515}$. Y_{440} was extracted from the images taken in attractant with 440 nm excitation light during the FRET measurements and after subtracting the CFP contribution; in our case, the presence of attractant eliminates the FRET contributions. Y_{515} was directly extracted from the corresponding image in attractant.

Images were analyzed using MATLAB (MathWorks, Natick, MA). Cells were considered for in-depth analysis only if they met the following criteria: the cyan and yellow images did not move during the experiment, fluorescence changes were consistent throughout a series of stimulations, and fluorescence changes were opposite for addition and removal of attractant and opposite in the yellow and cyan channels. The corrections embodied in Eq. (6) were done after alignment of the two images.

BRET Measurement of the Interaction Between YFP-CheY and -CheZ-RLUC in a Population of Bacteria Swimming in a Cuvette

Overview

The excitation light required for the FRET measurements described previously (especially the intense light required for single-cell work) can cause problems. These include damage to cells from photodynamic effects (generation of singlet oxygen), bleaching of the donor fluorophore, and high measurement background with samples that are autofluorescent or turbid. In experiments where these effects become prohibitive, bioluminescence resonance energy transfer (BRET) can provide a viable alternative to FRET. BRET depends upon the same principles of Förster resonance energy transfer as FRET, but the donor fluorophore is replaced with a bioluminescent luciferase enzyme, which catalyzes the light-producing oxidation reaction of a luciferin substrate. Many bioluminescent luciferases exist in nature, but a particularly convenient choice is *Renilla* luciferase (RLUC), which has an emission spectrum very similar to that of CFP. Its corresponding luciferin substrate, coelenterazine, is a small hydrophobic molecule that is permeable to the plasma membrane (Xu *et al.*, 1999, 2003).

FRET measurement systems designed for CFP/YFP FRET can thus be easily modified for RLUC/YFP BRET by simply replacing the CFP of the donor fusion protein with RLUC, adding coelenterazine at the beginning of the experiment, and aerating the sample.

To obtain the best possible BRET efficiency for measurements, we tested all four combinations of N- and C-terminal fusions between RLUC-CheZ, CheZ-RLUC, YFP-CheY, and CheY-YFP. Since fusions between CheY and YFP were available from previous FRET studies (Sourjik and Berg, 2000, 2002b), we only needed to construct the N- and C-terminal fusions between CheZ and RLUC. *Rluc* (the gift of C. H. Johnson, Vanderbilt University, Nashville, TN) was fused to *cheZ* and cloned into pBAD33 (Guzman *et al.*, 1995) under an arabinose-inducible promoter, by methods similar to those described for fusions between *ecfp* and *cheZ*. Both N- and C-terminal fusions were able to restore swarming in a *cheZ*-deleted strain, but the C/N combination (CheZ-RLUC/YFP-CheY) showed the largest BRET response upon addition of a saturating dose of MeAsp.

For the bacterial chemotaxis system, BRET proved particularly useful in measuring kinase activity in suspensions of swimming cells. Specifically, we were interested in determining whether mechanical perturbations to motor rotation are fed back in any way to kinase activity. A convenient way to apply such perturbations is to "jam" flagellar bundles by adding a bivalent anti-filament antibody (DiPierro and Doetsch, 1968; Greenbury and Moore, 1966; Meister *et al.*, 1987). However, suspensions of bacterial cells are turbid, due to the strong scattering properties of the cell itself (Wyatt, 1968), and FRET measurements in such suspensions are complicated by scattered light. This led us to develop a BRET-based assay for monitoring the kinase activity in cells suspended in a cuvette. Sample preparation and measurements follow the same basic principles as FRET, described previously, but with the notable additional requirement that coelenterazine and air be added to the sample.

A point of concern in switching from FRET to BRET is that maximum achievable emission levels are much lower for bioluminescence than for fluorescence. This is because the bioluminescence intensity per luciferase is limited by the turnover rate of the enzymatic reaction, each oxy-luciferin product producing, at most, one photon. The catalytic rate constant measured for RLUC is $\sim 2\ \mathrm{s}^{-1}$ or less (Hart *et al.*, 1979), so the emission intensity from each energy-transfer pair is much lower in BRET than in FRET. In general, therefore, a larger number of cells is required for experiments with BRET than with FRET. Since bioluminescence is sensitive to temperature, oxygen and salt concentrations, as well as pH, it is desirable to hold these parameters constant during the course of measurement.

With these concerns in mind, we developed a cuvette-based BRET measurement system, which does not require a microscope, but instead simply places emission filters and photomultiplier tubes (PMTs) in close proximity to the sample within a light-tight box. This apparatus allowed us to monitor changes in the kinase activity of swimming populations. The dose-response curves obtained for the attractant MeAsp closely matched

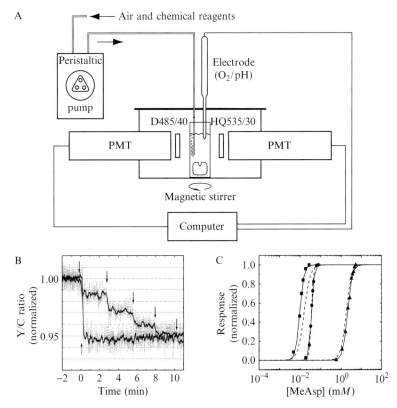

FIG. 5. BRET measurement of chemotactic signaling in swimming bacterial populations. (A) BRET measurement apparatus. Two PMTs collected light through bandpass filters (D485/40 and HQ535/30 for the cyan and yellow channels, respectively) mounted between the photocathodes and the cuvette, and output signals were analyzed online by a computer. A peristaltic pump provided constant aeration of the sample, as well as a convenient means of delivering other reagents required for experiments. In some experiments, an optional electrode for monitoring oxygen or pH was inserted into the cuvette through a hole in the lid of the box, which otherwise remained sealed. (B) Time course of a dose-response measurement to the attractant MeAsp in adaptation-deficient VS151 cells [*tsr::Tn5–1a* Δ(*cheR cheB cheY cheZ*)]. Superimposed is output from another experiment in which a single, saturating dose of attractant was added. Times at which attractant steps were applied are indicated by solid arrows for

those obtained in FRET, and no feedback loop was found that relays information from the flagellar motors to the kinase in the *E. coli* chemotaxis system (Shimizu *et al.*, 2006).

The BRET Apparatus and Measurement Protocol

A schematic illustration of the experimental apparatus is shown in Fig. 5A. Bioluminescence from a stirred suspension of swimming cells in a cuvette (1-cm path length) was measured by two photomultipliers (13-mm dia., R647P, Hamamatsu), each mounted with its photocathode ~1 cm away from the edge of the cuvette. Incoming light for each PMT passed through a 1-cm diameter aperture and a bandpass filter mounted between the photocathode and the cuvette (for the cyan and yellow channels, respectively, D485/40 and HQ535/30; Chroma Technology). The suspension of cells in the cuvette (2 ml) was stirred continuously at 300 rpm by a Teflon centrifugal stirrer (P-73, NSG Precision Cells, Farmingdale, NY) spinning just below the line of sight of the photomultipliers, driven by a synchronous motor and a small horseshoe magnet. To avoid signal perturbations by the magnetic stirrer, photomultiplier tubes were mounted in mu-metal shields. In the final design, the temperature of the box was controlled by a Peltier system, similar to one built previously (Khan and Berg, 1983). The photomultiplier gains (light intensity-to-voltage) were set arbitrarily by using 10^8-ohm feedback resistors in the current-to-voltage circuits and setting the cathode voltage at -800 v. The photomultiplier outputs, designated C and Y, were filtered with an 8-pole low-pass Bessel filter (2-Hz cutoff frequency, 3382, Krohn-Hite), and sampled at 2 Hz by a computer data-acquisition system controlled by LabView (National Instruments). The Y/C ratio was monitored during measurements; for offline analysis, it was smoothed further with a 10-point running average.

Since bioluminescence is dependent upon an oxidation reaction, aeration of the sample is essential for obtaining sufficiently intense signals. We found that bubbling the sample with room air results in more reproducible and stable results than those obtained when bubbling with pure oxygen. Air was bubbled continuously into the cuvette at a rate of ~30 μl/s with a small

---

the dose–response experiment and by a dashed arrow for the single-step experiment. Y/C ratio values are normalized to the pre-stimulus average for each experiment. (C) Dose–response curves obtained from measurements of the kind shown in (B) (solid lines) compared to those obtained for equivalent strains in FRET (dotted lines). Results for three strains known to have distinct sensitivities to [MeAsp] are shown: VS148 [*tsr::Tn5–1a* Δ(*tap cheR cheB cheY cheZ*)] (squares), VS150 [*tsr::Tn5–1a* Δ(*tap cheR cheB cheY cheZ*) *tar*(*QEQQ*)] (circles), VS179 [*tsr:: Tn5–1a* Δ*tap* Δ(*cheB cheY cheZ*)] (triangles). Response amplitudes are normalized to the maximal value for each experiment. Adapted from Figs. 1 and 4 of Shimizu *et al.* (2006).

peristaltic pump (RA034, Holter, Bridgeport, PA, but no longer made: 3 pinions on a 4.4-cm dia. circle rotating 33 rpm, stretch-occluding 1 mm i.d. by 2.2 mm o.d. silicone tubing, replaced daily) connected via 22-gauge stainless-steel tubing to 0.58-mm i.d. polyethylene tubing. The feed inside the cuvette also was 22-gauge stainless-steel tubing. Coelenterazine (10110-1, Biotium, Hayward, CA) was resuspended from freeze-dried stocks in a small amount of ethanol and then diluted in buffer to yield a concentrated solution (250 μM) delivered to the cuvette at the beginning of each experiment to initiate the bioluminescence reaction. The final concentration of coelenterazine in the cell suspension was 7.5 μM. This and other reagent solutions were pumped into the cuvette by dipping the input line into a droplet of the requisite volume pipetted onto a clean piece of parafilm. Each such aliquot was chased by an additional 30 to 60 μl of buffer to rinse the tubing.

Preparation of Cells

Overnight and day cultures are prepared by an identical procedure to that for population FRET, described previously. To determine an optimal cell density, we measured RLUC emission intensities at a variety of densities and found that an OD_{600} of $\sim$0.47, corresponding to a concentration of $\sim$10^8 cells/ml gives optimal emission at room temperature. Motility medium for cell suspensions was also the same as that used for FRET. We note that the addition to this medium of NaCl to 67 mM resulted in a sudden and large ($\sim$50%) decrease in RLUC emission. Media with high concentrations of NaCl are therefore not advisable for BRET measurements.

Stability of Luminescence Intensity and Spectrum

RLUC emission intensity typically peaked immediately and then decreased rapidly over the first $\sim$15 min after addition of coelenterazine, but then leveled off. The Y/C ratio changed by a smaller amount, reaching steady state in about 30 min at room temperature (in less time at higher temperatures). Over the course of the measurements, an increase in luminescence could sometimes be observed upon addition of certain reagents (e.g., serine) or changes in temperature, so the intensity decay was not necessarily monotonic; however, the Y/C ratio remained nearly constant after its initial relaxation to a steady-state value and provided a robust readout of BRET efficiency.

Remarkably, we found that this stability of the Y/C ratio persists over much longer times, on the order of hours. For example, we could measure nearly invariant responses to the attractant serine over periods greater than 4 h (Shimizu *et al.*, 2006). Although chemotactic responses occur on time scales much faster than this, the long-time stability of the BRET system might be useful for real-time monitoring of slower processes in the cell that involve, for example, changes in gene expression.

Dose-Response Curves Measured with BRET

In adaptation-deficient cells deleted for the *cheR* and *cheB* genes, we can apply stepwise concentration increments over time to obtain a complete dose–response curve in a single BRET experiment (Fig. 5B). We find that these dose–response curves give nearly identical $K_{1/2}$ values to those determined for the same background strain using FRET (Fig. 5C). Unlike the FRET experiments described previously, where the attachment of cells to a cover slip mounted on a flow cell allows easy addition and removal of stimuli, we cannot easily remove the reagents delivered to the cuvette in the BRET apparatus. It is, therefore, difficult to measure accurate dose–response curves for cells that adapt. In principle, one could still measure dose responses from such samples by conducting many experiments in which only one stimulus is applied, but this would be much more laborious than setting up an equivalent FRET measurement in a flow cell.

Comparison of Different Approaches and Application to Other Two-Component Systems

We have described three methods based on FRET and BRET for monitoring *in vivo* the activity of the two-component signal transduction pathway of bacterial chemotaxis. All depend on the same basic principles, but the specific design of each confers advantages and disadvantages that make them suitable for certain applications and less so for others. The first, population FRET performed on a microscope cover slip, gives the highest sensitivity and temporal resolution (down to the ms-range) in measuring the averaged behavior of populations. Because we collect fluorescence from many cells with sensitive detectors, we can use very low intensities of excitation light that strongly reduces photobleaching and photodamage of our samples while achieving quite high signal-to-noise ratios. But as a natural consequence of population averaging, information about individual cell behavior is lost. The second, single-cell FRET imaging, overcomes this shortcoming of population FRET, and can also provide a degree of intra-cellular spatial resolution. However, this requires much stronger excitation light as well as longer integration times, so the achievable temporal resolution and the duration over which measurements can be made are both decreased. Finally, BRET has the advantage that no excitation light is required at all, making it possible to measure signaling in turbid, strongly autofluorescent or photosensitive samples. Problems arising from photo-bleaching are also effectively eliminated, and when measurements are carried out in a cuvette, as we have done here, there is also the advantage of relatively simple instrumentation. But the achievable signal levels per cell are, in general, much lower in BRET, so signals must be collected from an even larger number of cells than in population FRET. In addition, we found that the chemical nature of bioluminescent light production required more care in controlling environmental variables and longer times for signals to stabilize.

Effectively applying FRET/BRET to studies of *in vivo* signaling depends critically on identifying the appropriate internal controls. In population FRET/BRET, monitoring changes in the ratio of donor/acceptor emissions, rather than the raw channel intensities, eliminated complications arising from light intensity fluctuations and sample-to-sample variations in cell density. In addition for these population measurements, it was important that we could simultaneously determine the value of donor/acceptor ratios in the absence of energy transfer (by attractant saturation or bleaching), another sample-dependent but critical parameter in converting the measured ratio changes back to units of energy transfer. In achieving spatial resolution in single-cell FRET, the coupling of CheY-YFP movement to FRET changes required the development of an additional control, namely, the measurement of fluorescence changes under YFP-specific excitation.

The existence of the specific phosphatase CheZ in *Escherichia coli* chemotaxis is fortuitous, since using FRET/BRET to monitor its interaction with the active response regulator, P-CheY, provides a readout directly proportional to CheA kinase activity. However, most two-component signaling pathways, including the chemotaxis pathways of many other organisms, do not have a dedicated phosphatase for the response regulator (Szurmant and Ordal, 2004; Wadhams and Armitage, 2004). This does not, by any means, preclude useful application of FRET/BRET methods analogous to those described here. As long as an energy-transfer pair that gives a measurable, stimulus-dependent change in emission intensities can be identified, there are ample opportunities for obtaining important information regarding intracellular signal processing. For example, within the chemotaxis system, we have used FRET between CheY-YFP and CFP-FliM to study the motor response to P-CheY (Sourjik and Berg, 2002a), and, more recently, homo-FRET between Tsr-YFP fusions to study the mechanical response of chemoreceptors (Vaknin and Berg, 2006).

Prior to the development of the methods described here, input–output approaches to the study of bacterial chemotaxis signaling required observations of motor switching or swimming behavior. The former generally require averaging over a very large number of measurements because of the stochastic nature of the motor response to P-CheY (Block *et al.*, 1982, 1983; Segall *et al.*, 1986), whereas the latter require many assumptions in bridging the observable swimming behavior to intracellular signaling (Alon *et al.*, 1998; Khan *et al.*, 1995; Kim *et al.*, 2001). By contrast, the methods described here enable direct and real-time monitoring of protein–protein interactions and allow us to peer into the workings of intracellular signaling at an unprecedented resolution and efficiency. This has, in turn, reinvigorated efforts to develop theories and models to explain the underlying mechanisms (Albert *et al.*, 2004; Endres and Wingreen, 2006; Keymer *et al.*, 2006; Mello and Tu, 2003, 2005; Rao *et al.*, 2004; Shimizu *et al.*, 2003; Skoge *et al.*, 2006). Given the many similarities between the bacterial chemotaxis pathway and other two-component systems, it is expected that adaptations of these methods developed for the former could serve as vital tools for the latter.

References

Albert, R., Chiu, Y. W., and Othmer, H. G. (2004). Dynamic receptor team formation can explain the high signal transduction gain in *Escherichia coli. Biophys. J.* **86**, 2650–2659.

Alon, U., Camarena, L., Surette, M. G., Aguera y Arcas, B., Liu, Y., Leibler, S., and Stock, J. B. (1998). Response regulator output in bacterial chemotaxis. *EMBO J.* **17**, 4238–4248.

Banno, S., Shiomi, D., Homma, M., and Kawagishi, I. (2004). Targeting of the chemotaxis methylesterase/deamidase CheB to the polar receptor–kinase cluster in an *Escherichia coli* cell. *Mol. Microbiol.* **53**, 1051–1063.

Block, S. M., Segall, J. E., and Berg, H. C. (1982). Impulse responses in bacterial chemotaxis. *Cell* **31**, 215–226.

Block, S. M., Segall, J. E., and Berg, H. C. (1983). Adaptation kinetics in bacterial chemotaxis. *J. Bacteriol.* **154**, 312–323.

Bourret, R. B., and Stock, A. M. (2002). Molecular information processing: Lessons from bacterial chemotaxis. *J. Biol. Chem.* **277**, 9625–9628.

Bren, A., and Eisenbach, M. (2000). How signals are heard during bacterial chemotaxis: Protein–protein interactions in sensory signal propagation. *J. Bacteriol.* **182**, 6865–6873.

Cantwell, B. J., Draheim, R. R., Weart, R. B., Nguyen, C., Stewart, R. C., and Manson, M. D. (2003). CheZ phosphatase localizes to chemoreceptor patches via CheA-short. *J. Bacteriol.* **185**, 2354–2361.

DiPierro, J. M., and Doetsch, R. N. (1968). Enzymatic reversibility of flagellar immobilization. *Can. J. Microbiol.* **14**, 487–489.

Endres, R. G., and Wingreen, N. S. (2006). Precise adaptation in bacterial chemotaxis through "assistance neighborhoods." *Proc. Natl. Acad. Sci. USA* **103**, 13040–13044.

Förster, T. (1948). Zwischenmolekulare Energiewanderung und Fluoreszenz. *Ann. Phys.* **2**, 55–75.

Gestwicki, J. E., and Kiessling, L. L. (2002). Inter-receptor communication through arrays of bacterial chemoreceptors. *Nature* **415**, 81–84.

Greenbury, C. L., and Moore, D. H. (1966). The mechanism of bacterial immobilization by anti-flagellar IgG antibody. *Immunology* **11**, 617–625.

Guzman, L. M., Belin, D., Carson, M. J., and Beckwith, J. (1995). Tight regulation, modulation, and high-level expression by vectors containing the arabinose pBAD promoter. *J. Bacteriol.* **177**, 4121–4130.

Hart, R. C., Matthews, J. C., Hori, K., and Cormier, M. J. (1979). *Renilla reniformis* biolumi-nescence: Luciferase-catalyzed production of nonradiating excited states from luciferin analogues and elucidation of the excited state species involved in energy transfer to *Renilla* green fluorescent protein. *Biochemistry* **18**, 2204–2210.

Keymer, J. E., Endres, R. G., Skoge, M., Meir, Y., and Wingreen, N. S. (2006). Chemosensing in *Escherichia coli*: Two regimes of two-state receptors. *Proc. Natl. Acad. Sci. USA* **103**, 1786–1791.

Khan, S., and Berg, H. C. (1983). Isotope and thermal effects in chemiosmotic coupling to the flagellar motor of *Streptococcus*. *Cell* **32**, 913–919.

Khan, S., Spudich, J. L., McCray, J. A., and Trentham, D. R. (1995). Chemotactic signal integration in bacteria. *Proc. Natl. Acad. Sci. USA* **92**, 9757–9761.

Kim, C., Jackson, M., Lux, R., and Khan, S. (2001). Determinants of chemotactic signal amplification in *Escherichia coli*. *J. Mol. Biol.* **307**, 119–135.

Lai, R. Z., Manson, J. M., Bormans, A. F., Draheim, R. R., Nguyen, N. T., and Manson, M. D. (2005). Cooperative signaling among bacterial chemoreceptors. *Biochemistry* **44**, 14298–14307.

Li, M., and Hazelbauer, G. L. (2004). Cellular stoichiometry of the components of the chemotaxis signaling complex. *J. Bacteriol.* **186**, 3687–3694.

Maddock, J. R., and Shapiro, L. (1993). Polar location of the chemoreceptor complex in the *Escherichia coli* cell. *Science* **259**, 1717–1723.

Meister, M., Lowe, G., and Berg, H. C. (1987). The proton flux through the bacterial flagellar motor. *Cell* **49**, 643–650.

Mello, B. A., and Tu, Y. (2003). Quantitative modeling of sensitivity in bacterial chemotaxis: The role of coupling among different chemoreceptor species. *Proc. Natl. Acad. Sci. USA* **100**, 8223–8228.

Mello, B. A., and Tu, Y. (2005). An allosteric model for heterogeneous receptor complexes: Understanding bacterial chemotaxis responses to multiple stimuli. *Proc. Natl. Acad. Sci. USA* **102**, 17354–17359.

Miyawaki, A., and Tsien, R. Y. (2000). Monitoring protein conformations and interactions by fluorescence resonance energy transfer between mutants of green fluorescent protein. *Methods Enzymol.* **327**, 472–500.

Miyawaki, A., Sawano, A., and Kogure, T. (2003). Lighting up cells: Labeling proteins with fluorophores. *Nat. Cell. Biol.* **Suppl.**, S1–S7.

Parkinson, J. S., Ames, P., and Studdert, C. A. (2005). Collaborative signaling by bacterial chemoreceptors. *Curr. Opin. Microbiol.* **8**, 116–121.

Rao, C. V., Frenklach, M., and Arkin, A. P. (2004). An allosteric model for transmembrane signaling in bacterial chemotaxis. *J. Mol. Biol.* **343**, 291–303.

Segall, J. E., Block, S. M., and Berg, H. C. (1986). Temporal comparisons in bacterial chemo-taxis. *Proc. Natl. Acad. Sci. USA* **83**, 8987–8991.

Selvin, P. R. (1995). Fluorescence resonance energy transfer. *Methods Enzymol.* **246**, 300–334.

Shimizu, T. S., Aksenov, S. V., and Bray, D. (2003). A spatially extended stochastic model of the bacterial chemotaxis signaling pathway. *J. Mol. Biol.* **329**, 291–309.

Shimizu, T. S., Delalez, N., Pichler, K., and Berg, H. C. (2006). Monitoring bacterial chemo-taxis by using bioluminescence resonance energy transfer: Absence of feedback from the flagellar motors. *Proc. Natl. Acad. Sci. USA* **103**, 2093–2097.

Shiomi, D., Zhulin, I. B., Homma, M., and Kawagishi, I. (2002). Dual recognition of the bacterial chemoreceptor by chemotaxis-specific domains of the CheR methyltransferase. *J. Biol. Chem.* **277**, 42325–42333.

Skoge, M. L., Endres, R. G., and Wingreen, N. S. (2006). Receptor–receptor coupling in bacterial chemotaxis: Evidence for strongly coupled clusters. *Biophys. J.* **90**, 4317–4326.

Sourjik, V. (2004). Receptor clustering and signal processing in *E. coli* chemotaxis. *Trends Microbiol.* **12**, 569–576.

Sourjik, V., and Berg, H. C. (2000). Localization of components of the chemotaxis machinery of *Escherichia coli* using fluorescent protein fusions. *Mol. Microbiol.* **37**, 740–751.

Sourjik, V., and Berg, H. C. (2002a). Binding of the *Escherichia coli* response regulator CheY to its target measured *in vivo* by fluorescence resonance energy transfer. *Proc. Natl. Acad. Sci. USA* **99**, 12669–12674.

Sourjik, V., and Berg, H. C. (2002b). Receptor sensitivity in bacterial chemotaxis. *Proc. Natl. Acad. Sci. USA* **99**, 123–127.

Sourjik, V., and Berg, H. C. (2004). Functional interactions between receptors in bacterial chemotaxis. *Nature* **428**, 437–441.

Studdert, C. A., and Parkinson, J. S. (2004). Crosslinking snapshots of bacterial chemoreceptor squads. *Proc. Natl. Acad. Sci. USA* **101**, 2117–2122.

Szurmant, H., and Ordal, G. W. (2004). Diversity in chemotaxis mechanisms among the bacteria and archaea. *Microbiol. Mol. Biol. Rev.* **68**, 301–319.

Ting, A. Y., Kain, K. H., Klemke, R. L., and Tsien, R. Y. (2001). Genetically encoded fluorescent reporters of protein tyrosine kinase activities in living cells. *Proc. Natl. Acad. Sci. USA* **98**, 15003–15008.

Tsien, R. Y. (1998). The green fluorescent protein. *Annu. Rev. Biochem.* **67**, 509–544.

Vaknin, A., and Berg, H. C. (2004). Single-cell FRET imaging of phosphatase activity in the *Escherichia coli* chemotaxis system. *Proc. Natl. Acad. Sci. USA* **101**, 17072–17077.

Vaknin, A., and Berg, H. C. (2006). Osmotic stress mechanically perturbs chemoreceptors in *Escherichia coli*. *Proc. Natl. Acad. Sci. USA* **103**, 592–596.

Wadhams, G. H., and Armitage, J. P. (2004). Making sense of it all: Bacterial chemotaxis. *Nat. Rev. Mol. Cell. Biol.* **5**, 1024–1037.

Wouters, F. S., Verveer, P. J., and Bastiaens, P. I. (2001). Imaging biochemistry inside cells. *Trends Cell Biol.* **11**, 203–211.

Wright, S., Walia, B., Parkinson, J. S., and Khan, S. (2006). Differential activation of *Escherichia coli* chemoreceptors by blue-light stimuli. *J. Bacteriol.* **188**, 3962–3971.

Wyatt, P. J. (1968). Differential light scattering—A physical method for identifying living bacterial cells. *Applied Optics* **7**, 1879–1896.

Xu, Y., Piston, D. W., and Johnson, C. H. (1999). A bioluminescence resonance energy transfer (BRET) system: Application to interacting circadian clock proteins. *Proc. Natl. Acad. Sci. USA* **96**, 151–156.

Xu, Y., Kanauchi, A., von Arnim, A. G., Piston, D. W., and Johnson, C. H. (2003). Bioluminescence resonance energy transfer: Monitoring protein–protein interactions in living cells. *Methods Enzymol.* **360**, 289–301.

Zaccolo, M., Filippin, L., Magalhaes, P., and Pozzan, T. (2001). Heterogeneity of second messenger levels in living cells. *Novartis Found. Symp.* **239**, 85–93; discussion 93–95, 150–159.

Zhang, J., Ma, Y., Taylor, S. S., and Tsien, R. Y. (2001). Genetically encoded reporters of protein kinase A activity reveal impact of substrate tethering. *Proc. Natl. Acad. Sci. USA* **98**, 14997–15002.

Zhao, R., Collins, E. J., Bourret, R. B., and Silversmith, R. E. (2002). Structure and catalytic mechanism of the *E. coli* chemotaxis phosphatase CheZ. *Nat. Struct. Biol.* **9**, 570–575.

[18]   *In Vivo* and *In Vitro* Analysis of the *Rhodobacter sphaeroides* Chemotaxis Signaling Complexes

By STEVEN L. PORTER, GEORGE H. WADHAMS, and
JUDITH P. ARMITAGE

Abstract

This chapter describes both the *in vivo* and *in vitro* methods that have been successfully used to analyze the chemotaxis pathways of *R. sphaeroides*, showing that two operons each encode a complete chemosensory pathway with each forming into independent signaling clusters. The methods used range from *in vitro* analysis of the chemotaxis phosphorylation reactions to protein localization experiments. *In vitro* analysis using purified proteins shows a complex pattern of phosphotransfer. However, protein localization studies show that the *R. sphaeroides* chemotaxis proteins are organized into two distinct sensory clusters—one containing transmembrane receptors located at the cell poles and the other containing soluble chemoreceptors located in the cytoplasm. Signal outputs from both clusters are essential for chemotaxis. Each cluster has a dedicated chemotaxis histidine protein kinase (HPK), CheA. There are a total of eight chemotaxis response regulators in *R. sphaeroides*, six CheYs and two CheBs, and each CheA shows a different pattern of phosphotransfer to these response regulators. The spatial separation of homologous proteins may mean that reactions that happen *in vitro* do not occur *in vivo,* suggesting great care should be taken when extrapolating from purely *in vitro* data to cell physiology. The methods described in this chapter are not confined to the study of *R. sphaeroides* chemotaxis but are applicable to the study of complex two-component systems in general.

Introduction

Bacteria use a biased random swimming pattern to reach better environments for growth, moving either away from repellents or toward attractants. The chemosensory pathway of *E. coli*, described in detail in other chapters of this volume, is probably the best understood sensory system in biology. In *E. coli*, chemoeffectors bind to specific chemoreceptors clustered at the poles of the cell (Maddock and Shapiro, 1993). There are five different chemoreceptors responding to a limited number of effectors. A change in chemoeffector concentration is signaled across the membrane, regulating the activity of an associated histidine protein kinase dimer, CheA. A decrease in attractant binding to the receptor results in trans-autophosphorylation of the

METHODS IN ENZYMOLOGY, VOL. 423   0076-6879/07 $35.00
Copyright 2007, Elsevier Inc. All rights reserved.   DOI: 10.1016/S0076-6879(07)23018-6

CheA dimer (Borkovich *et al.*, 1989; Ninfa *et al.*, 1991). Two response regulators compete for binding to CheA, CheY, a single domain response regulator, and CheB, a methyl esterase activated via a response regulator domain (Hess *et al.*, 1988a,b). When phosphorylated, CheY-P has reduced affinity for CheA but increased affinity for FliM on the flagellar motor, where it induces switching in the direction of motor rotation and, hence, a change in swimming direction (Welch *et al.*, 1993). The activity of CheB increases about 100-fold on phosphorylation (Anand *et al.*, 1998). It serves, with the constitutive methyl transferase, CheR, to reset the signaling state of the chemoreceptors in a process termed adaptation (Lupas and Stock, 1989; Springer and Koshland, 1977). Signaling occurs in about 100 ms whereas adaptation takes about 1 s, bringing memory into the system. The rate of spontaneous CheY-P dephosphorylation is increased by CheZ, to allow signal termination within the time course required for spatial gradient sensing (Lukat and Stock, 1993).

For many years, it has been apparent that chemotaxis in the photoheterotrophic species *Rhodobacter sphaeroides* is somewhat different from that of *E. coli*. *R. sphaeroides* is a photoheterotrophic bacterium, a member of the alpha proteobacteria. Sequencing of the *R. sphaeroides* genome revealed three loci, each encoding a complete putative chemosensory pathway (Mackenzie *et al.*, 2001). Our research over the past decade has centered on identifying why a bacterium with a single, unidirectional stop/start flagellum should apparently require three related pathways to control the stopping frequency of that flagellum (Armitage and Macnab, 1987). Two of these putative pathways have been shown to be essential for normal chemo (and photo) sensory behavior (Hamblin *et al.*, 1997; Porter *et al.*, 2002; Shah *et al.*, 2000), although the level of their expression changes under different growth conditions (Martin *et al.*, 2006; Shah *et al.*, 2000).

Our research has combined *in vitro* analysis of the phosphotransfer behavior of the different purified protein components with *in vivo* studies examining both the cellular localization of these proteins and the effects of mutating the genes encoding these proteins. The *in vitro* analysis has identified similarities and differences in phosphotransfer kinetics between components, and also the fact that some but not all components can cross-talk, despite sequence similarities (Porter and Armitage, 2002, 2004; Porter *et al.*, 2006). Although there is cross-talk between some phosphotransfer proteins in the test-tube, these proteins did not complement deletions of apparently equivalent proteins from the other loci *in vivo*. One possible reason for this became clear when we examined the localization of the proteins of two pathways; one pathway localizes to the cell poles with the transmembrane chemoreceptors and the other pathway localizes to a cytoplasmic midcell position with the soluble chemoreceptors (Martin *et al.*, 2003; Wadhams *et al.*, 2000, 2002, 2003).

This chapter illustrates the techniques we used to identify the pattern of cross phosphorylation, and also highlights why relying on this alone may give a false model of the chemosensory pathways (Fig. 1). Combining the different approaches outlined here has identified not only what can phosphotransfer, but what phosphotransfer reactions probably do occur *in vivo* and where these interactions happen. The chemosensory pathway is just one of multiple two-component pathways in bacteria, some species having over 100 putative pathways (Ashby, 2004; Galperin, 2006). Dissecting these interaction networks is a major challenge for bacteriology. The similarities among the different systems mean that the techniques described here are generally applicable to analyzing the phosphorylation and localization of two other component pathways.

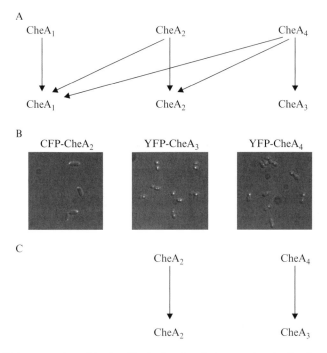

Fig. 1. Reinterpretation of *in vitro* CheA phosphorylation reactions in *R. sphaeroides* after *in vivo* imaging. (A) Phosphorylation reactions that have been detected *in vitro* using purified proteins. (B) The localization of $CheA_2$, $CheA_3$, and $CheA_4$. $CheA_1$ does not appear to be expressed, so its localization could not be determined. (C) The auto-/heterophosphorylation reactions that are likely to occur *in vivo*. These reactions are the same as those in panel (A) except that all reactions involving $CheA_1$ have been removed because $CheA_1$ is not expressed and is not required for chemotaxis; also, the phosphorylation of $CheA_2$ by $CheA_4$ has been removed because these proteins localize to different regions of the cell. (See color insert.)

In Vitro Analysis of Signaling by the Kinase Cluster

R. sphaeroides has loci that could encode the proteins for three different chemosensory pathways (Mackenzie *et al.*, 2001). Sequence comparisons suggest there are genes for 6 putative CheY, 2 CheB, 3 CheR, 4 CheA, but no CheZ proteins. Some have high levels of homology while others show variations on the *E. coli* theme. For example, $CheA_1$ and $CheA_2$ are very similar to the *E. coli* CheA, but $CheA_3$ and $CheA_4$ lack some of the usual domains (Porter and Armitage, 2004). To identify whether the HPKs and RRs of the putative *R. sphaeroides* chemosensory pathway show similar phosphotransfer kinetics to the *E. coli* pathway and how the proteins might relate in a signaling pathway, they were all purified and their phosphorylation kinetics measured.

Protein Purification

In vitro analysis of the chemotaxis signaling pathway requires modest quantities of purified proteins. These can be produced by overexpressing the proteins in *E. coli*, followed by purification. Histidine protein kinases and response regulators have been successfully purified using His and GST tagging strategies (Jimenez-Pearson *et al.*, 2005; Porter and Armitage, 2002; Sourjik and Schmitt, 1998).

A complete description of protein purification methodologies is beyond the scope of this chapter. Some HPKs are transmembrane proteins and may prove difficult to purify in sufficient quantity for analysis. The following protocol works for many of the cytoplasmic *R. sphaeroides* chemotaxis proteins when overexpressed using the His-tagging pQE expression vectors (Qiagen) and may be a useful starting point for designing purification strategies for related proteins from other species (Porter and Armitage, 2002):

1. Transform the pQE-based expression plasmid into the *E. coli* expression strain *M*15pREP4 (Qiagen). The transformed cells are resistant to both ampicillin (100 μg/ml) and kanamycin (25 μg/ml); both antibiotics should be used throughout the induction.

2. Grow a 25 ml overnight starter culture at 37° in Luria broth containing appropriate antibiotics.

3. Dilute overnight culture in 500 ml of 2YT medium containing appropriate antibiotics. Grow cells at 37° with shaking (225 rpm) to an $OD_{600nm} = 0.8$.

4. Add IPTG to the cultures to a final concentration of 100 μg/ml. Incubate cultures for 20 h at 18° with shaking at 225 rpm.

5. Harvest cells by centrifugation at 6000g for 15 min. Resuspend the cell pellet in 30 ml of lysis buffer (10% glycerol, 50 mM Tris-HCl, 150 mM

NaCl, 10 mM imidazole, 1 mM DTT, pH 8). Freeze cells at $-20°$. For the *R. sphaeroides* chemotaxis proteins, frozen resuspended cell pellets can be stored for up to 6 months without reduction in yield of purified protein.

6. Thaw the resuspended cell pellet and lyze cells by sonication on ice. Cell lysis can be followed by microscopic examination. For the Vibracell sonicator (Sonics & Materials Incorporated), 6×20 s full power bursts are sufficient for full lysis.

7. Centrifuge the lysate at 35000g for 15 min to remove insoluble material and cell debris.

8. Pour 1 ml of Ni-NTA Agarose slurry into a chromatography column (for example, catalog number 731–1550, Bio-Rad). Allow the column to settle for at least 10 min.

9. Equilibrate the nickel column with lysis buffer.

10. Apply the cleared lysate to the equilibrated nickel column.

11. Wash the column with at least 60 ml of lysis buffer.

12. Elute the protein using lysis buffer containing 500 mM imidazole. Collect 1 ml fractions of the eluate. Protein usually elutes in the first three fractions.

13. Assay the fractions for protein using, for example, a Bradford assay (BioRad).

14. Pool the fractions containing significant quantities of protein and dialyze overnight against lysis buffer lacking imidazole.

15. Assess protein purity by SDS-PAGE. Single-step His-tag purification is usually sufficient for phosphorylation assays. However, if significant levels of contaminating proteins are present, these can be removed by other standard protein purification techniques, for example, gel filtration or ion exchange chromatography.

16. Measure the protein concentration. Use a protein concentrator to increase the concentration if necessary (for typical phosphorylation/phosphotransfer assays, protein concentrations in excess of 20 μM are adequate).

17. Store the protein at $-20°$ in single-use aliquots. Although freezing does not affect the activity of the *R. sphaeroides* chemotaxis proteins (frozen proteins retain their activity for in excess of 1 year), it is possible that freezing may affect the activity of other proteins. Short-term storage at $4°$ is an alternative.

Measuring Kinase Activity

Autokinase activity can be detected by incubating the putative histidine protein kinase with $[\gamma\text{-}^{32}P]$ ATP for various time periods; this allows the kinase to autophosphorylate, which, depending on the kinase, can require time-periods ranging from seconds to hours. The reaction mixtures are then analyzed on SDS-PAGE gels. If autophosphorylation has occurred, then radioactive bands at a size corresponding to the protein of interest will be detectable by either autoradiography or phosphorimaging.

If no radiolabeling of the kinase has occurred, it could be due to a lack of kinase activity under the assay conditions. Kinase activity is usually controlled by a sensory domain; the inclusion/exclusion of the stimulus sensed by the sensory domain in the assay mixture may be necessary to increase kinase activity to a detectable level. Another possible reason for an apparent lack of kinase activity is that the kinase is a hybrid kinase (contains both kinase and receiver domains). It is possible that the kinase autophosphorylates slowly but then rapidly phosphotransfers to the receiver domain, which then rapidly dephosphorylates; in such a scenario, no phosphoprotein would accumulate so it would appear that the protein has no kinase activity (Rasmussen *et al.*, 2006). In such cases, there are two options: (1) Perform an ATPase assay on the protein (Ninfa *et al.*, 1991). Even though no phosphoprotein accumulates, there will still be conversion of ATP to ADP if the putative hybrid kinase has kinase activity; (2) Inactivate the receiver domain—the phosphorylatable aspartate within the receiver domain can be mutated to alanine to prevent phosphotransfer to the receiver domain. Any phosphoryl groups should then remain on the histidine phosphorylation site of the hybrid kinase (Rasmussen *et al.*, 2005).

The following protocol has been used to analyze the autophosphorylation of *R. sphaeroides* CheA$_1$ and CheA$_2$ (Porter and Armitage, 2002); it should form a suitable starting protocol for any histidine protein kinase, which can then be optimized to the system under study.

1. Dilute the HPK to a final concentration of 5 μM in TGMNKD buffer (final concentrations: 50 mM Tris HCl, 10% (v/v) glycerol, 5 mM MgCl$_2$, 150 mM NaCl, 50 mM KCl, 1 mM DTT, pH 8.0).

2. Incubate the reaction mixtures at 20° for 1 h.

3. Add 2 mM [γ-^{32}P] ATP (specific activity 14.8 GBq mmol^{-1}; this specific activity is prepared by mixing unlabeled ATP with a commercially available [γ-^{32}P] labeled ATP of higher specific activity, e.g., 220 TBq mmol^{-1}).

4. Following addition of ATP, sample the reaction mixtures at intervals (15, 30, 90, 240, 480, 1800, and 3600 s are good starting points but these can be varied to suit the rate of the kinase) by removing a 10 μl aliquot of the reaction mixture and mixing with 5 μl of the quenching reagent (3× SDS-PAGE loading dye (7.5% (w/v) SDS, 90 mM EDTA, 37.5 mM Tris HCl, 37.5% glycerol, 3% (v/v) β-mercaptoethanol, 0.05% (w/v) bromophenol blue, pH 6.8). Store the quenched samples on ice until all the samples have been collected.

5. Heat the quenched samples to 65° for 30 s, then return to ice. Phosphorylated histidine and aspartate residues are heat labile. Do not boil samples prior to SDS/PAGE.

6. Load samples onto an SDS-PAGE gel. To minimize hydrolysis of phosphoproteins, run gels at 500 V for as short a time period as possible at 4° (preferably less than 1 h). Use prechilled SDS-PAGE running buffer and a refrigerating circulating water bath to cool the gel core during the electrophoresis run.

7. To keep radioactive waste contained, it is preferable not to run the dye front off the end of the gel. Following the electrophoresis run, cut the dye front off the gel and rinse the gel briefly in deionized water. Sandwich the gel between two sheets of OHP (overhead projector) acetate.

8. Serially dilute $[\gamma\text{-}^{32}P]$ ATP (specific activity 14.8 GBq mmol^{-1}) in water to generate standards ranging in concentration from 0.05 to 5 μM. Spot 10 μl of each of these standards onto a strip of 3 MM Whatman paper. Allow these to dry and sandwich between two sheets of OHP acetate.

9. Expose the gel and the standard strip to a phosphorimaging screen or X-ray film. Due to their linear responses, phosphorimaging screens are recommended for quantification purposes; a 30 min exposure time is sufficient for detecting CheA kinase activity.

10. Bands corresponding to phosphorylated protein should be visible on either the X-ray film or the phosphorimage. To quantify the extent of phosphorylation, it is necessary to use image analysis software capable of outputting total pixel intensity for a defined region (e.g., ImageQuant TL, Amersham). Using this software, the regions containing the standards and the phosphoprotein bands are selected and the total pixel intensity measured. Following background subtraction, the total band intensity can be converted to amount of phosphorylated protein by comparison with the standard curve.

11. The autophosphorylation reaction should follow pseudofirst order rate kinetics. Plot the timecourse ([HPK-P] versus time) and fit to the following pseudofirst order rate equation using a mathematical analysis package (e.g., Origin, Microcal): $[\text{HPK-P}] = [\text{HPK-P}]_{\text{final}}\ (1 - e^{-k\text{obs}*t})$ where k_{obs} is the pseudofirst order rate constant, and $[\text{HPK-P}]_{\text{final}}$ is the final (maximal) concentration of HPK-P. Both k_{obs} and $[\text{HPK-P}]_{\text{final}}$ should be allowed to vary during the fitting procedure.

12. Repeat the entire experiment at different [ATP] ranging from 0.01 mM to 2 mM. This allows k_{obs} to be determined for a range of ATP concentrations. Plot a graph of k_{obs} vs [ATP]. Fit this curve to the following equation: $k_{\text{obs}} = k_{\text{cat}}[\text{ATP}]/(K_{\text{m}} + [\text{ATP}])$ where k_{cat} is the turnover rate for the HPK and K_{m} is the Michaelis constant for ATP. Both k_{cat} and K_{m} should be allowed to vary during the fitting procedure.

13. HPKs are functional only as dimers. At 5 μM, CheAs are almost completely dimeric; other HPKs may be different. It is advisable to measure k_{cat} under a range of different [HPK]; k_{cat} will increase as [HPK] is raised because a greater fraction of the HPK present will be dimerized. However, once the HPK concentration is high enough that all of HPK is dimerized, then no further increases in k_{cat} will be seen as the [HPK] is increased.

This protocol is flexible and will need to be modified to take into account the HPK under study; the incubation temperature can be varied and the assay buffer can be modified to include/exclude ligands that affect kinase activity. Mg^{2+} is essential for all the activity of all known HPKs, so it should always be included in any assay buffer.

Heterophosphorylation Reactions

Typically, HPKs are homodimeric proteins in which the kinase domain of one monomer phosphorylates a histidine residue within the other monomer. This process is referred to as autophosphorylation. Heterophosphorylation occurs when the kinase domain of one protein phosphorylates a histidine residue within another non-identical protein.

A variation of the measuring kinase activity protocol was used to analyze heterophosphorylation between the atypical CheAs, $CheA_3$ and $CheA_4$, from *R. sphaeroides*. Both proteins are missing some of the domains normally found in CheAs and, using the previous protocol, it was shown that neither protein could autophosphorylate. However, when a mixture of $CheA_3$ and $CheA_4$ was analyzed using the protocol, $CheA_3$ became phosphorylated, indicating that $CheA_4$ is able to (hetero)phosphorylate $CheA_3$ (Porter and Armitage, 2004).

A further variation of the measuring kinase activity protocol was used to examine potential heterophosphorylation reactions among the other *R. sphaeroides* CheAs, for example, the phosphorylation of $CheA_1$ by $CheA_2$. However, this analysis was complicated by the ability of $CheA_1$ and $CheA_2$ to autophosphorylate. For this reason, a $CheA_1$ mutant defective in autophosphorylation (containing the G501K mutation within its kinase domain) was used as the phosphoacceptor and a $CheA_2$ mutant (H46Q) lacking a phosphorylatable histidine residue as the phosphodonor. Since neither $CheA_1$(G501K) nor $CheA_2$(H46Q) can autophosphorylate when incubated alone, the phosphorylation of $CheA_1$(G501K) that was observed when these proteins were mixed can be assumed to be the result of the kinase domain of $CheA_2$(H46Q) phosphorylating $CheA_1$(G501K) (Porter and Armitage, 2004). Similar heterophosphorylation experiments on the other CheAs revealed the phosphorylation network shown in Fig. 1A.

Phosphotransfer from HPKs to Response Regulators

Phosphotransfer from a phosphorylated HPK to response regulators can be detected by mixing prephosphorylated ^{32}P labeled HPK with the response regulators. If phosphotransfer occurs, then radioactive label will be transferred from the kinase to the response regulator; this can be followed using SDS-PAGE followed by autoradiography or phosphorimaging.

The prephosphorylation of the HPK is achieved by incubating the HPK with $[\gamma\text{-}^{32}\text{P}]$ ATP for a period of time that allows almost complete phosphorylation. After this time, some investigators prefer to remove the unincorporated ATP from the reaction mixture prior to use in phosphotransfer assays; this prevents rephosphorylation of the HPK following phosphotransfer to response regulator and ensures that each phosphorylated kinase can only participate in a single phosphotransfer reaction. Alternatively, the unincorporated ATP can be left in the reaction mixture. However, this means that, following the initial phosphotransfer to the response regulators, the kinase can rephosphorylate and thus participate in further phosphotransfer reactions. If the ATP is left in the phosphotransfer reactions, then it is necessary to perform a control reaction in which the HPK is omitted from the reaction to demonstrate that the response regulator does not autophosphorylate.

Some HPKs (e.g., NtrB) have phosphatase activity on their cognate response regulators. This phosphatase activity can prevent the accumulation of phosphorylated response regulator in the phosphotransfer assay. This may result in no phosphorylated response regulator being detected; however, providing that the excess $[\gamma\text{-}^{32}\text{P}]$ ATP has been removed from the reaction, a reduction in the level of phosphorylated HPK should still be measurable if phosphotransfer has occurred.

The following protocol was designed for the analysis of phosphotransfer among the four *R. sphaeroides* CheAs and the six CheYs, and should be adaptable to the study of any two-component system:

1. Follow steps 1 through 3 of the measuring kinase activity protocol.

2. Following addition of $[\gamma\text{-}^{32}\text{P}]$ ATP to the kinase, incubate the reaction mixture for a period of time that allows almost complete phosphorylation of the kinase (30 min is sufficient for CheA).

3. Optional: Remove the unincorporated ATP from the reaction using a centrifugal protein concentrator; the ATP will pass through the membrane, while the protein will be retained. The concentrated protein solution should be diluted with 1 ml of TGMNKD buffer and concentrated; repeat at least twice to ensure the removal of the ATP. Protein recovery from this procedure is usually less than 100%; therefore, it is necessary to determine the concentration of the recovered protein using, for example, a Bradford assay (BioRad).

4. Immediately before addition of response regulator, remove a 10 μl aliquot of the reaction mixture and quench by mixing with 5 μl of the quenching reagent (3× SDS-PAGE loading dye (7.5% (w/v) SDS, 90 mM EDTA, 37.5 mM Tris HCl, 37.5% glycerol, 3% (v/v) β-mercaptoethanol, 0.05% (w/v) bromophenol blue, pH 6.8). Store the quenched samples on ice until all of the samples have been collected.

5. Add response regulator to a final concentration of 10 μM (the concentration of kinase and response regulator can be varied in this assay if, for example, their cellular concentrations have been determined).

6. Following addition of the response regulator, sample the reaction mixtures at intervals (15, 30, 90, 240, 480, 1800, and 3600 s are good starting points but these can be varied to suit the rate of the phosphotransfer) by removing a 10 μl aliquot of the reaction mixture and mixing with 5 μl of the quenching reagent. Store the quenched samples on ice until all of the samples have been collected.

7. Process the samples and analyze them by SDS-PAGE followed by autoradiography/phosphorimaging, as described in steps 5 through 10 of the measuring kinase activity protocol. If phosphotransfer has occurred, radioactive bands corresponding to the response regulators should be visible, which will be accompanied by a reduction in the amount of phosphorylated HPK (an example is shown in Fig. 2A).

8. Plot the concentration of HPK-P and RR-P at each time point to obtain the timecourse (see Fig. 2B for an example). This timecourse is determined by three processes: the rate of autophosphorylation of the HPK (if ATP was left in the reaction mixture), the rate of phosphotransfer from the HPK-P to the RR, and the rate of dephosphorylation of the RR-P. Estimation of the RR-P dephosphorylation rate constant has been described in detail elsewhere (Porter and Armitage, 2002). Obtaining estimates for the rate constants describing the phosphotransfer reaction requires further experimentation, curve fitting, and mathematical modeling (Li *et al.*, 1995; Stewart, 1997; Stewart *et al.*, 2000).

Response Regulator Phosphatase Assays

In addition to their intrinsic autophosphatase activity, the dephosphorylation of some response regulators is augmented by additional proteins, for example, CheZ, CheC, FliY, CheX, and RapA (Motaleb *et al.*, 2005; Szurmant *et al.*, 2004). The activity of these phosphatase proteins can be detected by performing parallel phosphotransfer experiments between the HPK and its cognate response regulator (as has been described; removal of the ATP following phosphorylation of the HPK is recommended) in the

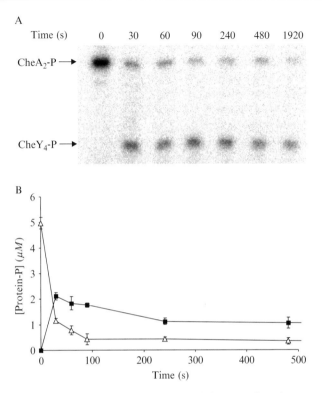

FIG. 2. Example of a phosphotransfer timecourse. (A) CheA$_2$ (5 μM) was preincubated together with 0.5 mM [γ-^{32}P] ATP for 30 min. CheY$_4$ (10 μM) was then added to the reaction mixture (the final volume was 100 μl). Ten microliter samples were removed at the indicated time-points and quenched immediately by addition of 5 μl of 3$\times$ SDS/EDTA loading dye. The quenched samples were analyzed by SDS-PAGE and detected by phosphorimaging. (B) The experiment shown in Panel (A) was repeated three times and the radioactive band intensity quantified by phosphorimaging. Graphs show the mean concentrations of the phosphoproteins $\pm$ the standard error of the mean. $\triangle$, CheA$_2$-P; $\blacksquare$, CheY$_4$-P.

presence and absence of the putative phosphatase. Phosphatase activity is suggested where there are reduced levels of RR-P in the presence of the putative phosphatase. However, phosphatase activity is not the only explanation for reduced levels of RR-P; an alternative possibility is that the putative phosphatase reduces the rate of phosphotransfer between the HPK and the RR. These two explanations are easily distinguished; if the rate of phosphotransfer were reduced, there would be an increase in HPK-P in the reactions containing the putative phosphatase, whereas if the putative phosphatase were a phosphatase for the RR, then HPK-P levels would either decrease or remain constant.

Site-Directed Mutagenesis of Kinases/RR to Confirm Role in Signaling

While phenotypic analysis of a gene deletion/disruption mutant for a HPK or RR can demonstrate that the encoded protein is required for a particular cellular process, this gives no information on the role of phosphorylation in the control of the pathway. For example, CheA$_2$ is essential for chemotaxis in *R. sphaeroides* (Martin *et al.*, 2001), and protein localization studies have shown that CheA$_2$ is necessary for the polar localization of CheW$_2$, CheW$_3$, and the transmembrane chemoreceptors (Wadhams *et al.*, 2003, 2005). Is CheA$_2$ essential for chemotaxis solely because of its role in localizing other proteins or is its phosphosignaling role also important? Deletion analysis could not answer this question. Instead, a CheA$_2$ mutant was required that would support the localization of the other chemotaxis proteins but be defective in phosphorylation; one candidate was the phosphorylation site mutant CheA$_2$(H46Q). This mutant was unable to support chemotaxis but resulted in correct localization of associated proteins, indicating that phosphosignaling from CheA$_2$ is essential for chemotaxis (Porter and Armitage, 2004).

1. Identify putative phosphorylation sites by sequence alignment with other HPKs/RRs.

2. Mutate these putative phosphorylation sites. For response regulators, substitution of the phosphorylatable aspartate residue with alanine is recommended. Several cases of alternative site phosphorylation have been documented where the phosphorylatable aspartate was replaced with asparagine (Appleby and Bourret, 1999; Moore *et al.*, 1993; Porter *et al.*, 2006; Reyrat *et al.*, 1994).

3. Purify the mutant HPKs/RRs and confirm using *in vitro* phosphorylation assays (described previously) that the phosphorylation site has been removed.

4. Compare the functionality of the mutant HPKs/RRs with their wild-type version *in vivo*. The most elegant way of doing this is to use a genomic replacement strategy, where the wild-type gene is directly replaced with the mutant gene. If this is not possible in the organism of choice, then the ability of the wild-type and mutant genes (expressed off plasmids at wild-type levels) to complement a deletion strain can be compared.

5. If the wild-type and mutant genes behave similarly, phosphorylation is not important for function of the HPKs/RRs. However, if there is a difference, this indicates that phosphorylation is important for function.

Genomic Replacements with Fluorescent Protein Fusions for
Studying Protein Localization

The *in vitro* analysis provided a very complex possible network of phosphotransfer reactions. We therefore examined the localization of each protein in the cell by fluorescently tagging each protein. This was carried out by replacing the wild-type gene in the *R. sphaeroides* genome with a gene encoding a fusion protein. The gene was inserted in the normal operon position to ensure normal expression patterns.

Whether to express a fusion protein from an inducible expression plasmid or by replacing the wild-type gene in the genome with its fluorescent fusion version depends, to a certain extent, on the protein of interest and the organism being used. In general, expression from a plasmid is easier since it requires fewer genetic manipulations. However, it is imperative that the level of expression of the fusion protein is carefully checked to ensure that it is the same as that of the wild-type protein, particularly in systems where there are large macromolecular complexes and stoichiometry might be important. Western blotting with an antibody raised to the wild-type protein is the most common method for measuring expression levels. However, this only determines that the average expression level in a population of cells is similar to wild-type. As the amount of inducer transported into each cell will vary, individual cells may express significantly more or less of the fusion than they would wild-type protein.

Expression of the fusion protein by direct replacement of the wild-type gene in the genome requires the ability to make unmarked changes within the genome. The significant advantage of this technique is that the fluorescent fusion protein is expressed behind the native gene's promoter elements and therefore should be expressed at the same levels and in the same location as the wild-type protein. This can be especially important with genes whose expression may vary with growth conditions and for those usually expressed in stoichiometric quantities with other components of a system, for example, within an operon. However, when inserting fusion genes into operons, it is essential to ensure that the fusion protein does not interfere with any regulatory sequences for adjacent genes. As with expression from a plasmid, it is important to confirm wild-type expression levels of the fusion protein by western blotting or a similar technique. It should also be confirmed that the expression of downstream genes in an operon has not been affected by the introduction of the fusion gene.

There are numerous different fluorescent proteins available commercially and a detailed description of the various proteins and their properties is outside the scope of this chapter. In general, the proteins are derivatives of either GFP from the jellyfish *Aequoria victoria* or of coral reef proteins.

These proteins are available in a wide range of spectral variants and some of them change colors when illuminated with light of a particular wavelength. The properties of different fluorescent proteins have been reviewed (Shaner *et al.*, 2005).

The fluorescent fusion can be to either the N- or C-terminus of the protein and may involve the addition of flexible linkers between the fluorescent protein and the protein of interest to assist in the correct folding of the protein and thus retention of function. The choice of N- or C-terminus for the fluorescent protein is often empirical, although if one terminus is known to be essential for function or interaction with other proteins, this terminus is often best avoided, at least initially. In either case, it is essential that the gene of interest and the gene encoding the fluorescent protein are in frame and there is no stop codon or start codon in the region linking the two proteins. The following describes the protocols used to replace genes with their corresponding fluorescent protein fusions in the genome of *R. sphaeroides* and methods for image acquisition and the subsequent analysis of these data.

Fluorescent Protein Fusion Constructs for Integration into the Genome

The first step in introducing an *egfp* tag into the *R. sphaeroides* genome involves generating the desired sequence within the suicide plasmid pK18*mobsacB* (Schäfer *et al.*, 1994). The *egfp* sequence needs to be flanked by upstream and downstream sequences that are found within the *R. sphaeroides* genome, so that double homologous recombination will result in the introduction of *egfp* into the *R. sphaeroides* genome between these flanking sequences (Fig. 3).

N-Terminal Fusions

In general, for an N-terminal fusion a construct is generated that contains ~500 bp immediately upstream of the gene fused to *egfp*, followed by the first ~500 bp of the gene that must be in-frame with the *egfp* (Fig. 4). The stop codon from *egfp* and the start codon for the gene should be excluded from the gene fusion. To ensure that the native Shine-Dalgarno sequence is used to control expression of the fusion protein, overlap-extension PCR is used to position the start codon of *egfp* in exactly the same position as the start codon of the wild-type gene, providing that this does not interfere with any stop codons or other important features of an upstream gene in the operon. If the stop codon from a preceding gene overlaps the start codon of the gene, it may be necessary to position the first codon of *egfp* a few codons into the existing gene. If the fusion is to a membrane protein, any membrane targeting sequence must be included before *egfp* for correct protein localization. As GFP folds as an autonomous

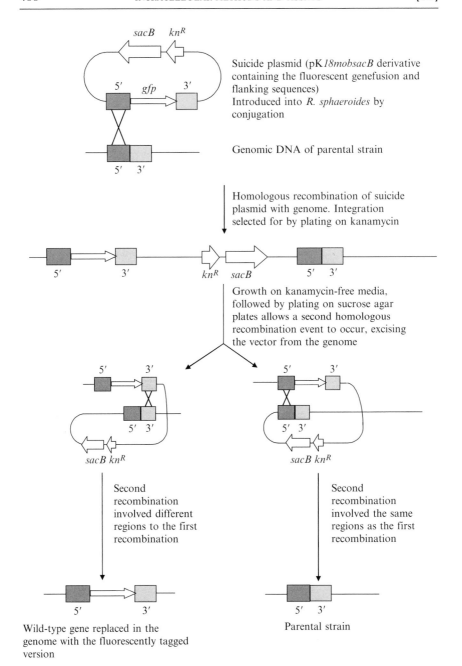

FIG. 3. Diagram outlining the steps involved in chromosomal gene replacement in *R. sphaeroides*.

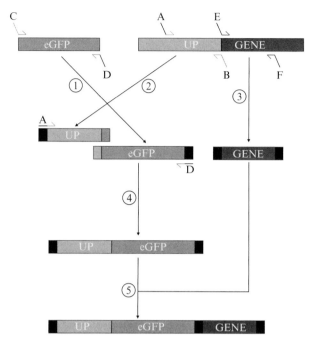

Fᴵɢ. 4. Diagram showing the production of a construct for introducing the gene encoding an N-terminal eGFP fusion into the *R. sphaeroides* genome. (1) Primers C and D are used to amplify the e*gfp* gene. (2) Primers A and B are used to amplify the ∼500 bp region immediately upstream of the gene to be tagged. (3) Primers E and F are used to amplify the first ∼500 bp of the gene to be tagged. (4) Primers B and C are exactly complementary to one another, facilitating an overlap extension reaction between the products of reactions 1 and 2. The overlap extension product is amplified using primers A and D. (5) The products of PCRs 3 and 4 are ligated into the suicide vector pK18*mobsacB* to generate the tagging construct. Restriction sites within primers are shown in black; the restriction sites used in primers D and E must be compatible. (See color insert.)

unit, extra amino acids prior to the GFP itself are fine, but the whole of the target protein (with the exception of the initiating methionine and any cleaved signal peptide) should always be included after the end of the GFP protein. The insert is ligated into a suicide vector, which integrates into the chromosome of the wild-type strain replacing the wild-type gene by double homologous recombination.

Example Protocol

1. A ∼500 bp region immediately upstream of the gene of interest is amplified by PCR to include a suitable 5′ restriction site.

2. *egfp* is amplified by PCR from pEGFP-N1 (BD Biosciences) to include a suitable 3′ restriction site and to exclude the *egfp* stop codon.

3. The central primers (B & C; Fig. 4) are designed so the 3' end of the upstream fragment is complementary to the 5' end of the *egfp* fragment and vice versa. The two PCR products are combined and used as templates in overlap-extension PCR with the 5' primer from the original upstream PCR and the 3' primer from the original *egfp* PCR. The product contains the region immediately upstream of the gene of interest fused to *egfp* with the start codon for *egfp* positioned exactly in the position of the start codon for the gene of interest.

4. This PCR product is cloned into the suicide vector pK18*mobsacB*.

5. The first ~500 bp of the gene of interest is amplified by PCR (excluding its initial ATG) to include an upstream restriction site compatible with that on the 3' end of the overlap-extension PCR product and a suitable downstream restriction site. This PCR product is cloned into the previous plasmid, generating a construct with the region upstream of the gene of interest fused to *egfp* and then ~500 bp of the gene of interest. It is important that the *egfp* and the gene of interest are in-frame.

C-Terminal Fusions

When constructing a C-terminal genomic fusion, similar principles apply as those for N-terminal fusions; however, a restriction site can be included between the end of the gene of interest and the beginning of *egfp*. The basic construct for a C-terminal fusion has ~500 bp immediately upstream from the stop codon of the gene of interest, followed in-frame by *egfp* and then the stop codon for the gene of interest and ~500 bp immediately downstream of that stop codon. If the gene of interest is part of an operon, it is important that the Shine-Dalgarno sequence or any other control elements for any downstream gene is not removed. In some cases, the start of a downstream gene overlaps the end of the gene of interest. The 3' region of the gene of interest can be repeated after *egfp* to ensure that potential control elements are retained. The insert is ligated into a suicide vector that integrates into the chromosome of the wild-type strain, replacing the wild-type gene by double homologous recombination. Confirm that expression of downstream genes in an operon is unaffected by the presence of the fusion using a technique such as western blotting.

Example Protocol

1. A ~500 bp fragment immediately preceding the stop codon of the gene of interest is amplified by PCR using primers that incorporate suitable restriction sites.

2. *egfp* without its initiating ATG is amplified by PCR using primers to introduce a restriction site at the 5' end compatible with the 3'

restriction site on the upstream fragment so that ligation results in the gene of interest and *egfp* being in-frame.

3. A ~500 bp fragment immediately downstream of and including the stop codon of the gene of interest is amplified by PCR using primers that include a 5' restriction site compatible with that on the 3' end of *egfp*.

4. The three fragments are ligated together into the suicide vector pK18*mobsacB*.

Integration of Gene Fusions into the R. sphaeroides *Genome*

The unmarked integration of genes into the *R. sphaeroides* genome is achieved by double homologous recombination (Fig. 3). Initially, the suicide plasmid pK18*mobsacB* integrates into the genome because of the homology between one of the flanking regions of the insert and the genomic DNA. The kanamycin resistance gene on the vector backbone allows selection of the recombinants. The cells are then grown in the absence of kanamycin selection and plated on sucrose-containing medium. Cells retaining the vector backbone are lost since the *sacB* gene on the vector backbone converts sucrose to toxic levansucrose. Southern blotting and PCR following the second recombination determines whether the two recombination events have happened within the same flanking region, resulting in reversion to wild-type, or in different flanking regions, resulting in the unmarked integration of the fusion gene into the genome.

Protocol

1. Transform the pK18*mobsacB* construct into *E. coli* S17–1λpir.
2. Grow *E. coli* S17-1 cells containing pK18*mobsacB* overnight at 37° in LB with kanamycin (25 μg/ml). Grow *Rhodobacter sphaeroides* WS8N for 2 days at 30° in succinate media with nalidixic acid (25 μg/ml).
3. On the day of the conjugation: Transfer 350 μl of S17-1 containing pK18*mobsacB* into 5 ml fresh LB with kanamycin. Grow at 37° with shaking until early log phase (faint cloudiness).
4. Harvest 1 ml of the S17-1 cells containing pK18*mobsacB* at 6000 rpm. Harvest 1 ml of *R. sphaeroides* cells.
5. Wash pellets very gently in 1 ml LB. Spin again at 6000 rpm.
6. Resuspend pellets very gently in 100 μl of LB.
7. Gently mix 10 μl of S17-1 containing pK18*mobsacB* and 100 μl of *R. sphaeroides*. Transfer gently to a sterile filter disc on a dry LB plate (aged 1 day). Incubate at 30° overnight.
8. Pick up the filter with sterile forceps and transfer it to an Eppendorf tube containing 800 μl LB. Vortex vigorously. Spread plate 100 μl

onto several LB agar plates containing nalidixic acid (25 μg/ml) and kanamycin (25 μg/ml).

9. Incubate plates at 30° for at least 2 days.
10. Pick a single colony into succinate media with nalidixic acid (25 μg/ml) and grow for 2 days at 30°.
11. Make serial dilutions into M22 minimal media and plate out 100 μl of neat, 1/10, 1/100, and 1/1000 dilution onto sucrose plates (M22 with 10% sucrose and 2% agar).
12. Incubate plates at 30° for 3 to 5 days.
13. Replicate plate by picking single colonies from the sucrose plates with a sterile toothpick and making crosses on an LB with nalidixic acid (25 μg/ml) and kanamycin (25 μg/ml) plate and on a LB with nalidixic acid (25 μg/ml) plate (48 colonies per pair of plates).
14. Incubate plates at 30° for 3 to 5 days.
15. Pick colonies from the LB with nalidixic acid plates that have not grown on the LB with nalidixic acid and kanamycin plates for genomic DNA extraction and further analysis.

Assessing the Functionality of the Fluorescent Protein Fusions

It is important to determine whether the fusion protein is functional and, if not, the level of reduction in functionality. This indicates probable correct localization and interaction. Comparison of the phenotype with a wild-type and deletion mutant provides a straightforward measure of function. If there is no deletion strain, or deletion does not produce an obvious phenotype, alternative methods of assessing the accuracy of the localization of the fusion protein should be investigated. Immuno-gold electron microscopy using antibodies to the wild-type protein show the usual localization pattern of the protein (Martin *et al.*, 2003; Wadhams *et al.*, 2003). Alternatively, co-localization studies with proteins whose fusions are functional and the ability to remove that localization on deletion of other proteins within the same system all help validate localization results (Wadhams *et al.*, 2005).

Image Acquisition and Data Analysis

A detailed description of fluorescence microscope systems suitable for visualizing the sub-cellular localization of fluorescent protein fusions has been reviewed elsewhere (Lichtman and Conchello, 2005). However, for most bacterial systems, a microscope with either phase contrast or DIC, a 100× oil immersion objective, and a mercury arc lamp or laser for excitation of the fluorescent protein are required. Sets of filters are also needed that are matched to the excitation and emission properties of the fluorophore being used. A cooled CCD camera is often used for higher resolution.

For any semi-quantitative data analysis it is important to ensure that the intensity of the excitation light and the exposure time are optimized to ensure that the fluorescence image is not saturating the camera's pixels, but is not so dim that fluorescence is lost in any background noise. Usually, trying to fill the pixel wells to about 90% of their maximum intensity is a good starting point. It is also important that the wild-type strain without any fluorescent fusion is imaged with the same microscope and camera settings on the same day as the strain containing the fluorescent fusion protein. This gives a background reading for cellular autofluorescence not attributable to the presence of the fluorescent protein fusion. If the objective of the experiment is simply to determine where within the cell the tagged protein is localized, then obtaining a bright field image showing the position of the cell bodies and a fluorescent image that can be superimposed on the first image will demonstrate protein localization (Fig. 1B). Many different software packages are available that will do this. The important principle is that any manipulation made to the images from the cells expressing the fluorescent protein fusion is also applied to images acquired with the same settings on the same day from cells without the fusion to ensure that observed fluorescence is directly attributable to the fusion protein.

Summary

Protein purification and analysis of phosphotransfer rates provide an accurate quantitative measure of the kinetics of phosphotransfer reactions. However, while the reactions tell you what can interact and phosphotransfer, they do not reveal which HPKs and RRs do really interact. If there are multiple homologues in a system, a combination of biochemical and microscopic approaches can provide a realistic model of the sensory network. For example, the CheAs of *R. sphaeroides* show complex heterophosphorylation patterns *in vitro* (Fig. 1A), but upon analysis of the *cheA* deletion phenotypes and the cellular localization of the CheAs (Fig. 1B), the story became much simpler (Fig. 1C). There are two separate signaling clusters; although $CheA_4$ can phosphorylate $CheA_2$ *in vitro*, $CheA_2$ is localized to the cell poles while $CheA_4$ is localized to the cytoplasmic chemotaxis cluster. This spatial separation of $CheA_2$ and $CheA_4$ would therefore make it unlikely that $CheA_4$ phosphorylates $CheA_2$ *in vivo*.

References

Anand, G. S., Goudreau, P. N., and Stock, A. M. (1998). Activation of methylesterase CheB: Evidence of a dual role for the regulatory domain. *Biochemistry* **37,** 14038–14047.

Appleby, J. L., and Bourret, R. B. (1999). Activation of CheY mutant D57N by phosphorylation at an alternative site, Ser-56. *Mol. Microbiol.* **34,** 915–925.

Armitage, J. P., and Macnab, R. M. (1987). Unidirectional intermittent rotation of the flagellum of *Rhodobacter sphaeroides*. *J. Bacteriol.* **169**, 514–518.

Ashby, M. K. (2004). Survey of the number of two-component response regulator genes in the complete and annotated genome sequences of prokaryotes. *FEMS Microbiol. Letts.* **231**, 277–281.

Borkovich, K. A., Kaplan, N., Hess, J. F., and Simon, M. I. (1989). Transmembrane signal transduction in bacterial chemotaxis involves ligand-dependent activation of phosphate group transfer. *Proc. Natl. Acad. Sci. USA* **86**, 1208–1212.

Galperin, M. Y. (2006). Structural classification of bacterial response regulators: Diversity of output domains and domain combinations. *J. Bacteriol.* **188**, 4169–4182.

Hamblin, P. A., Maguire, B. A., Grishanin, R. N., and Armitage, J. P. (1997). Evidence for two chemosensory pathways in *Rhodobacter sphaeroides*. *Mol. Microbiol.* **26**, 1083–1096.

Hess, J. F., Bourret, R. B., and Simon, M. I. (1988a). Histidine phosphorylation and phosphoryl group transfer in bacterial chemotaxis. *Nature* **336**, 139–143.

Hess, J. F., Oosawa, K., Kaplan, N., and Simon, M. I. (1988b). Phosphorylation of three proteins in the signaling pathway of bacterial chemotaxis. *Cell* **53**, 79–87.

Jimenez-Pearson, M. A., Delany, I., Scarlato, V., and Beier, D. (2005). Phosphate flow in the chemotactic response system of *Helicobacter pylori*. *Microbiology* **151**, 3299–3311.

Li, J. Y., Swanson, R. V., Simon, M. I., and Weis, R. M. (1995). The response regulators CheB and CheY exhibit competitive binding to the kinase CheA. *Biochemistry* **34**, 14626–14636.

Lichtman, J. W., and Conchello, J. A. (2005). Fluorescence microscopy. *Nat. Meth.* **2**, 910–919.

Lukat, G. S., and Stock, J. B. (1993). Response regulation in bacterial chemotaxis. *J. Cell. Biochem.* **51**, 41–46.

Lupas, A., and Stock, J. (1989). Phosphorylation of an N-terminal regulatory domain activates the CheB methylesterase in bacterial chemotaxis. *J. Biol. Chem.* **264**, 17337–17342.

Mackenzie, C., Choudhary, M., Larimer, F. W., Predki, P. F., Stilwagen, S., Armitage, J. P., Barber, R. D., Donohue, T. J., Hosler, J. P., Newman, J. E., Shapleigh, J. P., Sockett, R. E., *et al.* (2001). The home stretch, a first analysis of the nearly completed genome of *Rhodobacter sphaeroides* 2.4.1. *Photosynth. Res.* **70**, 19–41.

Maddock, J. R., and Shapiro, L. (1993). Polar location of the chemoreceptor complex in the *Escherichia coli* cell. *Science* **259**, 1717–1723.

Martin, A. C., Nair, U., Armitage, J. P., and Maddock, J. R. (2003). Polar localization of CheA(2) in *Rhodobacter sphaeroides* requires specific Che homologs. *J. Bacteriol.* **185**, 4667–4671.

Martin, A. C., Wadhams, G. H., and Armitage, J. P. (2001). The roles of the multiple CheW and CheA homologues in chemotaxis and in chemoreceptor localization in *Rhodobacter sphaeroides*. *Mol. Microbiol.* **40**, 1261–1272.

Martin, A. C., Gould, M., Byles, E., Roberts, M. A. J., and Armitage, J. P. (2006). Two chemosensory operons of *Rhodobacter sphaeroides* are regulated independently by sigma 28 and sigma 54. *J. Bacteriol.* **188**, 7932–7940.

Moore, J. B., Shiau, S. P., and Reitzer, L. J. (1993). Alterations of highly conserved residues in the regulatory domain of nitrogen regulator I (NtrC) of *Escherichia coli*. *J. Bacteriol.* **175**, 2692–2701.

Motaleb, M. A., Miller, M. R., Li, C. H., Bakker, R. G., Goldstein, S. F., Silversmith, R. E., Bourret, R. B., and Charon, N. W. (2005). CheX is a phosphorylated CheY phosphatase essential for *Borrelia burgdorferi* chemotaxis. *J. Bacteriol.* **187**, 7963–7969.

Ninfa, E. G., Stock, A., Mowbray, S., and Stock, J. (1991). Reconstruction of the bacterial chemotaxis signal transduction system from purified components. *J. Biol. Chem.* **266**, 9764–9770.

Porter, S. L., and Armitage, J. P. (2002). Phosphotransfer in *Rhodobacter sphaeroides* chemotaxis. *J. Mol. Biol.* **324**, 35–45.

Porter, S. L., and Armitage, J. P. (2004). Chemotaxis in *Rhodobacter sphaeroides* requires an atypical histidine protein kinase. *J. Biol. Chem.* **279,** 54573–54580.

Porter, S. L., Warren, A. V., Martin, A. C., and Armitage, J. P. (2002). The third chemotaxis locus of *Rhodobacter sphaeroides* is essential for chemotaxis. *Mol. Microbiol.* **46,** 1081–1094.

Porter, S. L., Wadhams, G. H., Martin, A. C., Byles, E. D., Lancaster, D. E., and Armitage, J. P. (2006). The CheYs of *Rhodobacter sphaeroides*. *J. Biol. Chem.* **281,** 32694–32704.

Rasmussen, A. A., Porter, S. L., Armitage, J. P., and Sogaard-Andersen, L. (2005). Coupling of multicellular morphogenesis and cellular differentiation by an unusual hybrid histidine protein kinase in *Myxococcus xanthus*. *Mol. Microbiol.* **56,** 1358–1372.

Rasmussen, A. A., Wegener-Feldbrugge, S., Porter, S. L., Armitage, J. P., and Sogaard-Andersen, L. (2006). Four signaling domains in the hybrid histidine protein kinase RodK of *Myxococcus xanthus* are required for activity. *Mol. Microbiol.* **60,** 525–534.

Reyrat, J. M., David, M., Batut, J., and Boistard, P. (1994). FixL of *Rhizobium meliloti* enhances the transcriptional activity of a mutant FixJD54N protein by phosphorylation of an alternate residue. *J. Bacteriol.* **176,** 1969–1976.

Schäfer, A., Tauch, A., Jäger, W., Kalinowski, J., Thierbach, G., and Pühler, A. (1994). Small mobilizable multipurpose cloning vectors derived from the *Escherichia coli* plasmids pK18 and pK19—Selection of defined deletions in the chromosome of *Corynebacterium glutamicum*. *Gene* **145,** 69–73.

Shah, D. S. H., Porter, S. L., Martin, A. C., Hamblin, P. A., and Armitage, J. P. (2000). Fine tuning bacterial chemotaxis: Analysis of *Rhodobacter sphaeroides* behavior under aerobic and anaerobic conditions by mutation of the major chemotaxis operons and *cheY* genes. *EMBO J.* **19,** 4601–4613.

Shaner, N. C., Steinbach, P. A., and Tsien, R. Y. (2005). A guide to choosing fluorescent proteins. *Nat. Meth.* **2,** 905–909.

Sourjik, V., and Schmitt, R. (1998). Phosphotransfer between CheA, CheY1, and CheY2 in the chemotaxis signal transduction chain of *Rhizobium meliloti*. *Biochemistry* **37,** 2327–2335.

Springer, W. R., and Koshland, D. E., Jr. (1977). Identification of a protein methyltransferase as the *cheR* gene product in the bacterial sensing system. *Proc. Natl. Acad. Sci. USA* **74,** 533–537.

Stewart, R. C. (1997). Kinetic characterization of phosphotransfer between CheA and CheY in the bacterial chemotaxis signal transduction pathway. *Biochemistry* **36,** 2030–2040.

Stewart, R. C., Jahreis, K., and Parkinson, J. S. (2000). Rapid phosphotransfer to CheY from a CheA protein lacking the CheY-binding domain. *Biochemistry* **39,** 13157–13165.

Szurmant, H., Muff, T. J., and Ordal, G. W. (2004). *Bacillus subtilis* CheC and FliY are members of a novel class of CheY-P-hydrolyzing proteins in the chemotactic signal transduction cascade. *J. Biol. Chem.* **279,** 21787–21792.

Wadhams, G. H., Martin, A. C., and Armitage, J. P. (2000). Identification and localization of a methyl-accepting chemotaxis protein in *Rhodobacter sphaeroides*. *Mol. Microbiol.* **36,** 1222–1233.

Wadhams, G. H., Martin, A. C., Porter, S. L., Maddock, J. R., Mantotta, J. C., King, H. M., and Armitage, J. P. (2002). TlpC, a novel chemotaxis protein in *Rhodobacter sphaeroides*, localizes to a discrete region in the cytoplasm. *Mol. Microbiol.* **46,** 1211–1221.

Wadhams, G. H., Martin, A. C., Warren, A. V., and Armitage, J. P. (2005). Requirements for chemotaxis protein localization in *Rhodobacter sphaeroides*. *Mol. Microbiol.* **58,** 895–902.

Wadhams, G. H., Warren, A. V., Martin, A. C., and Armitage, J. P. (2003). Targeting of two signal transduction pathways to different regions of the bacterial cell. *Mol. Microbiol.* **50,** 763–770.

Welch, M., Oosawa, K., Aizawa, S.-I., and Eisenbach, M. (1993). Phosphorylation-dependent binding of a signal molecule to the flagellar switch of bacteria. *Proc. Natl. Acad. Sci. USA* **90,** 8787–8791.

[19] *In Vivo* Crosslinking Methods for Analyzing the Assembly and Architecture of Chemoreceptor Arrays

By CLAUDIA A. STUDDERT and JOHN S. PARKINSON

Abstract

The chemoreceptor molecules that mediate chemotactic responses in bacteria and archaea are physically clustered and operate as highly cooperative arrays. Few experimental approaches are able to investigate the structure–function organization of these chemoreceptor networks in living cells. This chapter describes chemical crosslinking methods that can be applied under normal physiological conditions to explore physical interactions between chemoreceptors and their underlying genetic and structural basis. Most of these crosslinking approaches are based on available atomic structures for chemoreceptor homodimers, the fundamental building block for higher-order networks. However, the general logic of our *in vivo* crosslinking approaches is readily applicable to other protein–protein interactions and other organisms, even when high-resolution structural information is not available.

Introduction

Motile bacteria track gradients of attractant and repellent chemicals in the search for optimal living environments. This behavior, known as chemotaxis, involves a signal transduction pathway that has been extensively characterized in *E. coli* [reviewed by (Parkinson *et al.*, 2005; Sourjik, 2004; Szurmant and Ordal, 2004)]. The proteins responsible for sensing chemical gradients are known as methyl-accepting chemotaxis proteins, or MCPs (Zhulin, 2001). These chemoreceptors are homodimeric membrane proteins that typically contain a ligand-specific periplasmic sensing domain, flanked by two transmembrane regions, and a highly conserved cytoplasmic signaling domain (Fig. 1). MCPs form ternary signaling complexes with CheA, a histidine kinase, and CheW, an adaptor protein that couples CheA to receptor control. Information about receptor ligand occupancy is transmitted across the inner membrane to the cytoplasmic domain, which regulates autophosphorylation of CheA. CheA activity, in turn, regulates the phosphorylation state of the response regulator CheY, which modulates the direction of rotation of the flagellar motors, the final target of the signaling pathway.

METHODS IN ENZYMOLOGY, VOL. 423   0076-6879/07 $35.00
DOI: 10.1016/S0076-6879(07)23019-8

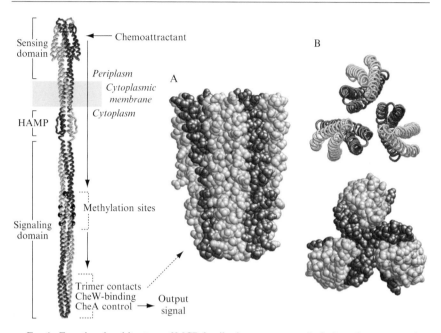

FIG. 1. Functional architecture of MCP-family chemoreceptors. Left: Atomic structure of an MCP homodimer assembled from structures for the Tar ligand-binding domain (Milburn *et al.*, 1991), the Tsr signaling domain (Kim *et al.*, 1999), and the HAMP domain from a non-MCP thermophile protein (Hulko *et al.*, 2006). MCP subunits also carry an unstructured segment at their C-terminus (not shown), whose function is not well understood. High-abundance receptors Tar and Tsr also carry a pentapeptide (not shown) at the end of the C-terminal linker that binds to and modulates the enzymatic activities of CheR and CheB, the MCP modifying enzymes. The backbone traces of the predominantly alpha-helical subunits are shaded differently to show the arrangements and structural interactions within the dimer. Chemoattractants trigger conformational changes in the periplasmic sensing domain that propagate across the cytoplasmic membrane, eventually influencing output signals generated at the cytoplasmic tip of the molecule. The tips of three MCP dimers are thought to associate in a trimer-of-dimers arrangement needed for CheA activation and control. (A) A space-filled side view of the trimer tip. Each dimer has one subunit (dark) that lies at the interdimer interface and contributes nine residues that stabilize the trimer arrangement. The other subunit in each dimer (light) lies mainly on the outside surface of the trimer, but contributes two additional residues to trimer packing interactions. (B) Backbone and space-filled structures of the trimer contact region viewed from the cytoplasmic tip. The shading scheme used in (A) also applies to these images. The structures in (A) and (B) are shown at the same relative scale.

MCPs sense temporal changes in chemoeffector levels by comparing their current occupancy state with that averaged over the past few seconds, recorded in the form of reversible methylation of 4 to 6 glutamic acid residues in the signaling domain. Cells adapt to unchanging chemical environments by adjusting their MCP methylation levels through the action of a methyltransferase CheR and a methylesterase CheB. CheR operates at a substrate-limited rate, whereas CheB is feedback-regulated through phosphorylation signals from the receptor complexes.

   E. coli has four MCPs with different detection specificities (Tsr, serine; Tar, aspartate and maltose; Tap, dipeptides; Trg, ribose and galactose) and a fifth MCP-like receptor (Aer) that mediates aerotactic behavior. These chemoreceptors and their associated signaling proteins are localized in clusters, often at the cell poles, in *E. coli* (Maddock and Shapiro, 1993; Sourjik and Berg, 2000) and in a variety of other bacteria (Gestwicki *et al.*, 2000). Clustering may enable receptor molecules to exchange sensory information and thereby act cooperatively to amplify small chemical stimuli into large flagellar-controlling output signals (Bray *et al.*, 1998; Sourjik and Berg, 2002, 2004). Although many genetic and biochemical studies have contributed to an understanding of the general organization of chemorecep-tor clusters and their underlying protein–protein interactions, the assembly and architecture of chemoreceptor clusters are not well understood.

   An interesting clue to the possible nature of receptor–receptor interac-tions was provided by the crystal structure of the Tsr signaling domain, which revealed a trimer-of-dimers arrangement (Kim *et al.*, 1999) (Fig. 1B). The eleven residues principally involved in dimer–dimer contacts at the trimer interface are identical in all five *E. coli* MCP-family transducers, suggesting that the trimer arrangement might be a structural and functional component of receptor signaling clusters.

   Crosslinking methods can be used to map protein–protein interactions. The availability of a large variety of crosslinkers, together with many analytical tools for characterizing the crosslinking products, makes it a versatile and reliable approach that permits inspection of macromolecular assemblies as they are found *in vivo*. This chapter focuses on the cross-linking methods that we have used to assess whether functional chemo-receptors are organized in trimers of dimers in intact cells, to characterize the interactions between receptors of different specificities, and to study the dynamics of those interactions under different cellular conditions.

Use of a Lysine-Targeted Crosslinker to Probe Receptor–Receptor Interactions in Cells

   DSP [di-thiobis(succinimidyl propionate)] is a bifunctional, lysine-reactive reagent with a relatively short spacer arm (11–12 Å). DSP crosslinks are reversible, owing to the presence of a disulfide bridge connecting the two succinimide moieties. DSP crosses biological membranes, so it can poten-tially trap any two proteins that reside close together in the cell, provided that they both contain lysine residues at an appropriate distance. Since lysine is a relatively common residue in proteins, DSP is a good candidate to use for a preliminary search of uncharacterized protein interactions. In fact, the first attempt to analyze interactions between chemotaxis proteins in intact cells made use of DSP (Chelsky and Dahlquist, 1980). Several oligomeric forms of different chemotaxis proteins were identified in that pioneering study. Notably, the MCPs Tar and Tsr formed DSP-crosslinked products of up to four subunits (Chelsky and Dahlquist, 1980).

We used DSP to show that Tar and Tsr molecules are physically associated in intact cells (Ames *et al.*, 2002). The experimental steps are depicted in Fig. 2. A Tar receptor carrying a 6× His tag (Tar-6× His) was coexpressed at physiological levels with a nontagged Tsr and cells in the midlog phase of growth were treated with DSP. After the treatment, membrane proteins were solubilized with a detergent, and all His-tagged molecules (together with any covalently attached proteins) were purified on a nickel resin column. His-tagged products were then treated with a reducing agent to break crosslinks and the individual protein components were analyzed by SDS-PAGE and visualized by immunoblotting with a polyclonal antiserum directed against a highly conserved segment of the MCP signaling domain.

To distinguish Tar and Tsr molecules, we used a gel system in which they had different mobilities. To simplify the resulting band patterns, we used a host strain that lacked the MCP-modifying CheR and CheB enzymes so that all receptor molecules were in the unmodified state. Comparing the protein

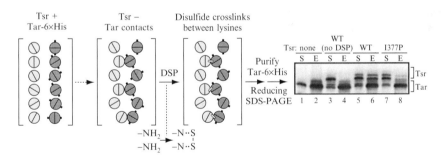

FIG. 2. Detection of interactions between receptor dimers with a lysine-reactive crosslinker. Cross-sections of receptor dimers are shown schematically as adjoining semicircles. Tsr (light) and Tar (dark) subunits do not form heterodimers, but do engage in dimer–dimer interactions detectable by DSP crosslinking. The Tar molecules carry a 6× His affinity tag (black rectangle) on each subunit that was used to purify Tar-containing crosslinking products before analyzing their composition by gel electrophoresis and immunoblotting. Samples labeled S were not affinity-purified; those labeled E were. Note that the DSP-mediated crosslinks were broken by reducing conditions before the gel analysis.

samples before (S) and after (E) elution from the nickel column, it is clear that the Tar-6× His molecules were efficiently retained by the nickel column (Fig. 2, lanes 2, 4, 6, and 8). The lower mobility Tsr band also appeared in the eluted samples, but only from DSP-treated cells (lane 6), not from untreated cells (lane 4). Thus, the presence of Tsr in the eluted material evidently depends on covalent connection(s) that DSP creates between His-tagged Tar and Tsr. Significantly, a mutant Tsr with a single amino acid replacement at a trimer contact residue (I377P) showed no DSP crosslinking to Tar-6× His (lane 8) even though the cellular level of the mutant protein was comparable to that of wild-type Tsr (compare lanes 5 and 7). These results suggested that DSP-mediated crosslinking between Tsr and Tar molecules could be based on a trimer-of-dimers interaction. This interaction was also detected in cells lacking all of the soluble chemotaxis proteins, including CheA and CheW, indicating that it was most likely a direct one between receptor molecules (Ames *et al.*, 2002).

A detailed description of the experiments (growth of cells, DSP treatment, cell disruption, membrane purification, solubilization of membrane proteins, purification of His-tagged proteins and crosslinked products on Ni-NTA matrix, gel electrophoresis, and Western blotting) can be found in Ames *et al.* (2002). Here, we emphasize three aspects of the experiments that are important to our conclusion that the observed crosslinking behavior reflects genuine interactions between different receptor dimers under normal physiological conditions. First, the 6× His tag at the C-terminus of Tar did not impair its chemotactic signaling ability, as judged by behavioral assays in soft agar plates, so Tar-6× His must be capable of the same protein–protein interactions as wild-type Tar. Second, Tar-6× His and wild-type Tsr were induced at levels optimal for wild-type function by each receptor. Thus, their interaction occurred under physiological conditions. Third, MCP molecules of different types do not seem to form heterodimers (Milligan and Koshland, 1988), implying that the observed crosslinking interactions occur between intact homodimers of both receptors.

Use of Cys-Targeted Crosslinking to Probe for the Trimer-of -Dimers Geometry in Cellular Chemoreceptor Assemblies

Tar and Tsr molecules have no native cysteine residues. We introduced cysteine reporters at positions that allowed us to exploit their unique sulfhydryl chemistry to look for trimer-of-dimer interactions in crosslinking experiments (Studdert and Parkinson, 2004). Inspection of the crystal structure of the Tsr cytoplasmic domain in which three different dimers closely interact at their hairpin tips (Kim *et al.*, 1999) (Fig. 1B) revealed several general structural features that needed to be taken into account when devising crosslinking strategies to test whether a similar trimer-based interaction occurs between dimers *in vivo*.

First, the basic structural units in the trimer are homodimers, with two identical subunits. Residues in one subunit can mediate close contact between dimers at the trimer interface, whereas their counterparts in the other subunit will reside in a very different structural environment at the trimer perimeter. Thus, a residue in the inner subunit of a trimer may serve to stabilize the trimer, whereas the same residue in the outer subunit could play a different functional role, for example, promoting an interaction with CheA or CheW, which are known to bind to the tip of the receptor signaling domain (Fig. 1). However, due to the coiled-coil nature of the interaction between the two subunits of a dimer, cross-sections at different distances from the tip show that the positions of residues in the "inner" and "outer" subunits change relative to the trimer interface. For example, the methylation site residues in both subunits have roughly comparable orientations in the trimer.

Second, in the trimer-of-dimers crystal structure, the distances between specific residues and the solvent exposure of those residues can be precisely determined. Accordingly, cysteine reporter positions can be chosen on the basis of their predicted and differential propensity to form disulfides in dimers versus trimers of dimers.

In the following two sections, we describe how we used this approach to test the trimer-of-dimers organization of different receptors in chemotaxis-proficient cells. Importantly, every position chosen for a cysteine replacement was tested for its effects on receptor function in soft agar chemotaxis assays. Only reporter sites that had little or no deleterious functional effects were used for crosslinking studies. Except where otherwise indicated, experiments were performed in strains lacking CheA, CheW, CheR, and CheB proteins and the Cys-containing receptors were expressed at levels required for optimal function.

Intracytoplasmic Disulfide Crosslinks

Residues V384 and V398 lie at solvent-exposed positions far from their counterparts in the other subunit of the Tsr dimer (Fig. 3A and B). Moreover, in the trimer-of-dimers structure, neither residue lies close to its counterparts in the other dimers (Fig. 3A and B). However, V384 from the inner subunit of one dimer interacts closely with V398 from the outer subunit of a neighboring dimer (6.2 Å between alpha carbons) (Fig. 3C). The V384–V398 pair does not seem to represent a critical trimer contact, because it tolerates some amino acid replacements without loss of function

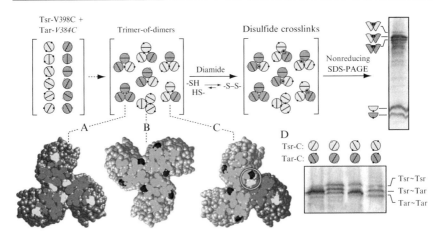

Fig. 3. Detection of trimer-based interactions between receptor dimers with cysteine-directed disulfide formation. Tsr and Tar dimers are depicted with the conventions described in Fig. 2. Their unique cysteine reporter sites are indicated by black (V398C) and white (*V384C*) circles. Cells expressing the two types of receptors at normal physiological levels were treated with diamide to render the cytoplasm more oxidizing and thereby promote disulfide bond formation. Crosslinked receptor subunits were analyzed by gel electrophoresis under nonreducing conditions and visualized by immunoblotting. (A) Expected arrangement of *V384C* reporter sites in a Tar trimer. (B) Expected arrangement of V398C reporter sites in a Tsr trimer. (C) Expected arrangement of V384C and V398C reporter sites in a mixed trimer containing both Tar and Tsr reporter molecules. Some of the reporter sites are much closer in the mixed trimer than in either of the pure trimers. (D) Crosslinked species generated by different combinations of Tsr and Tar reporter molecules. The mixed arrangements of reporter sites form disulfides much more efficiently than the others.

(Ames *et al.*, 2002). These structural and functional characteristics made V384 and V398 good candidates for probing higher-order chemoreceptor interactions *in vivo*. The trimer-of-dimers arrangement predicts that collisions between V384C and V398C reporters in different dimers should be much more frequent than V384–V384 or V398–V398 collisions. Some other higher-order receptor arrangements, for example, the "hedgerows" observed in a crystal structure of the signaling domain of an MCP molecule from *Thermotoga maritima* (Park *et al.*, 2006) make very different predictions about the proximity of the V384–V398 pair.

Based on these considerations, we created cysteine replacements at V384 and V398 in Tsr, and at the corresponding positions in Tar, and measured their ability to form interdimer disulfides by expressing different combinations of Cys-bearing receptors in the same cell, as summarized in Fig. 3 (Studdert and Parkinson, 2004). For simplicity, we refer to the Tar reporters by the (italicized) numbers of the corresponding residues in Tsr. (The actual Tar residue numbers are two less than their Tsr counterparts.) Three of the reporters (Tsr-V398C, Tar-*V384C*, Tar-*V398C*) retained good signaling function; the fourth reporter (Tsr-V384C) had somewhat impaired function, but nevertheless exhibited crosslinking behavior comparable to that of its fully functional Tar counterpart. Here, we discuss the rationale behind these experiments and some technical aspects that need to be addressed in order to test the proximity of cytoplasmic reporter sites in proteins using disulfides as structural probes.

To obtain an unequivocal readout of interdimer crosslinking (as opposed to intradimer crosslinking between subunits of the same dimer), we monitored disulfide formation between the Tsr and Tar reporter molecules, which do not form heterodimers (see preceding text). In addition, we used a polyacrylamide gel system (10% acrylamide, 0.05% bis-acrylamide, pH 8.2) that augmented the different electrophoretic mobilities of Tar and Tsr (Feng *et al.*, 1997). Thus, we could distinguish crosslinked products containing two Tar subunits, or two Tsr subunits, or one subunit of each type. Tar $\approx$ Tsr products should originate exclusively from interdimer collisions, whereas other crosslinked species could arise from either intra- or interdimer collisions.

As shown in the upper portion of Fig. 3, cells expressing Tar-*V384C* and Tsr-V398C formed three crosslinked receptor species, with the Tar $\approx$ Tsr product more prominent than the Tar $\approx$ Tar and Tsr $\approx$ Tsr products. This result could reflect a trimer-of-dimers geometry for inter-receptor interactions, but alternatively could be due to preferential collisions between heterologous receptors, no matter what their higher-order arrangement. Tests with other reporter combinations excluded this alternate explanation (Fig. 3D). In the Tar-*V384C* + Tsr-V398C and Tar-*V398C* + Tsr-V384C combinations, the residues expected to interact in a trimer-of-dimers assembly are cysteines and should crosslink efficiently (Fig. 3C), whereas in the Tar-*V398C* + Tsr-V398C and Tar-*V384C* + Tsr-V384C combinations, both receptors have cysteines at equivalent positions. The trimer-of-dimers geometry predicts that these reporter sites should crosslink less efficiently (Fig. 3A,B). As Fig. 3D shows, when both members of the 384–398 pair were cysteines, the Tar $\approx$ Tsr band predominated. When the same residue position in both Tar and Tsr carried the Cys reporter, the three possible crosslinking products formed with comparable efficiencies. This result is fully consistent with the trimer-of-dimers geometry of chemoreceptor assemblies, but cannot exclude other possible arrangements. However, any proposed alternative arrangement must be able to account for the preferential crosslinking of V398 and V384 residues. This is not the case for the hedgerow arrangement of the cytoplasmic domain of a *T. maritima* MCP (Park *et al.*, 2006), for example. In that structure, the distances between the *384* and *398* positions were not significantly less than the *398–398* or the *384–384* distances. Conceivably, the MCP arrangements could differ between *E. coli* and *T. maritima*, but it seems equally likely that the apparent disagreement reflects differences in crystal packing interactions rather than real physiological differences. Only *in vivo* crosslinking studies conducted in *Thermotoga*, as well as in other species, can resolve these issues.

Technical Considerations

The position of the bridging disulfide bond between receptor subunits can create small but consistent differences in mobility between various crosslinking products (Fig. 3D). Crosslinked products migrate faster with a disulfide near the subunit termini than with a connection near their centers, which creates a more "H"-shaped molecule (compare Tar-*V398C* ≈ Tsr-V398C with Tar-*V384C* ≈ Tsr-V384C in Fig. 3D).

The highly reducing nature of the cytoplasm opposes disulfide bond formation *in vivo* (Ritz and Beckwith, 2001). To perform the experiments just described, we had to create a less reducing cellular environment. To this end, cells were harvested at late-log phase of growth, resuspended in KEP buffer [10 mM potassium phosphate (pH 7), 0.1 mM EDTA] at an $OD_{600} = 2$, and incubated at 30° for 45 min in the presence of 0.5 mM diamide. Treatment with this thiol-specific oxidant (Kosower and Kosower, 1995) greatly enhanced the detection of intracytoplasmic disulfide bonds. Before lysis of the cells, the unreacted sulfhydryl groups were quenched by treatment with 10 mM NEM. Lysis buffer also contained NEM to avoid the formation of disulfides during processing of samples.

A Trifunctional Cys-Targeted Crosslinker

Above the trimer contact region, the three dimers of the trimer splay apart, forming a solvent-accessible space along the central axis of the trimer (Fig. 4A). We reasoned that, if the crystal structure represents the interactions that occur *in vivo*, it might be possible to capture all three axial subunits of a trimer with TMEA, a tri-functional thiol-specific crosslinking agent (Fig. 4C), provided that the target cysteines are appropriately positioned. The thiol-reactive maleimide groups in TMEA lie 10.3 Å apart, so we tested comparably spaced residues in the Tsr trimer as reporter sites. A cysteine replacement at S366 proved best. The alphacarbons at this position lie 12.3 Å from each other in the subunits facing the central space, whereas those on the outside of the trimer are very far apart and shielded from one another by the bulk of the trimer (Fig. 4B).

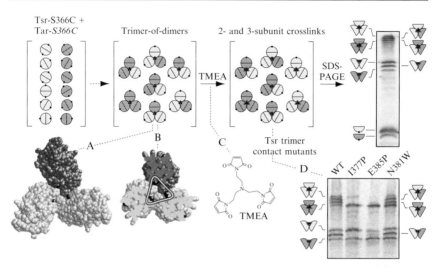

FIG. 4. Detection of trimer-based interactions between receptor dimers with a trifunctional cysteine-targeted crosslinker. Tsr and Tar dimers are depicted with the conventions described in Fig. 2 and Fig. 3. Cells expressing normal physiological levels of Tar and Tsr molecules with a cysteine reporter (S366C) were treated with TMEA (structure shown in [C]). TMEA-crosslinked receptor subunits were analyzed by gel electrophoresis and visualized by immunoblotting. (A) Arrangement of reporter sites in a mixed trimer, viewed from the cell membrane toward the cytoplasmic tip of the trimer. Reporter sites (black atoms) lie close together at the trimer axis, but far apart (not visible) in the outside subunits of the dimers. (B) Cross-section view of a mixed trimer showing the close-spaced triangular arrangement of reporter sites in the inside subunits of the dimers. (D) TMEA crosslinking patterns of wild-type (WT) and mutant Tsr reporter molecules when mixed with a wild-type Tar reporter. The mutant molecules carry single amino acid replacements at one of the trimer contact residues. The I377P and E385P lesions disrupt trimer formation; the N381W lesion does not.

Most importantly, Tsr and other receptors with a cysteine replacement at this position retained full signaling function.

Receptors with the S366C reporter generated two- and three-subunit crosslinking products upon treatment of whole cells with TMEA (Studdert and Parkinson, 2004). Cells were simply harvested, resuspended in KEP buffer, treated for 5 to 20 s with 50 μM TMEA, and lysed after quenching the reaction with 10 mM NEM. Then, the products were analyzed by SDS-PAGE and Western blotting with an anti-MCP antibody. A typical TMEA experiment with cells expressing both Tsr-S366C and Tar-*S366C* is shown in the upper portion of Fig. 4. Assuming that the 2- and 3-subunit crosslinking products represent TMEA-trapped inner subunits from trimers of different compositions (see supporting considerations in the following text), the distribution of crosslinked products was consistent with random formation of trimers from the available dimer pool. Thus, trimer composition would only depend on the relative abundance of the different receptor dimers (for theoretical predicted composition of trimers, see Fig. 5C). This assumption was experimentally tested by expressing Cys-containing Tar and Tsr receptors at different relative levels of expression and analyzing the distribution of crosslinked products upon TMEA treatment (Studdert and Parkinson, 2004). The results were clearly consistent with random mixing of dimers, with receptors expressed at a low level being found almost exclusively in mixed crosslinked species. Random assembly of trimers from the pool of receptor dimers implies that receptor types in relatively low cellular abundance should reside mainly in mixed trimers with higher abundance receptor types. The TMEA crosslinking behavior of the low abundance Trg (Studdert and Parkinson, 2004) and Aer (Gosink et al., 2006) receptors supported this view.

Technical Considerations About TMEA Treatment of Cells

TMEA treatments as short as 5 s and as long as 2 h produced similar extents and patterns of crosslinking. Higher TMEA concentrations had no enhancing effect on the yield of crosslinking products. Thus, TMEA seems to react rapidly with all receptor subunits that are available for crosslinking, yielding a "snapshot" of the higher-order structure of the receptor population.

Intact and broken cells exhibit very different TMEA behaviors. Cells expressing receptors at physiological levels yielded crosslinking products with high and reproducible efficiency, whereas membrane preparations from the same cells did not yield any crosslinking product (Studdert and Parkinson, 2004). TMEA treatment of membrane preparations was only effective if the receptors had been highly overexpressed. We hypothesize that cell disruption destabilizes interactions between receptors that can be restored or mimicked by dense crowding. Those interactions may be important for signaling because receptor crowding is also needed to activate CheA *in vitro*.

Evidence That TMEA Traps the Axial Subunits of Trimers

Formation of TMEA crosslinking products was not dependent on the presence of CheA and CheW (Studdert and Parkinson, 2004). In the absence of those proteins, receptor clusters are less tight (Kentner et al., 2006; Lybarger and Maddock, 2000), whereas direct interactions between receptors seem to be unaffected (Ames et al., 2002).

Tsr-S366C derivatives carrying loss-of-function lesions at trimer contact residues (e.g., I377P, E385P) were defective in forming 3-subunit TMEA products when expressed alone (Studdert and Parkinson, 2004) or mixed

TMEA products when coexpressed with Tar-*S366C* (Fig. 4D). In contrast, mutant Tsr molecules with epistatic trimer contact lesions (e.g., N381W), which impair the function of wild-type Tar and other receptors, still formed TMEA crosslinking products, both alone (Studdert and Parkinson, 2004) and when coexpressed with Tar-*S366C* (Fig. 4D). The crosslinking phenotypes of mutant Tsr receptors are fully consistent with their functional properties. Loss-of-function mutants evidently cannot form trimers, whereas epistatic mutants probably form defective trimers that impair the function of other receptors in the cluster (Ames *et al.*, 2002).

TMEA Competition Assay: A Tool for Assessing the Trimer-Forming Ability of Mutant Receptors

The trimer-based interpretation of our TMEA crosslinking results predicts that high relative levels of a receptor with no cysteine reporter (a competitor) should decrease the crosslinking products from a receptor with a TMEA reporter site. To simplify these competition experiments, we built a strain with a chromosomally encoded Tar-*S366C* (Tar·C) reporter and no other receptors (Studdert and Parkinson, 2005). In this host, we expressed different levels of wild-type or mutant Tsr molecules and examined the TMEA crosslinking pattern of Tar·C (Fig. 5A). We expected that Tsr molecules able to join mixed trimers of dimers would cause a corresponding decrease in interdimer-dependent TMEA products (Fig. 5A, trimer-proficient Tsr), whereas Tsr mutants with defects in interreceptor interactions would not interfere with the formation of any Tar·C crosslinking products (Fig. 5A, trimer-deficient Tsr). We found clear examples of both categories of mutants: For example, high levels of Tsr-N376W completely suppressed Tar·C crosslinking products, whereas comparable levels of Tsr-I377P had no effect on Tar·C crosslinking (Fig. 5A).

Several conclusions can be drawn from these competition experiments:

The competition results support our interpretation of direct TMEA crosslinking results. Conceivably, a mutant receptor could appear to be trimer-defective in direct TMEA tests if the mutation simply altered the orientation of the reporter site. However, this explanation can be excluded if the same mutant receptor fails to compete with Tar·C for mixed trimer formation, as proved to be the case for the prototypical trimer-defective lesion, I377P.

TMEA products mainly arise from *inter*dimer crosslinking events. Formation of crosslinks between the subunits of a dimer should not be subject to competition by a heterologous receptor because Tar and Tsr do not form heterodimers. Yet, trimer-proficient competitors blocked all Tar·C crosslinking products. Thus, both the 2- and 3-subunit TMEA products of

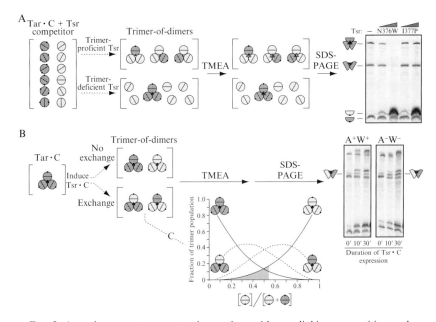

FIG. 5. Assessing receptor–receptor interactions with crosslinking competition and exchange assays. Tsr and Tar dimers are depicted with the conventions described in Figs. 2, 3, and 4. (A) Competition assay for evaluating the trimer-forming ability of Tsr molecules that do not bear a cysteine reporter site. Cells expressing a physiological level of Tar molecules with the *S366C* reporter (Tar·C) and different levels of a Tsr competitor were treated with TMEA and analyzed as described in Fig. 4. High levels of trimer-proficient Tsr molecules (e.g., N376W) prevent Tar crosslinking, whereas trimer-deficient Tsr mutants (e.g., I377P) do not. (B) Exchange assay for evaluating the stability of preformed trimers. Cells were allowed to express a Tar·C reporter at physiological levels for many generations, then induced for expression of a Tsr·C molecule, treated with TMEA, and the receptor crosslinking products analyzed by gel electrophoresis and immunoblotting. (C) Expected proportions of different trimer compositions if preexisting trimers freely exchange members with the pool of newly synthesized receptor molecules. The exchange assay is most sensitive at relatively low levels of newly made receptors (shaded portion of plot). Exchange efficiency was quantified by comparing the observed and expected proportions of mixed and pure two-subunit crosslinking products. Note that trimer exchanges occur much more readily in cells lacking the CheA and CheW proteins.

trimer-proficient receptors arise from *inter*dimer crosslinks. Some mutant receptors (e.g., Tsr-I377P) cannot form trimer-based crosslinks, but do form *intra*dimer crosslinks, due presumably to destabilization of the receptor tip where the mutation lies.

Because a competing receptor does not require a TMEA reporter site, the competition assay offers a simple method for assessing the trimer-forming potential of any mutant receptor. It could even be used to test heterologous receptors from other organisms, for example, *T. maritima* MCPs, for trimer-based interactions with the Tar·C reporter.

The competition assay should also serve to identify receptors with different types of trimer-forming defects. For example, some trimer-defective mutant receptors may be tolerated as a single member of a mixed trimer. Mutant receptors of this sort would be expected to cause a drastic decrease in the 3-subunit Tar·C crosslinking product and a concomitant increase in the 2-subunit crosslinking product.

Exchange Assay: Dynamic Changes in Trimer Composition as a
    Consequence of Changes in the Receptor Population

The TMEA crosslinking assay provides a snapshot of trimer groupings in the cell, which, in turn, reflects the composition of the entire receptor population. To explore the cellular stability of receptor trimers of dimers, we devised an exchange assay (Fig. 5B) for following the fate of established trimers upon a change in composition of the receptor population (Studdert and Parkinson, 2005). In this assay, a strain carrying a constitutively expressed, chromosomally encoded Tar·C reporter and an IPTG-inducible Tsr·C plasmid was grown to midlog phase and then induced for Tsr·C expression. The cells were treated with TMEA at different times after the onset of induction and their crosslinking products examined. If the preformed Tar·C trimers are highly dynamic, then the composition of trimers after Tsr·C induction should depend entirely on the relative cellular levels of the two receptor types, as depicted in Fig. 5B (exchange alternative). In this case, many of the newly made Tsr·C molecules should be found in mixed trimers, which would yield mixed crosslinking products. However, if preformed Tar·C trimers are stable, more of the newly made Tsr·C molecules should be found in pure crosslinking products (Fig. 5B, no exchange alternative). Thus, to assess the exchangeability of newly made Tsr·C dimers with the preexisting Tar·C population, we compared the measured levels of mixed crosslinking products with those predicted by purely random mixing (Fig. 5C). TMEA treatment was performed at short induction times, when the levels of Tsr·C were relatively low, to maximize the expected difference between the two exchange patterns (Fig. 5C, shaded region).

Although previous experiments showed that CheA and CheW were not required for trimer formation, exchange assays showed that these proteins made trimers exchange-resistant (Fig. 5B). In cells containing both CheA and CheW, pure 2- and 3-subunit crosslinking products were relatively more abundant than mixed crosslinking products. In cells lacking both proteins (or either one, not shown), mixed crosslinking products were relatively more abundant. These differences were most evident in the two-subunit products, for which we defined an "exchange factor" as the ratio of the observed to predicted levels of the Tar·C ≈ Tsr·C product. For the experiment shown in Fig. 5B, the exchange factor for cells lacking CheA and CheW was 0.95, indicative of nearly free exchange between new and old receptor molecules. In contrast, the exchange factor for cells that contained CheA and CheW was 0.35, indicating that the newly made Tsr·C receptors were not freely exchanging with the preexisting Tar·C population. The low-exchange crosslinking pattern persisted for up to 3 h when protein synthesis was inhibited before the TMEA treatment. Moreover, the presence of attractants (serine, aspartate, or both) in the incubation buffer did not increase receptor exchangeability. We conclude that receptors synthesized in the presence of both CheA and CheW assemble into an exchange-resistant complex based on trimers of dimers and that attractant stimuli do not alter the low exchangeability of preassembled receptor trimers.

Concluding Remarks

The experimental approaches described in this chapter were aimed at characterizing the higher-order organization of chemoreceptors in unperturbed cells. To that end, we reproduced normal *in vivo* stoichiometries as much as possible to minimize physiologically irrelevant collisional interactions between receptors. Our experimental designs were guided by the crystal structure of the cytoplasmic domain of Tsr and explored the premise that receptors form trimer-of-dimer arrangements in living cells. These *in vivo* crosslinking experiments have served to

• demonstrate direct interactions between both homologous and heterologous receptors that are not dependent on other chemotaxis signaling proteins (DSP, disulfides, TMEA).

• reveal crosslinking patterns consistent with the proposed trimer-of-dimers geometry for receptor–receptor interactions (disulfides, TMEA).

• demonstrate that amino acid replacements at trimer interface residues disrupt higher-order receptor interactions (DSP, disulfides, TMEA: direct and competition-based).

• define the conditions that promote or inhibit the exchangeability of dimers between trimers.

Disulfide crosslinking of cysteine reporters in the periplasmic domain of Tar has also been used to investigate higher-order receptor interactions in intact cells (Homma *et al.*, 2004; Irieda *et al.*, 2006). Those studies are consistent with a close interaction among three receptor dimers, but unlike the cytoplasmic contacts in trimers of dimers, the efficiency of periplasmic crosslinking was highly dependent on the presence of CheA and CheW and the crosslinking efficiency changed upon addition of attractant. It seems likely that these periplasmic interactions occur between members of different trimers of dimers, for example, through a receptor arrangement in which the periplasmic domains of each member of a trimer abut the periplasmic domains of dimers from other trimers in the receptor array (Kim *et al.*, 2002). In such an array arrangement, the periplasmic interactions between receptors, unlike the relatively static association of signaling domains in the cytoplasmic tip of a trimer, might depend on intertrimer bridging connections provided by CheA and CheW, which could be affected by the signaling state of the receptors. Indeed, using other periplasmic reporter sites, Irieda *et al.* (2006) observed changes in crosslinking efficiency that were consistent with stimulus-induced rotational movements of the individual receptor dimers in the array.

It should be possible to reconcile the structural information provided by periplasmic and cytoplasmic crosslinking approaches with additional *in vivo* studies that utilize both types of reporter sites. Other promising areas for investigation include crosslinking tests of the recently determined structure for a HAMP domain (Hulko *et al.*, 2006) and a search for cytoplasmic reporter sites that are sensitive to the receptor's signaling or modification state.

In vivo crosslinking approaches are one of the few experimental options available for exploring the structure of receptor signaling complexes in their native context. When combined with genetic and physiological studies, the crosslinking techniques described in this chapter should lead to a detailed molecular understanding of structure–function relationships in chemoreceptors. Similar approaches should be applicable to other protein–protein interactions.

Acknowledgments

Research in the authors' laboratories is funded by grant GM19559 from the National Institute of General Medical Sciences and by TW07216, a Fogarty International Research Collaboration Award.

References

Ames, P., Studdert, C. A., Reiser, R. H., and Parkinson, J. S. (2002). Collaborative signaling by mixed chemoreceptor teams in *Escherichia coli*. *Proc. Natl. Acad. Sci. USA* **99**, 7060–7065.
Bray, D., Levin, M., and Morton-Firth, C. (1998). Receptor clustering as a cellular mechanism to control sensitivity. *Nature* **393**, 85–88.
Chelsky, D., and Dahlquist, F. W. (1980). Chemotaxis in *Escherichia coli*: Associations of protein components. *Biochemistry* **19**, 4633–4639.

Feng, X., Baumgartner, J. W., and Hazelbauer, G. L. (1997). High- and low-abundance chemoreceptors in *Escherichia coli*: Differential activities associated with closely related cytoplasmic domains. *J. Bacteriol.* **179**, 6714–6720.

Gestwicki, J. E., Lamanna, A. C., Harshey, R. M., McCarter, L. L., Kiessling, L. L., and Adler, J. (2000). Evolutionary conservation of methyl-accepting chemotaxis protein location in Bacteria and Archaea. *J. Bacteriol.* **182**, 6499–6502.

Gosink, K. K., Buron-Barral, M., and Parkinson, J. S. (2006). Signaling interactions between the aerotaxis transducer Aer and heterologous chemoreceptors in *Escherichia coli*. *J. Bacteriol.* **188**, 3487–3493.

Homma, M., Shiomi, D., and Kawagishi, I. (2004). Attractant binding alters arrangement of chemoreceptor dimers within its cluster at a cell pole. *Proc. Natl. Acad. Sci. USA* **101**, 3462–3467.

Hulko, M., Berndt, F., Gruber, M., Linder, J. U., Truffault, V., Schultz, A., Martin, J., Schultz, J. E., Lupas, A. N., and Coles, M. (2006). The HAMP domain structure implies helix rotation in transmembrane signaling. *Cell* **126**, 929–940.

Irieda, H., Homma, M., and Kawagishi, I. (2006). Control of chemotactic signal gain via modulation of a pre-formed receptor array. *J. Biol. Chem.* **281**, 23880–23886.

Kentner, D., Thiem, S., Hildenbeutel, M., and Sourjik, V. (2006). Determinants of chemoreceptor cluster formation in *Escherichia coli*. *Mol. Microbiol.* **61**, 407–417.

Kim, K. K., Yokota, H., and Kim, S. H. (1999). Four-helical-bundle structure of the cytoplasmic domain of a serine chemotaxis receptor. *Nature* **400**, 787–792.

Kim, S. H., Wang, W., and Kim, K. K. (2002). Dynamic and clustering model of bacterial chemotaxis receptors: Structural basis for signaling and high sensitivity. *Proc. Natl. Acad. Sci. USA* **99**, 11611–11615.

Kosower, N. S., and Kosower, E. M. (1995). Diamide: An oxidant probe for thiols. *Methods Enzymol.* **251**, 123–133.

Lybarger, S. R., and Maddock, J. R. (2000). Differences in the polar clustering of the high- and low-abundance chemoreceptors of *Escherichia coli*. *Proc. Natl. Acad. Sci. USA* **97**, 8057–8062.

Maddock, J. R., and Shapiro, L. (1993). Polar location of the chemoreceptor complex in the *Escherichia coli* cell. *Science* **259**, 1717–1723.

Milburn, M. V., Prive, G. G., Milligan, D. L., Scott, W. G., Yeh, J., Jancarik, J., Koshland, D. E., Jr., and Kim, S. H. (1991). Three-dimensional structures of the ligand-binding domain of the bacterial aspartate receptor with and without a ligand. *Science* **254**, 1342–1347.

Milligan, D. L., and Koshland, D. E., Jr. (1988). Site-directed cross-linking. Establishing the dimeric structure of the aspartate receptor of bacterial chemotaxis. *J. Biol. Chem.* **263**, 6268–6275.

Park, S. Y., Borbat, P. P., Gonzalez-Bonet, G., Bhatnagar, J., Pollard, A. M., Freed, J. H., Bilwes, A. M., and Crane, B. R. (2006). Reconstruction of the chemotaxis receptor-kinase assembly. *Nat. Struct. Mol. Biol.* **13**, 400–407.

Parkinson, J. S., Ames, P., and Studdert, C. A. (2005). Collaborative signaling by bacterial chemoreceptors. *Curr. Opin. Microbiol.* **8**, 116–121.

Ritz, D., and Beckwith, J. (2001). Roles of thiol-redox pathways in bacteria. *Annu. Rev. Microbiol.* **55**, 21–48.

Sourjik, V. (2004). Receptor clustering and signal processing in *E. coli* chemotaxis. *Trends Microbiol.* **12**, 569–576.

Sourjik, V., and Berg, H. C. (2000). Localization of components of the chemotaxis machinery of *Escherichia coli* using fluorescent protein fusions. *Mol. Microbiol.* **37**, 740–751.

Sourjik, V., and Berg, H. C. (2002). Receptor sensitivity in bacterial chemotaxis. *Proc. Natl. Acad. Sci. USA* **99,** 123–127.

Sourjik, V., and Berg, H. C. (2004). Functional interactions between receptors in bacterial chemotaxis. *Nature* **428,** 437–441.

Studdert, C. A., and Parkinson, J. S. (2004). Crosslinking snapshots of bacterial chemoreceptor squads. *Proc. Natl. Acad. Sci. USA* **101,** 2117–2122.

Studdert, C. A., and Parkinson, J. S. (2005). Insights into the organization and dynamics of bacterial chemoreceptor clusters through *in vivo* crosslinking studies. *Proc. Natl. Acad. Sci. USA* **102,** 15623–15628.

Szurmant, H., and Ordal, G. W. (2004). Diversity in chemotaxis mechanisms among the bacteria and archaea. *Microbiol. Mol. Biol. Rev.* **68,** 301–319.

Zhulin, I. B. (2001). The superfamily of chemotaxis transducers: From physiology to genomics and back. *Adv. Microb. Physiol.* **45,** 157–198.

[20] A "Bucket of Light" for Viewing Bacterial Colonies in Soft Agar

By JOHN S. PARKINSON

Abstract

The morphologies of bacterial colonies in soft agar media can provide a wealth of information about a strain's locomotor and chemotactic abilities. Photographic images are often the simplest and most effective means of documenting these behavioral phenotypes. Uniform, indirect, transmitted illumination of the plates is essential for obtaining good colony images. This brief chapter describes a simple and relatively inexpensive illumination device for viewing and photographing bacterial colonies in soft agar.

Viewing Colonies Grown in Soft Agar

Soft agar assays for motility, chemotaxis, and aerotaxis are described in detail elsewhere in this volume (Ames and Parkinson, 2007; Taylor *et al.*, 2007). Briefly, in soft agar media, flagellated bacteria such as *E. coli* swim within the water-filled tunnels in the agar matrix. The expanding colony typically extends from the bottom of the plate to the top surface of the plate with the majority of the cells embedded in the agar rather than at the air interface. These colonies can contain one or more rings of cells that are concentrated near the outer boundaries of nutrient, energy, or electron-acceptor gradients generated through the cells' metabolic activities. The size, thickness, relative position, and depth of each ring provide important information about a strain's motility and tactic behavior.

Direct light from above is ill-suited for viewing or photographing soft agar colonies. Cells on the surface of the agar reflect much of the incident light, whereas those within the agar receive and reflect much less light. Consequently, this type of illumination obscures much of the detailed structure of the colonies (Fig. 1A). In contrast, indirect light from behind the agar more evenly illuminates the colonies and brings out their fine structure features, such as internal rings or fuzzy edges (Fig. 1B). However, the light should not pass from the source directly upward through the plate, but rather should enter only the bottom half of the plate and at an oblique angle, approximating a dark-field effect.

METHODS IN ENZYMOLOGY, VOL. 423   0076-6879/07 $35.00

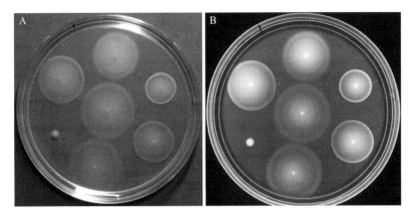

FIG. 1. Colonies in soft agar photographed with reflected (A) or transmitted (B) light. Isogenic *E. coli* strains with different chemotaxis phenotypes were inoculated into tryptone soft agar medium and incubated at 30° for 8 h. In panel (A), the plate was illuminated with incident fluorescent light from a desk lamp. In panel (B), the plate was illuminated with the bucket of light described in this chapter. Both photographs were taken with a digital camera at the same magnification and resolution and at an automatic exposure setting chosen by the camera.

Building a Bucket of Light

A dark-field illumination source for viewing bacterial colonies in soft agar plates can be built by a metalworking shop for less than $300. A design used by a number of research groups employs a circular fluorescent tube [22W Sylvania rapid-start cool white, model FC8T9 or equivalent] inside a cylindrical metal container (Fig. 2). Aluminum is the preferred material because of its relatively light weight, but other metals should also work. The cylinder is formed from a 1/8" (~2 mm) plate, approximately 18×83 cm, with the rolled-up edges joined by brazing. Enterprising researchers might be able to find ready-made circular containers with roughly the same dimensions, for example, a metal wastebasket. In our experience, none of the dimensions is very critical for good performance.

Both ends of the cylinder are covered by circular plates made from 1/4" (~6 mm) aluminum plate, but other materials could probably be substituted, particularly for the bottom plate. The top plate contains a circular hole ~82 mm in diameter that has a machined lip ~2 mm deep and 3 mm wide for supporting a standard-size plastic Petri dish (see enlarged cross-section view in Fig. 2). The dimensions of the viewing hole could presumably be changed to accommodate other sizes of culture plates, perhaps even square plates, although we have not tried them. The bottom plate is primarily intended to protect the electronics for the fluorescent tube and can

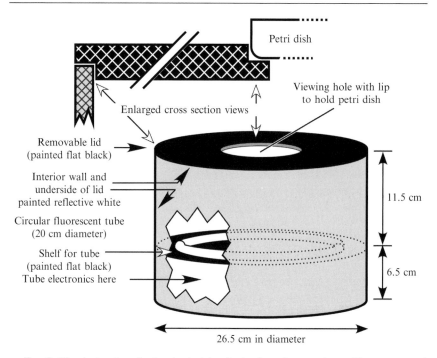

FIG. 2. Illuminator for viewing bacterial colonies in soft agar plates. The nature and dimensions of the stock material are provided in the text.

be mounted to the cylinder by brazing, machine screws, or some other means. The top plate should be removable for cleaning and maintenance access. This can be accomplished by machining a lip around its outside edge that allows it to nest on the top and against the inside edge of the cylinder wall (see enlarged cross-section view in Fig. 2).

The fluorescent tube is supported by clips mounted to a solid circular platform positioned approximately 1/3 of the overall height from the bottom of the cylinder. This platform can be mounted with machine screws (for maintenance access) to a few support posts brazed to the inside wall of the cylinder. The electronics for the fluorescent light are installed below this platform, which can be made from aluminum or any other suitable material. The upper surface of the platform is spray painted flat black to provide a dark, nonreflective viewing background. The inside wall of the cylinder and the underside of the lid should be spray painted with flat white paint to diffuse the light and to provide uniform intensity within the field of view.

In use, the fluorescent tube should not be visible when looking straight down through the viewing hole. If this proves to be a problem, a cylindrical shield (13–15 cm in diameter and 3–6 cm in height) can be made from black cardboard and placed on the tube support platform.

Acknowledgments

Thanks to Mark Johnson and Barry Taylor for helpful comments on the manuscript. Research in the author's laboratory is funded by grant GM19559 from the National Institute of General Medical Sciences.

References

Ames, P., and Parkinson, J. S. (2007). Phenotypic suppression methods for analyzing intra- and inter-molecular signaling interactions of chemoreceptors. *Methods Enzymol.* **423,** 436–457.
Taylor, B. L., Watts, K. J., and Johnson, M. S. (2007). Oxygen and redox sensing by two-component systems that regulate behavioral responses: Behavioral assays and structural studies of Aer using *in vivo* disulfide crosslinking. *Methods Enzymol.* **422,** 190–232.

[21] Phenotypic Suppression Methods for Analyzing Intra- and Inter-Molecular Signaling Interactions of Chemoreceptors

By PETER AMES and JOHN S. PARKINSON

Abstract

The receptors that mediate chemotactic behaviors in *E. coli* and other motile bacteria and archaea are exquisite molecular machines. They detect minute concentration changes in the organism's chemical environment, integrate multiple stimulus inputs, and generate a highly amplified output signal that modulates the cell's locomotor pattern. Genetic dissection and suppression analyses have played an important role in elucidating the molecular mechanisms that underlie chemoreceptor signaling. This chapter discusses three examples of phenotypic suppression analyses of receptor signaling defects. (i) Balancing suppression can occur in mutant receptors that have biased output signals and involves second-site mutations that create an offsetting bias change. Such suppressors can arise in many parts of the receptor and need not involve directly interacting parts of the molecule. (ii) Conformational suppression within a mutant receptor molecule occurs through a mutation that directly compensates for the initial structural defect. This form of suppression should be highly dependent on the nature of the structural alterations caused by the original mutation and its suppressor, but in practice may be difficult to distinguish from balancing suppression without high-resolution structural information about the mutant and pseudorevertant proteins. (iii) Conformational suppression between receptor molecules involves correction of a functional defect in one receptor by a mutational change in a heterologous receptor with which it normally interacts. The suppression patterns exhibit allele-specificity with respect to the compensatory residue positions and amino acid side chains, a hallmark of stereospecific protein–protein interactions.

Introduction

Motile bacteria exhibit sophisticated chemotactic behaviors that are good models for exploring the molecular mechanisms that proteins use to detect and process sensory information about their chemical environment (reviewed by Armitage, 1999; Parkinson *et al.*, 2005; Sourjik, 2004; Wadhams and Armitage, 2004). *Escherichia coli*, the best-studied

METHODS IN ENZYMOLOGY, VOL. 423
0076-6879/07 $35.00
DOI: 10.1016/S0076-6879(07)23021-6

chemotactic organism, tracks very shallow gradients of attractant and repellent chemicals by continuously scanning for temporal changes in chemoeffector concentrations as it swims about. Favorable stimuli—for example, an increasing attractant level—suppress the likelihood of a directional change, thereby prolonging cell movement in the favorable direction. *E. coli* senses temporal concentration changes by comparing current conditions to those averaged over the past few seconds in its travels. Concentration changes as small as 0.1% can produce much larger changes in the rotational behavior of the flagellar motors, corresponding to a roughly 50-fold amplification of the input stimulus. An adaptation system that tunes the sensory machinery to match ambient chemoeffector levels maintains sensitive gradient detection over a nanomolar to millimolar concentration range.

Chemoreceptors known as methyl-accepting chemotaxis proteins (MCPs) mediate these remarkable signaling feats (see reviews by Falke and Hazelbauer, 2001; Zhulin, 2001). MCPs are transmembrane homodimers with a periplasmic sensing domain and a cytoplasmic signaling domain (Fig. 1A). MCPs monitor chemoeffector levels through the occupancy state of their periplasmic ligand-binding domains. Past chemical conditions are recorded in the form of reversible covalent modifications at 4 to 6 specific glutamic acid residues in the MCP signaling domain. CheR, a methyltransferase, adds methyl groups to MCP molecules; CheB, a methylesterase, hydrolyzes MCP glutamyl-methyl esters back to glutamic acid. The MCP signaling domain forms stable ternary complexes with two cytoplasmic proteins, CheA, a histidine kinase, and CheW, which couples CheA to receptor control. The chemoreceptor signaling complex transmits phosphoryl groups to the response regulators CheB and CheY to control the cell's swimming behavior. Phospho-CheY interacts with the flagellar motor to enhance the probability of clockwise (CW) rotation, which causes random directional turns. Counterclockwise (CCW) rotation produces forward swimming, the default behavior. Phospho-CheB, the active form of the MCP methylesterase, is part of a feedback sensory adaptation circuit that adjusts the sensitive detection range of the receptors.

Although a receptor molecule can modulate the activity of its associated CheA partner over a 1000-fold range, this control alone is not responsible for the prodigious signal gain in the chemotactic signaling system. Rather, each receptor is able to regulate about 35 CheA molecules, most of which are probably physically associated with other receptor molecules (Sourjik and Berg, 2002). Signal amplification appears to arise through communication among receptor molecules in a cooperative signaling array (Sourjik and Berg, 2004). Receptor dimers form trimers of dimers that are, in turn, networked to other trimer-based receptor teams, possibly through shared

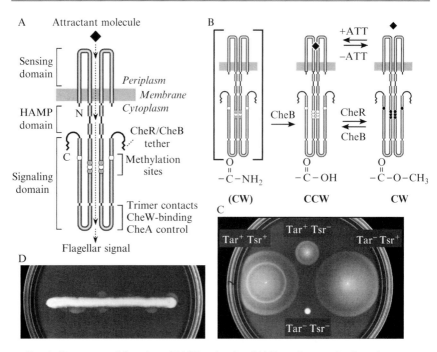

FIG. 1. Structure and function of MCP molecules. (A) Domain organization of transmembrane chemoreceptors of the MCP family. MCP subunits are ~550 residues in length; the native molecule is a homodimer. Thick segments denote polypeptide segments that have predominantly alpha-helical secondary structure. The periplasmic sensing domain contains two symmetric ligand-binding sites at the dimer interface (see Fig. 3D). Ligand-binding events trigger conformational changes that are transmitted through the membrane-spanning segments to the HAMP domain, which, in turn, regulates the activity of the signaling domain. The signaling domain helices of each subunit form anti-parallel coiled-coils that interact in the dimer to form a four-helix bundle. Conformational changes at the tip of the helix bundle regulate CheA activity, possibly by modulating receptor–receptor interactions in trimer-of-dimer signaling complexes. The modification status of the intervening methylation sites determines the sensitive range of the input–output connection. In *E. coli*, the high-abundance MCPs, Tar and Tsr, carry a pentapeptide sequence (NWETF) at their C-termini that interacts with the methylation (CheR) and deamidation/demethylation (CheB) enzymes. *E. coli* MCPs typically have two methylation sites (gray circles) that are synthesized as glutamine (Q) residues and subsequently converted to glutamic acid (E) by CheB-mediated deamidation. The remaining methylation sites (white circles) are synthesized as glutamic acid residues that are immediately competent to accept a methyl group, forming a glutamyl methyl ester. (B) Two-state model of receptor signaling and CheR/CheB-mediated modification reactions. MCP molecules can exist in CheA-activating (CW) or -deactivating (CCW) output states. The chemical structures of the methylation sites associated with each signaling state are shown beneath the molecules. Newly synthesized receptors (QEQE) have transient CW output until irreversible deamidation is complete (EEEE). In mature MCP molecules, the proportion of receptors in each signaling state is controlled by the interplay between ligand occupancy and modification state. (C) Chemotaxis phenotypes assayed on soft agar plates. The medium was tryptone broth

connections to their CheA and CheW signaling partners (Ames *et al.*, 2002; Studdert and Parkinson, 2004, 2005). The trimers can contain receptors of different detection specificities, any of which can modulate team signal output and relay sensory information to other receptor signaling teams (Parkinson *et al.*, 2005).

The signaling properties of a receptor team can be understood in terms of two alternative signaling states (Fig. 1B), a CW mode that activates CheA autophosphorylation and a CCW mode that deactivates CheA. The proportion of receptor molecules or signaling teams in each state and, consequently, the cell's behavior, reflect the interplay between ligand occupancy and methylation state averaged over the receptor population (Fig. 1B). Attractant increases, for example, drive receptor molecules toward the CCW signaling state, reducing CW motor rotation and initiating a signal-offsetting increase in MCP methylation state through feedback control of CheB activity. The cell "remembers" the chemoeffector change until sensory adaptation is complete.

Genetic Analyses of Chemoreceptors

Genetic methods and logic compose a powerful tool for dissecting structure–function relationships in proteins. In bacterial chemotaxis, genetic studies of the *E. coli* system over the past 40 years have served to identify the various components of the signaling pathway and their overall

---

(10 g/l Difco tryptone, 5 g/l NaCl) containing ~3 g/l Difco bacto-agar. The firmness of agar varies from lot to lot; each new lot must be empirically calibrated (typically over a range of 2–3 g/l in agar concentration). Colonies were inoculated with cells (usually from a fresh overnight colony, but liquid cultures also work) on a sterile toothpick stabbed to the bottom of the plate. Strains: RP437 (Tar$^+$ Tsr$^+$); RP8604 (Tar$^+$ Tsr$^-$); RP8606 (Tar$^-$ Tsr$^+$); RP8611 (Tar$^-$ Tsr$^-$). Plates were incubated at 32.5° for 8 h. (D) Selection for chemotactic revertants on tryptone soft agar. Approximately 50 μl of liquid culture containing nonchemotactic parental cells and rare chemotactic revertant or pseudorevertant cells (at spontaneous or mutagen-induced frequencies) were gently spread on the surface of a soft agar plate with a plastic pipette tip so as to avoid tearing the surface of the agar. The plate was incubated overnight (~17 h) at 32.5°. The number of cells in the inoculum should be adjusted to obtain ~10 discrete revertant "flares." If desired, nonmotile or nonchemotactic cells can be added to the inoculum to hasten formation of attractant gradients through growth in the cell streak. The "helper" cells should be incapable of reverting, for example, cells that have deletions of motility or chemotaxis genes. Note that the revertants in this example arose before placing the cells on the plate. This method can also be used to select spontaneous revertants that arise during growth on the soft agar plate, but the cell density of the inoculum and the overall incubation time will probably need to be increased to detect such reversion events. However, it is important to avoid prolonged incubation times (>24 h), which may yield undesirable multi-step revertants.

"wiring diagram" (Parkinson, 1977, 1993) Genetic approaches have also provided valuable insights into the functional architecture of individual chemoreceptor molecules (Ames *et al.*, 1988) and, more recently, of trimers of dimers and chemoreceptor clusters (Ames and Parkinson, 2006; Ames *et al.*, 2002). Chemoreceptor defects that eliminate chemotactic ability in *E. coli* include lesions that affect maturation and stability of the native protein (Buron-Barral *et al.*, 2006; Butler and Falke, 1998; Danielson *et al.*, 1997; Ma *et al.*, 2005), ligand-binding determinants (Gardina and Manson, 1996; Lee and Imae, 1990; Wolfe *et al.*, 1988; Yaghmai and Hazelbauer, 1992, 1993), transmembrane segments (Chen, 1992; Jeffery and Koshland, 1994, 1999; Maruyama *et al.*, 1995; Nishiyama *et al.*, 1999; Oosawa and Simon, 1986), the HAMP domain (Ames and Parkinson, 1988), input/output control (Coleman *et al.*, 2005; Trammell and Falke, 1999), CheA control (Ames and Parkinson, 1994), sensory adaptation determinants (Nara *et al.*, 1996; Nishiyama *et al.*, 1997; Shapiro *et al.*, 1995; Shiomi *et al.*, 2000, 2002; Starrett and Falke, 2005), and trimer formation (Ames and Parkinson, 2006; Ames *et al.*, 2002). Most of these functional defects reflect loss-of-function lesions, but gain-of-function lesions that lock receptor output can also abrogate chemotactic responses (Ames and Parkinson, 1988; Mutoh *et al.*, 1986).

Knowing the primary structure change in a mutant protein (inferred from the DNA sequence of the mutant gene) may not shed much light on the nature of its functional defect. A powerful genetic approach that can provide insight into the functional nature of a mutant defect involves the isolation and characterization of secondary mutational changes that restore some measure of function to the mutant protein (Manson, 2000). Such second-site suppressors can arise within the original mutant gene (intragenic suppressors) or in some other gene (extragenic suppressors) whose product functionally interacts with the mutant protein. This chapter discusses three suppression case studies of chemoreceptor mutants to illustrate both the power and the pitfalls of this approach. The subjects of these studies are Tsr, the serine receptor, and Tar, the aspartate receptor, the predominant receptor types in *E. coli*. However, the lessons learned from Tar and Tsr are generally applicable to the low-abundance chemoreceptors (Aer, Tap, Trg) as well. We begin with a general discussion of the basic genetic tools available in the chemotaxis system.

Soft Agar Chemotaxis Assays

E. coli chemoreceptor mutants exhibit distinctive colony morphologies on nutrient soft agar plates (Fig. 1C). At agar concentrations of ~3 g/l, cells can swim in the water-filled tunnels created by the agar matrix. As the colonies grow, the cells consume nutrients in the medium, such as aspartate and serine, that are chemotattractants. If the cells are able to detect and respond to those chemoeffectors, the colony expands rapidly, consuming and following the attractant gradients. On complex tryptone medium, serine is the first attractant to be exhausted, so strains capable of serine chemotaxis (Tsr$^+$) form a band of cells at the colony margin that expands outward as they consume the serine (Fig. 1C). Cells left behind the serine pioneers no longer have free serine to eat and so consume aspartate, leading to a second ring of cells doing aspartate chemotaxis. Tsr$^+$ strains that lack the aspartate receptor (Tar$^-$) expand equally fast, but lack the inner aspartate ring (Fig. 1C). In contrast, Tar$^+$ colonies lacking Tsr function expand more slowly than does the aspartate ring inside wild-type colonies because the Tsr$^-$ Tar$^+$ strain must still consume serine before it can establish an aspartate gradient (Fig. 1C). Strains lacking both of these major chemoreceptors (Tar$^-$ Tsr$^-$) are generally nonchemotactic (Che$^-$) because the remaining low-abundance receptors (Tap, dipeptides; Trg, ribose and galactose; Aer, aerotaxis) cannot by themselves sufficiently activate CheA to establish a suitable balance of running and tumbling behavior (Fig. 1C). Mutant strains lacking any of the shared components of the chemotaxis signaling pathway also exhibit Che$^-$ phenotypes. Those cells are motile, but either constantly running (CCW-biased: CheA$^-$, CheW$^-$, CheR$^-$, CheY$^-$) or constantly tumbling (CW-biased: CheB$^-$ and CheZ$^-$) and, consequently, cannot track chemoeffector gradients.

Colonies on soft agar plates are usually inoculated with a toothpick carrying cells from fresh colonies growing on hard agar plates, ideally composed of the same growth medium. The toothpicks are stabbed nearly to the bottom of the soft agar plate, which is then incubated until colonies are sufficiently large to score their chemotaxis phenotypes (typically, 6–10 hours at 30–35°). To select chemotactic revertants from a nonchemotactic parent, the mutant cells are generally inoculated in a stripe across the surface and incubated overnight or longer, depending on the frequency of reversion events. Chemotactic revertants appear as small "flares" that emanate from the border of the parental stripe (Fig. 1D).

The Pros and Cons of Plasmids

The case studies described in the following text were carried out with plasmid-borne chemoreceptor genes, an experimental approach that facilitates many aspects of suppression analyses: targeted mutagenesis, control of gene expression levels, large-scale revertant hunts, DNA sequencing analyses, and rapid transfer of mutant genes to new genetic backgrounds. However, plasmids can also complicate genetic analyses and it is important to understand and appreciate their limitations.

Expression level effects. Optimal chemotactic behavior depends on proper stoichiometry of the chemotaxis signaling components. Owing to relatively high copy numbers, plasmid-borne genes may express too much product for optimal performance. If the plasmid-borne gene is accompanied by its native promoter, it may titrate positive regulatory factors needed for transcription of other flagellar and chemotaxis genes, leading to a significant disparity in the expression levels of plasmid and chromosomal gene products. Foreign promoters that do not compete for shared transcription factors can alleviate this problem, but may result in high basal expression levels of the plasmid-borne gene(s), depending on the tightness of their regulatory controls. Our criterion for choosing plasmid constructs for genetic studies is that the plasmid-borne gene be able to produce optimal chemotactic ability at an intermediate induction level. This ensures that its uninduced expression level is below that of its chromosomally encoded counterpart and obviates aberrant stoichiometry effects.

Genotypic and phenotypic lag. A cell containing a multicopy plasmid is quite slow to produce mutant offspring following a loss-of-function mutation in one of the plasmid molecules. The long lag in mutant appearance is due to two factors: (1) the need to dilute away the functional nonmutant gene products present in the parental cell (phenotypic lag), and (2) the need to generate a progeny cell with all mutant plasmids (genotypic lag). Genotypic lag is the more severe problem because plasmids usually partition randomly, but more or less equally, into daughter cells at division. A single mutant plasmid among many nonmutant ones in a cell will need many generations of random segregation and genetic drift before it becomes fixed in the plasmid population of a descendant cell. Genotypic lag is further exacerbated by recombination or replication events that create multimeric plasmids with several tandem copies in the same molecule.

Segregation lag should not be a significant factor when selecting for gain-of-function revertants from a mutant parental plasmid. Any mutation that restores function should be expressed phenotypically as soon as it arises, provided that multiple copies of the mutant gene do not interfere with that function. However, subsequent analysis of the gain-of-function mutation may be confounded by the dominant nature of the revertant phenotype. The revertant cell will probably not carry a genetically homogeneous plasmid population. Plasmid dimers are especially insidious because they can remain heterozygous through single-plasmid transformations of new host cells.

Plasmid maintenance. Plasmids represent an added genetic load to their host cell and must be maintained by positive selection, most often for a plasmid-encoded antibiotic resistance trait. Growth in the presence of antibiotics seems to have a generalized dampening effect on chemotactic performance in soft agar plates, perhaps owing to slowed growth rates. To ameliorate this effect as much as possible, we routinely halve the usual concentration of an antibiotic for use in soft agar plates.

All of these plasmid-related problems can be overcome or circumvented. For example, plasmid dimers can be eliminated by linearizing a population of plasmid molecules with a restriction enzyme that cuts once per monomer and religating the linear products at low DNA concentration to "clone" monomeric forms of the plasmid. Similarly, two different copies of a mutant gene can be carried on compatible plasmids for complementation analyses, using recombination-deficient host cells to avoid recombination events between the multicopy parental plasmids. Thus, with foreknowledge and appropriate precautions, the technical advantages of manipulating chemo-receptor genes on plasmids vastly outweigh the potential drawbacks. The plasmid-specific details of these manipulations will not be explicitly discussed in the case studies to follow.

Balancing Suppression: Methylation-Independent Chemoreceptors

According to the two-state signaling model, chemoreceptor molecules with low methylation states have CCW output; those with high methylation states produce CW output (Fig. 1B). Receptors shift between these two signaling modes upon changes in ligand occupancy and upon subsequent compensatory changes in methylation state. Control of receptor methyla-tion level occurs in two ways, through stimulus-induced changes in the substrate properties of the receptor molecules for the CheR and CheB enzymes and through feedback regulation of CheB activity by phosphoryla-tion. Cells lacking CheR and/or CheB function are generally nonchemotac-tic because they cannot adjust their chemoreceptor methylation levels. The receptor molecules in CheB mutants are fully methylated (CW-signaling); those in CheR mutants are fully demethylated (CCW-signaling) (Fig. 2A). These defects, respectively, cause incessantly tumbling or constantly run-ning swimming behaviors that preclude gradient tracking in soft agar plates. Thus, in adaptation-defective cells, wild-type receptors have locked output signals and cannot mediate chemotaxis.

In CheR-deficient cells, the serine receptor (Tsr) readily acquires the ability to promote chemotaxis-like colony expansion on soft agar plates (Fig. 2B). Single amino acid replacements in many parts of the Tsr molecule can create CheR-independent receptors (designated Tsr!): near the ligand-binding sites in the sensing domain, in the membrane-spanning segments, in the HAMP domain, near the methylation site residues, and near the tip of the cytoplasmic signaling domain (Ames, unpublished results). In CheR$^+$ cells, most Tsr! receptors mediate robust serine chemotaxis, demonstrating

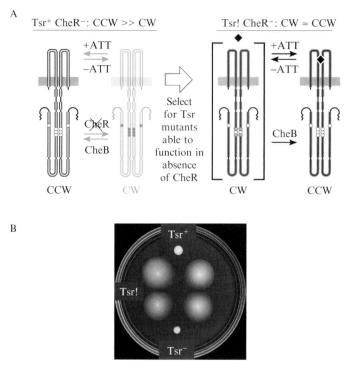

FIG. 2. Gain-of-function suppression of a methyltransferase defect. (A) Summary of mutant phenotypes and suppressor selection. Wild-type Tsr molecules cannot function properly in the absence of CheR function because the mature receptors are trapped in the unmethylated (CCW) signaling state. Selection for chemotactic ability yields Tsr! mutations that most likely extend the lifetime or enhance the CW-signaling activity of undeamidated Tsr molecules. (B) Examples of Tsr phenotypes in the absence of CheR function. Various Tsr-bearing plasmids were transferred to a Δtsr $\Delta cheR$ recipient strain and tested for chemotactic ability on tryptone soft agar. Plasmids expressing wild-type Tsr (Tsr⁺) or no Tsr (Tsr⁻) cannot support chemotaxis in a CheR-deficient host, whereas Tsr! receptors do. Tsr! alleles were, reading clockwise from upper left: L291F, E304K, A495T, A498T. The plate was incubated at 30° for 15 h.

that their signaling properties are quite similar to those of wild-type Tsr under adaptation-competent conditions. However, in CheR⁻ cells, Tsr! mutations probably shift the unmethylated receptor molecules toward the CW-signaling state, resulting in a more balanced CCW/CW distribution of signaling states in the receptor population (Fig. 2A). Evidently, subtle structural changes in many parts of the Tsr molecule can accomplish this balancing act, conceivably by a variety of mechanisms. Ligand-binding and HAMP alterations might attenuate the attractant-bound (CCW) signaling state or mimic a repellent-sensing (CW) condition. Structural alterations near the adaptation sites might mimic the methylated (CW) conformation. Alterations in the signaling tip might directly enhance the stability of the CheA-activating (CW) output state.

Do Tsr! suppressors restore true chemotaxis in the absence of a methylation system? Probably not. We know that colony expansion by Tsr! CheR$^-$ strains on soft agar plates involves serine sensing because serine-binding lesions abrogate that behavior. However, Tsr! receptors mediate rather slow colony expansion in CheR$^-$ strains relative to CheR$^+$ strains, implying an inefficient gradient-tracking strategy. Moreover, efficient chemotaxis most likely requires a system for sensory adaptation and it is difficult to imagine how CheB alone could *reversibly* modulate the CCW/CW equilibrium of a Tsr! receptor population in CheR$^-$ cells. Nevertheless, Tsr! CheR$^-$ "chemotaxis" remains CheB-dependent, suggesting that CheB-mediated deamidation reactions, the only ones possible in the absence of methylation, play some role in colony expansion. One behavioral scenario is that Tsr! CheR$^-$ colonies expand, albeit inefficiently, by modulating down-gradient rather than up-gradient cell movements. Owing to their intrinsic CW bias, unmethylated Tsr! molecules should be more sensitive than their wild-type counterparts to decreasing serine levels, which would tend to shift the receptor population toward the CW output state and elevated tumbling probability (Fig. 2A). Subsequent CheB-mediated deamidation of nascent MCP molecules might serve as a sensory adaptation mechanism by restoring a more balanced CCW/CW distribution in the receptor population. Thus, outward expansion of Tsr! CheR$^-$ colonies might occur through a difference in average cell path lengths between random up-gradient excursions and stimulus-enhanced tumbling during down-gradient travels.

Shiomi *et al.* (2002) demonstrated a similar type of balancing suppression in the aspartate receptor, Tar. The C-termini of Tsr and Tar molecules carry a pentapeptide sequence (NWETF) to which the CheR and CheB adaptation enzymes can bind (Fig. 1A). Binding tethers the adaptation enzymes to the receptor cluster and enhances their activities on Tar and Tsr molecules as well as on their low-abundance neighbors (Tap and Trg), which lack the NWETF sequence. Tar and Tsr mutants that lack the tethering sequence are inefficiently methylated and confer phenotypes very similar to those of wild-type receptors in a CheR-defective strain. Shiomi *et al.* isolated pseudorevertants of a Tar mutant lacking the NWETF sequence and showed that they contained second-site mutations in the Tar signaling domain that imparted a CW output bias.

The sensory adaptation capacity of MCP molecules makes balancing suppression a very common mechanism for restoring receptor function. Most loss-of-function receptor lesions shift the equilibrium between CCW and CW signaling states beyond the control range of the sensory adaptation system, resulting in excessively CCW or CW signal output in the absence of stimuli. Such receptor mutants can readily regain function through a variety of second-site mutations that create an offsetting bias change to bring the CCW-CW equilibrium back into the control range of the adaptation system. Because so many parts of the receptor molecule influence its signal output, the initial bias alteration and its suppressor need not affect the same step in signal production. For example, a receptor mutant with a binding site lesion that mimics ligand occupancy could most likely be phenotypically suppressed by a mutational change in one of the methylation sites. Obviously, the ligand-binding and methylation sites do not interact directly with one another, but rather both regions influence the receptor's signal output. In conclusion, balancing suppression is a good way to isolate mutations that cause a particular type of output bias, but reveals little about their structural and functional relationship to the initial signaling defect of the mutant receptor.

Conformational Suppression within Receptor Molecules

Second-site suppressors in a mutant receptor gene can also identify direct structural interactions within receptor molecules, but it may be difficult to distinguish a compensatory structural interaction from less direct balancing suppression mechanisms. Here, we describe a well-studied example of what appears to be conformational suppression between mutations in a receptor molecule, which nevertheless defies an entirely satisfactory mechanistic explanation.

MCP subunits have two membrane-spanning segments, one (TM1) leading to the periplasm and a second (TM2) that returns the polypeptide to the cytoplasm (Fig. 3A). In a seminal attempt to investigate the signaling role of these transmembrane segments, Oosawa and Simon (1986) engineered a loss-of-function lesion in TM1 of the aspartate receptor, Tar, and then isolated and characterized second-site suppressor mutations in the mutant gene. The starting mutation, A19K, in a residue thought to lie near the middle of TM1, eliminated aspartate chemotaxis on soft agar plates even though the mutant protein was membrane-localized and bound aspartate with normal affinity. Cells containing Tar-A19K as their sole receptor exhibited CCW-biased signal output, consistent with a substantial defect in activating the CheA kinase.

To explore the basis for the Tar-A19K signaling defect, Oosawa and Simon selected chemotactic pseudorevertants as spontaneous cell flares on soft agar medium (see Fig. 1D). A number of second-site mutations were found to restore function to Tar-A19K (Fig. 3B), including one in TM1 (V17E) and four in TM2 (W192R, A198E, V201E, V202L). Based on these

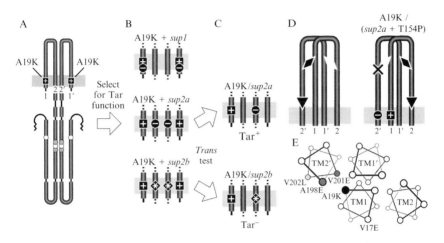

FIG. 3. Intragenic, second-site suppression of a transmembrane segment lesion. (A) A mutant Tar homodimer carrying the A19K mutation ("+" symbol) in the first transmembrane segment (TM1) of each subunit cannot mediate chemotactic responses to aspartate. (B) Selection for chemotactic pseudorevertants of Tar-A19K yielded second-site suppressor mutations in TM1 (*sup1*) or in the second transmembrane segment TM2 (*sup2*). Some of the suppressors introduced an acidic residue ("−" symbols); other suppressors had nonacidic amino acid replacements ("*" symbols). (C) *Trans* test of *sup2* action. A19K and *sup2* subunits were coexpressed to determine whether heterodimeric Tar molecules were functional. The *sup2* suppressors with acidic residue replacements were able to act in *trans*; the other *sup2* suppressors were not. To ensure that both types of mutant homodimers (A19K/A19K and *sup2* + T154P/*sup2* + T154P) were nonfunctional, the *sup2* mutations were combined with an aspartate binding site lesion (T154P), described in panel D. (D-left) The piston model of transmembrane signaling. The periplasmic sensing domains of Tar and Tsr have two symmetric, negatively cooperative ligand-binding sites at the dimer interface. Binding of one ligand molecule (black diamond) to either site induces a small downward movement of the TM2 segment in one of the subunits, determined by ligand orientation at the binding sites. (D-right) Use of a binding site lesion (T154P) to control the route of the transmembrane conformational signal in heterodimeric receptor molecules (Tar-A19K/*sup2a* + T154P). T154P subunits cannot function as homodimers, but contribute aspartate-binding determinants that allow heterodimers to sense aspartate ligands in one orientation, resulting in transmission of a transmembrane piston motion through the subunit that does not carry the T154P lesion (in this case, the A19K subunit). (E) Cross-section of the TM segments in a Tar dimer viewed from the periplasmic side. For each helix, the largest residue circles and thickest connecting lines are those closest to the periplasm. Note that A19 in one subunit is close to A198 and V201 in the other subunit. In contrast, V202L, which does not suppress A19K in *trans*, is not as closely aligned. Another suppressor that cannot act in *trans*, W192R, is not shown, but lies two turns above and just clockwise of the V202 position, even further out of line with residue 19 in TM1.

results, Oosawa and Simon suggested that the positively charged side-chain of the lysine replacement in Tar-A19K might distort the structure or position of TM1 through interaction with the negatively charged head groups in the membrane phospholipid. This TM1 structural change, in turn, must somehow alter the signaling state of the Tar output domain. Oosawa and Simon suggested that the V17E change in TM1 and the A198E and V201E changes in TM2 might suppress A19K through formation of a salt bridge that effectively satisfied the offending positive charge to restore correct TM1 structure. They offered no specific explanation for the suppressor mutations in TM2 with non-acidic amino acid replacements (W192R and V202L), but these residue changes could conceivably compensate the A19K lesion through a direct structural interaction of a different sort. This reversion study provided the first evidence for an important role of the TM segments in MCP signaling and suggested that TM1 and TM2 might interact structurally, as well.

In a subsequent study, Umemura *et al.* (1998) obtained strong evidence for TM1–TM2 interactions and demonstrated that the TM2 suppressors of Tar-A19K fell into two distinct functional groups (Fig. 3B). Through cleverly designed experiments, they asked whether, in heterodimeric Tar molecules, a TM2 alteration in one subunit could suppress an A19K lesion in the other. To perform the *trans* test, Umemura *et al.* exploited information about the structure of the Tar ligand-binding domain and the transmembrane signaling mechanism that had emerged since the original study by Oosawa and Simon.

X-ray structures of the aspartate-bound form of the Tar sensing domain had revealed two symmetric binding sites at the dimer interface, each relying on side chain contacts from residues in both subunits (Fig. 3D). The critical binding-site determinants involve residues in the helical extension (α1) adjoining TM1 from one subunit and the helical extension (α4) adjoining TM2 in the other subunit. Ligand binding at the two sites is negatively cooperative; except at very high ligand concentrations, aspartate can only occupy one site per dimer. Comparisons of the ligand-occupied and the unliganded X-ray structures of the Tar sensing domain had also revealed a small ($\sim$2 Å) downward displacement of the α4 helix in one subunit of the aspartate-bound receptor (Chervitz and Falke, 1996). Subsequent cysteine crosslinking studies of several MCPs confirmed that ligand binding induces a modest piston motion in the TM2 segment that shifts the receptor signaling domain toward the CCW output state (Falke and Hazelbauer, 2001). Thus, aspartate binds to either one of two symmetric sites in the Tar periplasmic domain, inducing a downward movement of one of the TM2 segments in the dimer to modulate its signal output.

To set up A19K/*sup2* heterodimers for the *trans* tests, Umemura *et al.* coexpressed A19K and *sup2* subunits in the same cell. However, at comparable subunit expression levels, both types of homodimers will also form. Although A19K homodimers have no Tar function, *sup2* homodimers exhibited Tar function. To ensure that only heterodimers were capable of furnishing Tar function, Umemura *et al.* introduced a binding site lesion (T154P) into the *sup2* subunit (Fig. 3D). Under these conditions, they found that the acidic (*sup2a*) suppressors (A198E, V201E) could act in *trans*, whereas the non-acidic (*sup2b*) suppressors (W192R, V202L) could not (Fig. 3C). In the test configuration, the T154P binding site lesion in the *sup2a* subunit constrains the ligand-induced piston motion to travel through the TM2 segment of the A19K subunit, demonstrating that the TM1 lesion does not interfere with the piston transmembrane signaling mechanism in the heterodimer.

Possible Mechanisms of Tar-A19K Suppression

The spatial arrangement of Tar TM segments in the membrane, established through cysteine-directed crosslinking and molecular modeling studies (Chervitz and Falke, 1995; Milburn *et al.*, 1991; Pakula and Simon, 1992), suggests a simple explanation for *trans* action of the *sup2a* suppressors (Fig. 3E). The four TM segments in a Tar dimer form a bundle in which the TM1 segments pack against one another at the dimer interface. The TM2 segments, in contrast, are far from one another and loosely associated with both TM segments. In the Tar TM bundle, residue A19 in TM1 lies close to residues A198 and V201 in TM2 of the opposing subunit (Fig. 3E). This spatial arrangement supports the idea that acidic replacements at the TM2 residues could form a salt bridge with the lysine replacement in TM1 of the opposing subunit to suppress the A19K defect in *trans*. Moreover, it implies that the productive TM interactions in A19K + *sup2a* homodimers also occur in *trans*, that is, between rather than within subunits. This predicts that a Tar heterodimer with one wild-type and one A19K + *sup2a* subunit might not function. The *cis* arrangement could be tested by introducing complementary binding-site lesions into the two different subunits to prevent both types of homodimers from functioning. This experiment has not, to our knowledge, been done. However, Umemura *et al.* provided strong support for the salt-bridge model by constructing additional acidic replacements at TM2 residues that were predicted to face residue 19 in TM1 of the opposing subunit (Umemura *et al.*, 1998). Replacements at residues most closely aligned with A19K (e.g., V201D, L205D, L205E) suppressed well in *trans*, whereas those with less optimal alignments suppressed poorly (I204D, I204E) or not at all (A208D, A208E). Interestingly, although A198E was a good suppressor, A198D failed to suppress, perhaps owing to its shorter side-chain.

In the context of the piston model for Tar transmembrane signaling, the A19K lesion most likely produces a downward movement of TM2 because it causes CCW-biased signal output. This could happen by any of several different mechanisms: (1) The A19K alteration might destabilize the TM1/TM1′ interface, leading to a conformational change in the periplasmic domain that mimics the ligand-occupied state. Indeed, basic replacements at nearby residues (G22R, S25R) in Tar-TM1 peptides have been shown to impair their dimerization (Sal-Man and Shai, 2005). (2) Interaction of the A19K side-chains with the lipid head groups at the cytoplasmic or periplasmic interface could shift the membrane position of the mutant TM1 segments, in turn triggering TM2 displacement through conformational changes in the periplasmic domain. (3) The mutant TM1 segment might perturb the membrane position of TM2 through a direct structural interaction. The trans-acting sup2a suppressors are consistent with all three mechanisms, but perhaps most compatible with a direct effect of TM1-A19K on the membrane position of TM2′, because the interacting TM segments are not covalently connected. In this scenario, it is easy to imagine how the postulated salt-bridge could resist the downward movement of TM2′ caused by the TM1-A19K lesion, thus restoring the ability to produce a downward piston movement in response to ligand-binding events.

It is less obvious how the acidic suppressor in TM1 (V17E) operates because this residue position is not close to A19 in either subunit (Fig. 3E) and has not been tested in trans. Conceivably, the TM1/TM1′ interaction is sufficiently malleable to allow salt bridge formation between V17E in one subunit and A19K in the other. This could serve to stabilize the TM1-TM1′ interface to alleviate a deleterious interaction of the TM1-A19K side-chain with TM2′. These suppression mechanisms predict that the V17E suppressor should be able to act in trans. If the acidic TM1 suppressor cannot act in trans, the suppression mechanism could be altogether different. For example, the A19K signaling defect might be caused by attraction of the basic side-chain toward the lipid head groups. The V17E acidic side-chain in cis might create a countering repulsive force that restores a more normal membrane position to the doubly mutant TM1.

The two cis-acting TM2 suppressors (W192R and V202L) probably work by different mechanisms. Owing to its aromatic side-chain, Tar-W192 is positioned at the lipid-head group interface at the periplasmic side of the membrane. Miller and Falke (2004) showed that Tar-W192R "superactivated" CheA in the absence of aspartate but was still able to inhibit kinase activity upon aspartate binding. They proposed that the arginine side-chain is attracted to the polar head groups, thereby shifting TM2 toward the periplasmic interface. This CW-biased conformational change should serve to offset the CCW-biased downward shift of TM2 caused by the A19K lesion. The failure of W192R to act on A19K in trans indicates that the compensatory membrane shift must occur in the subunit in which TM2 is displaced downward by the mutant TM1. The other cis-acting suppressor (V202L) may also counteract the downward piston displacement caused by the A19K lesion, but the underlying compensatory mechanism is less apparent. V202 lies midway in TM2, roughly equidistant from the periplasmic and cytoplasmic interfaces, so it is unclear how a valine to leucine change at this position would preferentially cause an upward shift of TM2. However, V202L in one subunit might be close enough to A19K in the other subunit for their side-chains to interact (Fig. 3E), which might serve to shift TM1 and/or TM2' toward the periplasmic side of the membrane. Although this is a *trans* interaction, perhaps it must occur in both subunits to achieve a significant suppression effect.

Further *cis-trans* tests of the A19K suppressors should serve to eliminate some of the proposed structural compensation mechanisms. However, the complex suppression behaviors in this relatively simple system illustrate the difficulty of reaching explicit molecular explanations for second-site reversion effects. To do so really requires accompanying high-resolution structural information for the original mutant protein, for its suppressor alteration, and for the doubly mutant revertant protein.

Conformational Suppression Between Receptor Molecules

Initial genetic evidence for collaborative signaling interactions between receptors of different types came from studies of Tsr trimer contact mutants that had amino acid replacements in residues thought to promote trimer-of-dimer formation (Ames *et al.*, 2002). The principal trimer contact residues are identical in all *E. coli* MCPs and most single amino acid changes at any of these positions abolished receptor function. However, some Tsr trimer contact mutants exhibited interesting functional effects when coexpressed with wild-type aspartate receptors: some regained the ability to mediate chemotactic responses to serine (rescuable Tsr defects); others blocked Tar function (epistatic Tsr defects) (Fig. 4A,B). The rescue and epistasis effects are consistent with the idea that receptors function in mixed, higher-order signaling teams. In this view, rescuable receptors benefit from association with normal team members, whereas epistatic members spoil the function of the entire team. Presumably, the conformations and/or dynamic motions of the mixed receptor teams are instrumental in both effects. Accordingly, we reasoned that it might be possible to find mutant forms of Tar (designated Tar^) that "rescued" the function of epistatic Tsr defects (designated Tsr*) by imparting a compensatory conformational change to a mixed Tar^/Tsr* signaling team.

To look for Tar^ suppressors of Tsr* defects, we induced random mutations in a wild-type Tar expression plasmid by passage through a *mutD* host. Independently mutagenized plasmid pools were transformed

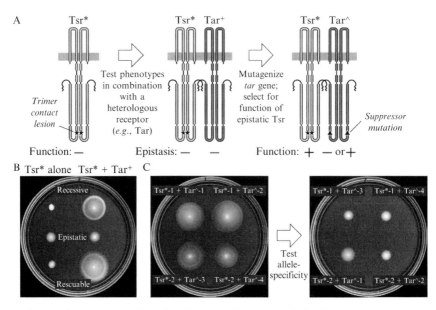

FIG. 4. Intergenic, conformational suppression of epistatic trimer contact lesions. (A) Scheme for selecting mutant Tar receptors (Tar^) that suppress epistatic Tsr trimer contact defects (Tsr*). Cells expressing Tsr* and Tar+ receptors cannot mediate chemotactic responses to either serine or aspartate. Tar^ receptors were selected from such cells as Tsr+ pseudorevertants on tryptone soft agar. Some Tar^ receptors exhibited Tar function in the presence of Tsr* receptors; others did not. (B) Phenotypes produced by Tsr* receptors alone and in combination with wild-type Tar receptors. Recessive Tsr* lesions (illustrated here by a vector control) have no effect on Tar function; rescuable Tsr* lesions (Tsr-R388A, in this example) regain Tsr function in the presence of wild-type Tar; epistatic lesions (Tsr-F373W, in this example) block the function of wild-type Tar molecules. The plate was incubated at 32.5° for 10 h. (C) Phenotypes and allele-specificity of Tar^ Tsr* suppression effects. Mutations: Tsr-F373W (Tsr*-1); Tsr-L380A (Tsr*-2); Tar-A380V (Tar^-1); Tar-V397M (Tar^-2); Tar-E389G (Tar^-3); and Tar-R386H (Tar^-4). The host strain (UU1250) carried chromosomal deletions of all MCP genes; Tar^ and Tsr* were supplied from compatible, independently regulatable expression plasmids. The plates were incubated at 32.5° for 10 h.

en masse into a receptorless recipient strain carrying a compatible, independently regulatable Tsr* plasmid and chemotactic cells in the transformation mix were selected as flares from a cell streak on a soft agar plate, as illustrated in Fig. 1D. The Tar^ plasmids were then purified from independent revertants and further characterized.

In all, we obtained 19 different Tar^ suppressors from four different Tsr* mutations. Ten more epistatic lesions at other Tsr trimer contact residues yielded no Tar^ suppressors. We subsequently showed that none

of the recalcitrant Tsr* mutations were suppressible by any of the Tar^
plasmids, implying that their structural defects were somehow different
from those of the suppressible Tsr* receptors. All Tar^ Tsr* revertants
had regained serine chemotaxis ability; some had also regained aspartate
responsiveness, whereas others were still defective for aspartate chemotaxis.
These findings implied, consistent with our working hypothesis of collaborative signaling in trimer-based receptor teams, that single-step mutational
changes cannot create "epistasis-resistant" Tar proteins, whereas they can
create Tar alterations that correct Tsr* functional defects, often with a
concomitant loss of Tar function.

The amino acid changes in all Tar^ receptors, inferred from DNA
sequence analysis of the mutant plasmids, occurred within or very close to
the trimer contact region of Tar, including three with alterations at a trimer
contact residue. Not surprisingly, all 19 Tar^ mutants exhibited functional
defects similar to those of their Tsr* counterparts. In the absence of their
Tsr* partner, most Tar^ receptors were defective in promoting aspartate
chemotaxis. Moreover, most of the nonresponsive Tar^ receptors exerted a
strong epistatic effect on wild-type Tsr function. Presumably, the similar
functional alterations of Tar^ and Tsr* receptors arise through similar
structural alterations of trimers of dimers.

To test the proposition that Tar^ suppressors act by creating a compensatory conformational change in a mixed Tar^ Tsr* signaling team, we
tested the allele specificity of Tar^-Tsr* interaction. Allele specificity is
the litmus test for conformational suppression between directly interacting proteins, reflecting the underlying stereospecificity of the interaction.
Accordingly, we tested each Tar^ receptor for ability to restore function to
each of the four suppressible Tsr* receptors. The suppression pattern
revealed six major interaction classes: three groups of Tar^ suppressors
acted productively with only a single, different Tsr* partner; another two
groups suppressed different pairs of partners; and one Tar^ group suppressed three of the four Tsr* receptors. Some examples of the test results
are shown in Fig. 4C. Tar-A397M (Tar^-2) and Tar-R386H (Tar^-4) represent two of the highly specific suppressor classes, acting, respectively, on
Tsr-F373W (Tsr*-1) and Tsr-L380A (Tsr*-2). Tar-A380V (Tar^-1) and
Tar-E389G (Tar^-3) represent the two group-specific suppressor classes.
Their Tsr* target groups are represented by Tsr-F373W and Tsr-L380A,
respectively.

The Tar^-Tsr* test matrix showed that many Tar^ suppressors discri-
minated among Tsr* changes at different residue positions (e.g., F373 and
L380). However, true allele-specific suppressors should also be able to
distinguish different amino acid changes at the same position in an interacting protein. To explore the side-chain specificity of Tar^-Tsr*

suppression, we isolated additional epistatic lesions by all-codon mutagenesis of Tsr residues F373 and L380, then compared the suppression patterns of the new epistatic alleles with those of the original Tsr* alleles (F373W and L380A) (Ames, unpublished results). For both residues, some of the Tar^ alleles that suppressed the Tsr* prototypes also suppressed some of the new epistatic lesions at the same residue position. The F373W suppressors fell into two groups, one that could suppress F373K and F373M and one that could not. The L380A suppressors fell into three groups, one that could suppress L380F, L380M, L380Q, and L380Y; a second that could suppress all of these except L380M; and a third group that could not suppress any of the new alleles. None of the new Tsr* alleles were suppressible by a Tar^ allele that could not act on the corresponding Tsr* prototype. However, if we were to repeat the suppressor selection by starting with the new Tsr* alleles, we would expect to find novel Tar^ alleles that would not suppress the original Tsr* mutations.

The highly specific pattern of Tar^-Tsr* suppression effects, both with respect to residue position and side-chain character, is fully consistent with the receptor team hypothesis. The location and nature of the various residue changes provide tantalizing clues about the underlying suppression mechanism(s). Productive Tar^-Tsr* combinations might arise through two types of compensatory structural interactions. Bulky amino acid replacements at a Tsr trimer contact site (e.g., F373W) that most likely distort the trimer interface might require a correspondingly reduced side-chain volume in their Tar^ partner to restore a functional trimer shape. In contrast, a small amino acid replacement at a Tsr trimer contact site (e.g., L380A), would most likely destabilize the trimer, perhaps increasing its dynamic motions. Tar^ alterations that enhance dimer–dimer packing interactions or that reduce receptor dynamics might correct these sorts of Tsr* defects. Only structural studies of the interacting proteins and their complexes can eliminate this guesswork.

General Guidelines for Intermolecular Conformational Suppression Studies

In theory, any direct protein–protein interaction should be amenable to conformational suppression analysis. One needs only a strong selection for restored function and methods for analyzing the genetic changes in the revertants. However, the success of such endeavors is critically dependent on starting with the right mutants; some of them must carry lesions that affect a contact surface for the interaction. Alterations that do not cause a structural change at the site of protein–protein interaction are not likely to be conformationally suppressed by a structural change in a partner protein. The key to a successful suppression study is to focus on the types of mutants that are most likely to have interaction site lesions. Most randomly isolated loss-of-function mutations will not fall within protein–protein interaction determinants, but rather at the many sites critical for proper protein folding, maturation, and stability. Thus, mutant proteins that do not have wild-type expression level and stability should be excluded from the reversion analysis. They may yield revertants that have a second-site suppressor mutation, but they will probably not identify an interacting partner protein.

The second key to a successful conformational suppression study is to work with a large enough set of suppressible mutations and suppressors to build a compelling case for allele-specificity. The mutation set can be expanded in several ways. A putative interaction determinant that has been identified by a suppressible mutation can be explored by constructing additional mutational changes that may affect the same docking surface and isolating suppressors of those mutations as well. Obviously, knowledge about the structure of the starting protein is helpful, but not essential, in guiding such approaches. It is also possible to expand the mutation set by using any suppressor mutations that are functionally defective with a wild-type partner as the starting point for additional rounds of reversion (Liu and Parkinson, 1991).

Acknowledgments

Research in the authors' laboratory is funded by grant GM19559 from the National Institute of General Medical Sciences.

References

Ames, P., Chen, J., Wolff, C., and Parkinson, J. S. (1988). Structure–function studies of bacterial chemosensors. *Cold Spring Harbor Symp. Quant. Biol.* **53,** 59–65.

Ames, P., and Parkinson, J. S. (1988). Transmembrane signaling by bacterial chemoreceptors: *E. coli* transducers with locked signal output. *Cell* **55,** 817–826.

Ames, P., and Parkinson, J. S. (1994). Constitutively signaling fragments of Tsr, the *Escherichia coli* serine chemoreceptor. *J. Bacteriol.* **176,** 6340–6348.

Ames, P., and Parkinson, J. S. (2006). Conformational suppression of inter-receptor signaling defects. *Proc. Natl. Acad. Sci. USA* **103,** 9292–9297.

Ames, P., Studdert, C. A., Reiser, R. H., and Parkinson, J. S. (2002). Collaborative signaling by mixed chemoreceptor teams in *Escherichia coli*. *Proc. Natl. Acad. Sci. USA* **99,** 7060–7065.

Armitage, J. P. (1999). Bacterial tactic responses. *Adv. Microb. Physiol.* **41,** 229–289.

Buron-Barral, M., Gosink, K. K., and Parkinson, J. S. (2006). Loss- and gain-of-function mutations in the F1-HAMP region of the *Escherichia coli* aerotaxis transducer Aer. *J. Bacteriol.* **188,** 3477–3486.

Butler, S. L., and Falke, J. J. (1998). Cysteine and disulfide scanning reveals two amphiphilic helices in the linker region of the aspartate chemoreceptor. *Biochem.* **37,** 10746–10756.

Chen, J. (1992). Genetic studies of transmembrane and intracellular signaling by a bacterial chemoreceptor. Ph.D. thesis. University of Utah, Salt Lake City.

Chervitz, S. A., and Falke, J. J. (1995). Lock on/off disulfides identify the transmembrane signaling helix of the aspartate receptor. *J. Biol. Chem.* **270,** 24043–24053.

Chervitz, S. A., and Falke, J. J. (1996). Molecular mechanism of transmembrane signaling by the aspartate receptor: A model. *Proc. Natl. Acad. Sci. USA* **93,** 2545–2550.

Coleman, M. D., Bass, R. B., Mehan, R. S., and Falke, J. J. (2005). Conserved glycine residues in the cytoplasmic domain of the aspartate receptor play essential roles in kinase coupling and on-off switching. *Biochem.* **44,** 7687–7695.

Danielson, M., Bass, R., and Falke, J. (1997). Cysteine and disulfide scanning reveals a regulatory a-helix in the cytoplasmic domain of the aspartate receptor. *J. Biol. Chem.* **272,** 32878–32888.

Falke, J. J., and Hazelbauer, G. L. (2001). Transmembrane signaling in bacterial chemoreceptors. *Trends. Biochem. Sci.* **26,** 257–265.

Gardina, P. J., and Manson, M. D. (1996). Attractant signaling by an aspartate chemoreceptor dimer with a single cytoplasmic domain. *Science* **274,** 425–426.

Jeffery, C. J., and Koshland, D. E., Jr. (1994). A single hydrophobic to hydrophobic substitution in the transmembrane domain impairs aspartate receptor function. *Biochem.* **33,** 3457–3463.

Jeffery, C. J., and Koshland, D. E., Jr. (1999). The *Escherichia coli* aspartate receptor: Sequence specificity of a transmembrane helix studied by hydrophobic-biased random mutagenesis. *Protein Eng.* **12,** 863–872.

Lee, L., and Imae, Y. (1990). Role of threonine residue 154 in ligand recognition of the Tar chemoreceptor in *Escherichia coli. J. Bacteriol.* **172,** 377–382.

Liu, J. D., and Parkinson, J. S. (1991). Genetic evidence for interaction between the CheW and Tsr proteins during chemoreceptor signaling by *Escherichia coli. J. Bacteriol.* **173,** 4941–4951.

Ma, Q., Johnson, M. S., and Taylor, B. L. (2005). Genetic analysis of the HAMP domain of the Aer aerotaxis sensor localizes flavin adenine dinucleotide-binding determinants to the AS-2 helix. *J. Bacteriol.* **187,** 193–201.

Manson, M. D. (2000). Allele-specific suppression as a tool to study protein–protein interactions in bacteria. *Methods* **20,** 18–34.

Maruyama, I. N., Mikawa, Y. G., and Maruyama, H. I. (1995). A model for transmembrane signaling by the aspartate receptor based on random-cassette mutagenesis and site-directed disulfide cross-linking. *J. Mol. Biol.* **253,** 530–546.

Milburn, M. V., Prive, G. G., Milligan, D. L., Scott, W. G., Yeh, J., Jancarik, J., Koshland, D. E., Jr., and Kim, S. H. (1991). Three-dimensional structures of the ligand-binding domain of the bacterial aspartate receptor with and without a ligand. *Science* **254,** 1342–1347.

Miller, A. S., and Falke, J. J. (2004). Side chains at the membrane-water interface modulate the signaling state of a transmembrane receptor. *Biochem.* **43,** 1763–1770.

Mutoh, N., Oosawa, K., and Simon, M. I. (1986). Characterization of *Escherichia coli* chemotaxis receptor mutants with null phenotypes. *J. Bacteriol.* **167,** 992–998.

Nara, T., Kawagishi, I., Nishiyama, S., Homma, M., and Imae, Y. (1996). Modulation of the thermosensing profile of the *Escherichia coli* aspartate receptor tar by covalent modification of its methyl-accepting sites. *J. Biol. Chem.* **271,** 17932–17936.

Nishiyama, S., Maruyama, I. N., Homma, M., and Kawagishi, I. (1999). Inversion of thermosensing property of the bacterial receptor Tar by mutations in the second transmembrane region. *J. Mol. Biol.* **286,** 1275–1284.

Nishiyama, S., Nara, T., Homma, M., Imae, Y., and Kawagishi, I. (1997). Thermosensing properties of mutant aspartate chemoreceptors with methyl-accepting sites replaced singly or multiply by alanine. *J. Bacteriol.* **179,** 6573–6580.

Oosawa, K., and Simon, M. (1986). Analysis of mutations in the transmembrane region of the aspartate chemoreceptor in *Escherichia coli. Proc. Natl. Acad. Sci. USA* **83**, 6930–6934.

Pakula, A. A., and Simon, M. I. (1992). Determination of transmembrane protein structure by disulfide cross-linking: The *Escherichia coli* Tar receptor. *Proc. Natl. Acad. Sci. USA* **89**, 4144–4148.

Parkinson, J. S. (1977). Behavioral genetics in bacteria. *Annu. Rev. Genet.* **11**, 397–414.

Parkinson, J. S. (1993). Signal transduction schemes of bacteria. *Cell* **73**, 857–871.

Parkinson, J. S., Ames, P., and Studdert, C. A. (2005). Collaborative signaling by bacterial chemoreceptors. *Curr. Opin. Microbiol.* **8**, 116–121.

Sal-Man, N., and Shai, Y. (2005). Arginine mutations within a transmembrane domain of Tar, an *Escherichia coli* aspartate receptor, can drive homodimer dissociation and heterodimer association *in vivo. Biochem. J.* **385**, 29–36.

Shapiro, M. J., Chakrabarti, I., and Koshland, D. E., Jr. (1995). Contributions made by individual methylation sites of the *Escherichia coli* aspartate receptor to chemotactic behavior. *Proc. Natl. Acad. Sci. USA* **92**, 1053–1056.

Shiomi, D., Homma, M., and Kawagishi, I. (2002). Intragenic suppressors of a mutation in the aspartate chemoreceptor gene that abolishes binding of the receptor to methyltransferase. *Microbiology* **148**, 3265–3275.

Shiomi, D., Okumura, H., Homma, M., and Kawagishi, I. (2000). The aspartate chemoreceptor Tar is effectively methylated by binding to the methyltransferase mainly through hydrophobic interaction. *Mol. Microbiol.* **36**, 132–140.

Sourjik, V. (2004). Receptor clustering and signal processing in *E. coli* chemotaxis. *Trends Microbiol.* **12**, 569–576.

Sourjik, V., and Berg, H. C. (2002). Receptor sensitivity in bacterial chemotaxis. *Proc. Natl. Acad. Sci. USA* **99**, 123–127.

Sourjik, V., and Berg, H. C. (2004). Functional interactions between receptors in bacterial chemotaxis. *Nature* **428**, 437–441.

Starrett, D. J., and Falke, J. J. (2005). Adaptation mechanism of the aspartate receptor: Electrostatics of the adaptation subdomain play a key role in modulating kinase activity. *Biochem.* **44**, 1550–1560.

Studdert, C. A., and Parkinson, J. S. (2004). Crosslinking snapshots of bacterial chemoreceptor squads. *Proc. Natl. Acad. Sci. USA* **101**, 2117–2122.

Studdert, C. A., and Parkinson, J. S. (2005). Insights into the organization and dynamics of bacterial chemoreceptor clusters through *in vivo* crosslinking studies. *Proc. Natl. Acad. Sci. USA* **102**, 15623–15628.

Trammell, M. A., and Falke, J. J. (1999). Identification of a site critical for kinase regulation on the central processing unit (CPU) helix of the aspartate receptor. *Biochem.* **38**, 329–336.

Umemura, T., Tatsuno, I., Shibasaki, M., Homma, M., and Kawagishi, I. (1998). Intersubunit interaction between transmembrane helices of the bacterial aspartate chemoreceptor homodimer. *J. Biol. Chem.* **273**, 30110–30115.

Wadhams, G. H., and Armitage, J. P. (2004). Making sense of it all: Bacterial chemotaxis. *Nat. Rev. Mol. Cell Biol.* **5**, 1024–1037.

Wolfe, A. J., Conley, M. P., and Berg, H. C. (1988). Acetyladenylate plays a role in controlling the direction of flagellar rotation. *Proc. Natl. Acad. Sci. USA* **85**, 6711–6715.

Yaghmai, R., and Hazelbauer, G. L. (1992). Ligand occupancy mimicked by single residue substitutions in a receptor: Transmembrane signaling induced by mutation. *Proc. Natl. Acad. Sci. USA* **89**, 7890–7894.

Yaghmai, R., and Hazelbauer, G. L. (1993). Strategies for differential sensory responses mediated through the same transmembrane receptor. *EMBO J.* **12**, 1897–1905.

Zhulin, I. B. (2001). The superfamily of chemotaxis transducers: From physiology to genomics and back. *Adv. Microb. Physiol.* **45**, 157–198.

[22] Single-Cell Analysis of Gene Expression by Fluorescence Microscopy

By TIM MIYASHIRO and MARK GOULIAN

Abstract

Gene regulation by two-component systems has traditionally been studied using assays that involve averages over large numbers of cells. Single-cell measurements of transcription offer a complementary approach that provides the distribution of gene expression among the population. This chapter focuses on methods for using fluorescence microscopy and fluorescent proteins to study gene expression in single cells.

Introduction

Single-cell measurements make it possible to explore potential variability in gene expression within a population. There are many reasons why a two-component system may give rise to gene expression that varies from cell to cell. A spatially heterogeneous distribution in the input signal, such as may occur in biofilms, host cells, or in any other non-uniform environment, will give rise to a broad range of cellular responses (Larrainzar *et al.*, 2005). Heterogeneous gene expression can also be inherent to the underlying two-component regulatory network (Smits *et al.*, 2006). For example, a bimodal population distribution can occur in networks that contain positive or double-negative feedback loops (Ferrell, 2002). In the *Bacillus subtilis* sporulation network, the bistable expression of the *spoIIA* gene was shown to depend on the positive autoregulation of the response regulator Spo0A (Veening *et al.*, 2005). Bimodal distributions have also been described for the competence network in *B. subtilis*, although it appears that the positive feedback loop that gives rise to this behavior is independent of the DegS-DegU two-component regulatory system (Maamar and Dubnau, 2005; Smits *et al.*, 2005; Suel *et al.*, 2006). In *Agrobacterium tumefaciens*, genes regulated by the VirA-VirG two-component system show a broad distribution of expression levels among cells for intermediate stimulus (Brencic *et al.*, 2005; Goulian and van der Woude, 2006). This may reflect a true bistable response (e.g., from positive autoregulation of VirA and VirG) or a bimodal distribution arising from fluctuations around a threshold response. Many other two-component systems are also positively autoregulated (Bijlsma and Groisman, 2003). However, such positive feedback

METHODS IN ENZYMOLOGY, VOL. 423
Copyright 2007, Elsevier Inc. All rights reserved.

0076-6879/07 $35.00
DOI: 10.1016/S0076-6879(07)23022-8

does not guarantee a bistable response. Indeed, in *Escherichia coli,* genes regulated by the PhoQ-PhoP system, which is positively autoregulated, show a monomodal population distribution (unpublished data and see following text). The *E. coli* EnvZ-OmpR system, which lacks autoregulation, also shows a monomodal distribution of gene expression (Batchelor *et al.,* 2004). Noise, or stochastic fluctuations, can also give rise to heterogeneity in gene expression (Elowitz *et al.,* 2002; Ozbudak *et al.,* 2002). Although the relevance of such fluctuations in two-component signaling has not been explored, a recent study has suggested that noise in the expression of certain regulatory proteins is important in the *B. subtilis* competence network (Suel *et al.,* 2006).

Besides heterogeneity in gene expression, there are other reasons to adopt a single-cell approach. Depending on the technique, measurements of single cells can give greater sensitivity and precision by minimizing background signal from the ambient environment and other sources of experimental error. By following individual cells over time, one can also study the dynamics of gene expression through the cell cycle or through developmental programs without taking steps to synchronize the population. In this chapter, we briefly describe approaches for measuring gene expression in single cells. We then describe specific protocols for quantifying single-cell fluorescence using microscopy and image analysis.

Transcriptional Reporters

Fluorescent Proteins

Fluorescent proteins, such as green fluorescent protein (GFP), its spectral variants, and more distantly related proteins, have become indispensable tools for single-cell studies in biology. There are numerous reviews of the properties of these proteins (Chudakov *et al.,* 2005; Schmid and Neumeier, 2005) and of their applications in bacteriology (Bongaerts *et al.,* 2002; Phillips, 2001; Southward and Surette, 2002). Despite their many advantages, fluorescent proteins also have some limitations, which are often underappreciated and should be considered when designing experiments. Fluorescent proteins lack enzymatic amplification, which significantly decreases their sensitivity for most applications when compared with enzyme-based reporters, such as beta-galactosidase. Fluorescence from the extracellular environment as well as endogenous autofluorescence also limit the lowest fluorescence levels that can be detected. Photobleaching, which is the destruction of fluorescent molecules by excitation light, limits the length of time over which fluorescence measurements can be made. Because GFP and related proteins require molecular oxygen for maturation into fluorescent forms, they cannot be used in anaerobic conditions. In some applications, the time required for maturation of the protein into a fluorescent form may be of concern. These issues of sensitivity, photostability, and maturation time can at least be partially addressed by choosing the appropriate fluorescent protein for the desired application. For guidance in selecting fluorescent proteins, tables listing many of the spectral variants and their properties are available (Shaner et al., 2005; Tsien, 1998). Unfortunately, many properties, such as expression efficiency and maturation time, are likely to depend on the organism. Some of the most popular genes for fluorescent proteins have been optimized in E. coli or eukaryotic cells (Cormack et al., 1996; Scholz et al., 2000; Shaner et al., 2005) and may require further improvements for efficient use in other bacteria (e.g., Veening et al., 2004). A good rule of thumb is to choose alleles that have been shown to work well in the organism of interest whenever possible.

The availability of many different colors of fluorescent proteins with distinct excitation/emission spectra makes it possible to quantify multiple colors of fluorescent proteins in the same cell (Shaner et al., 2005). To date, the most frequently used proteins for dual-fluorescence applications are cyan fluorescent protein (CFP) and yellow fluorescent protein (YFP). However, with improvements in proteins that have longer excitation and emission wavelengths, many more combinations of proteins are now possible (Shaner et al., 2005). For studies of gene expression, the precision of transcriptional measurements can be greatly increased by including a second color of fluorescent protein, which can be used for normalization. For example, if YFP is expressed from the promoter of interest and CFP is expressed from a constitutive promoter, then the ratio of YFP to CFP fluorescence eliminates many sources of experimental error. This is especially the case for single-cell measurements by fluorescence microscopy. Sources of cell-to-cell variability from non-uniformity in the excitation light intensity or in the plane of focus across the field of view, or variations in the identification of cell boundaries during image analysis steps (see later), will at least partially cancel in the YFP/CFP fluorescence ratio.

An example of a two-color fluorescent reporter strain for the magnesium-regulated PhoQ-PhoP two-component system in E. coli is shown in Fig. 1. CFP was expressed from a constitutive promoter and YFP was expressed from the promoter of the PhoP-regulated gene mgrB. Histograms of cellular fluorescence for various levels of magnesium are shown in Fig. 2A,B. Much of the cell-to-cell variability in the YFP fluorescence is correlated with corresponding variability in the CFP fluorescence. This is evident in the YFP-CFP scatter plot shown in Fig. 2D. One can also see this directly in Fig. 1B (e.g., compare bright and dim cells in the YFP and CFP images for the culture grown in 100 μM magnesium). The improvement in precision

A

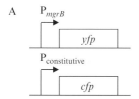

B

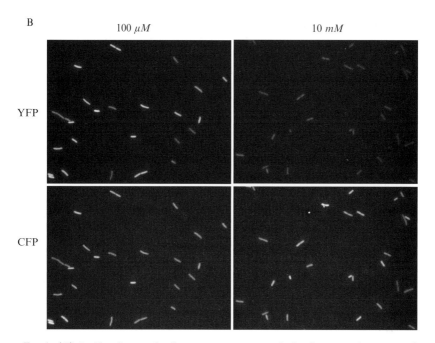

FIG. 1. (A) An *E. coli* two-color fluorescent reporter strain for the magnesium-responsive PhoQ-PhoP two-component system was constructed by inserting in the chromosome both a copy of *yfp*, which encodes yellow fluorescent protein (YFP), under control of the promoter for the PhoP-regulated gene *mgrB*, and a copy of *cfp*, which encodes cyan fluorescent protein (CFP), expressed from a constitutive promoter. The YFP and CFP proteins are variants of GFPmut3.1 (Clontech, Mountain View, CA), which have the substitutions V68L Q69K Q80R T203Y, and F64L G65T Y66W A72S Q80R N146I M153T V163A N164H, respectively. (B) Images of the PhoQ-PhoP reporter strain grown in low (100 μM) and high (10 mM) magnesium taken with YFP and CFP filter sets. Note the variability in fluorescence among cells, most of which is correlated between the CFP and YFP images (e.g., compare the CFP and YFP images in the left column). The reporter strain was grown overnight at 37° in minimal A medium (Miller, 1992) supplemented with 0.1% casamino acids, 0.2% glucose, and 1 mM MgSO$_4$. To measure steady-state transcription, the overnight culture was diluted 1:10^4 into prewarmed media containing either 100 μM or 30 mM MgSO$_4$ and grown at 37° for 3.5 h. Cultures were then quickly cooled using an ice-water slurry, and streptomycin was added to a concentration of 100 μg/ml. To concentrate cells, 1 ml samples were spun at 4° for 5 min and all but 10 μl of the supernatant was removed. The sample was resuspended in the residual volume.

from using the YFP/CFP ratio to measure transcription is evident in Fig. 2C and E.

The fluorescence ratio also provides a normalization for variations that affect overall protein expression levels. Such variations can emerge from extrinsic noise (Elowitz *et al.*, 2002; Ozbudak *et al.*, 2002) or when one compares cell populations grown under different conditions. For example, a decrease in cell growth rate can result in an increase in expression level per cell. Under most conditions, fluorescent proteins are stable in cells, unless the proteins have been engineered to be degraded (Andersen *et al.*, 1998). In this respect, they are similar to beta-galactosidase. Under conditions of balanced growth, the steady-state level of a stable protein will be equal to the product of the production rate and the specific growth rate (e.g., Warner and Lolkema, 2002). Hence, in comparing protein expression levels under two different environmental conditions that result in two different growth rates, one should take care to normalize by the growth rates. The ratio of YFP/CFP, however, will not have this growth-rate dependence, since the effects will cancel between the numerator and denominator.

Alternatively, to explore differential gene expression within a particular regulatory system, CFP and YFP can be expressed from different promoters within the regulon. For example, two-color fluorescent reporters for the *E. coli* EnvZ-OmpR system have been constructed using operon fusions of *yfp* to the porin gene *ompF* and *cfp* to the porin gene *ompC* (Batchelor *et al.*, 2004). At high levels of OmpR-P, *ompF* transcription is repressed and *ompC* expression is activated. Reporters with this form of reciprocal regulation have the additional advantage that the fluorescence ratio will effectively amplify the output since the ratio will show a larger change than will CFP or YFP alone.

Alternatives to Fluorescent Proteins

It is worth noting that there are a number of reporter systems based on the enzymatic conversion of substrates to fluorescent products that can be used for single-cell analysis. To be useful for single-cell measurements, the

---

To image cells, a 5-μl sample was placed on a cover glass and touched to a 1% agarose pad that was made using minimal A medium. Fluorescence measurements were obtained using an Olympus IX81 microscope with a 100W mercury lamp and 100× UPlanApo NA 1.35 objective lens. Filter sets were HQ500/20 excitation, Q5151p beam splitter, and HQ535/30 emission for YFP fluorescence and D436/20 excitation, 455dclp beam splitter, and D480/40 emission for CFP measurements (Chroma, Brattleboro, VT). Sixteen-bit images were acquired with a SensiCam QE cooled charge-coupled device camera (Cooke Corporation, Romulus, MI) and IPLab v3.7 software (Scanalytics, Fairfax, VA) with 2×2 binning. Exposure times were 250 ms for YFP images and 100 ms for CFP images.

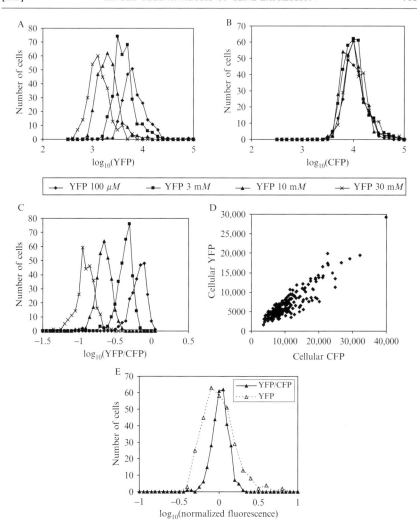

FIG. 2. Distributions of single-cell YFP fluorescence (A), CFP fluorescence (B), and ratio of YFP/CFP (C) for the PhoQ-PhoP reporter strain from Fig. 1 grown in 30 mM (crosses), 10 mM (triangles), 3 mM (squares), and 100 μM (diamonds) magnesium. Experimental procedures were essentially the same as described in the caption to Fig. 1, except for differences in the final magnesium concentration. (D) Scatter plot of single-cell YFP versus CFP shows a strong correlation between CFP and YFP fluorescence. The data is from the 100 μM magnesium culture shown in the previous figures. (E) Comparison of histograms of single-cell $\log_{10}$(YFP/CFP) fluorescence (closed triangles) and $\log_{10}$(YFP) fluorescence (open triangles) for cells grown in 10 mM magnesium. The histograms have been shifted so that both distributions have zero mean. The standard deviations of the $\log_{10}$(YFP) and $\log_{10}$(YFP/CFP) distributions are 0.2 and 0.1, respectively.

substrates must be cell-permeable and the fluorescent products must remain associated with the cell. Examples of enzymes for which such substrates are available include beta-galactosidase, beta-glucoronidase, and beta-lactamase (Winson and Davey, 2000)(http://probes.invitrogen.com/handbook/). Because these systems take advantage of enzymatic amplification, they can be used for extremely sensitive measurements of gene expression. Indeed, beta-galactosidase has been used to study gene expression in single cells with single-protein sensitivity (Cai *et al.*, 2006; Maloney and Rotman, 1973). These enzyme-based systems have not been extensively used for single-cell applications in bacteriology. At least in some cases, difficulties with substrate permeabilization or leakage of the fluorescent product out of the cell have limited their applicability.

The luciferases are another class of enzymes that are used as transcriptional reporters in bacteria and can, in principle, be used for single-cell analysis (Greer and Szalay, 2002; Wilson and Hastings, 1998). These enzymes catalyze reactions in which conversion of substrate to product results in the emission of photons (chemiluminescence). As a result, there is no need for the excitation light source that is required for fluorescence-based methods. This eliminates factors such as light leakage and auto-fluorescence, which limit the sensitivity of fluorescence measurements. Unfortunately, the luciferases that are currently available produce photons at a relatively slow rate. The photon fluxes from luciferase-expressing cells, therefore, tend to be quite low; quantification of light emission from individual cells requires expensive cameras or other specialized instrumentation (Greer and Szalay, 2002; Wilson and Hastings, 1998).

Measuring Cellular Fluorescence by Microscopy

Fluorescence of individual cells can be measured with a flow cytometer or a fluorescence microscope. Standard flow cytometers are relatively easy to use and can rapidly quantify the fluorescence of large numbers of cells—hundreds to tens of thousands of cells can be analyzed in seconds (Bongaerts *et al.*, 2002; Shapiro, 2000; Winson and Davey, 2000). The large sample sizes that can be acquired with these machines are particularly useful for studying rare, statistically significant events. In contrast, quantifying cellular fluorescence with a fluorescence microscope is technically more involved and is usually limited to much smaller sample sizes (a few hundred cells). However, with a fluorescence microscope equipped with a high numerical aperture lens and a standard cooled charge coupled device (CCD) camera, one can detect cellular fluorescence with much greater sensitivity and precision compared with a flow cytometer. This is partly due to efficient collection of emitted photons and to the long exposures that can be achieved for a static field of cells on the microscope stage. Using more sophisticated microscopy techniques, even single molecules of GFP can be detected in bacteria (Yu *et al.*, 2006). Fluorescence microscopy has the additional benefit that live cells can be followed for long periods of time. This makes it possible to follow gene expression across many generations of cells, or during the course of developmental or other regulatory programs.

To quantify cellular fluorescence with a microscope, cells are immobilized on a slide with an agarose pad, digital images are acquired, and the fluorescence in each cell in the images is determined with computer software. In the following text, we describe each of these steps in detail.

Agarose Pads

For quantifying fluorescence by microscopy, it is preferable to work with cells that are held immobile in a fixed plane. This ensures that all the cells in the field of view are in focus and eliminates background fluorescence from cells that might otherwise be above or below the cells of interest. In addition, immobilization ensures that the cells do not move during long exposures or between subsequent measurements of the same group of cells.

A relatively simple and gentle method to immobilize bacteria is to use an agarose pad. The pad presses the cells against the surface of a cover glass and forces the cells onto a single plane. The agarose pad also provides a reservoir of media and is sufficiently soft that, over long times, the cells are able to grow, divide, and form microcolonies (Elowitz *et al.*, 2002; Stewart *et al.*, 2005; Suel *et al.*, 2006). For live-cell work, the choice of media will depend on the physiological requirements of the experiment. However, whenever possible, media with minimal autofluorescence should be used. If an endpoint measurement of total cellular fluorescence is being made, so that the cells do not have to be kept alive on the microscope, then a nonfluorescent buffer that does not lyse or severely deform the cells can be used. For example, in experiments with *E. coli*, we often use minimal media without added supplements, or simply phosphate-buffered saline.

A number of steps may be taken to increase the reproducibility and sensitivity of fluorescence-based measurements. If the data analysis will require identifying cell boundaries, for example, to measure total fluorescence, then it is convenient to prepare slides from sufficiently dilute solutions of cells so that relatively few cells are touching. A concentration of roughly 10^8 cells/ml will accomplish this and still give a large number of cells in the field of view. Single-cell measurements from cultures with a lower density of cells may require centrifugation and re-suspension into a smaller volume to concentrate the cells. Another issue concerns uniformity across the field under the microscope. Only slides that contain agarose pads that present cells in the same plane of focus should be used. If all the cells in a field are not in focus, then this can introduce a corresponding cell-to-cell variability in the detected level of fluorescence.

The thin agarose pads that we will describe are particularly simple and work well for short measurements, such as less than an hour. They can be used with either an upright or an inverted microscope. For longer times, evaporation may be a problem. In this case, the edges of the cover glass can be sealed with silicone grease or paraffin. For some applications, microscope slides with cavities can be used to provide sufficient media and oxygenation (Stewart *et al.*, 2005). Alternatively, for long time courses, a Petri dish with a cover glass attached to the bottom can be used (Suel *et al.*, 2006), although this requires an inverted microscope.

Protocol to Prepare Microscope Slides with Agarose Pads

1. Prepare a molten solution of 1% agarose (e.g., Seakem LE, Cambrex Bioscience, Rockland) in appropriate growth media or buffer. Compounds that precipitate out of solution or are heat sensitive should be added after boiling the agarose. If necessary, low-melt agarose can be used to keep the agarose molten at lower temperatures.

2. While the agarose is molten, place $50\text{-}\mu l$ in the center of a standard 75 $\times$ 25 mm microscope slide. Immediately place a 22 mm square cover glass onto the molten agarose. The agarose should spread out evenly over most of the cover glass. If it does not, then it has cooled too much and should be reheated. There may be some small air bubbles in the agarose, which is not a problem. It is a good idea to prepare extra slides since it is likely that a few will fail to yield adequate pads due to mistakes during slide preparation (uneven or torn pads, dropped slides, etc.). It is important to leave enough time for the agarose to solidify well before attempting to pry off the cover glass. Thirty minutes at room temperature should be adequate. However, slides should not be left out for too long since they will dry out. The slides can be stored at 4° for short periods. If they need to be prewarmed, they should be placed at the appropriate temperature in a wet-box to reduce evaporation.

3. To use a slide, pry off the cover glass with a razor blade, starting at one corner. This step can be a little tricky—it is important to lift off the cover glass quickly but smoothly to ensure that the agarose stays with the slide and does not tear. If a bubble appears between the agarose pad and slide while removing the cover glass, pop it using the razor blade.

4. Clean the cover glass with a lint-free wipe to remove any residual agarose (or take a new cover glass). Place a 5 μl droplet of cells in the center

of the cover glass. Invert the slide with the agarose pad onto the cover glass. The droplet should spread over the majority of the pad. The cells should now be immobilized and spread out over the cover glass. Occasionally, some cells will not be fixed in place. This is most likely due to wrinkles or bumps on the agarose that prevent the cover glass from making a good seal. In most cases, these are in a localized region and one can simply move to a different field of view.

Fluorescence Microscopy and Image Acquisition

Before taking fluorescence images, it is advisable to check that the excitation light is aligned. Non-uniformity of the excitation light across the field of view will introduce heterogeneity in the fluorescence measurements. For measurements using a single fluorescence color, it is important for the excitation light intensity to be as uniform as possible. For two-color fluorescence measurements, small variations in excitation light intensity should cancel in the fluorescence ratio (see preceding discussion). However, large variations can still cause variability due to loss of sensitivity in regions of low excitation intensity. We recommend checking the excitation light using a uniform or at least well-defined fluorescent sample prior to any experiment. A convenient reference slide for such alignments can be made from fluorescent latex beads (such as those from Invitrogen, Carlsbad, CA) immobilized on a cover glass. Gross non-uniformities can easily be seen by eye and corrected by realigning the excitation lamp. The alignment procedure is straightforward for most fluorescence microscopes (http://www.microscopyu.com/tutorials/java/arclamp/index.html). Specific instructions can usually be found in the microscope manual or at the website of the manufacturer.

Once a sample is mounted on the stage and brought into focus, a field of cells is selected. The cell density will often be quite variable across the slide. To reduce the number of images needed to analyze a particular sample, one would like to have as many cells as possible in the field of view. However, if an automated program is going to be used for identifying the cells in images, selected fields should contain as few cells touching as possible. In addition, it is important to avoid fields that have dirt or cell debris that could be misinterpreted as intact cells. One can take images from many fields of cells on a single slide. However, one should also acquire images from additional slides for each culture condition to control for possible variability in fluorescence measurements due to variability in the slide preparation. Repeated measurements of the same cells should be avoided whenever possible since photobleaching will lower the fluorescence and result in inaccurate quantification of fluorescent protein levels. Therefore, before acquiring a fluorescence image, the field of cells should be viewed using nonfluorescence illumination, for example, bright-field, phase contrast, or differential interference contrast microscopy. Once a field is selected, the cells should be carefully focused (again, without using fluorescence). Before taking the fluorescence images, it is a good idea to take a nonfluorescence image. This provides a useful reference during analysis if one sees odd or difficult to interpret patterns of fluorescence in particular images. In addition, depending on the method of generating image contrast, one may be able to use the nonfluorescence image to identify cell boundaries. After bringing the cells into focus, the fluorescence images should be taken as quickly as possible to avoid drift of the stage (motion within the plane of view) and drift in the focus.

For acquiring fluorescence images, the choice of exposure or integration time depends on the expression level of the fluorescent protein, and on the details of the microscope and imaging equipment. Long exposures should be avoided whenever possible since they will lead to excessive photobleaching. In addition, the exposure time should be sufficiently short that none of the pixels in the image are saturated, that is, so that they are below the maximal pixel value. Otherwise, the intensity corresponding to a saturated pixel will underestimate the true fluorescence intensity. The maximal pixel value depends on the image type (255 for 8-bit images, 4095 for 12-bit images, 65,535 for unsigned 16-bit images, etc.) and can be determined empirically by deliberately overexposing an image. Since one can never be certain of the maximal fluorescence intensity for a new field of cells before the image is taken, one should choose an exposure setting that leaves a wide margin below the saturation value. In addition, if different cultures will be used, the exposure should be calibrated using the sample that is expected to have the highest fluorescence. At the other extreme, very short exposures will have limited sensitivity since the signal coming from the low number of photons from the fluorescence sample will be comparable to the read noise of the camera. For multiple samples that span a wide range of fluorescence levels, it may be necessary to use different exposures to capture the low and high ranges of fluorescence. Most modern cooled CCD cameras allow "binning," in which pixels on the CCD chip are combined or binned into larger pixels, often called superpixels. The most common binning settings are 2×2 (in which each superpixel is a 2×2 square of pixels on the chip), 4×4, and 8×8. Binning increases the number of photons incident on a superpixel without a comparable increase in the read noise per superpixel. This results in a significant increase in sensitivity but it comes at the price of a decrease in spatial resolution in the image. For example, for a chip with 1024×1024 pixels, 4×4 binning will result in a resolution of 256 × 256. If one is only interested in quantifying fluorescence from cells, then the loss of resolution from binning is often not a problem.

When two or more fluorescence colors will be quantified, the exposures for the different fluorescence channels do not have to be the same. However, one should take care that acquiring an image in one channel (e.g., CFP) does not result in significant photobleaching of the other fluorescent protein (e.g., YFP). The severity of this problem will depend on the extent to which the absorption spectrum of the second protein (YFP) overlaps the range of wavelengths used to excite the first protein (CFP). This can usually be minimized by acquiring images from the channel with the longer excitation wavelength first (YFP in our example).

In many modern high-end fluorescence microscopes, the fluorescence filters are computer-controlled, and one can readily take images in several different fluorescence channels without disturbing the field of view. With microscopes in which the filters are manually controlled, one should change the filter as gently as possible to avoid moving the stage or the plane of focus.

Image Analysis

Overview

Once digital images have been acquired, the fluorescence must be quantified for each cell. This requires identifying cells and the area that they occupy. This can be done manually, although there is a risk of introducing biases in the measurements. In addition, because of the considerable labor involved, the number of cells that can be processed will be relatively low. We therefore recommend using computer software that will automate the process. Many different software packages for analyzing images are available. Some may already have built-in functions that can be used. Others will require writing macros or short software routines. Unfortunately, many image processing programs are quite expensive. In some cases, the software used to control the CCD camera can also be used for image analysis. There are also several free software packages, such as NIH image (http://rsb.info.nih.gov/nih-image/) and ImageJ (http://rsb.info.nih.gov/ij/), with a large (and growing) number of macros or plug-ins written by the community.

Even when using computer software for image analysis, there is a risk of introducing biases (or other errors) into the quantification of cellular fluorescence. This is especially a concern when the algorithms used by the software are not available. Whether one uses third-party or in-house software, one should carefully test the performance of the program with control images.

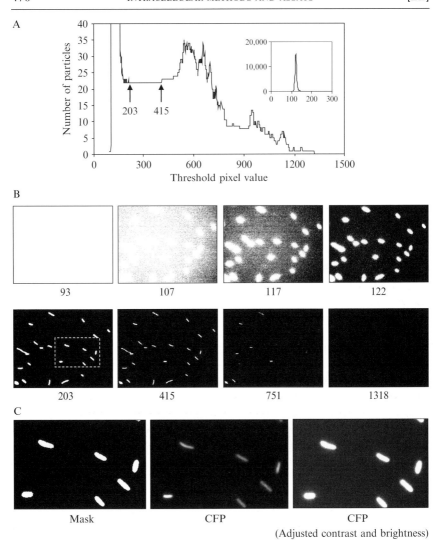

Fig. 3. (A) Number of particles versus threshold for a sequence of binary images. As discussed in the text, a particle is defined to be a maximal set of connected nonzero pixels in a binary image. The reference image used to construct the binary images was the CFP image in the lower left of Fig. 1B. For each threshold value, a binary image was constructed by setting pixels in the reference image to zero or one if they were greater than or less than the threshold, respectively. The range of threshold pixel values over which the number of particles remains constant lies between 203 and 415 (indicated by the arrows). The inset plot shows the maximum number of particles. (B) Selected binary images from the sequence of images used to generate the plot in (A). The corresponding threshold values are indicated below each image. The image

One approach for identifying cell boundaries is to use a nonfluorescence image. It is relatively easy to develop software to identify the boundaries in phase contrast images, which tend to have fairly uniform intensity differences between the cell interior and exterior. Identifying boundaries in brightfield and DIC images, however, may be more difficult. An alternative approach is to identify cell boundaries in fluorescence images. This may be problematic if a single fluorescence color is used since many algorithms will assign a larger area to a cell with bright fluorescence than to a cell with faint fluorescence. However, for two-color reporter strains, in which one of the colors provides a constant reference fluorescence (see preceding text), the reference fluorescence image can be used to identify the cells. Alternatively, if the two colors are reciprocally regulated, one can use the sum of the two fluorescence images as the reference image (Batchelor et al., 2004). For this approach to work, the fluorescence intensity of the cells in the reference image must be well above background.

To identify cells in the reference image, we use an image thresholding algorithm. The reference image is first converted into a series of binary images, in which the values of those pixels of the reference image at or above a particular threshold value are set to 1 and the values of the remaining pixels are set to zero. The number of particles (regions of adjoining nonzero pixels) within each binary image are then determined for a range of threshold values. An example of the resulting particle number versus threshold plot is shown in Fig. 3A. In such a plot, the particle number will show a plateau over a range of threshold values that are comparable to the pixel values of cells in the reference fluorescence image. The binary image resulting from the lowest threshold value within this range provides a mask in which the nonzero pixel values define the regions occupied by cells. In the example shown in Fig. 3, the mask is taken to be the binary image corresponding to the threshold value of 203.

The fluorescence of a particular cell can be calculated by simply adding the values of those pixels in the fluorescence image that correspond to the pixels that make up its area within the mask and then subtracting the background fluorescence.

The following protocol describes in detail how to quantify the fluorescence levels of single cells from two-color fluorescence images taken using fluorescence microscopy.

mask, which is used to identify particles in the CFP and YFP images, is taken to be the binary image with the lowest threshold in the plateau in (A). This corresponds to a threshold pixel value of 203. (C) A magnified region for the mask and for the original CFP image, corresponding to the dashed square in the lower left image in (B). A CFP image with increased brightness and contrast, which shows the full fluorescence of the cells, is shown on the right.

Image Analysis Algorithm

1. For each pair of images, make a reference image, either by adding the two images or by making a copy of the image that corresponds to the reference color, depending on the application.

2. Identify the minimum pixel value in the reference image and use this as the starting threshold value = T.

3. For a threshold value of T, construct a binary image from the reference image by setting each pixel to 0 or 1 if the corresponding pixel in the reference image is $<$T or $\geq$T, respectively. Note that for the initial threshold, which is set to the minimum pixel value (in step 2), all the pixels in the resulting binary image will be equal to 1.

4. Determine the number of particles (the number of maximally connected sets of nonzero pixels) in the binary image. In counting particles, one can choose to discard particles that have an area below a cutoff value A. The choice of A will depend on the properties of the reference image and should be determined empirically. A nonzero value for A is usually necessary only when there is not sufficient contrast between the cells and the background in the reference image.

5. Increment T (T $\rightarrow$ T $+$ 1) and return to step 3. Repeat this process until the number of particles has remained constant for N successive threshold increments. The choice of N will depend on the properties of the images being analyzed and should be determined empirically. Instead of requiring that the number of particles remain constant, one can require that the number of particles does not vary by more than an integer n over N successive thresholds. The choice of n will depend on the images. (For reference images with cells well above background, we generally find that we can take n = 0.) To monitor the progress of the program and to make sure that it is working properly, it is a good idea to display the binary image during each iteration. As the threshold value increases, the background should fade, revealing the single cells within the image (see Fig. 3B).

6. When the condition in step 5 is met for N successive thresholds, select the binary image associated with the lowest of these thresholds (a threshold of 203 in the example of Fig. 3). The particles in this binary image, or mask, identify the cells within the reference image. For each fluorescence image, the mask produced in step 6 is used to quantify the cellular fluorescence as follows:

7. For each particle in the mask, add up the corresponding pixels in the fluorescence image. The sums give the integrated fluorescence intensity for each cell.

8. To calculate the background signal per pixel in the fluorescence image, add up the pixels that correspond to zero pixels values in the mask and divide by the total number of pixels contributing to the sum.

9. To determine the background signal of each cell in the fluorescence image, calculate the area of each cell by adding up the number of pixels in the corresponding particle in the mask and then multiply this area by the background signal per pixel from step 8.

10. To calculate the background-subtracted fluorescence for each cell, subtract the background signal for the cell, determined in step 9, from the integrated fluorescence intensity, determined in step 8.

11. The fluorescence ratio for each cell is the ratio of the background-subtracted fluorescence in the two different channels (e.g., YFP and CFP).

Before combining the data from multiple images, it is a good idea to compare the histograms of number of cells versus fluorescence ratio for the various images. Although the number of cells will be small for each histogram, the histograms should show roughly similar distributions with similar means, at least for cell populations that are not too heterogeneous. If there are notable discrepancies, one should carefully inspect the images to make sure that there are no artifacts and that the software analysis was correct. When working with fluorescence ratios, it is convenient to construct the histogram of the logarithm of the ratio, for example, $\log_{10}(\text{YFP/CFP})$. In this way, the numerator and denominator are effectively weighted equally along the horizontal axis. Otherwise, large values in the denominator will bunch up at the origin whereas large values in the numerator will be far to the right on the horizontal axis.

Concluding Remarks

The methods described here provide a simple introduction to following gene expression in single bacterial cells. Most of the protocols can be readily generalized to other applications involving imaging and analysis of bacteria under the fluorescence microscope. Single-cell studies are increasingly playing an important role in the analysis of two-component systems. New techniques continue to be developed that, at least in specialized cases, enable single-cell studies of different steps in the signal transduction process by fluorescence microscopy (Batchelor and Goulian, 2006; Cluzel et al., 2000; Deich et al., 2004; Matroule et al., 2004; Suel et al., 2006; Vaknin and Berg, 2004). There is a growing appreciation for the importance of heterogeneity in cellular responses, for the importance of spatial localization within the cell, and for the need to follow the dynamics of intermediate steps in the signaling process with high sensitivity. We can expect then that more and more researchers will turn to the fluorescence microscope as an additional tool for studying two-component signaling and other regulatory systems in bacteria.

References

Andersen, J. B., Sternberg, C., Poulsen, L. K., Bjorn, S. P., Givskov, M., and Molin, S. (1998). New unstable variants of green fluorescent protein for studies of transient gene expression in bacteria. *Appl. Environ. Microbiol.* **64,** 2240–2246.

Batchelor, E., and Goulian, M. (2006). Imaging OmpR localization in *Escherichia coli. Mol. Microbiol.* **59,** 1767–1778.

Batchelor, E., Silhavy, T. J., and Goulian, M. (2004). Continuous control in bacterial regulatory circuits. *J. Bacteriol.* **186,** 7618–7625.

Bijlsma, J. J., and Groisman, E. A. (2003). Making informed decisions: Regulatory interactions between two-component systems. *Trends Microbiol.* **11,** 359–366.

Bongaerts, R. J., Hautefort, I., Sidebotham, J. M., and Hinton, J. C. (2002). Green fluorescent protein as a marker for conditional gene expression in bacterial cells. *Methods Enzymol.* **358,** 43–66.

Brencic, A., Angert, E. R., and Winans, S. C. (2005). Unwounded plants elicit Agrobacterium vir gene induction and T-DNA transfer: Transformed plant cells produce opines yet are tumor free. *Mol. Microbiol.* **57,** 1522–1531.

Cai, L., Friedman, N., and Xie, X. S. (2006). Stochastic protein expression in individual cells at the single molecule level. *Nature* **440,** 358–362.

Chudakov, D. M., Lukyanov, S., and Lukyanov, K. A. (2005). Fluorescent proteins as a toolkit for *in vivo* imaging. *Trends Biotechnol.* **23,** 605–613.

Cluzel, P., Surette, M., and Leibler, S. (2000). An ultrasensitive bacterial motor revealed by monitoring signaling proteins in single cells. *Science* **287,** 1652–1655.

Cormack, B. P., Valdivia, R. H., and Falkow, S. (1996). FACS-optimized mutants of the green fluorescent protein (GFP). *Gene* **173,** 33–38.

Deich, J., Judd, E. M., McAdams, H. H., and Moerner, W. E. (2004). Visualization of the movement of single histidine kinase molecules in live Caulobacter cells. *Proc. Natl. Acad. Sci. USA* **101,** 15921–15926.

Elowitz, M. B., Levine, A. J., Siggia, E. D., and Swain, P. S. (2002). Stochastic gene expression in a single cell. *Science* **297,** 1183–1186.

Ferrell, J. E., Jr. (2002). Self-perpetuating states in signal transduction: Positive feedback, double-negative feedback, and bistability. *Curr. Opin. Cell. Biol.* **14,** 140–148.

Goulian, M., and van der Woude, M. (2006). A simple system for converting lacZ to gfp reporter fusions in diverse bacteria. *Gene* **372,** 219–226.

Greer, L. F., 3rd, and Szalay, A. A. (2002). Imaging of light emission from the expression of luciferases in living cells and organisms: A review. *Luminescence* **17,** 43–74.

Larrainzar, E., O'Gara, F., and Morrissey, J. P. (2005). Applications of autofluorescent proteins for in situ studies in microbial ecology. *Annu. Rev. Microbiol.* **59,** 257–277.

Maamar, H., and Dubnau, D. (2005). Bistability in the Bacillus subtilis K-state (competence) system requires a positive feedback loop. *Mol. Microbiol.* **56,** 615–624.

Maloney, P. C., and Rotman, B. (1973). Distribution of suboptimally induced -D-galactosidase in *Escherichia coli.* The enzyme content of individual cells. *J. Mol. Biol.* **73,** 77–91.

Matroule, J. Y., Lam, H., Burnette, D. T., and Jacobs-Wagner, C. (2004). Cytokinesis monitoring during development: Rapid pole-to-pole shuttling of a signaling protein by localized kinase and phosphatase in Caulobacter. *Cell* **118,** 579–590.

Miller, J. H. (1992). "A Short Course in Bacterial Genetics: A Laboratory Manual and Handbook for *Escherichia coli* and Related Bacteria." Cold Spring Harbor Laboratory Press, Plainview, NY.

Ozbudak, E. M., Thattai, M., Kurtser, I., Grossman, A. D., and van Oudenaarden, A. (2002). Regulation of noise in the expression of a single gene. *Nat. Genet.* **31,** 69–73.

Phillips, G. J. (2001). Green fluorescent protein—A bright idea for the study of bacterial protein localization. *FEMS Microbiol. Lett.* **204,** 9–18.

Schmid, J. A., and Neumeier, H. (2005). Evolutions in science triggered by green fluorescent protein (GFP). *Chembiochem.* **6,** 1149–1156.

Scholz, O., Thiel, A., Hillen, W., and Niederweis, M. (2000). Quantitative analysis of gene expression with an improved green fluorescent protein. p6. *Eur. J. Biochem.* **267,** 1565–1570.

Shaner, N. C., Steinbach, P. A., and Tsien, R. Y. (2005). A guide to choosing fluorescent proteins. *Nat. Methods* **2,** 905–909.

Shapiro, H. M. (2000). Microbial analysis at the single-cell level: Tasks and techniques. *J. Microbiol. Methods* **42,** 3–16.

Smits, W. K., Eschevins, C. C., Susanna, K. A., Bron, S., Kuipers, O. P., and Hamoen, L. W. (2005). Stripping Bacillus: ComK auto-stimulation is responsible for the bistable response in competence development. *Mol. Microbiol.* **56,** 604–614.

Smits, W. K., Kuipers, O. P., and Veening, J. W. (2006). Phenotypic variation in bacteria: The role of feedback regulation. *Nat. Rev. Microbiol.* **4,** 259–271.

Southward, C. M., and Surette, M. G. (2002). The dynamic microbe: Green fluorescent protein brings bacteria to light. *Mol. Microbiol.* **45,** 1191–1196.

Stewart, E. J., Madden, R., Paul, G., and Taddei, F. (2005). Aging and death in an organism that reproduces by morphologically symmetric division. *PLoS Biol.* **3,** e45.

Suel, G. M., Garcia-Ojalvo, J., Liberman, L. M., and Elowitz, M. B. (2006). An excitable gene regulatory circuit induces transient cellular differentiation. *Nature* **440,** 545–550.

Tsien, R. Y. (1998). The green fluorescent protein. *Annu. Rev. Biochem.* **67,** 509–544.

Vaknin, A., and Berg, H. C. (2004). Single-cell FRET imaging of phosphatase activity in the *Escherichia coli* chemotaxis system. *Proc. Natl. Acad. Sci. USA* **101,** 17072–17077.

Veening, J. W., Hamoen, L. W., and Kuipers, O. P. (2005). Phosphatases modulate the bistable sporulation gene expression pattern in *Bacillus subtilis. Mol. Microbiol.* **56,** 1481–1494.

Veening, J. W., Smits, W. K., Hamoen, L. W., Jongbloed, J. D., and Kuipers, O. P. (2004). Visualization of differential gene expression by improved cyan fluorescent protein and yellow fluorescent protein production in Bacillus subtilis. *Appl. Environ. Microbiol.* **70,** 6809–6815.

Warner, J. B., and Lolkema, J. S. (2002). LacZ-promoter fusions: The effect of growth. *Microbiology* **148,** 1241–1243.

Wilson, T., and Hastings, J. W. (1998). Bioluminescence. *Annu. Rev. Cell. Dev. Biol.* **14,** 197–230.

Winson, M. K., and Davey, H. M. (2000). Flow cytometric analysis of microorganisms. *Methods* **21,** 231–240.

Yu, J., Xiao, J., Ren, X., Lao, K., and Xie, X. S. (2006). Probing gene expression in live cells, one protein molecule at a time. *Science* **311,** 1600–1603.

Section IV

Genome-Wide Analyses of Two-Component Systems

[23] Two-Component Systems of *Mycobacterium tuberculosis*—Structure-Based Approaches

By Paul A. Tucker, Elzbieta Nowak, and Jens Preben Morth

Abstract

Mycobacterium tuberculosis contains few two-component systems compared to many other bacteria, possibly because it has more serine/threonine signaling pathways. Even so, these two-component systems appear to play an important role in early intracellular survival of the pathogen as well as in aspects of virulence. In this chapter, we discuss what has been learned about the mycobacterial two-component systems, with particular emphasis on knowledge gained from structural genomics projects.

Introduction

To survive in an inhospitable word, microbes, like all life forms, must be able to adapt to changing environmental conditions. Extracellular parasites must attach to and enter the host cells to which they are adapted, avoiding potentially dangerous host responses before they are internalized. After becoming intracellular, these pathogens are faced with a new environment and must adapt to the host cell. In response, pathogens must be able to perceive and overcome, or neutralize, the direct and indirect threats triggered by the host cell.

Bacterial pathogens frequently use two-component systems to recognize and respond to the changing environmental conditions within the host and they normally do this by means of a phosphotransfer reaction between a membrane-localized histidine kinase sensor protein (SK) and a cytoplasmic response regulator (RR), which usually functions at the level of transcriptional activation or repression. Genomic analysis indicates that *Mycobacterium tuberculosis* (MtB) encodes 12 complete two-component systems (TCS) and five remaining (see the following text) potential orphan RR or SK proteins. The TCS are summarized in Table I with their genomic location tags and their literature nomenclature, which usually refers to their supposed function based on close homologs of known or postulated function. The number of TCS is rather low when compared with *Escherichia coli*, which has more than 30. It is unlikely that the reason for the low number of TCS currently identified in MtB is due to limitations in the bioinformatics analysis but rather is directly related to a larger number,

METHODS IN ENZYMOLOGY, VOL. 423
0076-6879/07 $35.00
DOI: 10.1016/S0076-6879(07)23023-X

TABLE I
THE TCS IDENTIFIED IN *MYCOBACTERIUM TUBERCULOSIS* H37RV STRAIN

ORF annotation	Name	References	*M. leprae* #
Rv0490/Rv0491	SenX3/RegX3	(Himpens *et al.*, 2000)	+/+
Rv0600c/Rv0601c/ Rv0602c[+]	U/U/TcrA	(Haydel and Clark-Curtiss, 2006)	−/−/−
Rv0757/Rv0758	PhoP/PhoR	(Zahrt and Deretic, 2001)	P/P
Rv0844c[+]/Rv0845	NarL/NarS		−/−
Rv0902c/Rv0903c	PrrB/PrrA	(Ewann *et al.*, 2004)	+/+
Rv0981/Rv0982	MprA/MprB	(Zahrt *et al.*, 2003)	+/+
Rv1027c/Rv1028c	KdpE/KdpD	(Parish *et al.*, 2003b)	−/−
Rv1032c/Rv1033c	TrcS/TrcR	(Haydel *et al.*, 1999)	P/P
Rv3132c/Rv3133c	DevS/DevR (DosS/DosR)	(Saini *et al.*, 2004)	−/−
Rv3245c/Rv3246c	MtrB/MtrA	(Zahrt and Deretic, 2000)	+/+
Rv3764c/Rv3765c	TcrY/TcrX	(Parish *et al.*, 2003b)	P/P
Rv1626/Rv3220c	PdtaR/PdtaS	(Morth *et al.*, 2005)	+/+

Possible orphan TCS

Rv0195[+]	U	D but apparently no REC domain	
Rv0260c[+]	U	D with N-terminal HEM4 domain. Regulation by phosphorylation unlikely.	−
Rv0818[+]	U	D with low probability N-terminal REC domain	P
Rv2027c	DosT (DevS homolog)	Contains GAF/GAF/H-ATPase domains	−
Rv2884[+]	U	D	−
Rv3143[+]	U	Contains REC domain only	−

[+] No detectable expression in infected macrophages (Haydel and Clark-Curtiss, 2004).
#, regarded as the minimum gene requirement for Mycobacteria.
U, unassigned name.
P, pseudogene of Mtb (www.sanger.ac.uk).
D, Contains a C-terminal DNA binding domain, suggesting a possible RR.

relative to other prokaryotes, of the eukaryotic like serine/threonine kinase signaling families (Cole *et al.*, 1998).

Five TCS—namely, SenX3-RegX3, PrrA-PrrB, MprA-MprB, MtrA-MtrB, and PdtaR-PdtaS—are conserved in all mycobacterial species (Tyagi and Sharma, 2004) sequenced to date. Five more TCS (PhoP-PhoR, KdpD-KdpE, TrcR-TrcS, DevR-DevS, and TrcX-TrcY, NarL) as well as three orphan proteins (Rv2060c, Rv0818, and Rv3143) are conserved in all mycobacteria except for *M. leprae*, which is thought to represent the minimal TCS requirement for mycobacteria.

The histidine to aspartate phosphotransfer-based signaling has been established for six of the twelve putative TCS, namely, RegX3-SenX3 (Himpens *et al.*, 2000), TrcR-TrcS (Haydel *et al.*, 1999), MprA-MprB (Zahrt *et al.*, 2003), PrrA-PrrB (Ewann *et al.*, 2004), DevR-DevS (Saini *et al.*, 2004), and PdtaR-PdtaS (Morth *et al.*, 2005). One apparently unique feature has been observed in the TCS proteins of MtB, where two proteins (with ORFs Rv0600c and Rv0601c) are annotated to contain the ATPase and histidine kinase domains that are normally found in a single protein and that would be expected to be necessary in order to phosphorylate the response regulator TcrA (with ORF Rv0602c). The two open reading frames overlap. Shirivastava *et al.* (2006) suggest the Rv0601c is an HPt domain, rather than a classical SK dimerization domain, whereas the SMART server identifies a HAMP domain. Clearly, further work is required on this unusual system.

The functions of many mycobacterial TCS proteins remain unknown but several studies have analyzed the expression profiles during MtB growth in human macrophages (Haydel and Clark-Curtiss, 2004; Zahrt and Deretic, 2001) and have begun to ascertain the biological role for these signal transduction systems. Constitutive expression of *pdtaR, dosT,* and *mtrA* during intercellular growth indicates that these genes are likely to be involved in MtB adaptation to life within macrophages. The other genes, *pdtaS, regX3, phoP, prrA, mprA, kdpA, trcR, devR,* and *trcX,* exhibit differential expression patterns in the presence of changing intracellular environments. This indicates that these genes are likely to be involved in a temporal pattern of regulation in response to intracellular adaptation. The studies describing high-density mutagenesis support the role of TCS in growth and survival (Sassetti *et al.*, 2001, 2003a,b). Based on these investigations, it appears that the *senX3, kdpD,* and *mtrA* (Sassetti *et al.*, 2003a) gene products are required for survival in mice and that the response regulators PhoP, KdpE, PdtR, and MtrA, as well as the sensor kinases MprB, DevS, and MtrB (Sassetti *et al.*, 2003b) are required for optimal growth *in vitro*.

A number of site-directed and deletion mutagenesis experiments have indicated that several TCS proteins are important for aspects of virulence of the tubercle bacillus. The RR- or SK-containing mutations do not grow well in macrophages relative to the wild-type MtB. This applies to DevR (Malhotra *et al.*, 2004), RegX3 (Parish *et al.*, 2003a), PhoP (Perez *et al.*, 2001), SenX3 (Rickman *et al.*, 2004), MprA (Zahrt and Deretic, 2001), and PrrA (Ewann *et al.*, 2002). MtrA-MtrB is believed to be essential for survival because, so far, it has not been possible to obtain *mtrA* knockout strains of MtB (Zahrt and Deretic, 2000), but most TCS genes do not appear to be essential for *in vitro* growth of MtB.

As a result of various genetic and biochemical analyses, PhoP has been suggested as a regulator of the synthesis of methyl-branched fatty acid-containing acetyltrehaloses, which are known to be restricted to pathogenic species of the *M. tuberculosis* (Asensio *et al.*, 2006; Walters *et al.*, 2006). However, specific signals sensed by the SK PhoR must be identified in order to determine under which conditions, and with what goal, the production of methyl-branched fatty acids is regulated. The hypothesis that MtB utilizes fatty acids as carbon sources during *in vivo* growth and persistence suggests that genes involved in metabolism of fatty acids are important in the mycobacterial infection process. The mycolic acids, important for the impervious nature of the bacterial cell wall, are synthesized by elongation using a type-II fatty acid synthetase and, primarily, palmitoyl acyl carrier protein (Kremer *et al*, 2002).

The best characterized of the MtB response regulators, DevR, has been shown to modulate its own expression, and that of a 47-gene regulon in MtB, in response to hypoxia and nitric oxide exposure (Ohno *et al.*, 2003; Sherman *et al.*, 2001; Voskuil *et al.*, 2003). It has also been shown that the *devR* gene responds to different stress situations, such as exposure to S-nitroglutathione, ethanol, and H_2O_2 (Kendall *et al.*, 2004). DevR activity is modulated by hypoxia via the TCS phosphorelay from DevS but might also be regulated through DosT (see Table I) in response to other, as yet unidentified environmental signals (Roberts *et al.*, 2004; Saini *et al.*, 2004).

The MprA/B TCS is required by MtB for persistent infection (Zahrt and Deretic, 2001) and participates in a multifaceted stress response in the tubercle bacillus. In particular, MprA directly regulates two key sigma factors, SigB and SigE (He *et al.*, 2006). SigB encodes a σ^A-like sigma factor that is dispensable for the growth of *M. tuberculosis* and *M. smegmatis* (Manganelli *et al.*, 2004), while SigE is one of 10 alternate sigma factors present in the *M. tuberculosis* genome (Cole *et al.*, 1998) that are believed to be activated in response to various stress situations. MprA-regulated SigB expression, during growth of MtB, was observed only under physiological conditions, whereas regulation of SigE expression depends upon the absence or presence of stress such as treatment with SDS, alkaline pH, or the presence of ionic or non-ionic detergents. MprA regulates eight genes (*mprA* (*rv0891*) , *rv1057, sigE* (*rv1221*), *rv1813c, rv2053c, rv2626c, rv2627c,* and *rv2628*) and some of these genes (*rv2053c, rv2626c, rv2627c,* and *rv2628*) that are induced under hypoxia are also under control of DevR (Park *et al.*, 2003). Although the majority of these genes can be up-regulated or repressed in response to a specific stress signal, many have been shown to be regulated in response to a multiplicity of environmental stresses, such as hypoxia, nutrient starvation, heat shock, antibiotic treatment, low pH, or detergent such as SDS (He *et al.*, 2006).

Studies on the TrcR response regulator have shown that it binds and regulates its own promoter via an AT-rich region (Haydel *et al.*, 2002). A similar AT-rich region was identified within the intergenic region located upstream of the *rv1057* gene, which codes for a hypothetical protein of unknown function. Transcriptional analysis revealed that the TrcR response regulator represses expression of the ORF *rv1057* by binding to the AT-rich region located 347bp upstream of the ATG codon (Haydel and Clark-Curtiss, 2006). It was shown that TrcR generates specific contacts on one side of the DNA helix and wraps around the ends of the DNA binding region. According to a computational analysis of *rv1057* (Haydel and Clark-Curtiss, 2006) the protein, which is annotated as being related to surface antigens in other bacteria, belongs to a family that possesses a β-propeller fold. The function and biological role of other TCS are unknown and require a considerable amount of further investigation at a functional and gene expression level.

Orphan TCS Proteins

Annotation as a response regulator or as a sensor kinase is usually made on the basis of the sequence signal of the receiver domain for RRs and the histidine kinase and ATPase domains for SKs. Identification of a RR/SK pair is then usually straightforward because the ORFs are adjacent on the genome. For the orphan proteins, this is not the case and the pairs, and consequently the nature of either output or input signal, are harder to identify. In addition the diversity of output domains for RR proteins (reviewed by Galperin, 2006) results sometimes in uncertainty in the identification of an RR when it itself has, for example, in the case of a hybrid SK, a histidine kinase domain. In MtB, there are isolated postulated open reading frames that could, based on sequence comparisons, be one or another component of a two-component system. Five of these are possibly RRs and one is possibly an SK (Table I). Thanks to the increasing number of complete genome sequences, the number and efficacy of bioinformatics tools has developed dramatically. The genome databases have also allowed the classification of protein families, especially for such modular systems as the TCS. An example is an analysis of the domain combinations and diversity of the bacterial RRs (Galperin, 2006).

To identify the putative SK corresponding to the, at that time, apparent orphan RR PdtaR (genomic location tag Rv1626), we adopted a systems biology approach using the program STRING, which makes a total genomic comparison from a current database of 179 species (von Mering *et al.*, 2003). In this case, the program suggests one putative SK (since named PdtaS, with genomic location tag Rv3220c) as a possible protein interaction partner based on gene co-occurrence and neighborhood of PdtaR orthologs. By using STRING, it was possible to quickly check putative orthologs and their genomic arrangement; in the case of PdtaR and PdtaS, a distant ortholog was found in *Listeria monocytogenes* (*lmo1172* and *lm01173*), still located next to each other on the genome. To establish the generality of this putative RR/HK pair, a PHI and PSI-BLAST (Altschul *et al.*, 1997) search was performed through the NCBI web server (http://www.ncbi.nlm.nih.gov/ BLAST). In our example, when using PdtaS as the query sequence and with three iterations, we found likely orthologs in the *Actinobacteria, Fusobacterium, Chloroflexus, and Firmicutes* phyla. These are the same phyla where we had previously found orthologs of PdtaR, thus substantiating the likely pairing. As a further control, we performed a sequence alignment on the identified HKs. To avoid redundancy, only sequences with less than 70% sequence identity were aligned. These sequences are identifiable by the highly conserved motifs corresponding to PFAM (Bateman *et al.*, 2004) HisKA_2 and HATPase domains. It was then possible to look for hints as to the domain organization of the rest of the protein. In our example, the N-terminus is not annotated for orthologs of PdtaS except for Cac2720 from *Clostridium acetobutylicum*, where the SMART server (http://smart.embl-heidelberg.de) suggests the presence of the ubiquitous GAF and PAS domains. Both the SMART and PFAM (http://pfam.wustl.edu/) servers have options to change their search criteria with the default cutoff value of both servers being sufficiently stringent to ensure a minimum of false positives. By increasing this value, it is possible to get a set of weak domain predictions, which can be evaluated individually. For each of the aligned gene products, both domain types are often found associated with HisKA_2 and HATPase domains and are often involved in signal sensing. To further validate whether we had found the true orthologs, we performed a secondary structure prediction of all the aligned sequences and docked them to our aligned putative orthologs. In our example, this indicated that all the aligned sequences had the same domain topology and that a GAF and PAS domain organization at the N-terminus would fit this secondary structure profile (Morth *et al.*, 2005).

At this point, we believed there was enough evidence to indicate a real TCS pair, so it was then necessary to clone and purify PdtaS in order to obtain biochemical verification of the hypothesis. The most informative biochemical assay is to check the ability of the putative SK to specifically phosphorylate the putative RR. The assay based on the detection of ^{32}P labeled products derived from radiolabeled ATP, although in many ways similar to that used for serine/threonine/tyrosine kinases, is less robust due to the instability of the phosphoamidate bond, especially at acidic pH, and to hydrolysis of the phosphoaspartate of the activated RR. The assay is effective because there were substantial amounts of both RR and SK to work with and so, by using these relatively large amounts of protein to perform a phosphorylation experiment, we were able to detect whether self and cross phosphorylation reactions had taken place. In our example, self-phosphorylation assays of Rv3220c, as well as the phosphotransfer between Rv3220c and the putative response regulator PdtaR and other RRs of MtB (PrrA;Rv0903c, NarL;Rv0844, and RegX3;Rv0491) were performed. In addition, we investigated the ability of another SK (PrrB;Rv0902c) to phosphorylate PdtaR (Morth *et al.*, 2005).

Information from Crystal Structures

The medical consequences of tuberculosis cannot be underestimated. The fact is that one-third of the world's population is infected by MtB and that there are over two million deaths per year. Only recently has increasing focus been placed on developing MtB treatments, in part, because it is a disease that is predominant in the third world and among the poor and the immunocompromised. Little is known about the multidrug resistant strains of MtB that result, to a great extent, from patient noncompliance to treatment. MtB treatment normally involves a 2-month-long intensive phase followed by 4 months of continuation (Hatfull and Jacobs, 2000). This treatment is expensive but necessary due to the persistence of the bacteria and its ability to avoid the host immune system. These considerations have driven a number of structural genomics initiatives in a number of countries/continents (see following text). The general goal of all of these initiatives is to provide the basis for structure-based drug design (SBDD), although it has never been clear whether these efforts could have any commercial success, given the economic circumstances of those requiring treatment. Even so, the TCS of MtB are ideal targets for SBDD because these signal transduction systems—at least in the sense of phosphotransfer from histidine to aspartic acid—have not been found in higher eukaryotes (the host) other than in plants (Mizuno, 2005).

Structural Genomics as a Driving Force

The TCS of MtB have been targets in a number of structural genomics consortia. The big advantage of these consortia is the open (web) availability of information and anyone interested in MtB structural genomics has a variety of (cross-linked) resources to consult (http://www.doe-mbi.ucla.edu/TB/, www.pasteur.fr/recherche/X-TB/, http://xmtb.org/, http://genolist.pasteur.fr/TubercuList/, http://pedant.gsf.de/). This list may not be, by oversight only, exhaustive, but it does reflect the real concern regarding the spread of tuberculosis in and outside the structural biology community.

Domain Boundary Definitions

As a first step in the analysis of the TCS—on a structural basis—is the realization that these are, of necessity multidomain proteins. Multi-domain proteins often will not crystallize as such, especially where a non-structured linker has been predicted by programs such as Globplot (Linding *et al.*, 2003) and DIsEMBL (Iakoucheva and Dunker, 2003). We believe it is still worth the attempt. There are a number of two-domain RR structures in the literature, for example from thermophiles such as *Thermotoga maritima* (Buckler *et al.*, 2002; Robinson *et al.*, 2003) but also from MtB (Figs. 3 and 4). There is just one multiple domain SK structure in the literature (Marina *et al.*, 2005) and, to date, our attempts to crystallize multidomain constructs of MtB SKs have been unsuccessful. It is interesting to note that we have observed in two cases (PrrB and SenX3) that, when the region N-terminal of the dimerization (HK domain) is removed, we obtain proteins that are soluble, dimeric, and which appear to specifically phosphory-late their cognate RR when magnesium and ATP are present. When the multidomain approach fails, we have attempted to define domain bound-aries using extensive sequence alignments. The domain boundaries of the RR receiver domain and the SK HisKA_2 and HATPase domains are relatively well defined. This is not true of the so-called HAMP domain (Appleman *et al.*, 2003; Aravind and Ponting, 1999) that is described for some SKs and is thought to be involved in the signal transduction pathway from the extracellular region. Domain prediction programs (for a summary, visit the CAFASP4 web site) also fail to give clear answers. When this is the case, a multiconstruct approach should always be used, where the (in this case) C- and N-terminal extents must be explored experimentally. High-throughput, rational approaches to this problem have been developed (e.g., Dahlroth *et al.*, 2006; Hart and Tarendeau, 2006; see http://psb.esrf.fr/), but this does not mean that a simpler experimental approach testing some 6 to 12 different constructs may not be a productive way to proceed.

Protein Production as a Source of Material for Structural Studies and *In Vitro* Inhibition Assays

One major limitation to the progress of any structural biology project will always be the requirement for large amounts (<5 mg) of pure, soluble protein. Many approaches are possible that depend, to a large extent, on the degree of high throughput technologies available (see, e.g., OPPF;

http://www.oppf.ox.ac.uk). For a medium-sized laboratory such as ours, a small-scale expression and solubility methodology is appropriate and we chose to adapt this from the MtB structural genomics program at UCLA (Goulding and Perry, 2003; Goulding *et al.*, 2003).

For MtB, the expression strain BL21(DE)RP proved particularly useful. The *E. coli* strain has an extra plasmid, which encodes tRNA for the rare codons AGA, AGG, and CCC, which code for arginine and proline, respectively. These particular rare codons are often found in the open reading frames of MtB. Our first choice of expression was, therefore, always the RP cell strain, but we also had success with BL21(DE3)RIL, BL21(DE) pLysS, and Rosetta (DE3). An effective procedure that we have used is:

Day 1: Plasmid transformed into the chosen cell strain.

Day 2: Overnight (O/N) cultures were prepared in common Luria Betani (LB) media, with the appropriate antibiotic resistance. O/N cultures were performed with 5 ml LB media in 10 ml falcon tubes at 37°.

Day 3: One ml O/N culture was used to inoculate 100 ml LB media. Inoculating with approximately 1% of the desired expression media normally requires 2 to 3 h at 37° before an $OD_{600} = 0.6$ to 0.8 is reached. Before induction, 3×1 ml samples were taken for later analysis and the OD_{600} was carefully measured. The remaining culture was induced with IPTG to a final concentration of 0.5 to 1.0 mM and left to induce for 4 to 5 h at 37°. At the end of induction, OD_{600} was measured again and 10×1 ml samples where taken. The samples taken for analysis were spun down for 3 min at 13,000 rpm and the supernatant carefully removed with a pipette. The LB medium used contains a large amount of soluble protein, which will increase the background on the SDS-PAGE when testing for expression/solubility. It is, therefore, necessary to remove as much as possible by washing the cells. Larger-scale purifications were normally washed with Tris buffer containing 1 mM PMSF, to inhibit the excreted *E. coli* proteases. The remaining 80 to 90 ml LB was spun down and stored in a single pellet in a 50 ml NUNC tube, which was normally kept to make a quick off-column Ni-NTA purification, to see how the protein behaved.

Day 4: Expression (Ex) and Solubility (SB) tests.

Buffers

   Ex: 50 mM Tris pH 8.0, 1% SDS, 30 mM DTT, 1mM EDTA
   SB1. Native: 50 mM Tris pH 8.0, 150 KCl, 4 mM DTT
   SB2. Glycerol: 50 mM Tris pH 8.0, 150 KCl, 10% Glycerol, 4 mM DTT
   SB3. Urea: 50 mM Tris pH 8.0, 150 KCl, 0.5 M Urea, 4 mM DTT
   SB4. Bugbuster: 50 mM Tris pH 8.0, 150 KCl, 1% BugBuster®
       10× Protein Extraction Reagent (Novagen), 4 mM DTT

Protocol

Two hundred microliters of **Ex** is added to samples taken both before and after induction. The cells are resuspended and lysis is performed in a water ultrasound bath for 30 to 45 min at 4°. Samples are spun for 10 min at 13,000 rpm and the supernatant carefully transferred to a clean Eppendorf tube. The supernatant is precipitated with cold acetone and left on ice for 15 min, then centrifuged at 13,000 rpm for 10 min. The bulk acetone is removed and the tubes left open until the remaining acetone has evaporated. The remaining pellet is resuspended in $1\times$ SDS-loading buffer. The quantity of loading buffer added depends on the OD_{600} measured earlier. Fifty microliters of $1\times$ SDS-Loading Buffer is added per unit OD_{600} measured. This pellet is highly concentrated with protein, and loading 5 to 10 μl of each sample to the subsequent SDS-PAGE 12% polyacrylamide will be sufficient.

For SDS-PAGE analyses, it is important that samples from before and after induction are placed next to each other (for comparison).

Solubility

Using the same procedure, the four given buffers are commonly used as a first test for solubility. The protocol for solubility studies is identical to the expression protocol with regard to cell lysis and sample preparation, except that both pellet and supernatant are kept for analysis. The pellet remaining from the acetone precipitation of the supernatant should be resuspended in 15 μl $1\times$ SDS loading buffer per unit OD_{600} and the pellet in 25 μl $1\times$ SDS loading buffer. Loading a 5 to 10 μl sample to the SDS-PAGE gel is sufficient. The pellet will often contain large amounts of genomic DNA, which should not be disturbed when taking a sample for the SDS-PAGE. Consequently, the SDS samples should be spun for 5 min at 13,000 rpm before applying them to the SDS-PAGE, making it easier to remove the sample without disturbing the pellet.

For structural studies, it is very important to have some preliminary information about protein solubility and homogeneity. In one of the example cases we are using in this chapter, both PdtaR (Rv1626) and PdtaS (Rv3220c) were found to be most soluble in low salt buffers. The expression temperature had no effect on PdtaR, but was a very important factor for PdtaS (Fig. 1), where a temperature dependence on the solubility of the protein ongoing from 37 to 24° upon induction is clear (24 to 16° brings no significant improvement). The OD_{600} measurements allow one to optimize the total wet cell weight produced at the different induction temperatures. From the initial solubility study, it was estimated that the urea buffer would give most soluble protein when cells were grown at 16°. (The estimate is

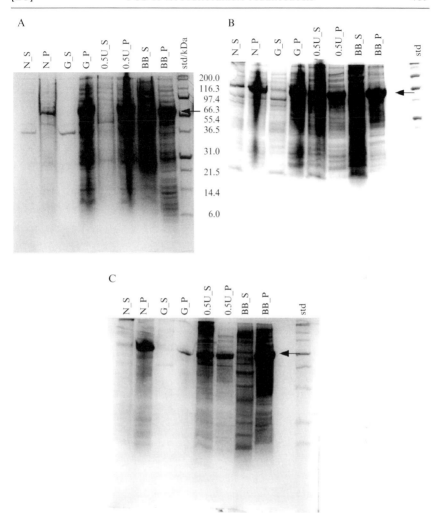

FIG. 1. PdtaS induced at A: 37°, B: 24°, and C: 16° and tested for solubility, supernatant (S), and pellet (P) are loaded next to each other. Four different buffers are tested, Native (N), Glycerol (G), Urea (0.5U), and Bugbuster (BB).

made from the SDS-PAGE gel [Fig. 1B]) by comparing supernatant:pellet ratios. At 16°, it is approximately 50:50 while at 24°, it is 40:60 and at 37° 0:100. We next test for salt dependency on solubility using a range from 0 to 450 mM KCl (in addition to the urea buffer). The result (Fig. 2) is atypical (but, therefore, justifies the protocol) in that PdtaS shows a significant improvement in solubility at *low* salt concentrations. To validate this

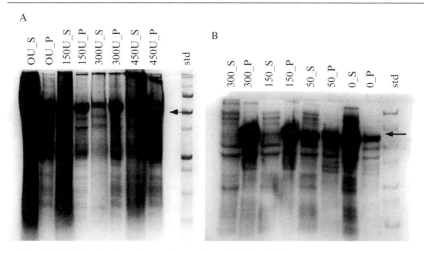

FIG. 2. PdtaS induced at 16° and tested for solubility. (A) Urea buffer with KCl ranging from 0 to 450 mM. (B) Native buffer with KCl ranging from 0 to 300 mM KCl.

unexpected result, we repeated the test in native buffer (no urea) with a KCl range between 0 and 300 mM (see Fig. 2B), which indicated that the trend also was significant in the native buffer. Despite the fact that the solubility was greater in urea buffer (Fig. 2), we preferred to continue with the native buffer. In this case, because PdtaS contains putative GAF and PAS domains with unknown cofactors/ligands and by minimizing chaotropes, we enhance the possibility that the recombinant protein will pick up its (unknown) partner from the expression host and carry it along through the following purification steps. We have rarely been lucky in this respect.

Crystallographic Studies

Having produced soluble, and hopefully conformationally and chemically monodisperse, protein solutions, structural biology offers several, usually complementary, techniques. In this contribution, we do not consider NMR methods (rarely employed for multidomain MtB proteins), although there are many important benefits with NMR—not least because it explores structure in solution. High-throughput crystallization facilities (e.g., http://www.embl-hamburg.de/service/crystallization/) are becoming more widespread and 300-odd crystallization experiments can be conducted with less than 100 μl of pure protein at around 10 mg/ml concentration. There is a general consensus that if these experiments fail to give a lead, then the law of diminishing returns suggests that it would be more productive to change the construct, or, where one has the freedom to do so, the organism.

The process of crystal structure determination is beyond the scope of this chapter—indeed, there are four volumes in this series (Vols. 276, 277, 368, and 374, all edited by C. W. Carter, Jr., and R. M. Sweet) covering macromolecular crystallography. We would simply observe that the ready availability of tunable synchrotron beamlines (allowing the determination of phase information from the measurement of anomalous scattering effects (Hendrickson, 1991), the development of automated procedures ranging from sample mounting through crystal centering (e.g., XREC, Pothineni *et al.*, 2006), and expert systems for data collection and processing (e.g., DNA, Leslie *et al.*, 2002), as well as some attempts to automate the structure determination and model building (see, for example, Panjikar *et al.*, 2005), have made the process much simpler for the novice than it once was.

Although we do not wish to cover the methods involved in crystal structure determination, we do wish to summarize the structural results on MtB TCS that are either in the literature or have been deposited in the Protein Data Bank (Berman *et al.*, 2000). The structures are shown in Figs. 3 and 4.

Information on Solution Structure from Small-Angle X-Ray Scattering

If a multidomain protein, containing one or more flexible linkers between domains, yields to crystal structure determination, the result will, in general, be one (or more) of the possible lower energy conformers. Given that the crystallization conditions are far from those found in the microbial cytosol, there is no guarantee that this (crystal) structure has any physiological significance or, indeed, that it even represents a major conformer in solution. As a consequence, it is a good idea to perform X-ray small angle scattering experiments to check the generality of the crystal structure results. Such experiments have produced some interesting and/ or unexpected results that we will illustrate from some of our own work (see later).

The technique has developed enormously over the last few years both because of improvements in equipment and, more importantly, developments in the software used to process and interpret the data. The experimental methodology has been reviewed (Svergun and Koch, 2003) but for completeness, a simplified experimental setup is shown in Fig. 5. The requirement for good signal-to-noise ratio, especially at higher scattering angles, requires a stable, well-collimated X-ray beam from a synchrotron source and a low-noise, well-calibrated X-ray detector. The sample requirements are modest by structural biology standards, typically, a few tens of microliters at protein concentrations between 5 and 20 mg/ml.

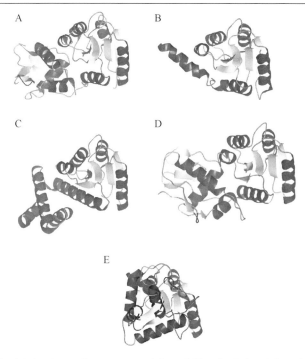

FIG. 3. Crystal structures of response regulators of *Mycobacterium tuberculosis* displayed using the receiver domain to fix the orientation. (A) The two domain construct of MtrA (PDB code 2GWR; Friedland *et al.*, to be published). (B) The receiver domain of NarL (Nowak *et al.*, to be published). (C) PdtaR (PDB code 1SN8; Morth *et al.*, 2004). (D) PrrA (PDB codes 1YS6,1YS7; Nowak *et al.*, 2006a), and (E) RegX3 (PDB code 1ZKV; King-Scott *et al.*, 2007).

The samples should have a high-speed spin ($>$10,000 rpm at $4°$) prior to usage to remove aggregates which otherwise dominate the scattering, especially at small scattering angles. The analysis requires a reasonably accurate determination of the protein concentration, for which measurement at 280 nm in a nanodrop spectrophotometer (Peqlab) using a calculated extinction coefficient is adequate. This instrument may not be suitable to measure very high protein concentrations ($>$10 mg/ml) but this can be easily remedied by accurate dilution prior to measurement. Attention needs to be paid to the protein buffer, first, because the buffer scattering needs to be subtracted to obtain the scattering due to the molecules of interest and second, because it should not contain high concentrations of glycerol or other chemicals that would have high affinity for the protein surface. The experimental procedure normally requires measurement of the scattering from the sample, from the buffer (both before and after the sample measurement) and from a standard such as a known concentration

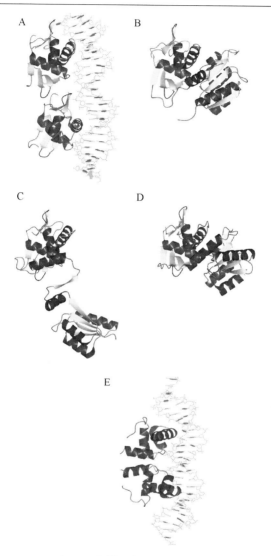

FIG. 4. The response regulators of *Mycobaterium tuberculosis* shown with the output domains overlapped. (A) Winged HTH of RegX3 modeled onto the tandem repeat DNA by homology with the PhoB DNA structure (PDB code 1GXP; Blanco *et al.*, 2002), (B) MtrA (PDB code 2GWR), (C) RegX3 (PDB code 1ZKV; King-Scott *et al.*, 2007), (D) PrrA (Nowak *et al.*, 2006a), (E) the DosR-DNA complex (PDB codes 1ZLK, 1ZLJ; Wisedchaisri *et al.*, 2005). Here, the DNA binding region is a palindromic sequence and the dimer has two-fold symmetry.

(around 6 mg/ml) of bovine serum albumin. The measurements need to be repeated, not only if statistics need to be improved, but also to check reproducibility and that radiation damage has not adversely affected the scattering curves.

The scattering curves, scattered intensity as a function of the scattering vector **s** (where $|s| = 2\sin\theta/\lambda$, where θ is the scattering angle and λ is the X-ray wavelength), can be analyzed with a variety of software, most of which can be found in the ATSAS package (Konarev *et al.*, 2006; see also www.embl-hamburg.de/ExternalInfo/Research/Sax/software.html). The types of analysis can be divided into a number of areas, depending on the degree of a *priori* knowledge, as follows:

 i. *Ab initio* structure analysis. These programs attempt to determine the shape of the molecule using no, or very few, simple assumptions, resulting in an envelope, a dummy atom, or a dummy residue model (Svergun, 1999; Svergun *et al.*, 1996, 2001).
 ii. Testing for mixtures to yield, for example, volume fractions of different components (Konarev *et al.*, 2003).
 iii. Comparing the solution scattering with that calculated from an atomic model (Svergun *et al.*, 1995).
 iv. Assembling different subunits of a protein or protein complex, for example, domains of known structure, into a larger structural assembly (Konarev *et al.*, 2001; Petoukhov and Svergun, 2005). This could also include determining the shapes of additional domains in a multidomain protein.

SAXS has become especially relevant to studies of TCS, but the first paper of real significance, relating to the response regulator FixJ of *Sinorhizobium meliloti* (Birck *et al.*, 2002), is relatively recent. This work was important in determining the relative position and orientations of the two domains. The $\alpha4$-$\beta5$-$\alpha5$ surface, the linker region, and the C-terminal helix of the effector domain were identified as being important for signal transduction. In the same paper, the SAXS results were supported by biochemical investigations.

In our own work on the MtB TCS, we have used the method to:

1. Show that for PdtaR (Rv1626), the structure in the crystal is, in solution, dependent on the ionic strength of the solution but apparently not on the pH (Morth *et al.*, 2004). However, it was clear that the molecule, at least in the inactivated form, was monomeric and did not form the helical coiled-coil dimer interface seen in the structure of the homologous *Pseudomonas aeruginosa* AmiR protein in the AmiS-AmiR complex (O'Hara *et al.*, 1999).

Schematic X33 SAXS/WAXS setup

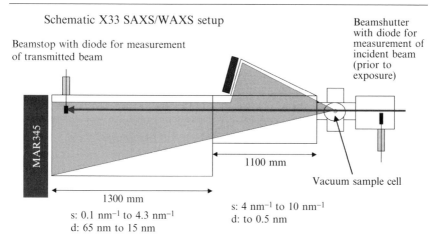

FIG. 5. A schematic view of the EMBL Hamburg X33 SAXS beamline. The small angle scattering regime is covered by the MAR345 image plate detector and the wider angle region by a gas detector. Figure courtesy of Manfred Roessle.

2. Show that the position of the HAMP domain of the dimeric sensor kinase PrrB (Rv0902c) was spatially between the membrane anchor and the histidine-kinase (dimerization) domain–ATPase domain (Nowak *et al.*, 2006b) pair. Interestingly, the only high-resolution X-ray structure of such a two-domain sensor kinase construct (Marina *et al.*, 2005) shows a slightly different orientation of the two domains. This reflects the importance of the hinge between the dimerization and ATPase domains, which is necessary in order that the latter domain can transfer the phosphate to the conserved histidine of the dimerization domain. A possible mechanism is that one of the two ATPase domains swings around the axis of the dimer formed by the dimerization domain to bring the interfaces in proximity for phosphotransfer to take place.

3. Show that the response regulator PrrA (Rv0903c) exists as a conformational mixture in solution where, in the absence of the sensor kinase, the recognition helix of the effector domain is buried, and therefore could not bind DNA, whereas upon activation by the cognate sensor kinase (PrrB, Rv0902c), the molecule adopts a more open conformation, similar to that observed in *Thermotoga maritima* DrrD, which would allow the effector domain to bind DNA (Nowak *et al.*, 2006a). The experiments agree with the established view that the signal transduction pathway involves changes in orientation of a serine (or threonine) close to the phosphorylation site, which, in turn, allows a reorientation of a tyrosine residue (Tyr105 in PrrA).

This tyrosine in PrrA is part of the interdomain interface—which does not involve the interdomain linker—thus explaining how the interdomain interface can be destabilized upon phosphorylation, generating a higher concentration of a species that is able to bind DNA.

4. Show that there is some doubt about the physiological significance of the domain swapped dimer of RegX3 (King-Scott *et al.*, 2007), despite the fact that the domain swapping leaves a structurally intact phosphorylation site with the arginine coming from another protomer.

SAXS has also been used in combination with low-resolution X-ray diffraction experiments to establish the structural basis of the signaling pathway of the ThkA/TrrA system in *T. maritima* (Yamada *et al.*, 2006).

Structural Information Relating to Regulation Mechanisms

We embarked on our work on the TCS of MtB in order to identify putative inhibitors of the phosphotransfer pathway either *in silico*, from the modeling of compounds that would interfere with the surface complementarity of the RR receiver domain and the SK HisKA_2 domain, or *in vitro* from high-throughput screening assays. The former has not been possible because of the current paucity of information we have on the structure of the MtB SKs or their domains, with only one HATPase domain structure published (Nowak *et al.*, 2006b). Clearly, the structures of RR/SK complexes would provide the most useful source of this information; however, we would expect that such complexes would be, in general, transitory in nature and we have failed to detect any. The latter studies are under way and are facilitated by the previously mentioned observation that deletion mutants of the SKs appear constitutively active for phosphotransfer.

Thus, for MtB, we have a good basis for understanding the structural biology but there is a notable lack of information on the trigger for the signaling pathway and, in very many cases, on the downstream response. To illustrate the latter point, we searched for a PhoB-like (Makino *et al.*, 1996) tandem repeat that might be bound by the winged HTH domain of the RR, RegX3. It appears that the RR binds to a tandem repeat (Himpens *et al.*, 2000; King-Scott *et al.*, 2007) in the promoter region of SenX3/RegX3, thus apparently regulating the production of the signaling system itself. Furthermore, it has been shown that RegX3 binds to DNA in both phosphorylated and unphosphorylated forms (Himpens *et al.*, 2000), suggesting that regulation is via interaction of the RR with the sigma factor of the RNA polymerase rather than with the DNA itself, the binding of which only acts as an anchor. The MtB operon *rv0096-rv0101* is strongly upregulated in *senX3/regX3* knockouts (Parish *et al.*, 2003), suggesting that the observed effect is a direct one, with RegX3 acting as a transcriptional repressor of this operon.

To date, only two structures (Blanco *et al.*, 2002; Wisedchaisri *et al.*, 2005) of MtB response regulators bound to their cognate DNA have been published (Fig. 4). One belongs to the PhoB/OmpR family (Galperin, 2006) where the binding is to a tandem repeat DNA sequence, the other to a member of the FixJ/NarL family where binding is to a palindromic, or pseudo-palindromic, DNA sequence. These structures, however, only contain the effector domain and so provide scant information relating to the activation mechanism itself.

References

Altschul, S. F., Madden, T. L., Schäffer, A. A., Zhang, J., Zhang, Z., Miller, W., and Lipman, D. J. (1997). Gapped BLAST and PSI-BLAST: A new generation of protein database search programs. *Nucleic Acids Res.* **25**, 3389–3402.

Appleman, J. A., Chen, L. L., and Steward, V. (2003). Probing conservation of HAMP linker structure and signal transduction mechanism through analysis hybrid sensor kinases. *J. Bacteriol.* **185**, 4872–4882.

Aravind, L., and Ponting, C. P. (1999). The cytoplasmic helical linker domain of receptor histidine kinase and methyl-accepting proteins is common to many prokaryotic signaling proteins. *FEMS Microbiol. Lett.* **176**, 111–116.

Asensio, J. G., Maia, C., Ferrer, N. L., Barilone, N., Laval, F., Soto, C. Y., Winter, N., Daffe, M., Gicquel, B., Martin, C., and Jackson, M. (2006). The virulence-associated two-component PhoP-PhoR system controls the biosyntesis of polyketide-derived lipids in *Mycobacterium tuberculosis. J. Biol. Chem.* **281**, 1313–1316.

Bateman, A., Coin, L., Durbin, R., Finn, R. D., Hollich, V., Griffiths-Jones, S., Khanna, A., Marshall, M., Moxon, S., Sonnhammer, E. L., *et al.* (2004). The Pfam protein families database. *Nucleic Acids Res.* **32**, D138–D141.

Berman, H. M., Westbrook, J., Feng, Z., Gilliland, G., Bhat, T. N., Weissig, H., Shindyalov, I. N., and Bourne, P. E. (2000). The Protein Data Bank. *Nucleic Acids Res.* **28**, 235–242.

Birck, C., Malfois, M., Svergun, D., and Samma, J. (2002). Insights into signal transduction revealed by the low resolution structure of FixJ response regulator. *J. Mol. Biol.* **16**, 447–457.

Blanco, A. G., Sola, M., Gomis-Ruth, F. X., and Coll, M. (2002). Tandem DNA recognition by two-component signal transduction transcriptional activator PhoB. *Structure* **10**, 701–713.

Buckler, D. R., Zhou, Y., and Stock, A. M. (2002). Evidence of intradomain and interdomain flexibility in an OmpR/PhoB homolog from *Thermotoga maritima. Structure* **10**, 153–164.

Cole, S. T., Brosch, R., Parkhill, J., Garnier, T., Churcher, C., Harris, D., Gordon, S. V., Eiglmeier, K., Gas, S., Barry, C., III, Tekaia, F., Badcock, K., *et al.* (1998). Deciphering the biology of *Mycobacterium tuberculosis* from the complete genome sequence. *Nature* **393**, 537–544.

Dahlroth, S.-L., Nordlund, P., and Cornvik, T. (2006). Colony filtration blotting for screening soluble expression in *Escherichia coli. Nature Protocols* **1**, 253–258.

Ewann, F., Locht, C., and Supply, P. (2004). Intracellular autoregulation of the *Mycobacterium tuberculosis* PrrA response regulator. *Microbiology* **150**, 241–246.

Ewann, F., Jackson, M., Pethe, K., Cooper, A., Mielcarek, N., Ensergueix, D., Gicquel, B., Locht, C., and Supply, P. (2002). Transient requirement of PrrA-PrrB two-component system for early intracellular multiplication of *Mycobacterium tuberculosis*. *Infect. Immun.* **70,** 2256–2263.

Galperin, M. Y. (2006). Structural classification of bacterial response regulators: Diversity of output domains and domain combinations. *J. Bacteriol.* **188,** 4169–4182.

Goulding, C. W., and Perry, L. J. (2003). Protein production in *Escherichia coli* for structural studies by X-ray crystallography. *J. Struct. Biol.* **142,** 133–143.

Goulding, C. W., Perry, L. J., Anderson, D., Sawaya, M. R., Cascio, D., Apostol, M. I., Chan, S., Parseghian, A., Wang, S. S., Wu, Y., *et al.* (2003). Structural genomics of *Mycobacterium tuberculosis*: A preliminary report of progress at UCLA. *Biophys. Chem.* **105,** 361–370.

Hart, D. J., and Tarendeau, F. (2006). Combinatorial library approaches for improving soluble protein expression in *Escherichia coli*. *Acta Crystallogr.* **D62,** 19–26.

Hatfull, G. F., and Jacobs, W. R. (2000). "Molecular Genetics of Mycobacterium," 1st ed. ASM Press, Washington DC.

Haydel, S. E., Benjamin, W. H., Jr., Dunlap, N. E., and Cark-Curtiss, J. E. (2002). Expression, autoregulation, and DNA binding properties of the *Mycobacterium tuberculosis* TrcR response regulator. *J. Bacteriol.* **184,** 2192–2203.

Haydel, S. E., and Clark-Curtiss, J. E. (2004). Global expression analysis of two-component system regulator genes during *Mycobacterium tuberculosis* growth in human macrophages. FEMS *Microbiol. Lett.* **236,** 341–347.

Haydel, S. E., and Clark-Curtiss, J. E. (2006). The *Mycobacterium tuberculosis* Trc response regulator represses transcription of the intracellularly expressed Rv1057 gene, encoding a seven-bladed β-propeller. *J. Bacteriol.* **188,** 150–159.

Haydel, S. E., Dunlap, N. E., and Benjamin, W. H., Jr. (1999). *In vitro* evidence of the two-component system phosphorylation between the *Mycobacterium tuberculosis* TrcR/TrcS proteins. *Microb. Pathog.* **26,** 195–206.

He, H., Hovey, R., Kane, J., Singh, V., and Zhart, T. C. (2006). MprAB is a stress-responsive two-component system that directly regulates expression of sigma factors SigB and SigE in *Mycobacterium tuberculosis*. *J. Bacteriol.* **188,** 2134–2143.

Hendrickson, W. A. (1991). Determination of macromolecular structures from anomalous diffraction of synchrotron radiation. *Science* **254,** 51–58.

Himpens, S., Locht, C., and Supply, P. (2000). Molecular characterization of the mycobacterial SenX3-RegX3 two-component systems: Evidence for autoregulation. *Microbiology* **146,** 3091–3098.

Iakoucheva, L. M., and Dunker, A. K. (2003). Order, disorder, and flexibility: Prediction from protein sequence. *Structure* **11,** 1316–1317.

Kendall, S. L., Movahedzadeh, F., Rison, S. C. G., Wernisch, L., Parish, T., Duncan, K., Betts, J. C., and Stoker, N. G. (2004). The *Mycobacterium tuberculosis* dosRS two-component system is induced by multiple stresses. *Tuberculosis* **84,** 247–255.

King-Scott, J., Nowak, E., Mylonas, E., Svergun, D. I., and Tucker, P. A. (2007). The crystal and solution structure of the response regulator RegX3 from *Mycobacterium tuberculosis*. In preparation.

Konarev, P. V., Petoukhov, M. V., and Svergun, D. I. (2001). MASSHA—A graphics system for rigid-body modeling of macromolecular complexes against solution scattering data. *J. Appl. Crystallogr.* **34,** 527–532.

Konarev, P. V., Volkov, V. V., Sokolova, A. V., Koch, M. H. J., and Svergun, D. I. (2003). PRIMUS: A Windows PC-based system for small-angle scattering data analysis. *J. Appl. Crystallogr.* **36,** 1277–1282.

Konarev, P. V., Petoukhov, M. V., Volkov, V. V., and Svergun, D. I. (2006). ATSAS 2.1, a program package for small-angle scattering data analysis. *J. Appl. Crystallogr.* **39,** 277–286.

Kremer, L., Dover, L. G., Carrere, S., Nampoothiri, K. M., Lesjean, S., Brown, A. K., Brennan, P. J., Minnikin, D. E., Locht, C., and Besra, G. S. (2002). Mycolic acid biosynthesis and enzymic characterization of the beta-ketoacyl-ACP synthase A-condensing system of *Mycobacterium tuberculosis. Biochem. J.* **364,** 423–430.

Leslie, A. G. W., Powell, H. R., Winter, G., Svensson, O., Spruce, D., McSweeney, S., Love, D., Kinder, S., Duke, E., and Nave, C. (2002). Automation of the collection and processing of X-ray diffraction data—A generic approach. *Acta Crystallogr.* **D58,** 1924–1928.

Linding, R., Russell, R. B., Neduva, V, and Gibson, T. J. (2003). GlobPlot: Exploring protein sequences for globularity and disorder. *Nucleic Acids Res.* **31,** 3701–3708.

Makino, K., Amemura, M., Kawamoto, T., Kimura, S., Shinagawa, H., Nakata, A., and Suzuki, M. (1996). DNA binding of PhoB and its interaction with RNA polymerase. *J. Mol. Biol.* **259,** 15–26.

Malhotra, V., Sharma, D., Ramanathan, V. D., Shakila, H., Saini, D. K., Chakravorty, S., Das, T. K., Li, Q., Silver, R. F., Narayanan, P. R., and Tayagi, J. S. (2004). Disruption of response regulator gene, devR, leads to attenuation in virulence of *Mycobacterium tuberculosis. FEMS Microbiol. Lett.* **231,** 237–245.

Manganelli, R., Provvedi, S., Rodrigue, S., Beaucher, J., Gaudreau, L., and Smith, I. (2004). Sigma factors and global gene regulation in *Mycobacterium tuberculosis. J. Bacteriol.* **186,** 895–902.

Marina, A., Waldburger, C. D., and Hendrickson, W. A. (2005). Structure of the entire cytoplasmic portion of a sensor histidine-kinase protein. *EMBO J.* **24,** 4247–4259.

Mizuno, T. (2005). Two-component phosphorelay signal transduction systems in plants: From hormone responses to circadian rhythms. *Biosci. Biotechnol. Biochem.* **69,** 2263–2276.

Morth, J. P., Feng, V., Perry, L. J., Svergun, D. I., and Tucker, P. A. (2004). The crystal and solution structure of a putative transcriptional antiterminator from *Mycobacterium tuberculosis. Structure* **12,** 1595–1605.

Morth, J. P., Gosmann, S., Nowak, E., and Tucker, P. A. (2005). A novel two-component system found in *Mycobacterium tuberculosis. FEBS Lett.* **579,** 4145–4148.

Nowak, E., Panjikar, S., Konarev, P., Svergun, D. J., and Tucker, P. A. (2006a). The structural basis of singal transduction for the response regulator PrrA from *Mycobacterium tuberculosis. J. Biol. Chem.* **281,** 9659–9666.

Nowak, E., Panjikar, S., Morth, J. P., Jordanova, R., Svergun, D. J., and Tucker, P. A. (2006b). Structural and functional aspects of the sensor histidine kinase PrrB from *Mycobacterium tuberculosis. Structure* **14,** 275–285.

O'Hara, B. P., Norman, R. A., Wan, P. T. C., Roe, S. M., Barrett, T. E., Drew, R. E., and Pearl, L. H. (1999). Crystal structure and induction mechanism of AmiC-AmiR: A ligand-regulated transcription antitermination complex. *EMBO J.* **18,** 5175–5186.

Ohno, H., Zhu, G., Mohan, V. P., Chu, D., Kohno, S., Jacobs, W. R., Jr., and Chan, J. (2003). The effects of reactive nitrogen intermediates on gene expression in *Mycobacterium tuberculosis. Cell Microbiol.* **5,** 637–648.

Panjikar, S., Parthasarathy, V., Lamzin, V. S., Weiss, M. S., and Tucker, P. A. (2005). *Auto-Rickshaw*—An automated crystal structure determination platform as an efficient tool for the validation of an X-ray diffraction experiment. *Acta Cryst.* **D61,** 449–457.

Parish, T., Smith, D. A., Roberts, G., Betts, J., and Stoker, N. G. (2003a). The senX3-regX3 two-component regulatory system of *Mycobacterium tuberculosis* is required for virulence. *Microbiology* **149,** 1423–1435.

Parish, T., Smith, D. A., Kendall, S., Casali, N., Bancroft, G. J., and Stoker, N. G. (2003b). Deletion of two-component regulatory systems increases the virulence of *Mycobacterium tuberculosis*. *Infect. Immun.* **71,** 1134–1140.

Park, H. D., Guinn, K. M., Harrell, M. I., Liao, R., Voskuil, M. I., Tompa, M., Schoolnik, G. K., and Shermann, D. R. (2003). Rv3133c/dosR is a transcription factor that mediates the hypoxic response of *Mycobacterium tuberculosis*. *Mol. Microbiol.* **48,** 833–843.

Perez, E., Samper, S., Bordas, Y., Guilhot, C., Gicquel, B., and Martin, C. (2001). An essential role for phoP in *Mycobacterium tuberculosis* virulence. *Mol. Microbiol.* **41,** 179–187.

Petoukhov, M. V., and Svergun, D. I. (2005). Global rigid body modeling of macromolecular complexes against small-angle scattering data. *Biophys. J.* **89,** 1237–1250.

Pothineni, S. B., Strutz, T., and Lamzin, V. S. (2006). Automated detection and centring of cryo-ooled protein crystals. *Acta Crystallogr.* **D62,** 1358–1368.

Rickman, L., Saldanha, J. W., Hunt, D. M., Hoar, D. N., Colston, M. J., Millar, J. B., and Buxton, R. S. (2004). A two-component signal transduction system with a PAS domain-containing sensor is required for virulence of *Mycobacterium tuberculosis* in mice. *Biochem. Biophys. Res. Commun.* **314,** 259–267.

Roberts, D. M., Liao, R. P., Wisedchaisri, G., Hol, W. G., and Sherman, D. R. (2004). Two sensor kinases contribute to the hypoxic response of *Mycobacterium tuberculosis*. *J. Biol. Chem.* **279,** 23082–23087.

Robinson, V. L., Wu, T., and Stock, A. M. (2003). Structural analysis of the domain interface in DrrB, a response regulator of the OmpR/PhoB subfamily. *J. Bacter.* **185,** 4186–4194.

Saini, D. K., Malhotra, V., Dey, D., Pant, N., Das, T. K., and Tyagi, J. S. (2004). DevR-DevS is a bona fide two-component system of *Mycobacterium tuberculosis* that is hypoxia-responsive in the absence of DNA-binding domain of DevR. *Microbiology* **150,** 865–875.

Saini, D. K., Malhotra, V., and Tyagi, J. S. (2004). Cross talk between DevS sensor kinase homologue, Rv2027c, and DevR response regulator of *Mycobacterium tuberculosis*. *FEBS Lett.* **565,** 75–80.

Sassetti, C. M., Boyd, D. H., and Rubin, E. J. (2001). Comprehensive identification of conditionally essential genes in mycobacteria. *Proc. Natl. Acad. Sci. USA* **98,** 12712–12717.

Sassetti, C. M., Boyd, D. H., and Rubin, E. J. (2003a). Genetic requirements for mycobacterial survival during infection. *Proc. Natl. Acad. Sci. USA* **100,** 12989–12994.

Sassetti, C. M., Boyd, D. H., and Rubin, E. J. (2003b). Genes required for mycobacterial growth defined by high density mutagenesis. *Mol. Microbiol.* **41,** 179–187.

Sherman, D. R., Voskuil, M., Schnappinger, D., Liao, R., Harrell, M. I., and Schoolnik, G. K. (2001). Regulation of the *Mycobacterium tuberculosis* hypoxic response gene encoding alpha-crystallin. *Proc. Natl. Acad. Sci. USA* **98,** 7534–7539.

Shrivastava, R., Das, D. R., Wiker, H. G., and Das, A. K. (2006). Functional insights from the molecular modeling of a novel two-component system. *Biochem. Biophys. Res.* **344,** 1327–1333.

Svergun, D. I., Barberato, C., and Koch, M. H. J. (1995). CRYSOL, a program to evaluate X-ray solution scattering of biological macromolecules from atomic coordinates. *J. Appl. Crystallogr.* **28,** 768–773.

Svergun, D. I., and Koch, M. H. J. (2003). Small angle scattering studies of biological macromolecules in solution. *Rep. Progr. Phys.* **66,** 1735–1782.

Svergun, D. I., Volkov, V. V., Kozin, M. B., and Stuhrmann, H. B. (1996). New developments in direct shape determination from small-angle scattering. 2. *Acta Crystallogr.* **A52,** 419–426.

Svergun, D. I. (1999). Restoring low resolution structure of biological macromolecules from solution scattering using simulated annealing. *Biophys. J.* **76,** 2879–2886.

Svergun, D. I., Petoukhov, M. V., and Koch, M. H. (2001). Determination of domain structure of proteins from X-ray solution scattering. *Biophys. J.* **80,** 2946–2953.

Tyagi, J. S., and Sharma, D. (2004). Signal transduction systems of mycobacteria with special reference to *M. tuberculosis*. *Curr. Science* **86,** 93–102.

von Mering, C., Huynen, M., Jaeggi, D., Schmidt, S., Bork, P., and Snel, B. (2003). STRING: A database of predicted functional associations between proteins. *Nucleic Acids Res.* **31,** 258–261.

Voskuil, M. I., Schanppinger, D., Viscotni, K. C., Harrel, M. I., Dolganov, G. M., Sherman, D. R., and Schoolnik, G. K. (2003). Inhibition of respiration by nitric oxide induces a *Mycobacterium tuberculosis* dormancy program. *J. Exp. Med.* **198,** 705–713.

Walters, S. B., Dubnau, E., Kolesnikowa, I., Laval, F., Daffe, M., and Smith, I. (2006). The *Mycobacterium tuberculosis* PhoPR two-component system regulates genes essential for virulence and complex lipid biosynthesis. *Mol. Microbiol.* **60,** 312–330.

Wisedchaisri, G., Wu, M., Rice, A. E., Roberts, D. M., Sherman, D. R., and Hol, W. G. (2005). Structures of *Mycobacterium tuberculosis* DosR and DosR-DNA complex involved in gene activation during adaptation to hypoxic latency. *J. Mol. Biol.* **354,** 630–641.

Yamada, S., Akiyama, S., Sugimoto, H., Kumita, H., Ito, K., Fujisawa, T., Nakamura, H., and Shiro, Y. (2006). The Signaling pathway in histidine kinase and the response regulator complex revealed by C-ray crystallography and solution scattering. *J. Mol. Biol.* **362,** 123–139.

Zahrt, T. C., and Deretic, V. (2000). An essential two-component signal transduction system in *Mycobacterium tuberculosis*. *J. Bacteriol.* **182,** 3832–3838.

Zahrt, T. C., and Deretic, V. (2001). *Mycobacterium tuberculosis* system is required for persistent infections. *Proc. Natl. Acad. Sci. USA* **98,** 12706–12711.

Zahrt, T. C., Wozniak, C., Jones, D., and Trevett, A. (2003). Functional analysis of the *Mycobacterium tuberculosis* MprAB two-component signal tunsductions system. *Infect. Immun.* **71,** 6962–6970.

[24]   Transcriptomic Analysis of ArlRS Two-Component
Signaling Regulon, a Global Regulator,
in *Staphylococcus aureus*

By Yinduo Ji, Chuanxin Yu, and Xudong Liang

Abstract

The two-component signal transduction system plays an important role for bacteria to adapt to diverse niches by sensing the environmental stimuli and modulating gene expression. In *Staphylococcus aureus*, at least 16 pairs of two-component systems have been discovered and some of them coordinate with different regulators to modulate the expression of virulence factors. The availability of complete genome sequences, and transcriptome and proteome, enables us to identify the genes mediated by different regulators. The RT-PCR-based method and microarray technology have made it feasible for high throughput screening of genomewide transcription profiles. These techniques have been used to investigate different TCS in *S. aureus* and to identify the regulons of regulators in different bacterial systems. Therefore, combined with the inactivation of gene expression, microarray technology should be more useful to identify genes transcriptionally controlled by the TCS system. We propose that similar approaches can be used to understand the regulon of ArlRS two-component signaling by the comparison of gene expression profiles between wild type and *arlR* mutant. This chapter provides detailed protocols for identification of *arlRS* regulon using Affymetrix *S. aureus* chips and describes general considerations of microarray assay.

Introduction

S. aureus is an important community- and hospital-acquired pathogen, causing a wide variety of infections, ranging from superficial skin to life-threatening systemic infections, including endocarditis, septic arthritis, and toxic shock syndrome (Lowy, 1998). The emergence of multiple-antibiotic resistant *S. aureus*, especially methicillin-resistant *S. aureus* (MRSA) and vancomycin-intermediate resistant *S. aureus* (VIRSA), is causing increased public health concerns (Jones *et al.*, 1999). Compared to the period of 1994 to 1998, there was a 43% increase of MRS in 1999 alone. Data from the December 2000 report of the National Nosocomial Infection Surveillance (NNIS) System showed that about 47% of *S. aureus* isolates from intensive

METHODS IN ENZYMOLOGY, VOL. 423   0076-6879/07 $35.00
                        DOI: 10.1016/S0076-6879(07)23024-1

care units were methicillin-resistant (Srinivasan *et al.*, 2002). The threat of vancomycin resistance in *S. aureus* has been the topic of intensive research and discussion, since it is the last drug of choice for MRSA infections. Therefore, there is an urgent need for new classes of antibiotics and potent vaccines to fight infections caused by *S. aureus*.

S. *aureus* expresses more than 30 virulence factors, such as surface-associated adhesins, a polysaccharide capsule, a range of extracellular cytotoxins, proteases, DNases, and enterotoxins (Foster and Hook, 1998), most of which have been found to be differentially controlled by different regulators including two-component signal transduction systems (TCSs), such as *agr* (Novick, 2003), *srrAB* (Yarwood *et al.*, 2001), *arlRS*, and *saeRS* (Liang *et al.*, 2005, 2006). At least 16 different two-component signal transduction systems in *S. aureus* have been revealed. TCSs have been implicated together with other regulators to play a role in pathogenesis or biofilm formation in a wide range of bacterial species (Bronner *et al.*, 2004; Li *et al.*, 2002; Novick, 2003). Therefore, TCS has been explored as potential targets for new antimicrobials (Barrett and Hoch, 1998; Ji *et al.*, 1995).

A typical TCS is composed of a membrane-associated histidine kinase, which acts as a sensor protein extending through the cytoplasmic membrane to monitor environmental changes, and a cognate response regulator existing in the cytoplasm that modulates gene expression (Stock *et al.*, 2000). TCSs have evolved into a variety of organisms and regulate metabolic processes, cell cycles, communication, and virulence factor expression by sensing changes in the microenvironment (Beier and Gross, 2006; Loomis *et al.*, 1998).

The well-known accessory gene regulator (*agr*) is a positive regulator of staphylococcal exoproteins, including proteases, hemolysins, and toxins, and is a repressor of surface proteins, such as protein A, coagulase, and some adhesins expressed in late exponential phase of growth *in vitro* (Novick, 2003). Numerous studies indicate that *agr* activity plays an important role in the progression of infections in the rabbit model of osteomyetitis and the rat model of endophthalmitis. Mutations in *agr* decrease but do not eliminate the infection (Abelnour *et al.*, 1993; Booth *et al.*, 1995). Several studies suggest that Agr positively regulates *cap5* expression both *in vitro* and *in vivo* (in the rabbit endocarditis model, van Wamel *et al.*, 2002) and plays a key role in abscess formation (Wright *et al.*, 2005). Data indicate that Agr regulatory activity is not critical for the expression of exotoxins important for *S. aureus* to induce toxic shock syndrome (TSS) associated with an abscess or bacteremia (Yarwood *et al.*, 2002). The *srrAB* TCS is involved in adaptation to anaerobic growth of *S. aureus* (Throup *et al.*, 2001) and in regulation of virulence factors such as toxic shock syndrome toxin 1 and protein A (Yarwood *et al.*, 2001). Other TCS loci, such as *sae* (Giraudo *et al.*, 1997; Liang *et al.*, 2006) and *arlRS* (Fournier and Hooper, 2000; Fournier *et al.*, 2001), also regulate the expression of proteins, including α- and β-hemolysins (*hla*, *hlb*), coagulase (*coa*), and DNase. It has been found that together with ArlRS, another TCS, LytSR, is involved in regulation of autolysis. LytSR positively regulates the transcription of an operon responsible for the synthesis and transport of a cell wall murein hydrolase (Brunskill and Bayles, 1996).

The availability of microarray technology enables us to measure changes of gene expression in a whole genome on a single chip and see the big picture of the interactions among thousands of genes simultaneously. Regulatory systems are part of important networks modulating the expression of *S. aureus* genes, including genes that control antibiotic resistance. Therefore, the elucidation of the regulons of these regulatory systems is important for us to better understand the molecular mechanisms of pathogenesis. Using a microarray-based approach, different *S. aureus* regulons including Agr, ArlRS, SeaRS, Sar, SigB, Rot, and Mgr have been revealed (Bischoff *et al.*, 2004; Dunman *et al.*, 2001; Liang *et al.*, 2005, 2006; Luong *et al.*, 2006; Saïd-Salim *et al.*, 2003). We constructed an *arl* deletion mutant and compared the transcriptional profile of an *arl* deletion mutant to that of the parental strain by using *S. aureus* Affymetrix arrays. By identifying genes whose expression changed significantly, we have been able to demonstrate that ArlRS is a global regulator. It was found that the ArlRS system positively regulates *agr*, *lytSR*, and *rots* regulators, and negatively regulates SarH2 transcription regulator. In this chapter, we describe a comprehensive genetic approach to studying the regulon of *arlRS* and provide the general consideration in performing microarray assay.

Construction of an *arlR* Allelic Replacement Mutant in S. *aureus*

The *S. aureus* vector pSA7755 is used to generate the gene replacement mutant, as described (Fan *et al.*, 2001). The plasmid pSA7755 containing an erythromycin-resistant gene is a derivative of the pE194 temperature-sensitive mutant pRN5101 (Novick *et al.*, 1984), carrying the pUC19 poly-linker at its unique *Pst*I site and a fragment with the pT181 origin of replication at the *Cla*I site. A cassette containing the *tetA* gene, flanked sequences from upstream and downstream of the *alrR* to be replaced, is constructed in pBluescript vector and inserted into pSA7755. An *S. aureus* strain carrying this hybrid grows at a restrictive temperature (40°) for pSA7755 replication in the presence of 5 μg/ml tetracycline. These conditions allow the growth of only cells with *tetA* inserted into the chromosome by homologous recombination. In addition, the presence of the pT181 origin of replication on pSA7755 provides a way for the enrichment of clones with gene replacement. This approach is based on the deleterious effect on the cell of an excess of the pT181 replication initiator, RepC, only in the presence of a functional pT181 origin (Iordanescu, 1995). Based on this rationale, a plasmid overproducing RepC is introduced into clones, carrying the entire hybrid inserted in the chromosome and the cells screened for a tetracycline-resistant, erythromycin-sensitive phenotype. To obtain gene replacement, the mutated locus is transformed into the wild-type strain WCUH29 using phage11 transduction. Selection for tetracycline resistance and screening for the loss of the erythromycin resistance marker carried by the vector indicated that allelic replacement occurred and resulted in the *arlR* mutant strain.

The mutation of *arlR* mutation is verified by PCR and a Southern blot analysis.

Purification of Total RNA From Wild Type and *arlR* Mutant Strains

Overview

In order to identify genes whose expression is affected by the mutation of *arlRS* system, we choose the medium-exponential-growth phase because the RNA, at this time point, is more stable and the results are more reproducible. Overnight cultures of *S. aureus* are inoculated in 5% in Tryptic Soy broth (TSB) (Difco) medium and grow to the medium-exponential (3 h) phase of growth. Cells are harvested by centrifugation, and the RNA is isolated by the RNAPrep Kit (Promega). Contaminating DNA is removed with a DNA-*free* Kit (Ambion), and the RNA yield is determined spectrophotometrically at 260 nm.

Total RNA Purification Protocol

1. *S. aureus* isolates are incubated overnight at 37° in TSB with appropriate antibiotics and with shaking (225 rpm). The following day, inoculate the overnight cultures in 5% until the OD_{600nm} reaches 0.4. This should take only a few hours. If growth is too slow, reduce the dilution factor.

2. Transfer 3 ml of culture to a 10 ml tube. Centrifuge for 2 min at $14,000 \times g$. Remove the supernatant, leaving the pellet as dry as possible.

3. Resuspend the pellet in 100 μl of freshly prepared TE, add 6.5 μl of lysostaphin (2 mg/ml) and 6 μl of lysozyme (50 mg/ml), mix, and incubate the tube at 37° for 10 min. Add 75 μl of SV RNA Lysis buffer and 350 μl of RNA dilution buffer. Gently mix by inversion until the content become clear. Then, add 200 μl of 95% ethanol to the cleared lysate and mix by pipetting to cut the genomic DNA until the content becomes clear.

4. Transfer the transparent content into a spin column that has been put in a collection tube, and centrifuge the spin column assembly at $14,000 \times g$

for 1 min. Discard the follow-through solution, and add 600 μl of SV RNA wash solution to the spin column. Centrifuge the spin column assembly at 14,000×g for 1 min.

5. Empty the collection tube as before and prepare the DNase1 incubation mix by combining (in this order) 40 μl yellow core buffer, 5 μl of 0.09 M MnCl$_2$, and 5 μl of DNase1 enzyme per sample in a sterile tube. Apply 50 μl of this freshly prepared DNase1 incubation mix directly to the membrane inside the spin basket, making sure that the solution is in contact with and thoroughly covering the membrane. Incubate for 15 min at room temperature, and add 200 μl of SV DNase stop solution to the spin column and centrifuge at 14,000×g for 1 min. Add 600 μl of SV RNA wash solution and centrifuge at 14,000×g for 1 min. Empty the collection tube; add 250 μl of SV RNA wash solution. Centrifuge at high speed for 2 min.

6. Transfer the spin column to a sterilized 1.5 ml centrifuge tube, remove the cap of the tube, and apply 100 μl of nuclease-free water to the column membrane. Be sure to completely cover the surface of the membrane with the water. Centrifuge at 14,000×g for 1 min. Collect and label the elution tube containing the purified RNA.

7. Add 11 μl of 10× DNase1 buffer and 2 μl of DNase1(2 U/μl) to the RNA solution to digest the contaminated genomic DNA. Mix and spin briefly to collect the solution at the bottom of the tube, and keep the tube at 37° for 30 min.

8. Add 1/10 volume of DNase1 inactivator (Ambion) to the RNA solution to remove the DNase1. Mix and keep it at room temperature for 2 min (mixing once during this period). Centrifuge the reaction at 12,000×g for 2 min, and carefully transfer the supernatant containing the RNA to a sterilized clean tube.

9. Measure the quality and quantity of purified RNA by using a photometer ($1OD_{260} = 40$ μg, the ratio of OD_{260}/OD_{280} should be among 1.8 to 2.1). The purified RNA should be stored at $-70°$.

cDNA Synthesis, cDNA Fragmentation, and Labeling

Overview

In order to ensure the integrity of the RNA, the quality of total RNA is analyzed by electrophoresis in 1.2% agarose-0.66 M formaldehyde gels. The 23S and 16S rRNA bands should be clear without any obvious smearing patterns. For microarray assay, a total of 10 μg of RNA is reverse-transcribed to generate cDNA using Superscript III reverse transcriptase and random primers (Invitrogen). Then, the RNA is removed by treatment with NaOH. The cDNA is purified by using the QIAquick PCR Purification

Kit (Qiagen). The purified cDNA is digested with DNase I. The 3' termini of the fragmented cDNA are then directly labeled with Biotin-ddUTP. The efficiency of the labeling procedure is assessed using the Gel-shift assay. The absence of a shift pattern indicates poor biotin labeling. This quality control protocol prevents the hybridization of a poorly labeled target onto the probe array.

Synthesis of First Strand cDNA Protocol

1. Put 10 μg total RNA and 750 ng random primers in a sterilized 0.2 ml tube. Add nuclease-free water to make the total volume 30 μl; mix and centrifuge briefly.
2. Incubate at 70° for 10 min and keep at 25° for 10 min. Finally, put the reaction on ice for at least 2 min.
3. Add the following components (12 μl 5× first strand cDNA synthesis buffer, 6 μl 0.1 M DTT, 3 μl 10 mM dNTP, 1.5 μl SuperRNase In, 5 μl SuperScriptII, and 2.5 μl nuclease-free H_2O) into the RNA/primer mixture in the indicated order. The total volume of the reaction is 60 μl. Mix and incubate at 25° for 10 min, at 37° for 60 min, then at 42° for 60 min.
4. Terminate the reactions by keeping the tube at 70° for 10 min, then hold at 4°. Centrifuge the reaction tube briefly to collect products.
5. Add 20 μl of 1N NaOH; mix and keep at 65° for 30 min to remove the RNA. Then, add 10 μl of 1N HCl to neutralize the solution.
6. Use the QIAquick PCR Purification Kit to clean up the cDNA synthesis product following the protocol provided by the supplier (Qiagen). The synthesizing cDNA products can be stored at −20°.

cDNA Fragmentation and Terminal Labeling Protocol

1. Prepare the reaction mixture 5 μl 10× DNase 1 Buffer, 6 μg cDNA, 0.18 μl DNase 1(2 U/μl), and dH_2O up to 50 μl on ice. Mix and incubate the reaction at 37° for 10 min. Inactivate the enzyme at 98° to 100° for 10 min.

2. For quality control, pre- and post-fragmented cDNA (~500 ng) samples are analyzed by 2% agarose gel electrophoresis. The desired result should yield a majority of the DNA fragments within a distribution of 50 to 200bp (Fig. 1).

3. The 3' termini of the fragmentized cDNA are labeled using the Biotin-ddUTP by terminal transferase (Roche). Prepare the reaction mixtures 14 μl 5× Reaction Buffer, 14 μl 5× $CoCl_2$ (25 mM), 1 μl Biotin ddUTP, 2 μl Terminal deoxyribonucleotide transferase, and 4–5 μg fragmentized cDNA in a total volume 70 μl. Incubate the reaction for 1 h at 37°.

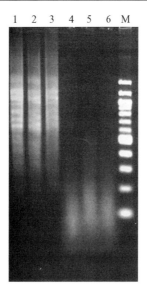

Fig. 1. Two percent agarose gel analysis of pre- and post-fragmented cDNA. Lane 1–3, total cDNA; Lane 4–6, fragmented cDNA; M:100bp DNA marker. (See color insert.)

Add 1.5 μl of 0.5 M EDTA to terminate the reaction. This labeled fragmented cDNA can be used for microarray directly.

4. Analyze the labeling efficiency by gel shift assay. Add 150 to 200 ng aliquots of fragmented and biotinylated sample in a fresh tube, and add 5 μl of 2 mg/ml NeutrAvidin to each sample, mix, and incubate at room temperature for 5 min. Run the RNA sample in a 4 to 20% TBE polyacrylamide gel at 150 volts until the front dye (red) almost reaches the bottom. Stain the gel with 1× solution of SYBR GreenII or 0.5 μg/ml of EB solution. Place the gel on the ultraviolet (UV) light box and produce an image.

Microarray Analysis

Overview

The *S. aureus* array (Affymetrix) contains probe sets to over 3300 *S. aureus* open reading frames based on the updated *S. aureus* genomic sequences of N315, Mu50, NCTC 8325, and COL. Additionally, the array also contains probes to study both the forward and reverse orientation of over 4800 intergenic regions throughout the *S. aureus* genome. A subsequent analysis suggests that these probes represent approximately 2738, 2668, 2773, and 2810 individual genes of the *S. aureus* COL, N315, NTC8325, and Mu50 genome, respectively, on the chip (Liang *et al.*, 2005). To address the issue of reproducibility, two separate RNA samples from each strain are prepared from two different experiments. For each sample, biotinylated cDNA is hybridized to at least two separate GeneChips.

Hybridization and Scanning Protocol

1. After determining that the fragmented cDNA is labeled with biotin, the fragmented biotinylated cDNA were hybridized to *S. aureus* chips (Affymetrix) that contain probe sets for *S. aureus* genomic ORFs. Three micrograms of fragmented RNA and the appropriate controls are hybridized to the chips at 45° for 16 h with constant rotation at 60 rpm.

2. The GeneChips are then washed and subjected to a series of staining procedures that included the successive binding of streptavidin, biotinylated anti-streptavidin, and phycoerythrin-conjugated streptavidin. Each GeneChip is washed and scanned at a 570-nm wavelength and a 3-μm resolution in an Affymetrix GeneChip scanner. Hybridization intensities for each of the genes/transcripts are collected from the scanned images.

3. The Affymetrix Microarray Suite 4.0 algorithms are used to calculate the signal intensities (average differences) and the present or absent determinations for each ORF. The scatter plot of signal intensities for each ORF is obtained to differentiate genes up- and down-regulated by *arlR* (Fig. 2). The GeneChips are then normalized, and their backgrounds are defined by using GeneSpring 4.0 (Silicon Genetics). The GeneSpring software is used to further analyze the transcription patterns of genes.

4. To identify genes with significantly altered expression, a series of statistical analyses (filtering) was performed: cutoff values for ratio of expression levels 1.80 and 0.55 are used to filter genes with expression level fold changes greater than + 1.8 in all three independent samples. Genes with fold change variations > 1.5 across the three samples are excluded. Furthermore, a "statistical group comparison" using the Student t-test/ANOVA is conducted to compare the mean expression levels between the control and the *arlR* mutant samples. The genes with significant differential expression (*p*-value < 0.05) are selected. Those genes negatively regulated by ArlR are identified as ORFs with transcript titers at least 1.8-fold higher in 316ko (*arlR* negative) than in WCUH29. The genes whose transcript levels are at least 1.8-fold higher in WCUH29 than in Sa316ko (*arlR* negative) are categorized as being positively regulated by *arlR*.

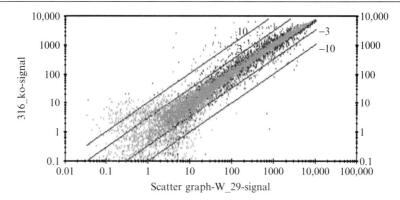

FIG. 2. Scatter plot of signal intensities from two GeneChips for *arlR* mutant and wild type strains. X axis, wild type strain; Y axis, *arlR* mutant. −3 and −10, fold of decrease after the mutation of arlR; 3 and 10, fold of increase after the mutation of *arlR*. (See color insert.)

Quantitative Real-Time RT-PCR Analysis

Overview

In order to confirm the results obtained from the microarray analyses, we employ quantitative real-time reverse transcription (RT)-PCR to compare the RNA levels for genes located in different operons (5 up-regulated and 6 down-regulated), showing a significant change of expression on the microarray assay. Briefly, the total RNA isolated from *S. aureus* is further processed by additional DNA removal using the DNA-free Kit (Ambion). The first strand cDNA is synthesized using reverse transcriptase with the SuperScript III Platinum Two-Step qRT-PCR Kit (Invitrogen). For each RNA sample, duplicate reactions of reverse transcription are performed, as well as a control without reverse transcriptase, in order to determine the levels of DNA contamination. PCR reactions are set up in triplicate by using the SYBR Green PCR Master Mix (Stratagene). Real-time sequence-specific detection and relative quantization are performed with the Stratagene Mx3000P Real Time PCR System. Gene-specific primers are designed to yield ~100 bp of specific products. Relative quantification of the product will be calculated using the Comparative C_T method, as described for the Stratagene Mx3000P system. The housekeeping gene 16s rRNA is used as an endogenous control. All samples are analyzed in triplicate and normalized against 16 s rRNA gene expression. The results are statistically analyzed for correlation to the microarray results.

cDNA Synthesis and qReal-Time PCR Protocol

1. Set up a 20 μl reaction by adding the following components: 1 to 2 μl of random primers (100ng, included in the cDNA synthesis kit), 1 μl of dNTP mix (10 mM each), 1 μg of total RNA, and sterile, distilled water to 13 μl of total volume.

2. Heat the mixture at 65° for 10 min, then chill it on ice immediately. Collect the contents of the tube by brief centrifugation. Add 4 μl of 5× First-Strand Buffer and 2 μl of 0.1 M DTT to the tube.

3. Incubate the reaction at 25° for 2 min; add 1 μl of SuperScript^{TM} III Reverse transcriptase (200 units), and incubate the reaction at 25° for 10 min, followed by 42° for 50 min. Inactivate the enzymes by heating the reaction to 70° for 15 min.

4. Thaw the Brilliant SYBR Green qPCR master mixture at room temperature, store it on ice, and keep the unused portion at 4° in a dark container. Dilute the reference dye 1:500 in nuclease-free water if it will be used in the reaction. Dilute the primers to a concentration of 3 μM. Dilute the template cDNA in 20 ng/μl.

5. Set up the reaction mixture. For a single reaction mixture of 15 μl, add 7.5 μl of 2× master mixture, 1.0 μl of upstream and downstream primer (3.0 μM each), respectively, 0.225 μl of diluted preference dye (optional, final concentration: 30 nM), and 1.0 μl of diluted cDNA template into each well of the 96-well plate. Adjust the final volume to 15 μl with nuclease-free water. Gently mix the reactions without creating bubbles, and centrifuge the reactions briefly. The amount of cDNA can vary from 1 to 1000 ng depending on the abundance of the specific mRNA in the cells or the diluted genomic DNA in control experiment.

6. Seal the cover of the 96-well plates with thermo-stable film. Put the plate with the reaction mixtures onto the RT-PCR instrument.

7. Run the PCR program according to the following procedure (Table I) being applied to the amplification of 100 to 300 bp DNA fragment.

TABLE I
REAL-TIME PCR PROGRAM

Cycle(s)	Temperature	Duration of cycle
1 (denature)	95°	5 min
40 (amplification)	95°	30 s
	55–60°	1 min
	72°	30 s
1 (dissociation)	95°	1 min
	53–58°	1 min

After the program is finished (about 2.5 h), save and then analyze the data on the computer using the software provided by the manufacturer.

Acknowledgments

We thank Aaron Becker for his assistance with microarray assay. This work was supported in part by NIH Grant AI057451 (Y. Ji) and AHC Faculty Research Development Grant 03-02 (Y. Ji) at the University of Minnesota.

References

Abelnour, A., Arvidson, S., Bremell, T., Ryden, C., and Tarkowski, A. (1993). The accessory gene regulator (*agr*) controls *Staphylococcus aureus* virulence in a murine arthritis model. *Infect. Immun.* **61,** 3879–3885.

Barrett, J., and Hoch, J. (1998). Two-component signal transduction as a target for microbial anti-infective therapy. *Antimicrob. Agents Chemother.* **42,** 1529–1536.

Beier, D., and Gross, R. (2006). Regulation of bacterial virulence by two-component systems. *Curr. Opin. Microbiol.* **9,** 143–152.

Bischoff, M., Dunman, P., Kormanec, J., Macapagal, D., Murphy, E., Mounts, W., Berger-Bächi, B., and Projan, S. (2004). Microarray-based analysis of the *Staphylococcus aureus s* regulon. *J. Bacteriol.* **186,** 4085–4099.

Booth, M. C., Atkuri, R. V., Nanda, S. K., Iandolo, J. J., and Gilmore, M. S. (1995). Accessory gene regulator controls *Staphylococcus aureus* virulence in endophthalmitis. *Invest. Ophthalmol. Vis. Sci.* **36,** 1828–1836.

Bronner, S., Monteil, H., and Prevost, G. (2004). Regulation of virulence determinants in *Staphylococcus aureus:* Complexity and applications. *FEMS Microbiol. Rev.* **28,** 183–200.

Brunskill, E. W., and Bayles, K. W. (1996). Identification and molecular characterization of a putative regulatory locus that affects autolysis in *Staphylococcus aureus. J. Bacteriol.* **178,** 611–618.

Dunman, P. M., Murphy, E., Haney, S., Palacios, D., Tucker-Kellogg, G., Wu, S., Brown, E. L., Zagursky, R. J., Shlaes, D., and Projan, S. J. (2001). Transcription profiling-based identification of *Staphylococcus aureus* genes regulated by the *agr* and/or *sarA* loci. *J. Bacteriol.* **183,** 7341–7353.

Fan, F., Lunsford, R. D., Sylvester, D., Fan, J., Celesnik, H., Iordanescu, S., Rosenberg, M., and McDevitt, D. (2001). Regulated ectopic expression and allelic-replacement mutagenesis as a method for gene essentiality testing in *Staphylococcus aureus. Plasmid* **46,** 71–75.

Foster, T. J., and Höök, M. (1998). Surface protein adhesins of *Staphylococcus aureus. Trends Microbiol.* **6,** 484–488.

Fournier, B., and Hooper, D. C. (2000). A new two-component regulatory system involved in adhesion, autolysis, and extracellular proteolytic activity of *Staphylococcus aureus. J. Bacteriol.* **182,** 3955–3964.

Fournier, B., Klier, A., and Rapoport, G. (2001). The two-component system ArlS-ArlR is a regulator of virulence gene expression in *Staphylococcus aureus. Mol. Microbiol.* **41,** 247–261.

Giraudo, A. T., Cheung, A. L., and Nagel, R. (1997). The *sae* locus of *Staphylococcus aureus* controls exoprotein synthesis at the transcriptional level. *Arch. Microbiol.* **168,** 53–58.

Iordanescu, S. (1995). Plasmid pT181 replication is decreased at high levels of RepC per plasmid copy. *Mol. Microbiol.* **16,** 477–484.

Ji, G., Beavis, R., and Novick, R. (1995). Cell density control of staphylococcal virulence mediated by an octapeptide pheromone. *Proc. Natl. Acad. Sci. USA* **92,** 12055–12059.

Jones, R. N., Erwin, M. E., Biedenbach, D. J., Johnson, D. M., and Pfaller, M. A. (1999). Epidemiological trends in nosocomial and community-acquired infections due to antibiotic resistant Gram-positive bacteria. *Diagn. Microbiol. Infect. Dis.* **33,** 101–112.

Li, Y. H., Lau, P. C. Y., Tang, N., Svensater, G., Ellen, R. P., and Cvitkovitch, D. G. (2002). Novel two-component regulatory system involved in biofilm formation and acid resistance in *Streptococcus mutans*. *J. Bacteriol.* **184,** 6333–6342.

Liang, X., Zheng, L., Landwehr, C., Lunsford, D., Holmes, D., and Ji, Y. (2005). Global regulation of gene expression by ArlRS, a two-component signal transduction regulatory system of *Staphylococcus aureus*. *J. Bacteriol.* **187,** 5486–5492.

Liang, X., Yu, C., Sun, J., Liu, H., Landwehr, C., Holmes, D., and Ji, Y. (2006). Inactivation of a two-component signal transduction system, SaeRS, eliminates adherence and attenuates vivulence of *Staphylococcus aureus*. *Infect. Immun.* **74,** 4655–4665.

Loomis, W. F., Kuspa, A., and Shaulsky, G. (1998). Two-component signal transduction systems in eukaryotic microorganisms. *Curr. Opin. Microbiol.* **1,** 643–648.

Lowy, F. D. (1998). *Staphylococcus aureus* infections. *N. Engl. J. Med.* **339,** 520–532.

Luong, T., Dunman, P., Murphy, E., Projan, S., and Lee, C. (2006). Transcription profiling of the *mgrA* regulon in *Staphylococcus aureus*. *J. Bacteriol.* **188,** 1899–1910.

Novick, R. P. (2003). Autoinduction and signal transduction in the regulation of staphylococcal virulence. *Mol. Microbiol.* **48,** 1429–1449.

Novick, R. P., Adler, G. K., Projan, S. J., Carleton, S., Highlander, S., Gruss, A., Khan, S. A., and Iordanescu, S. (1984). Control of pT181 replication I. The pT181 copy control function acts by inhibiting the synthesis of a replication protein. *EMBO J.* **3,** 2399–2405.

Saïd-Salim, B., Dunman, P. M., McAleese, F. M., Macapagal, D., Murphy, E., McNamara, P. J., Arvidson, S., Foster, T. J., Projan, S. J., and Kreiswirth, B. N. (2003). Global regulation of *Staphylococcus aureus* genes by Rot. *J. Bacteriol.* **185,** 610–619.

Srinivasan, A., Dick, J. D., and Perl, T. M. (2002). Vancomycin resistance in staphylococci. *Clin. Microbiol. Rev.* **15,** 430–438.

Stock, A. M., Robinson, V. L., and Goudreau, P. N. (2000). Two-component signal transduction. *Annu. Rev. Biochem.* **69,** 183–215.

Throup, J. P., Zappacosta, F., Lunsford, R. D., Annan, R. S., Carr, S. A., Lonsdale, J. T., Bryant, A. P., McDevitt, D., Rosenberg, M., and Burnham, M. K. R. (2001). The *srhSR* gene pair from *Staphylococcus aureus*: Genomic and proteomic approaches to the identification and characterization of gene function. *Biochemistry* **40,** 10392–10401.

van Wamel, W., Xiong, Y. Q., Bayer, A. S., Yeaman, M. R., Nast, C. C., and Cheung, A. L. (2002). Regulation of *Staphylococcus aureus* type 5 capsular polysaccharides by *agr* and *sarA in vitro* and in an experimental endocarditis model. *Microb. Pathogene.* **23,** 73–79.

Wright, J. S., Ji, R., and Novick, R. P. (2005). Transient interference with staphylococcal quorum sensing blocks abscess formation. *Proc. Natl. Acad. Sci. USA* **102,** 1691–1696.

Yarwood, J. M., McCormick, J. K., Paustian, M. L., Kapur, V., and Schlievert, P. M. (2002). Repression of the *Staphylococcus aureus* accessory gene regulator in serum and *in vivo*. *J. Bacteriol.* **184,** 1095–1101.

Yarwood, J. M., McCormick, J. K., and Schlievert, P. M. (2001). Identification of a novel two-component regulatory system that acts in global regulation of virulence factors of *Staphylococcus aureus*. *J. Bacteriol.* **183,** 1113–1123.

[25] Global Analysis of Two-Component Gene Regulation in *H. pylori* by Mutation Analysis and Transcriptional Profiling

By BIJU JOSEPH and DAGMAR BEIER

Abstract

The human gastric pathogen *Helicobacter pylori* was among the first microorganisms whose genome sequence was determined. It has a remarkably small repertoire of two-component regulators comprising three histidine kinases and five response regulators involved in transcriptional regulators as well as a bifunctional histidine kinase and four response regulators which build up the chemotaxis regulatory system. However, the two-component systems of *H. pylori* proved to play an important role for both *in vitro* growth of the organism and its ability to colonize its host. Here, we describe the experimental approaches applied to characterize the two-component systems of *H. pylori*, which were mostly based on the availability of the *H. pylori* genome sequence. These approaches comprise conventional techniques including mutation analysis as well as sophisticated methods like whole genome transcriptional profiling.

Introduction

Helicobacter pylori is a human pathogen that thrives in the mucous layer overlaying the stomach epithelium. Infection with *H. pylori*, which affects more than half of the world's population, is the major cause of chronic superficial gastritis and peptic ulcer disease. In addition, persistent infection which, if untreated, lasts for the lifetime of the infected individual, predisposes to gastric malignancies like adenocarcinoma and mucosa-associated lymphoid tissue (MALT) lymphoma (reviewed in Kusters *et al.*, 2006). In 1997 and 1999, the genome sequences of two independent clinical isolates of *H. pylori*, 26695 and J99, isolated from a patient with gastritis in the United Kingdom and a patient with duodenal ulcer in the United States, respectively, became available (Alm *et al.*, 1999; Tomb *et al.*, 1997). From the genomes of *H. pylori* 26695 and J99 comprising 1667867 and 1643831 bp encoding 1590 and 1495 open reading frames, respectively, a very restricted repertoire of regulatory proteins was predicted. This is consistent with the human gastric mucosa's being the only known niche of *H. pylori*, which is believed to provide a relatively stable environment lacking competition

METHODS IN ENZYMOLOGY, VOL. 423   0076-6879/07 $35.00
     DOI: 10.1016/S0076-6879(07)23025-3

from other microorganisms. Containing only the chemotaxis proteins CheAY2 (HP0392), CheY1 (HP1067), CheV1 (HP0019), CheV2 (HP0616), and CheV3 (HP0393) as well as three histidine kinases and five response regulators involved in transcriptional regulation, *H. pylori* is among the organisms with the lowest number of two-component regulatory systems whose genome sequences have been characterized so far. However, it was shown that important virulence traits of *H. pylori* which are required for the successful colonization of the stomach, that is, motility and urease-dependent acid resistance, are regulated by two-component systems. The two-component proteins of *H. pylori* and their respective characteristics are listed in Table I.

The two-component system HP0703-HP0244 (FlgRS) is part of the regulatory cascade governing flagellar gene expression (Niehus *et al.*, 2004) while the two-component system HP0166-HP0165 (ArsRS) mediates the pH-responsive transcriptional control of the urease gene cluster and of other genes involved in acid adaptation (Pflock *et al.*, 2005, 2006; Wen *et al.*, 2006). The urease gene cluster consisting of the genes *ureABIEFGH* encodes the major acid resistance system of *H. pylori*. *ureAB* encode the subunits of the nickel-containing dodecameric heterodimer urease that hydrolyzes urea, leading to the formation of ammonia and carbon dioxide,

TABLE I
TWO-COMPONENT PROTEINS OF *H.PYLORI*

Designation	Characteristics	Function
HP0703-HP0244 (FlgRS)	TCS	Regulation of class II flagellar genes
HP0166-HP0165 (ArsRS)	TCS; RR essential	pH-responsive regulation of genes involved in acid resistance
HP1365-HP1364 (CrdRS)	TCS	Regulation of genes encoding Cu^{2+} resistance determinant regulation of acid resistance genes?
HP1043	Orphan RR; essential	Unknown
HP1021	Orphan RR	Unknown
HP0392-HP1067 (CheAY2-CheY)	TCS; bifunctional HK	Chemotaxis, regulation of flagellar motion
HP0019 (CheV1), HP0616 (CheV2), HP0393 (CheV3)	RRs (cognate HK: HP0392)	Chemotaxis, precise function unknown

TCS, two-component system; HK, histidine kinase; RR, response regulator.

which buffer the cytoplasm and periplasm of the bacteria. *ureI* encodes a pH-gated urea channel protein and *ureEFGH* encode accessory proteins required for subunit assembly and incorporation of nickel ions into the urease apoenzyme (reviewed in Stingl and De Reuse, 2005). The third complete two-component system of *H. pylori*, HP1365-HP1364 (CrdRS), was demonstrated to positively regulate the expression of the copper resistance determinant CrdAB-CzcAB in response to increasing concentrations of copper ions (Waidner *et al.*, 2005).

In addition, *H. pylori* contains two orphan response regulators with atypical receiver sequences, HP1021 and HP1043, which, to current knowledge, lack a cognate histidine kinase (Beier and Frank, 2000; Schär *et al.*, 2005). According to the sequences of their output domains, the transcriptional regulators HP0166 (ArsR), HP1365 (CrdR), and HP1043 are grouped into the OmpR family of response regulators, while HP0703 (FlgR) is an NtrC-like protein. Orthologs of HP1021 are present in all other members of the ε-proteobacteria for which genomes have been sequenced, that is, *H. hepaticus*, *H. acinonychis*, *Campylobacter jejuni*, *Wolinella succinogenes*, and *Thiomicrospira denitrificans* (Baar *et al.*, 2003; Eppinger *et al.*, 2006; Parkhill *et al.*, 2000; Suerbaum *et al.*, 2003; http://cmr.tigr.org). However, its output domain, which contains a helix–turn–helix motif, cannot be grouped into any of the known response regulator subclasses. The receiver domains of the orphan response regulators HP1043 and HP1021 differ from the consensus sequence at highly conserved positions which are crucial for the phosphotransfer reaction from the transmitter domain of the histidine kinase to the receiver. In HP1043, the highly conserved aspartic acid residue D8 corresponding to position 13 in the sequence of CheY from *E. coli* is replaced by lysine. Furthermore, comparison with the homologous response regulator proteins of *H. hepaticus*, *C. jejuni*, and *W. succinogenes* shows that HP1043 has a deletion of four amino acids, which affects the canonical receiver phosphorylation site (D57 in *E. coli* CheY). Interestingly, in the HP1021 orthologs of the ε-proteobacteria, the highly conserved phosphate-accepting aspartic acid residue is replaced by serine. Though serine phosphorylation has been observed in response regulator mutants lacking an aspartic acid residue at the canonical phosphorylation site (Appleby and Bourret, 1999; Moore *et al.*, 1993), until now there are no reports on serine phosphorylation as a means of modulating the activity of wild-type response regulator proteins. Despite their atypical receiver sequences, the response regulators HP1043 and HP1021 serve important functions in *H. pylori* which, however, are unknown so far. Here, we review the approaches that were made to characterize the two-component systems of *H. pylori*.

Functional Analysis of Essential Response Regulators of *H. pylori*

In attempts to construct null mutants of the *H. pylori* two-component genes, it proved impossible to obtain deletion mutants of the response regulator genes hp0166 (*arsR*) and hp1043 by allelic exchange mutagenesis in the *H. pylori* strains 26695 and G27. Strain G27 is a clinical isolate that was collected at the hospital of Grosseto, Italy (Xiang *et al.*, 1995). Null mutants of hp1021 exhibited a small colony phenotype (Beier and Frank, 2000; McDaniel *et al.*, 2001). When a second copy of the response regulator genes hp0166 (*arsR*), hp1043, or hp1021 together with a chloramphenicol resistance cassette from *C. coli* (Wang and Taylor, 1999) was introduced into the chromosome of *H. pylori* G27 by substituting parts of the divergently transcribed genes *cagA* and *cagBC* as well as their intergenic region, the respective response regulator gene could be deleted from its wild-type locus in the merodiploid strain, yielding mutants that were indistinguishable from the wild-type regarding their *in vitro* growth. *cagA* and *cagBC* are components of the *cag* pathogenicity island (PAI) encoding a type IV secretion system and its effector protein, the immunodominant antigen A (CagA) (Censini *et al.*, 1996). The *cag* PAI, which is present in the so-called type I isolates but is missing in type II strains, renders *H. pylori* more virulent, resulting in more severe clinical outcomes of an infection; however, it is dispensable for host colonization (reviewed in Kusters *et al.*, 2006). Hence, the *cag* PAI is a suitable site for the insertion of extra DNA into the *H. pylori* chromosome. These complementation experiments proved that hp0166 (*arsR*) and hp1043 are essential genes (Schär *et al.*, 2005) and that the growth retardation of the hp1021 null mutant is, indeed, a consequence of the deletion of the response regulator gene. The *rdxA* gene encoding an oxygen-insensitive NADPH nitroreductase has also been used as insertion site in complementation experiments relying on the chromosomal integration of the gene copy of interest (Croxen *et al.*, 2006; Loh and Cover, 2006). Disruption of the *rdxA* gene confers a metronidazole-resistant phenotype to *H. pylori* (Goodwin *et al.*, 1998), which can be exploited for the selection of transformants. The pHel plasmid vector shuttle system developed by Heuermann and Haas (1998) can also be used for the complementation of *H. pylori* mutants with cloned genes of homologous or heterologous origin.

We exploited the *in vivo* complementation system based on the chromosomal integration of additional gene copies into the *cag* PAI of *H. pylori* to investigate the impact of the atypical receiver sequences of HP1043 and HP1021 on response regulator function by constructing mutants that express solely mutated derivatives of the respective response regulator proteins. Suicide plasmids containing a mutated copy of the respective response regulator gene as well as its promoter region and a chloramphenicol acetyl transferease (*cat*) gene flanked by DNA fragments derived from the *cagA* and *cagCD* genes were transformed into *H. pylori* G27 via natural transformation. Selection for chloramphenicol resistance acquired by homologous recombination yielded merodiploid *H. pylori* strains which were subsequently transformed with suicide plasmids designed for the allelic replacement of the wild-type response regulator gene by a kanamycin resistance cassette from *C. coli* (Labigne-Roussel *et al.*, 1988). When kanamycin- and chloramphenicol-resistant transformants exhibiting a normal growth phenotype could be obtained, these strains were then analyzed for the exclusive presence of the mutated response regulator gene. The analysis of various mutants of HP1043 and HP1021 for their capability to functionally substitute for the respective wild-type response regulator protein led to the conclusion that the atypical amino acids (K8 in HP1043, S47 in HP1021) in the receiver sequences of HP1043 and HP1021 are not crucial and that phosphorylation of these response regulators is not a prerequisite for their cell growth-associated functions. Furthermore, by replacing wild-type hp0166 (*arsR*) with an allele encoding a response regulator with a D52N mutation at the canonical phosphorylation site, we could demonstrate that phosphorylation of HP0166 (ArsR) is not required for its essential function (Schär *et al.*, 2005). This is in line with the observation that the cognate histidine kinase HP0165 (ArsS) can be inactivated without affecting the growth of *H. pylori* under standard laboratory culture conditions (Beier and Frank, 2000) and indicates that two sets of target genes of HP0166 (ArsR) do exist: one group of genes, at least one of which is essential for viability, is controlled by the unphosphorylated HP0166 (ArsR), while the second group consists of nonessential genes that are regulated by the phosphorylated HP0166 (ArsR).

The complementation experiments described previously take advantage of the easy practicability of allelic exchange mutagenesis in *H. pylori*. *H. pylori* is naturally competent for the uptake of foreign DNA and is extremely prone to recombining its chromosomal DNA (reviewed in Kraft and Suerbaum, 2006). Therefore, double crossover resulting in allelic exchange occurs readily and is achieved in one step after transformation with an appropriate suicide plasmid. In case unmarked mutations of a particular gene are desired, a sucrose-based counterselection system is available (Copass *et al.*, 1997). In a first step, the target gene is replaced via allelic exchange by a gene cassette consisting of a kanamycin resistance module and the *sacB* gene of *Bacillus subtillis* that was fused to the promoter of the *H. pylori flaA* gene. Following transformation with a mutated allele, strains that have integrated this allele can be selected by growing the transformants on sucrose-containing medium, which requires the loss of the *sacB* gene. In the following, we provide protocols for the natural transformation and electroporation of *H. pylori* strains. The latter method (Ferrero *et al.*, 1992) is appropriate for isolates that are only poorly competent.

Natural Transformation of H. pylori *G27*

1. Grow *H. pylori* G27 recovered from a frozen stock for 2 days at 37° in an anaerobic jar providing a microaerophilic atmosphere (6% oxygen, 10% carbon dioxide; Oxoid Gas Generating Kits for Campylobacter) on Columbia agar plates containing Dent's *Helicobacter pylori* selective supplement (Oxoid) and 5% horse blood.

2. Passage the bacteria on a fresh blood agar plate and grow them under the same conditions until a confluent bacterial layer is formed.

3. Harvest the bacteria, spot (diameter of the spot = 2 cm) them onto a fresh agar plate and incubate the plate for 5 h at 37° in a microaerophilic atmosphere.

4. Precipitate 5 μg of plasmid DNA, wash with 70% ethanol, and dissolve the dried DNA pellet in 20 μl of Brucella broth. An appropriate suicide plasmid contains an antibiotic-resistance cassette from *C. coli* (kanamycin or chloramphenicol) flanked by approximately 500 bp DNA fragments derived from the regions that on the *H. pylori* chromosome flank the gene targeted by allelic exchange mutagenesis in any standard cloning vector.

5. Add the DNA solution dropwise to the spotted bacteria, mix using a sterile inoculation loop, and incubate the bacteria for about 15 h at 37° in a microaerophilic atmosphere.

6. Using an inoculation loop, spread the bacteria onto the surface of a Columbia blood agar plate containing the antibiotic of choice in a concentration of 20 μg/ml. Incubate the plates at 37° in a microaerophilic atmosphere for 5 days.

7. Pick single colonies and grow the bacteria on plates until there are enough cells to prepare frozen stocks in Brucella broth containing 20% glycerol.

8. Recover the bacteria, prepare chromosomal DNA, and check for the desired recombination event by PCR analysis using appropriate primers.

Electrotransformation of H. pylori *26695*

1. Harvest the cells from a fresh confluent Columbia blood agar plate, wash them in 500 μl of a cold (4°) 15% glycerol–9% sucrose solution, and suspend them in 50 μl 15% glycerol–9% sucrose solution.

2. Add 1 μl of plasmid DNA (500 μg/μl) and incubate 1 min on ice.

3. Transfer the mixture to a prechilled 0.2-cm electroporation cuvette placed in a Gene Pulser apparatus (BioRad MicroPulser) and apply a pulse with settings of 25 F, 2.5 kV, and 200 Ω giving a time constant ranging from 4 to 5 ms.

4. Suspend the pulsed cells in 100 μl of SOC medium (2% Bacto Tryptone, 0.5% Bacto Yeast Extract, 10 mM NaCl, 2.5 mM KCl, 10 mM MgCl$_2$, 10 mM MgSO$_4$, 20 mM glucose) and spread them on a Columbia blood agar plate. Incubate the plate for 48 h at 37° in a microaerophilic atmosphere.

5. Harvest the bacteria, suspend them in 500 μl of Brucella broth, and plate aliquots of 100 μl onto Columbia blood agar plates containing the antibiotic appropriate for the selection of transformants.

6. Check for the presence of transformants after 72 h of incubation at 37° in a microaerophilic atmosphere.

Characterization of the Regulons Controlled by the H. pylori Two-Component Systems

Whole genome transcriptional profiling is a powerful tool for the comprehensive analysis of bacterial stimulons and regulons and has been applied for the characterization of *H. pylori* null mutants of the two-component genes hp0244 (*flgS*), hp0703 (*flgR*), hp0165 (*arsS*), and hp1365 (*crdS*) (Niehus *et al.*, 2004; Pflock *et al.*, 2006, Loh and Cover, 2006). Using the *H. pylori* strains N6 (Ferrero *et al.*, 1992) and 88–3887, which is a motile derivative of strain 26695 (Josenhans *et al.*, 2000), the HP0703-HP0244 (FlgRS) two-component system was shown to regulate, together with the sigma factor RpoN, the transcription of the class 2 flagellar genes encoding the flagellar hook protein (FlgE1), hook-filament adapter proteins (FlgK, FlgL), and the minor flagellin subunit FlaB. In addition, hp0869 (*hypA*) encoding a nickel-binding protein involved in hydrogenase and urease maturation, hp1155 (*murG*) encoding a glycosyltransferase, and six genes (hp0114, hp0906, hp1076, hp1120, hp1233, and hp1154) encoding proteins of unknown function are regulated by the HP0244-HP0703 (FlgRS) two-component system (Niehus *et al.*, 2004). Since inactivation of hp1067 and hp1233 resulted in altered flagellar morphology, it is likely that the respective gene products have a role in flagellar biosynthesis (Niehus *et al.*, 2004). However, the stimulus governing the autokinase activity of the cytoplasmic protein HP0244 remains unknown.

It has been shown that the stimulus sensed by the histidine kinase HP0165 (ArsS) is acidic pH (Pflock *et al.*, 2004). Transcriptional profiling of an hp0165 (*arsS*)-deficient mutant of *H. pylori* G27 that was grown at pH 5.0 revealed that the expression of 109 genes differed at least two-fold as compared to the wild-type strain. Seventy-five genes were found to be activated by the phosphorylated response regulator HP0166 (ArsR) at low pH, while 34 genes were repressed. The acid-responsive HP0166 (ArsR) regulon includes the urease gene cluster, the amidase genes *amiE* and *amiF* encoding ammonia-producing enzymes, and genes encoding outer membrane proteins, Ni^{2+}-storage proteins, and detoxifying enzymes involved in oxidative stress responses (Pflock *et al.*, 2006). In a similar study by Loh and Cover (2006) that was performed with an isogenic hp0165 (*arsS*) null mutant of *H. pylori* J99, 68 genes were found to be differentially expressed. However, it should be noted that only eight genes, including *ureAB*, *amiE*, and *amiF*, were detected in both the study of Pflock *et al.* (2006) and Loh and Cover (2006), indicating that subtle differences in the experimental design and the use of different strains may strongly influence the outcome of a microarray experiment. Binding of phosphorylated HP0166 (ArsR) to the promoters of the urease gene cluster and the amidase genes was confirmed by DNase I footprint experiments that revealed extended HP0166 (ArsR) binding regions overlapping the -35 box of the respective promoters (Pflock *et al.*, 2005, 2006).

The HP1365-HP1364 (CrdRS) two-component system was shown to positively regulate the expression of the copper resistance determinant CrdAB-CzcAB in response to increasing concentration of copper ions via the direct binding of the response regulator HP1365 (CrdR) to the promoter of the *crdA* gene (Waidner *et al.*, 2005). However, Loh and Cover (2006) also reported that the histidine kinase HP1364 (CrdS) is required for pH-responsive gene regulation since transcriptional profiling of an isogenic hp1364 (*crdS*) null mutant of *H. pylori* J99 revealed that transcription of 63 genes, including *ureAB*, *amiE*, and *amiF*, did no longer respond to low pH. Surprisingly, this pH-responsive HP1364 (CrdS)-dependent regulon was largely identical to the HP0165 (ArsS)-dependent regulon identified by the same authors. Interestingly, no requirement of the HP1365-HP1364 (CrdRS) two-component system for acid-induced transcription of the *ureAB*, *amiE* and *amiF* genes was observed when an isogenic hp1364 (*crdS*) null mutant of *H. pylori* G27 was analyzed in our laboratory (Pflock *et al.*, 2007), suggesting profound strain-specific differences in the regulation of important virulence traits.

The target genes regulated by the orphan response regulators HP1021 and HP1043 are largely unknown. Transcriptome analysis of an hp1021 null mutant in *H. pylori* 26695 is currently performed in our laboratory. On the basis of *in vitro* DNA binding experiments, it was hypothesized that the essential response regulator HP1043 regulates its own expression as well as transcription of the *tlpB* gene encoding a methyl-accepting chemotaxis protein (Delany *et al.*, 2002). In an attempt to identify target genes of HP1043, *H. pylori* strains transcribing the response regulator gene under control of the strong constitutive promoter of the *cagA* gene (P_{cagA}) and the promoter of the *pfr* gene (P_{pfr}) encoding bacterioferritin, respectively, were constructed (Delany *et al.*, 2002; Müller *et al.*, 2006). Transcription from the P_{pfr} promoter is strongly derepressed in the presence of Fe^{2+}

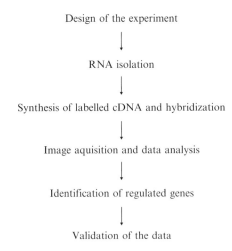

Fɪɢ. 1. Work flow in a microarray experiment.

(Delany *et al.*, 2001). It was supposed that overexpression of the response regulator protein causes changes in the transcription profiles of its target genes, which could be detected using microarray technology. However, it turned out that although transcription of hp1043 was strongly increased in the mutants, the response regulator was not overexpressed on the protein level, suggesting posttranscriptional and/or posttranslational regulation of the expression of HP1043 (Delany *et al.*, 2002; Müller *et al.*, 2006).

Here, we describe the design and realization of a typical microarray hybridization experiment performed with cDNA prepared from RNA extracted from *H. pylori* (Fig. 1).

Design of the Experiment for Transcriptional Profiling

As is the case with any experiment, design of the experimental setup for whole genome transcriptional analysis is of utmost importance to obtain meaningful data. The setup could be of two types:

a. Loop design: Here, one strain/condition is compared directly to another strain/condition by hybridizing the differentially labeled cDNAs and regulated genes are identified. This design is ideal when comparing one or two mutant strains to the wild-type strain. But if the candidates to be analyzed are more in number, this design becomes tedious.

b. Common reference design: This is a more robust design in which one strain/condition could be compared to several strains/conditions. The advantage of this design is that one strain or condition is a common reference and all other strains or conditions are compared to this common reference, which allows comparisons among the various experimental strains/conditions. It is almost impossible to analyze large numbers of experimental conditions/mutant strains by a loop design and, in such cases, this design is well suited.

Based on the issues to be addressed, one could choose an appropriate design and perform the experiment with sufficient biological replicates (different RNA preparations) and technical replicates (dye swap experiments, replicates of the probes on the slide, different probes for a gene on the slide, etc.) (for a detailed review, see Yang and Speed, 2002).

Substrates and Arrays

Microarrays consist of several thousand probes, each representing an ORF, which are biological samples (cDNA, PCR products, or oligonucleotides) immobilized on a glass slide.

The other substrate on which the probes are spotted are nylon membranes, termed macroarrays. Both high-density macroarrays and microarrays have been used to study whole genome transcriptional profiling in *H. pylori*. In the transcriptome studies performed by Niehus *et al.* (2004) and Pflock *et al.* (2006), a whole-genome microarray containing 1649 PCR products generated with specific primer pairs derived from the genome sequences of *H. pylori* 26695 (Tomb *et al.*, 1997) and J99 (Alm *et al.*, 1999) and comprising 98% of the coding sequences present in both genomes (Gressmann *et al.*, 2005) was used. A PCR product-based DNA microarray comprising 1578 genomic fragments and representing 97% of the open reading frames (ORFs) in the genomes of *H. pylori* 26695 and J99 (Eurogentec), as well as a 50-mer oligonucleotide-based DNA microarray covering 98% of the coding regions of both genomes (Ocimum Biosolutions), are commercially available. Furthermore, a nylon Panorama *H. pylori* array based on PCR-amplified ORFs is available (Sigma-Genosys), which was used in transcriptome studies by Loh and Cover (2006) and Kim *et al.* (2004). Presently, Operon Inc. (http://www.operon.com) is developing an oligonucleotide set covering the entire genome of *H. pylori*.

RNA Isolation and cDNA Synthesis

Preparation of *H. pylori* RNA:

1. Inoculate 40 ml of BHI broth containing 5% fetal calf serum and Dent's antibiotic supplement (Oxoid) to a starting OD_{550} of 0.1 with a suspension of *H. pylori* cells harvested from a fresh Columbia blood agar plate prepared in 1 ml of BHI broth. Transfer the culture to a 250 ml cell

culture flask with filter cap (Greiner Bio-One) placed into an anaerobic jar with a microaerophilic atmosphere and incubate at 37° with shaking until an OD_{550} of 0.8 is reached.

2. Transfer the culture to a precooled 50 ml Falcon tube containing 125 ml phenol and 2 ml 96% ethanol, centrifuge (4000 rpm, 10 min, 4°), and freeze the sedimented cells immediately in liquid nitrogen.

3. Resuspend the pellet in 1 ml Trizol reagent (Invitrogen) and proceed according to the manufacturer's recommendations.

4. Incubate the RNA in a final volume of 500 μl 10 mM $MgCl_2$, 40 mM Tris-HCl, pH 7.5 with 100 units RNase-free DNase I (Amersham Biosciences) for 45 min at 25°. Precipitate the RNA by the addition of 1/10 Vol. 10 mM LiCl and 2.5 Vol. 96% ethanol, wash with 70% ethanol, and dissolve the dried RNA in an appropriate volume of DEPC-treated water.

5. Perform a further purification step using the RNeasy Mini Kit (Qiagen) according to the manufacturer's RNA-cleanup protocol.

Purity of the RNA isolated is one of the most critical parameters that will determine the outcome of a microarray experiment and assessment of the quality of isolated RNA using a bioanalyzer is highly recommended to ensure that one has good starting material for a successful microarray experiment. In all the published work on transcriptional profiling of *H. pylori* so far, total RNA, which was transcribed into cDNA and further processed, has been used. However, it is also possible to enrich the RNA sample isolated by any of the traditional methods using commercial kits (Ambion) to remove ribosomal RNA from the sample, leaving behind pure mRNA for downstream processing.

cDNA Synthesis, Labeling, and Hybridization

Here, we describe a typical protocol for the preparation of Cy3-dCTP and Cy5-dCTP labeled cDNA for a microarray experiment.

1. Equal amounts (20–40 μg) of total RNA from the two strains or conditions to be tested are mixed either with random hexamers (Invitrogen) or gene-specific reverse primers and allowed to anneal by heating the probes to 70° for 5 min, with incubation on ice for another 5 min.

2. Components for the reverse transcription reaction then are added to both the samples, namely, dATP, dTTP, dGTP (0.5 mM), and dCTP (0.2 mM), dithiothreitol (1 mM; Invitrogen), RNase inhibitor (40 units; Invitrogen), and reverse transcriptase (200 units; Invitrogen) with an appropriate buffer.

3. Two nanomoles of Cy3 dCTP is added to one sample and 2 nmoles of Cy5 dCTP is added to the other. The probes are then incubated at room

temperature for 10 mins and at 42° for 2 hrs for reverse transcription to proceed.

4. Following the inactivation of the RNase inhibitor at 70° for 15 mins, the RNA in both the samples is degraded by addition of 2 μg of RNase (Roche) and incubation at 37° for 45 mins.

5. The Cy3 and Cy5 labeled cDNAs are now purified using a PCR purification kit (Qiagen) or the CyScribe^{TM}GFX^{TM} purification kit (Amersham Biosciences) and pooled together. The target (biological sample labeled with the fluorescent dyes) is now ready to be hybridized on the microarray.

6. The microarray slides, in the meantime, are preprocessed according to manufacturer's protocols (Quantifoil) appropriate for the given microarray type (oligonucleotide-based or PCR product-based).

7. The labeled cDNAs are mixed with an appropriate hybridization buffer (50% formamide, 6% SSC, 0.5% SDS, 50 mM NaPO4, pH 8.0, 5% Denhardt's solution, and hybridized onto the preprocessed microarray slide at 50° for 16 h or, as an alternative, the target is mixed with 2 to 4× SSC and 0.1% SDS) and hybridized at 65° for 16 h.

Image Acquisition and Data Analysis

The slides are then washed according to the appropriate manufacturer's protocol and scanned for the two fluorescent dyes at their respective wavelengths, that is, 543 nm for Cy3-dCTP and 633 nm for Cy5-dCTP, to yield two images in TIFF format, which are then analyzed. Several commercial microarray scanners produced by, for example, Agilent Technologies, Perkin Elmer, or Genaxon, are available. During the process of image acquisition, scanner settings including laser power and photo multiplier tube (pmt) settings are adjusted in such a way as to yield images in which the overall intensities are balanced to keep in line with the basic assumption of any microarray experiment that most of the genes are not regulated and only a small subset of the genome is differentially regulated, which are then identified based on the difference in the intensities of the spots in the two channels.

The generated images are then quantified using image analysis software generally provided along with the scanner or specific image analyis software applications such as Imagene (Biodiscovery). Normally, each image analysis software has its own quality control criteria of the spots, namely, signal to noise ratio, total background, etc. It is important at this point to also visually examine the images and manually flag the spots in cases of stray fluorescent signals or obvious cases of high background, which will then be excluded from further analysis. Several options for in-slide normalization of the data to be generated are available in the image analysis software which will remove the bias produced by the differential dye incorporation in the probes. The commonly used methods for in-slide normalization are total or linear normalization, wherein a factor derived based on the total intensities of the two dyes is applied to the experimental sample. The other more robust method of in-slide normalization is the global locally weighted scattered plot smoothing (LOWESS), which takes into consideration differences in the intensities of the individual spots (Yang *et al.*, 2002). Once data from individual slides are generated, the data are then pooled together and further processed using any preferred platforms, such as the open source BIOCONDUCTOR package (http://www.bioconductor.org/) or EMMA (Dondrup *et al.*, 2003), and differentially regulated genes that are statistically significant are identified. The other commonly used method for identifying significantly regulated genes in a microarray experiment is the Microsoft Excel add-in Significance Analysis of Microarray (SAM), based on Tusher *et al.* (2001).

Validation of the Data

The significantly regulated genes which are identified have now to be further validated by independent techniques. The most commonly used methods for the validation of microarray data are quantitative real time reverse transcriptase PCR (qRT PCR), Northern blot analysis, or semi-quantitative primer extension analysis. Furthermore, phenotypic characterization of mutants generated for the interesting genes identified can be performed. Both β-galactosidase (*lacZ*) and chloramphenicol acetyl transferase (*cat*) have been shown to be appropriate reporter genes for gene expression analysis in *H. pylori* (Bijlsma *et al.*, 2002; van Vliet *et al.*, 2001).

The Chemotaxis System of H. pylori

The chemotactic behavior of *H. pylori* aids the efficient colonization of the gastric mucosa. The chemotaxis system consists of three classical MCPs (HP0082, HP0099, HP0103), which are likely to sense ligands external to the cell, a soluble MCP orthologue (HP0599), which might respond to internal physiological signals, a coupling protein CheW (HP0391), a histidine kinase CheA containing a C-terminal CheY-like receiver domain (CheAY2/HP0392), a separate CheY response regulator (CheY1/HP1067), and three CheV (CheV1/HP0019; CheV2/HP0616; CheV3/HP0393) proteins consisting of an N-terminal CheW-like domain and a C-terminal receiver domain. CheV was first described in *B. subtilis* and was shown to be involved in the adaptation to attractants in this organism (Karatan *et al.*, 2001). Inactivation of *cheV1* (HP0019) significantly reduced swarming of *H. pylori* on semisolid agar plates while swarming was unchanged in insertion mutants of *cheV2* (HP0616) and *cheV3* (HP0393), respectively, and in a *cheV2/cheV3* double mutant (Pittman *et al.*, 2001). In a 2006 study, ORF HP0170 was identified as a remote homolog of the CheY~P-specific CheZ phosphatase (Terry *et al.*, 2006). Analysis of the phosphotransfer reactions between purified two-component chemotaxis signaling modules *in vitro* demonstrated that both CheY1 and CheY2 are phosphorylated by CheA~P and that the three CheV proteins mediate the dephosphorylation of CheA~P, but with a clearly reduced efficiency as compared to CheY1 and CheY2. Interestingly, retrophosphorylation of CheAY2 by CheY1~P was observed suggesting a role of CheY2 as a phosphate sink to modulate the half-life of CheY~P (Jiménez-Pearson *et al.*, 2005). Therefore, it is tempting to speculate that the newly identified CheZ-homolog HP0170 might be a CheY2-specific phosphatase.

Concluding Remarks

H. pylori is an interesting organism to investigate regulatory networks due to its quite uniform habitat, which allowed the streamlining of its environmental response repertoire, and due to the ease of genetic manipulation. The few signal transduction systems present are mainly involved either in essential life functions that so far are not well understood or in accessory functions required for the establishment of a stable infection in the hostile environment of the host. Modern genome technology such as whole genome transcriptional profiling had an enormous impact on the elucidation of the regulatory network governing the expression of virulence traits in *H. pylori*. However, it has to be kept in mind that transcriptome technology is very sensitive to minor differences at various steps starting from the experimental design to analysis of the data and, therefore, compliance to globally acceptable standards and techniques is required to facilitate comparison of data generated from different laboratories.

References

Alm, R. A., Ling, L.-S. L., Moir, D. T., King, B. L., Brown, E. D., Doig, P. C., Smith, D. R., Noonan, B., Guild, B. C., deJonge, B. L., Carmel, G., Tummino, P. J., *et al.* (1999). Genomic-sequence comparison of two unrelated isolates of the human gastric pathogen *Helicobacter pylori*. *Nature* **397,** 176–180.

Appleby, J. L., and Bourret, R. B. (1999). Activation of CheY mutant D57N by phosphorylation at an alternative site, Ser-56. *Mol. Microbiol.* **34,** 915–925.

Baar, C., Eppinger, M., Raddatz, G., Simon, J., Lanz, C., Klimmek, O., Nandakumar, R., Gross, R., Rosinus, A., Keller, H., Jagtap, P., Linke, B., *et al.* (2003). Complete genome sequence and analysis of *Wolinella succinogenes*. *Proc. Natl. Acad. Sci. USA* **100,** 11690–11695.

Beier, D., and Frank, R. (2000). Molecular characterization of two-component systems of *Helicobacter pylori*. *J. Bacteriol.* **182,** 2068–2076.

Bijlsma, J. J. E., Waidner, B., van Vliet, A. H. M., Hughes, N. J., Häg, S., Bereswill, S., Kelly, D. J., Vandenbroucke-Grauls, C. M. J. E., Kist, M., and Kusters, J. G. (2002). The *Helicobacter pylori* homologue of the ferric uptake regulator is involved in acid resistance. *Infect. Immun.* **70,** 606–611.

Censini, S., Lange, C., Xiang, Z., Crabtree, J. E., Ghiara, P., Borodovsky, M., Rappuoli, R., and Covacci, A. (1996). *cag*, a pathogenicity island of *Helicobacter pylori*, encodes type I-specific and disease-associated virulence factors. *Proc. Natl. Acad. Sci. USA* **93,** 14648–14653.

Copass, M., Grandi, G., and Rappuoli, R. (1997). Introduction of unmarked mutations in the *Helicobacter pylori vacA* gene with a sucrose sensitivity marker. *Infect. Immun.* **65,** 1949–1952.

Croxen, M. A., Sisson, G., Melano, R., and Hoffman, P. S. (2006). The *Helicobacter pylori* chemotaxis receptor TlpB (HP0103) is required for pH taxis and for colonization of the gastric mucosa. *J. Bacteriol.* **188,** 2656–2665.

Delany, I., Spohn, G., Rappuoli, R., and Scarlato, V. (2001). The Fur repressor controls transcription of iron-activated and -repressed genes in *Helicobacter pylori*. *Mol. Microbiol.* **42,** 1297–1309.

Delany, I., Spohn, G., Rappuoli, R., and Scarlato, V. (2002). Growth phase-dependent regulation of target gene promoters for binding of the essential orphan response regulator HP1043 of *Helicobacter pylori*. *J. Bacteriol.* **184,** 4800–4810.

Dondrup, M., Goesmann, A., Bartels, D., Kalinowski, J., Krause, L., Linke, B., Rupp, O., Sczyrba, A., Pühler, A., and Meyer, F. (2003). EMMA: A platform for consistent storage and efficient analysis of microarray data. *J. Biotechnol.* **106,** 135–146.

Eppinger, M., Baar, C., Linz, B., Raddatz, G., Lanz, C., Keller, H., Morelli, G., Gressmann, H., Achtman, M., and Schuster, S. C. (2006). Who ate whom? Adaptive *Helicobacter* genomic changes that accompanied a host jump from early humans to large felines. *PLoS Genet.* **2,** e120.

Ferrero, R. L., Cussac, V., Courcoux, P., and Labigne, A. (1992). Construction of isogenic urease-negative mutants of *Helicobacter pylori* by allelic exchange. *J. Bacteriol.* **174,** 4212–4217.

Goodwin, A., Kersulyte, D., Sisson, G., Veldhuyzen van Zanten, S. J. O., Berg, D. E., and Hoffman, P. S. (1998). Metronidazole resistance in *Helicobacter pylori* is due to null mutations in a gene (*rdxA*) that encodes an oxygen-insensitive NADPH nitroreductase. *Mol. Microbiol.* **28,** 383–393.

Gressmann, H., Linz, B., Ghai, R., Pleissner, K.-P., Schlapbach, R., Yamaoka, Y., Kraft, C., Suerbaum, S., Meyer, T. F., and Achtmann, M. (2005). Gain and loss of multiple genes during the evolution of *Helicobacter pylori*. *PLoS Genet.* **1(4),** e43.

Heuermann, D., and Haas, R. (1998). A stable shuttle vector system for efficient genetic complementation of *Helicobacter pylori* strains by transformation and conjugation. *Mol. Gen. Genet.* **257,** 519–528.

Jiménez-Pearson, M.-A., Delany, I., Scarlato, V., and Beier, D. (2005). Phosphate flow in the chemotactic response system of *Helicobacter pylori*. *Microbiology* **151,** 3299–3311.

Josenhans, C., Eaton, K. A., Thevenot, T., and Suerbaum, S. (2000). Switching of flagellar motility in *Helicobacter pylori* by reversible length variation of a short homopolymeric sequence repeat in *fliP*, a gene encoding a basal body protein. *Infect. Immun.* **68,** 4598–4603.

Karatan, E., Saulmon, M. M., Bunn, M. W., and Ordal, G. W. (2001). Phosphorylation of the response regulator CheV is required for adaptation to attractants during *Bacillus subtilis* chemotaxis. *J. Biol. Chem.* **276,** 43618–43626.

Kim, N., Marcus, M. A., Wen, Y., Weeks, D. L., Scott, D. R., Jung, H. C., Sung, I. S., and Sachs, G. (2004). Genes of *Helicobacter pylori* regulated by attachment to AGS cells. *Infect. Immun.* **72,** 2358–2368.

Kraft, C., and Suerbaum, S. (2006). Mutation and recombination in *Helicobacter pylori*: Mechanisms and role in generating strain diversity. *Int. J. Med. Microbiol.* **295,** 299–305.

Kusters, J. G., van Vliet, A. H. M., and Kuipers, E. J. (2006). Pathogenesis of *Helicobacter pylori* infection. *Clin. Microbiol. Rev.* **19,** 449–490.

Labigne-Roussel, A., Courcoux, P., and Tompkins, L. (1988). Gene disruption and replacement as feasible approach for mutagenesis of *Campylobacter jejuni. J. Bacteriol.* **170,** 1704–1708.

Loh, J. T., and Cover, T. L. (2006). Requirement of histidine kinases HP0165 and HP1364 for acid resistance in *Helicobacter pylori. Infect. Immun.* **74,** 3052–3059.

McDaniel, T. K., DeWalt, K. C., Salama, N. R., and Falkow, S. (2001). New approaches for validation of lethal phenotypes and genetic reversion in *Helicobacter pylori. Helicobacter* **6,** 15–23.

Moore, J. B., Shiau, S.-P., and Reitzer, L. J. (1993). Alterations of highly conserved residues in the regulatory domain of nitrogen regulator I (NtrC) of *Escherichia coli. J. Bacteriol.* **175,** 2692–2701.

Müller, S., Pflock, M., Schär, S., Kennard, S., and Beier, D. (2006). Regulation of expression of atypical orphan response regulators of *Helicobacter pylori. Microbiol. Res.* **162,** 1–14.

Niehus, E., Gressmann, H., Ye, F., Schlapbach, R., Dehio, M., Dehio, C., Stack, A., Meyer, T. F., Suerbaum, S., and Josenhans, C. (2004). Genome-wide analysis of transcriptional hierarchy and feedback regulation in the flagellar system of *Helicobacter pylori. Mol. Microbiol.* **52,** 947–961.

Parkhill, J., Wren, B. W., Mungall, K., Ketley, J. M., Churcher, C., Basham, D., Chillingworth, T., Davies, R. M., Feltwell, T., Holroyd, S., Jagels, K., Karlyshev, A. V., *et al.* (2000). The genome sequence of the food-borne pathogen *Campylobacter jejuni* reveals hypervariable sequences. *Nature* **403,** 665–668.

Pflock, M., Dietz, P., Schär, J., and Beier, D. (2004). Genetic evidence for histidine kinase HP165 being an acid sensor of *Helicobacter pylori. FEMS Microbiol. Lett.* **234,** 51–61.

Pflock, M., Kennard, S., Delany, I., Scarlato, V., and Beier, D. (2005). Acid-induced activation of the urease promoters is mediated directly by the ArsRS two-component system of *Helicobacter pylori. Infect. Immun.* **73,** 6437–6445.

Pflock, M., Finsterer, N., Joseph, B., Mollenkopf, H., Meyer, T. F., and Beier, D. (2006). Characterization of the ArsRS regulon of *Helicobacter pylori*, involved in acid adaptation. *J. Bacteriol.* **188,** 3449–3462.

Pflock, M., Müller, S., and Beier, D. (2007). The CrdRS (HP1365-HP1364) two-component system is not involved in pH-responsive gene regulation in the *Helicobacter pylori* strains 26695 and G27. *Curr. Microbiol.* **54,** 320–324.

Pittman, M. S., Goodwin, M., and Kelly, D. J. (2001). Chemotaxis in the human gastric pathogen *Helicobacter pylori*: Different roles for CheW and the three CheV paralogues, and evidence for CheV2 phosphorylation. *Microbiology* **147,** 2493–2504.

Schär, J., Sickmann, A., and Beier, D. (2005). Phosphorylation-independent activity of atypical response regulators of *Helicobacter pylori. J. Bacteriol.* **187,** 3100–3109.

Stingl, K., and De Reuse, H. (2005). Staying alive overdosed: How does *Helicobacter pylori* control urease activity? *Int. J. Med. Microbiol.* **295,** 307–315.

Suerbaum, S., Josenhans, C., Sterzenbach, T., Drescher, B., Brandt, P., Bell, M., Dröge, M., Fartmann, B., Fischer, H.-P., Ge, Z., Hörster, A., Holland, R., *et al.* (2003). The complete genome sequence of the carcinogenic bacterium *Helicobacter hepaticus*. *Proc. Natl. Acad. Sci. USA* **100,** 7901–7906.

Terry, K., Go, A. C., and Ottemann, K. M. (2006). Proteomic mapping of a suppressor of non-chemotactic *cheW* mutants reveals that *Helicobacter pylori* contains a new chemotaxis protein. *Mol. Microbiol.* **61,** 871–882.

Tomb, J.-F., White, O., Kerlavage, A. R., Clayton, R. A., Sutton, G. G., Fleischmann, R. D., Ketchum, K. A., Klenk, H. P., Gill, S., Dougherty, B. A., Nelson, K., Quackenbush, J., *et al.* (1997). The complete genome sequence of the gastric pathogen *Helicobacter pylori*. *Nature* **388,** 539–547.

Tusher, V. G., Tibshirani, R., and Chu, G. (2001). Significance analysis of microarrays applied to the ionizing radiation response. *Proc. Natl. Acad. Sci. USA* **98,** 5116–5121.

van Vliet, A. H. M., Kuipers, E. J., Waidner, B., Davies, B. J., De Vries, N., Penn, C. W., Vandenbroucke-Grauls, C. M. J. E., Kist, M., Bereswill, S., and Kusters, J. G. (2001). Nickel-responsive induction of urease expression in *Helicobacter pylori* is mediated at the transcriptional level. *Infect. Immun.* **69,** 4891–4897.

Waidner, B., Melchers, K., Stähler, F. N., Kist, M., and Bereswill, S. (2005). The *Helicobacter pylori* CrdRS two-component regulation system (HP1364/HP1365) is required for copper-mediated induction of the copper resistance determinant CrdA. *J. Bacteriol.* **187,** 4683–4688.

Wang, Y., and Taylor, D. E. (1999). Chloramphenicol resistance in *Campylobacter coli*: Nucleotide sequence, expression, and cloning vector construction. *Gene* **94,** 23–28.

Wen, Y., Feng, J., Scott, D. R., Marcus, E. A., and Sachs, G. (2006). Involvement of the HP0165-HP0166 two-component system in expression of some acidic-pH-upregulated genes of *Helicobacter pylori*. *J. Bacteriol.* **188,** 1750–1761.

Xiang, Z., Censini, S., Bayeli, P. F., Telford, J. L., Figura, N., Rappuoli, R., and Covacci, A. (1995). Analysis of expression of CagA and VacA virulence factors in 43 strains of *Helicobacter pylori* reveals that clinical isolates can be divided into two major types and that CagA is not necessary for expression of the vacuolating cytotoxin. *Infect. Immun.* **63,** 94–98.

Yang, Y. H., Dudoit, S., Luu, P., Lin, D. M., Peng, V., Ngai, J., and Speed, T. P. (2002). Normalization for cDNA microarray data: A robust composite method addressing single and multiple slide systematic variation. *Nucleic Acids Res.* **30,** e15.

Yang, Y. H., and Speed, T. (2002). Design issues for cDNA microarray experiments. *Nat. Rev. Genet.* **3,** 579–588.

[26]   Phosphotransfer Profiling: Systematic Mapping of Two-Component Signal Transduction Pathways and Phosphorelays

By Michael T. Laub, Emanuele G. Biondi, and Jeffrey M. Skerker

Abstract

Two-component signal transduction systems, composed of histidine kinases and response regulators, enable bacteria to sense, respond, and adapt to changes in their internal and external conditions. The importance of these signaling systems is reflected in their widespread distribution and prevalence in the bacterial kingdom, with some organisms encoding as many as 250 two-component signaling proteins. In many cases, a histidine kinase and a response regulator are encoded in the same operon and, in such cases, the two molecules usually interact in an exclusive one-to-one fashion. However, in many organisms, the vast majority of two-component signaling genes are encoded as orphan genes, precluding the mapping of signaling pathways based on sequence information and genome position alone. There is also a growing number of examples of two-component signaling pathways with more complicated topologies, including one-to-many and many-to-one relationships, which cannot be inferred from sequence. To address these problems, we have developed an *in vitro* technique called phosphotransfer profiling, which enables the systematic identification of two-component signaling pathways. Purified histidine kinases are tested for their ability to transfer a phosphoryl group to each response regulator encoded in a genome of interest. As histidine kinases typically exhibit a strong kinetic preference *in vitro* for their *in vivo* cognate substrates, this technique allows the rapid mapping of cognate pairs and is applicable to any organism containing two-component signaling genes. The technique can be further extended to mapping phosphorelays and the cognate partners of histidine phosphotransferases. Here, we describe protocols and strategies for the successful implementation of this system-level technique.

Overview

Two-component signal transduction systems, comprising histidine kinases and response regulators, are the single most prevalent paralogous family of signaling proteins in the bacterial kingdom and are also found in archaea, plants, yeasts, and other lower eukaryotes. These signaling systems

METHODS IN ENZYMOLOGY, VOL. 423   0076-6879/07 $35.00

enable cells to sense and respond to a wide range of stimuli, both intracellular and extracellular, and thus help coordinate a wide range of adaptive
responses. Both histidine kinases and response regulators are easily identified by sequence homology and can be systematically enumerated in any
fully sequenced genome. Some organisms encode more than 200 two-
component signaling proteins, with an average bacterial genome containing
50 to 100.

The canonical two-component signaling pathway (Fig. 1A,B) contains a
histidine kinase, which, in response to a stimulus, autophosphorylates on a
conserved histidine residue. The phosphorylated histidine kinase then binds
and transfers its phosphoryl group to a conserved aspartate residue on the
response regulator. The phosphorylation of a response regulator, which
occurs on its receiver domain, typically activates an output domain that
can trigger changes in gene expression, protein–protein interactions, or
enzymatic activity (for a comprehensive review of two-component signaling
pathways, see Stock *et al.*, 2000). In addition to this canonical case involving
two proteins, one histidine kinase and one response regulator, these classes
of signaling proteins can be arranged into more complicated pathways. In
some cases, a histidine kinase can have multiple phosphotransfer targets or
a regulator can receive input from multiple kinases. Another common
arrangement is the so-called phosphorelay (Fig. 1C,D). In these pathways,
the top-level histidine kinase is usually a hybrid histidine kinase such that
the N-terminal portion of the molecule is similar to that of a canonical
histidine kinase and the C-terminus contains a receiver domain, similar to
that found on response regulators. In response to an input stimulus, these
hybrid kinases catalyze a histidine autophosphorylation and then transfer
the phosphoryl group intramolecularly to their receiver domain. The phosphoryl group can then be transferred to a different type of protein, called a
histidine phosphotransferase, which subsequently donates the phosphoryl
group to a soluble response regulator, leading to an output response.

Identifying the connectivity of two-component signal transduction proteins and phosphorelays remains a major challenge and is the focus of this
chapter. In *E. coli*, nearly all histidine kinases and response regulators occur
in operon pairs and, most probably, form exclusive one-to-one pairs. Operon pairs in other organisms also usually define exclusive phosphotransfer
relationships. In many bacteria, however, histidine kinases are often
encoded as orphans and the cognate response regulator(s) cannot be predicted based on sequence alone. Moreover, orphan two-component signaling proteins frequently have one-to-many or many-to-one connectivity or
are involved in phosphorelays. In some cases, phenotypic analysis of
mutants can help delineate these pathways. However, for some complicated
pathways, biochemical connectivity is difficult to infer from mutant

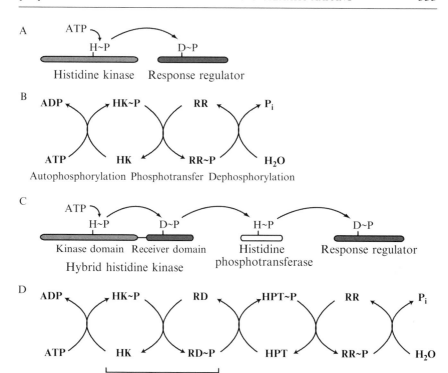

FIG. 1. Schematic overview of two-component signaling pathways. (A) In a canonical two-component signaling system, receipt of an input stimulates a histidine kinase to catalyze autophosphorylation on a conserved histidine residue. The phosphoryl group is subsequently passed to a cognate response regulator which, after phosphorylation, triggers changes in cellular physiology. (B) Schematic of the flow of phosphate through a two-component signal transduction system. Autophosphorylation involves the transfer of the gamma phosphoryl group from ATP to the histidine kinase (HK). Phosphotransfer shuttles the phosphoryl group to a cognate response regulator (RR). Dephosphorylation of the response regulator occurs by hydrolysis of the aspartyl-phosphate bond in a reaction stimulated by either an intrinsic autophosphatase activity in the regulator or by the phosphatase activity of a bifunctional histidine kinase. (C) In a canonical phosphorelay, receipt of an input stimulates autophosphorylation of a hybrid histidine kinase that harbors both a histidine kinase domain and a C-terminal receiver domain equivalent to those found in soluble response regulators. After autophosphorylation, the phosphoryl group is passed intramolecularly to the receiver domain, then to a histidine phosphotransferase, and finally to a soluble, terminal response regulator. (D) Schematic of the flow of phosphate through a canonical phosphorelay. After autophosphorylation of the kinase domain (HK) of a hybrid histidine kinase, the phosphoryl group is passed intramolecularly to the receiver domain (RD), then to a histidine phosphotransferase (HPT), and finally to a terminal response regulator (RR).

phenotypes and, for genetically intractable organisms, such analyses are simply not possible. To address these problems, we have developed an *in vitro* technique, called phosphotransfer profiling, which enables the systematic mapping of *in vivo* connectivity of two-component signal transduction proteins and phosphorelays (Biondi *et al.*, 2006; Skerker *et al.*, 2005). This technique leverages the fact that histidine kinases exhibit a kinetic preference *in vitro* for their *in vivo* cognate response regulator substrate(s). Kinetic preference was first demonstrated with subsets of two-component signaling proteins (Fisher *et al.*, 1996; Grimshaw *et al.*, 1998), and subsequently extended to a genome-wide level (Skerker *et al.*, 2005). The latter observation now enables the efficient mapping of two-component pathways in any organism with a sequenced genome.

In a typical phosphotransfer profiling experiment (Fig. 2A), the purified cytoplasmic, soluble kinase domain of a histidine kinase is first autophosphorylated *in vitro* with $[\gamma^{-32}P]ATP$. The radiolabeled kinase is then tested, in parallel, for phosphotransfer to each purified, full-length response regulator encoded in the genome of interest. Each phosphotransfer reaction is incubated for an identical period of time, with reaction products resolved by SDS-PAGE and analyzed by phosphor-imaging. Autophosphorylated kinase alone is included as a control and forms a single intense band. Efficient phosphotransfer to a response regulator can be manifested in one of two ways. First, a high-intensity band corresponding to the phosphorylated response regulator will appear at a position corresponding to the regulator's molecular weight. Alternatively, efficient phosphotransfer can deplete radiolabel from both the histidine kinase and response regulator, resulting in a blank lane. The latter case can result from a high autophosphatase activity intrinsic to some response regulators. Some histidine kinases are also bifunctional and can act as phosphatases for their cognate regulators (Igo *et al.*, 1989b). In either case, autophosphatase activity of the response regulator or phosphatase activity from the histidine kinase, the net result is nearly complete depletion of radiolabel from the reaction components (Fig. 1B). Thus, to identify kinase-regulator phosphotransfer relationships, each reaction in a profile experiment is inspected for (i) a band corresponding to the response regulator or (ii) a decrease in intensity of the kinase band relative to the kinase-only control. Because this profiling method relies on the comparison, in parallel, of all potential phosphotransfer substrates for a given kinase, it is independent of the specific activity of the kinase being tested.

Although phosphotransfer profiling is an *in vitro* technique, the *in vivo* targets can be identified because histidine kinases exhibit a kinetic preference for their *in vivo* cognate response regulators. In other words, the substrate specificity of a histidine kinase appears to be intrinsic to the

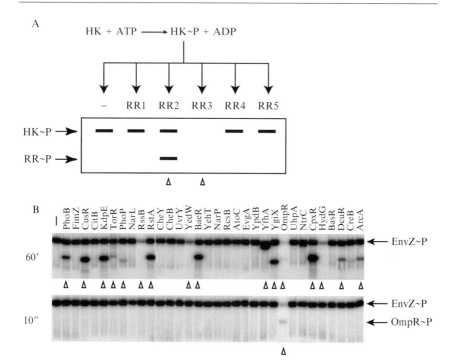

FIG. 2. Phosphotransfer profiling of two-component signaling proteins. (A) Diagram of the phosphotransfer profiling assay. A purified histidine kinase domain (HK) is incubated in the presence of radiolabeled ATP and then incubated with each of the purified response regulators (RR1–RR5) encoded in a genome of interest. Reactions are analyzed by SDS-PAGE and phosphorimaging. Each lane, including a control lane containing no response regulator, shows a band corresponding to the autophosphorylated histidine kinase. Phosphotransfer (lanes indicated by open arrowheads) is manifested as a second band at a position corresponding to the response regulator or by the depletion of radiolabel. The latter results from a high rate of phosphotransfer followed by rapid dephosphorylation of the response regulator. (B) An example of using phosphotransfer profiling to identify the kinetically preferred phospho-acceptor of a histidine kinase. The cytoplasmic kinase domain of EnvZ was profiled against each of the 32 *E. coli* response regulators with phosphotransfer reaction times of either 10 s or 60 min. Whereas multiple regulators are phosphorylated at the long timepoint, only the in vivo cognate substrate OmpR is phosphorylated at the short timepoint. Profiling data taken from Skerker *et al.*, 2005.

protein and not dependent on cellular context. Hence, by examining the phosphotransfer profile of a histidine kinase at multiple timepoints, one can quickly ascertain the preferred substrate. This *in vitro* kinetic preference of a histidine kinase for phosphotransfer to its *in vivo* cognate regulator has been demonstrated for kinase-regulator pairs in the phylogenetically distant organisms *E. coli* and *Caulobacter crescentus* (Skerker *et al.*, 2005).

This suggests that kinetic preference is a general property of two-component signal transduction systems, and that the technique can be applied to any organism that utilizes these signaling proteins.

An example of a typical phosphotransfer profile is shown in Fig. 2B. In *E. coli*, the histidine kinase EnvZ, upon stimulation, autophosphorylates and then transfers a phosphoryl group exclusively to OmpR (Forst *et al.*, 1989; Igo *et al.*, 1989a; Skerker *et al.*, 2005). As seen in Fig. 2B, the systematic profiling of EnvZ~P at a long timepoint of 1 h identifies many different response regulator partners, including OmpR. Although this may seem to suggest that EnvZ has multiple targets, a profile using just 10 s phosphotransfer incubation times demonstrates that EnvZ has a single, preferred target, the known cognate regulator OmpR. Additional kinetic analysis showed that EnvZ has at least a 2000-fold preference, in terms of relative k_{cat}/K_M ratios for phosphotransfer to OmpR relative to a noncognate substrate, CpxR (Skerker *et al.*, 2005). Similar profile experiments have identified the preferred substrates for CheA, CpxA, PhoQ, and PhoR in *E. coli*. Detailed kinetic studies of a small number of two-component signaling pathways have also demonstrated *in vitro* kinetic preference of a histidine kinase for its *in vivo* cognate substrate (Burbulys *et al.*, 1991; Fisher *et al.*, 1996; Grimshaw *et al.*, 1998; Igo *et al.*, 1989b). For example, in *B. subtilis*, the careful measurement of kinetic parameters showed that the kinase KinA has a 57,000-fold preference for phosphorylating SpoOF relative to SpoOA (Grimshaw *et al.*, 1998).

The phosphotransfer profiling technique is an extended, scaled-up version of the standard kinase assays that have been used for many years to study two-component signaling systems; for an earlier *Methods in Enzymology* chapter, see Hakenbeck and Stock, 1996. The protocols that will be described are derived from these earlier studies, but have been modified and, in some cases, streamlined to facilitate the scaling of these assays to a system-wide level. At the end of the chapter, we discuss the interpretation and pitfalls specific to the system-wide phosphotransfer profiling approach.

Detailed Protocols

Preparation of Sequence-Verified Clones and Expression Vectors

Response regulators and histidine kinases in a genome of interest can be identified by consulting databases such as SMART (http://smart.embl-heidelberg.de/) or PFAM (http://www.sanger.ac.uk/Software/Pfam/). However, the automated annotation procedures used by these databases often incorrectly identify some genes as encoding two-component signaling proteins and fail to identify others. A more tedious but reliable approach is to use BLAST analysis of a complete genome followed by manual inspection of each hit. The query sequences used should contain only the receiver domain of a known response regulator and the kinase domain of a histidine kinase to avoid finding proteins with homology to other parts of the response regulator, such as the DNA-binding domain, or the histidine kinase, such as a PAS domain. During manual inspection of the BLAST results, each potential response regulator and histidine kinase should be evaluated for the presence of all major conserved residues involved in phosphotransfer (for a detailed analysis of the conserved sequence features of two-component signaling proteins, see Hoch and Silhavy, 1995, or Grebe and Stock, 1999).

Once a comprehensive list of two-component signaling genes has been established, genes corresponding to each response regulator must be cloned into expression vectors for protein purification. Expression vectors for histidine kinases of interest must also be generated. For organisms with large numbers of response regulators and/or histidine kinases, the generation of expression vectors can be greatly facilitated by using the Invitrogen Gateway high-throughput recombinational cloning system (for a detailed description, see http://www.invitrogen.com/ and Walhout *et al.*, 2000). In this system, clones generated by PCR can be rapidly inserted into expression vectors using highly efficient site-specific recombination between sites present in the vector and sites incorporated into the clone. This system does not require restriction enzymes or ligase and so eliminates much of the time-consuming subcloning that would otherwise be necessary. In the following text, we briefly outline the use of the Gateway system to generate expression clones for the protein purifications required for phosphotransfer profiling.

1. For each regulator, amplify the entire open reading frame by PCR with a proofreading DNA polymerase such as Pfu. Ideally, primers should be purified by reverse-phase cartridge, PAGE, or HPLC because unpurified (desalted-only) primers often contain a significant fraction of incorrect sequences, such as single base deletions.

2. Clone PCR amplicons into the pENTR/D-TOPO vector (available from Invitrogen) according to the manufacturer's protocol and transform into competent *E. coli* cells.

3. Screen 5 to 10 kanamycin-resistant colonies by PCR using M13F (5'-TGTAAAACGACGGCCAGT-3') and M13R (5'-TCACACAGG-AAACAGCTATGAC-3') primers to verify vectors with inserts of the correct size. Even using a proofreading polymerase, positive clones should be further confirmed by sequencing to ensure that the protein that is ultimately produced will have no mutations that affect phosphotransfer behavior.

Once a set of sequence-verified pENTR clones is generated, they can be rapidly and efficiently mobilized into a wide range of expression vectors, called destination vectors, using the Gateway system's LR reaction. The products of these LR reactions are expression vectors in which each gene is fused to an affinity tag and under the control of an IPTG-inducible promoter. A wide range of destination vectors is commercially available for producing fusions to His_6, GST, and other common affinity tags. Alternatively, one can easily convert customized expression vectors into destination vectors (see http://www.invitrogen.com/). The protocol described here uses a customized destination vector that produces thioredoxin-His_6-TEV-tagged proteins (details on the tag used will be described later).

4. For each clone, set up a 5 μl LR reaction containing 75 ng destination vector, 30 ng pENTR plasmid DNA (equimolar ratio of destination:pENTR DNA), $1\times$ LR buffer, 0.75 U topoisomerase I, and 0.5 μl LR clonase enzyme mix (Invitrogen).

5. Incubate each LR reaction overnight at room temperature, then transform 2 μl into chemically competent *E. coli* DH5α cells, and plate on LB with the appropriate antibiotic, depending on the resistance cassette carried by the chosen destination vector.

6. Colonies recovered should be retested, by patching onto appropriate plates, for presence of the antibiotic resistance marker carried on the destination vector and for kanamycin sensitivity to ensure no carryover of pENTR DNA. Positive colonies bear the desired expression vector and are ready for protein purification.

As an alternative to cloning full-length response regulators, the phospho-accepting receiver domains alone can be cloned and used for phosphotransfer profiling. Since some response regulators are large, multi-domain proteins, the purification of receiver domains alone can often produce higher yields and more soluble protein. Moreover, the specificity of kinase interaction is conferred by the receiver domain alone and, hence, is sufficient for the mapping of kinase–regulator pairings. The receiver domains of hybrid histidine kinases can also be individually cloned and purified for profiling the specificity of histidine phosphotransferases (see later).

It should be noted that once sequence-verified clones are generated in pENTR vectors, these clones can be rapidly moved into a wide range of destination vectors using the highly efficient LR reaction. Destination vectors exist that create fusions to a number of popular affinity tags or fluorescent reporters such as GFP. The generation of pENTR clones thus makes possible a range of systematic studies (Walhout *et al.*, 2000).

Protein Purification by Affinity Chromatography

Response regulators can be easily purified in the native state by over-expression in *E. coli* followed by affinity chromatography. There are a wide variety of affinity tags (and corresponding Gateway destination vectors) available, but the protocol described here uses a thioredoxin-His$_6$-TEV tag. The N-terminal thioredoxin domain improves folding and solubility in *E. coli* and, hence, substantially increases the success rate and yield of the protein purifications. The inclusion of a TEV protease site allows for removal of the affinity tag, if necessary or desired.

1. Transform expression vector DNA for each response regulator independently into *E. coli* BL21-Tuner cells and select on LB plates supplemented with the appropriate antibiotic.

2. Inoculate 4 to 5 colonies into 500 ml of LB supplemented with the appropriate antibiotic and grow at 37° to an OD$_{600}$ ~ 0.6 (~4–5 h).

3. To induce expression of the His$_6$-tagged constructs, add IPTG to 300 μM and grow the culture at 30° for 4 h.

4. Harvest cells by centrifugation at 10,800g for 5 min and store cell pellets at −80° until needed.

5. Resuspend each cell pellet in 10 ml lysis buffer (20 mM Tris-HCl, pH 7.9, 0.5 M NaCl, 10% glycerol, 20 mM imidazole, 0.1% Triton X-100, 1 mM PMSF, 1 mg/ml lysozyme, 125 units benzonase nuclease (Novagen)), transfer to a 50 ml conical tube, and incubate at room temperature for 20 min to allow cell lysis.

6. Sonicate cells (2 × 30 s in a Fisher Model 550 with microtip) to complete cell lysis, add fresh PMSF, and then centrifuge for 60 min at 30,000g to generate a cleared lysate. Transfer the lysate to a clean 50 ml conical tube, add 1 ml of Ni-NTA agarose slurry (Qiagen), which has been pre-equilibrated in lysis buffer, and incubate at 4° for 30 min.

7. Wash (centrifuge and resuspend) the Ni-NTA beads twice with 50 ml wash buffer (20 mM HEPES-KOH, pH 8.0, 0.5M NaCl, 10% glycerol, 20 mM imidazole, 0.1% Triton X-100, 1 mM PMSF) and then load the slurry onto an Econo-column (Bio-Rad).

8. Add 2.5 ml elution buffer (20 mM HEPES-KOH, pH 8.0, 0.5M NaCl, 10% glycerol, 250 mM imidazole) and collect the eluate which contains the purified protein. Then, load the eluate directly onto a PD-10 column (Amersham Biosciences) that had been pre-equilibrated with kinase buffer (10 mM HEPES-KOH, pH 8.0, 50 mM KCl, 10% glycerol, 0.1 mM EDTA, 1 mM DTT). The purified protein is then eluted with 3.5 ml kinase buffer and ready for concentration and storage.

9. Concentrate eluted samples to ~1 to 10 mg/ml using Amicon Ultra 30K or 10K columns (Millipore), depending on the protein size. Filter each

sample through an Ultrafree-MC (0.22 μm) spin filter (Millipore) and aliquot for storage at $-80°$.

10. Estimate protein concentration using Coomassie Plus Protein Assay Reagent and a BSA standard (Pierce). An equal amount (500 ng) of each protein sample should also be analyzed by SDS-PAGE to verify molecular weight and purity. Prior to phosphotransfer profiling, normalize all response regulator concentrations against a 500 ng BSA standard using a ChemiImager 5500 and densitometry (Alpha Innotech).

Starting from cell pellets (step 5), one person can purify 8 to 10 proteins in parallel in one day. Hence, for most genomes, the entire set of response regulators can be purified in about a week. This protocol typically generates from 100 μg to 1 mg of protein, depending on its solubility, which is enough for at least 100 phosphotransfer profiling experiments.

Histidine kinases can be cloned, expressed, and purified as done for the response regulators, but there are several additional considerations. First, the phosphotransfer profiling technique is predicated on having a comprehensive set of response regulators, but histidine kinases can be cloned and tested as desired, not necessarily as a complete set. Second, it can be helpful to use a His_6-MBP affinity tag for the kinases rather than a thioredoxin-His_6 tag. The larger MBP tag helps to ensure that the molecular weight of the histidine kinase is greater than each of the TRX-His_6-tagged response regulators. This facilitates resolution of reaction components by SDS-PAGE and, hence, easier interpretation of the profile by phosphor-imaging, as will be described. Third, the forward primer used for PCR amplification of a histidine kinase should be designed to omit any known or predicted N-terminal transmembrane domains to allow the native purification of the soluble kinase domain. In addition, some kinases have large intervening portions between the end of the last transmembrane domain and the beginning of the conserved kinase domain. The inclusion of all or part of this region may or may not affect the autophosphorylation activity of the purified kinase. There is not yet, to the best of our knowledge, a simple way to determine *a priori* whether to include or omit this intervening region. Instead, the simplest strategy is to produce several versions with different N-terminal start points and empirically identify an active construct. For histidine kinases with a structure similar to EnvZ, active kinases can often be obtained with an N-terminal start point after the last transmembrane domain (I179) or just prior to the H-box domain that contains the active site histidine (M223) (Park *et al.*, 1998). Histidine kinases have been purified as full-length constructs and reconstituted in membrane vesicles (Jung *et al.*, 1997). Such full-length kinases could, in principle, be used for phosphotransfer profiling, although this has not yet been reported. Some histidine kinases, however, may not be active in any form *in vitro*, since they may require additional factors *in vivo* for activity. These additional factors, if known, could be purified and may be sufficient, if added to the purified kinase, to drive autophosphorylation. This would enable phosphotransfer profiling, but such a scenario has not yet been demonstrated.

Phosphotransfer Profiling

Before starting a phosphotransfer profile, the histidine kinase of interest should first be tested for autophosphorylation and the time required for producing radiolabeled kinase optimized. To accomplish this, set up a reaction containing 5 μM histidine kinase, 2 mM DTT, 5 mM MgCl$_2$, 500 μM ATP, 5 μCi [γ^{32}P]ATP ($\sim$6000 Ci/mmol; Amersham Biosciences), and enough kinase buffer to bring the reaction volume to 30 μl. Premix the radiolabeled and cold ATP and add last to initiate the reaction. Immediately begin taking samples every 15 min for up to 3 h while the reaction is incubated at room temperature or 30°. Stop the reaction by adding SDS-PAGE loading buffer and place on ice. After all samples have been collected, the autophosphorylation kinetics can be examined by SDS-PAGE and phosphorimaging. Once the optimal autophosphorylation time has been optimized in this fashion, the reaction can be scaled up to provide enough material for phosphotransfer profiling. The optimal autophosphorylation time for histidine kinases can vary from a few minutes to a few hours.

1. Dilute each purified response regulator to 5 μM in kinase buffer supplemented with 5 mM MgCl$_2$.

2. For each phosphotransfer reaction, mix 5 μl of autophosphorylated, radiolabeled kinase, with 5 μl of response regulator prepared and diluted as has been described. A control reaction containing kinase alone must also be included. Each 10 μl phosphotransfer reaction thus contains a final concentration of 2.5 μM of both a response regulator and the histidine kinase of interest.

3. Incubate the reactions at 30° for a defined period of time and then stop by adding 3.5 μl of 4× sample buffer (500 mM Tris-HCl pH 6.8, 8% SDS, 40% glycerol, 400 mM β-mercaptoethanol), and place the reaction on ice.

4. Load the entire reaction for each regulator and the control reaction into a 10% Tris-HCl polyacrylamide gel followed by electrophoresis at 150 volts at room temperature for 50 to 60 min. For typical 15 lane gels, several gels may be necessary to run out every reaction, depending on the number of response regulators.

5. Following electrophoresis, disassemble the gel tank and remove one of the glass plates sandwiching the gel. Then, using a razor blade, carefully remove the dye front which includes unincorporated ATP and place the wet gel (still on the back glass plate) into a sealed plastic bag.

6. Expose gels to a storage phosphor screen for 1 to 3 h at room temperature. If necessary, gels can be frozen at $-80°$ or dried and exposed for longer periods of time. Longer exposures are often necessary for kinases with low autophosphorylation activity. After exposure, scan the phosphor screen and stitch the gel images together for analysis and presentation using any standard image processing software.

Interpretation and Analysis

A typical profile is shown in Fig. 2B. Phosphotransfer, as noted earlier, is indicated by either the appearance of a band corresponding to the phosphorylated response regulator or by the disappearance of the histidine kinase band. To identify the kinetically preferred substrate often requires only running a comprehensive phosphotransfer profile at two timepoints. Using the reaction conditions described previously, an incubation time of 10 to 30 s usually reveals the kinetically preferred substrate(s). This may not always be the case and multiple profiles may be necessary. As each profile is labor-intensive, a more practical strategy is to run only a single profile at a long timepoint, such as 1 h, and then perform detailed kinetics only on the response regulators that are phosphotransfer targets at the long time-point. Although histidine kinases are typically quite promiscuous after 1 h phosphotransfer incubations, this approach will dramatically reduce the number of regulators being examined at shorter timepoints.

In general, detailed kinetic analysis should be performed after phosphotransfer profiling to confirm the kinetic preference of a substrate and to quantify the specificity. For enzymes, the specificity constant is often defined as k_{cat}/K_M and the ratio of specificity constants for two substrates quantifies the relative preference. For many histidine kinases, this ratio is on the order of 10^3 or 10^4 for the *in vivo* cognate response regulator relative to the next best substrates. For example, with EnvZ the k_{cat}/K_M ratio is at least 2000-fold higher for phosphotransfer to OmpR relative to one of the next best substrates, CpxR (Skerker *et al.*, 2005). Histidine kinases, such as EnvZ, with such large preferences probably exclusively phosphorylate the kinetically preferred substrate *in vivo*. However, the gap may not always be as large and certain histidine kinases may have multiple targets, each phosphorylated to a quantitatively different extent (Mika and Hengge, 2005). Detailed kinetic analysis can quantify the relative preferences and thus guide additional, *in vivo* experiments.

The autophosphorylated histidine kinase can be purified away from ATP before examining phosphotransfer, either in profiling experiments or during detailed kinetic characterization. This additional purification is often desirable so that one can examine phosphotransfer from a kinase to a regulator without the confounding effects of additional autophosphorylation of the histidine kinase. For such purifications, the entire autophosphorylation reaction can be loaded onto a Nanosep-10K column (Pall Corporation) and washed 3 to 4 times with kinase buffer by repeated dilution and centrifugation. The phospho-histidine is relatively stable compared to a phospho-aspartate so the histidine kinase will retain most of the initially incorporated radiolabel, and can be stored at $-20°$ for several days before use.

Although the protocol described previously uses equal concentrations of histidine kinase and response regulator in each phosphotransfer reaction, the concentration of phosphorylated histidine kinase is typically lower than the concentration of total histidine kinase. Autophosphorylation reactions are unlikely to go to completion, and during the steps prior to phosphotransfer, some of the phosphate incorporated can be lost to hydrolysis. The final ratio of HK~P:RR is thus likely much less than one. This does not, however, impact the ability to assess the preferred substrate as long as the same preparation of kinase is used for each phosphotransfer reaction. In fact, a ratio of HK~P:RR less than one may better mimic the *in vivo* stoichiometries. For the EnvZ-OmpR pathway, quantitative Western blotting estimates that EnvZ is present at approximately 100 molecules per cell ($\sim$100 nM) and OmpR at approximately 3500 molecules per cell ($\sim$3.5 μM) (Cai and Inouye, 2002). Therefore, even if all of the EnvZ in an *E. coli* cell were phosphorylated, the ratio of HK~P:RR would be approximately 1:20. Estimates of protein abundance for other two-component proteins are of similar orders of magnitude.

The phosphotransfer profiling technique is designed to test each response regulator in isolation. This is not, however, essential. Multiple response regulators can be combined and incubated with histidine kinase in a single reaction (J. M. S. M. T. L., unpublished data). By pooling regulators into small groups, a complete phosphotransfer profiling experiment can be done with a single gel. This pooling strategy will also set up a competition among the response regulators as phospho-acceptors because, as has been described, the concentration of phosphorylated histidine kinase is much lower than that of the response regulators. This competition will lead to an apparent enhancement of the kinase's specificity since phosphotransfer to the kinetically preferred substrate will deplete the reaction of phosphorylated histidine kinase; this, in turn, leads to even less phosphotransfer to noncognate substrates than would be observed if examined in isolation. The major barrier to performing phosphotransfer profiling in this manner is the inability to resolve, by SDS-PAGE and phosphorimaging, multiple response regulators, which are often of similar molecular weights. This fact can be circumvented by carefully choosing which response regulators to combine. Alternatively, response regulators can be grouped into sets of 3 to 5 and, if there is evidence of phosphotransfer to a set, the individual regulators in that group could be subsequently examined in detail.

Phosphorelays and Histidine Phosphotransferases

Phosphotransfer profiling can also be used to analyze hybrid histidine kinases and to map phosphorelays (Fig. 3) (Biondi *et al.*, 2006). For hybrid kinases, the kinase domain can be purified and profiled against the entire set of soluble response regulators, exactly as done with canonical, non-hybrid histidine kinases. In addition, the kinase domain can be profiled against the entire set of receiver domains from the hybrid kinases in a genome of interest. Receiver domains can be individually cloned, expressed, purified, and used for profiling exactly as has been described for full-length, soluble response regulators. Such profiles typically reveal that the kinase domain of a hybrid kinase preferentially phosphorylates its own receiver domain, reflecting what is usually an intramolecular phosphotransfer reaction.

In a typical phosphorelay, a histidine phosphotransferase shuttles phosphate from the receiver domain of a hybrid histidine kinase to the receiver domain of a diffusible response regulator. Unlike histidine kinases and response regulators, histidine phosphotransferases are difficult to identify by sequence homology. However, once identified by other means, they are easily purified since they are usually small, highly soluble proteins. All of the protocols that have been described can be easily modified to enable the profiling of histidine phosphotransferases (HPTs). Radiolabeling of an HPT for profiling first requires purification of a kinase domain and a receiver domain from a hybrid kinase. A fused version of the hybrid kinase containing both domains can be purified and used, but we have found that for a number of hybrid kinases, the activity of the kinase domain increases significantly when it is separated from its receiver domain (J. M. S. M. T. L., unpublished observations). A reaction similar to that described for histidine kinase autophosphorylation is then set up with three components: 1 μM hybrid kinase domain, 1 μM hybrid receiver domain, and 10 μM of the HPT of interest. For uncharacterized histidine phosphotransferases, the hybrid kinase used for this reaction does not need to be the cognate kinase. The purpose of this initial reaction is simply to "load" the HPT with radioactive phosphate, and a significant level of phosphorylated HPT can be achieved

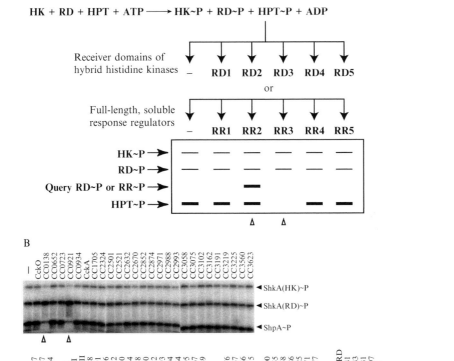

FIG. 3. Mapping phosphorelays. (A) Diagram of the phosphotransfer profiling assay for a histidine phosphotransferase (HPT). The kinase (HK) and receiver domains (RD) of a hybrid kinase are incubated with the HPT of interest and radiolabeled ATP, leading to phosphorylation of the HPT. The phosphorylated HPT is then mixed individually with the purified receiver domains from each hybrid histidine kinase or full-length, soluble response regulator from a genome of interest. As with phosphotransfer profiling of a histidine kinase (Fig. 2), reactions are analyzed by SDS-PAGE and phosphorimaging. The kinase and receiver domains of the hybrid kinase used for phosphorylating the HPT appear as bands in each lane. Phosphotransfer to a queried response regulator or receiver domain is manifested as a band at the appropriate position or by depletion of radiolabeled HPT (see Fig. 2 and text for details). (B) An example of using phosphotransfer profiling to identify the cognate partners of an HPT from *C. crescentus*, ShpA. Phosphorylated ShpA was profiled against the 27 receiver domains of *C. crescentus* hybrid kinases (top) and the 44 full-length response regulators (bottom). In each case, only a single timepoint of 2 min is shown. Open arrowheads indicate lanes with evidence of phosphotransfer. Data taken from Biondi *et al.*, 2006.

with noncognate substrates by using an extended incubation time. Since all known histidine phosphotransferases share a common structural fold, extended incubation eventually leads to their phosphorylation by the hybrid kinase components. The exact concentrations, ratios of components, and reaction times can be empirically optimized. Moreover, for a given HPT, some hybrid kinases will be better than others at phosphorylating an HPT; as of now, this can only be determined empirically.

To profile an HPT:

1. Dilute the HPT phosphorylation reaction described previously 1:10 in kinase buffer.

2. Add 5 μl of the diluted reaction to 5 μl of each purified response regulator to be profiled. Final concentrations are thus 0.5 μM HPT and 2.5 μM of the query response regulator.

3. Subsequent steps are exactly as those described for histidine kinase phosphotransfer profiles. Phosphorimaging will reveal bands corresponding to the hybrid kinase components used to generate phosphorylated HPT. Phosphotransfer, however, is still easily assessed by examining gel images for a band corresponding to the query response regulator or the depletion of radiolabel from the HPT relative to a control lane containing no response regulator.

Because a histidine kinase domain and ATP are required for the initial phosphorylation of an HPT, the phosphorylation of a query response regulator during HPT profiling could, in principle, result from direct phosphotransfer by the kinase rather than from the HPT. This is unlikely because the kinase is present, after dilution, at a concentration of only 0.05 μM. Nevertheless, for each regulator identified in an HPT phosphotransfer profile, a control reaction should be run in which the histidine kinase domain is tested alone (at 0.05 μM) for phosphotransfer to the response regulator.

In vivo, HPTs can shuttle phosphate from the receiver domain of a hybrid kinase to a soluble response regulator, or vice versa, with the directionality probably dictated largely by mass-action (Georgellis *et al.*, 1998; Uhl and Miller, 1996). However, *in vitro*, HPTs can be examined for their ability to transfer to each soluble response regulator, as has been noted, or to the receiver domains of each hybrid histidine kinase encoded in a genome of interest. Receiver domains can be examined as phosphotransfer targets of an HPT just as with the full-length, soluble response regulators. The phosphotransfer profiling of an HPT against both receiver domains and soluble response regulators can thus allow the rapid mapping of complete phosphorelays (Biondi *et al.*, 2006).

Concluding Remarks

The phosphotransfer profiling technique enables the rapid, systematic mapping of two-component signaling pathways and phosphorelays. It should be emphasized, however, that phosphotransfer profiling is an *in vitro* technique and therefore only *suggests* the most probable *in vivo* target. All potential substrates identified by this *in vitro* methodology should be confirmed through the various tools and approaches of molecular biology. Thus far, no discrepancy between the preferred *in vitro* substrates and the *in vivo* relationships has been reported, but results should nevertheless be interpreted with caution and used as a precursor to further experimentation.

References

Biondi, E. G., Skerker, J. M., Arif, M., Prasol, M. S., Perchuk, B. S., and Laub, M. T. (2006). A phosphorelay system controls stalk biogenesis during cell cycle progression in *Caulobacter crescentus*. *Mol. Microbiol.* **59,** 386–401.

Burbulys, D., Trach, K. A., and Hoch, J. A. (1991). Initiation of sporulation in *B. subtilis* is controlled by a multicomponent phosphorelay. *Cell* **64,** 545–552.

Cai, S. J., and Inouye, M. (2002). EnvZ–OmpR interaction and osmoregulation in *Escherichia coli*. *J. Biol. Chem.* **277,** 24155–24161.

Fisher, S. L., Kim, S. K., Wanner, B. L., and Walsh, C. T. (1996). Kinetic comparison of the specificity of the vancomycin resistance VanS for two response regulators, VanR and PhoB. *Biochemistry* **35,** 4732–4740.

Forst, S., Delgado, J., and Inouye, M. (1989). Phosphorylation of OmpR by the osmosensor EnvZ modulates expression of the *ompF* and *ompC* genes in *Escherichia coli*. *Proc. Natl. Acad. Sci. USA* **86,** 6052–6056.

Georgellis, D., Kwon, O., De Wulf, P., and Lin, E. C. (1998). Signal decay through a reverse phosphorelay in the Arc two-component signal transduction system. *J. Biol. Chem.* **273,** 32864–32869.

Grebe, T. W., and Stock, J. B. (1999). The histidine protein kinase superfamily. *Adv. Microb. Physiol.* **41,** 139–227.

Grimshaw, C. E., Huang, S., Hanstein, C. G., Strauch, M. A., Burbulys, D., Wang, L., Hoch, J. A., and Whiteley, J. M. (1998). Synergistic kinetic interactions between components of the phosphorelay controlling sporulation in *Bacillus subtilis*. *Biochemistry* **37,** 1365–1375.

Hakenbeck, R., and Stock, J. B. (1996). Analysis of two-component signal transduction systems involved in transcriptional regulation. *Methods Enzymol.* **273,** 281–300.

Hoch, J. A., and Silhavy, T. J. (1995). "Two-Component Signal Transduction." ASM Press, Washington, DC.

Igo, M. M., Ninfa, A. J., and Silhavy, T. J. (1989a). A bacterial environmental sensor that functions as a protein kinase and stimulates transcriptional activation. *Genes Dev.* **3,** 598–605.

Igo, M. M., Ninfa, A. J., Stock, J. B., and Silhavy, T. J. (1989b). Phosphorylation and dephosphorylation of a bacterial transcriptional activator by a transmembrane receptor. *Genes Dev.* **3,** 1725–1734.

Jung, K., Tjaden, B., and Altendorf, K. (1997). Purification, reconstitution, and characterization of KdpD, the turgor sensor of *Escherichia coli. J. Biol. Chem.* **272,** 10847–10852.

Mika, F., and Hengge, R. (2005). A two-component phosphotransfer network involving ArcB, ArcA, and RssB coordinates synthesis and proteolysis of sigmaS (RpoS) in *E. coli. Genes Dev.* **19,** 2770–2781.

Park, H., Saha, S. K., and Inouye, M. (1998). Two-domain reconstitution of a functional protein histidine kinase. *Proc. Natl. Acad. Sci. USA* **95,** 6728–6732.

Skerker, J. M., Prasol, M. S., Perchuk, B. S., Biondi, E. G., and Laub, M. T. (2005). Two-component signal transduction pathways regulating growth and cell cycle progression in a bacterium: A system-level analysis. *PLoS Biol.* **3,** e334.

Stock, A. M., Robinson, V. L., and Goudreau, P. N. (2000). Two-component signal transduction. *Annu. Rev. Biochem.* **69,** 183–215.

Uhl, M. A., and Miller, J. F. (1996). Central role of the BvgS receiver as a phosphorylated intermediate in a complex two-component phosphorelay. *J. Biol. Chem.* **271,** 33176–33180.

Walhout, A. J., Temple, G. F., Brasch, M. A., Hartley, J. L., Lorson, M. A., van den Heuvel, S., and Vidal, M. (2000). GATEWAY recombinational cloning: Application to the cloning of large numbers of open reading frames or ORFeomes. *Methods Enzymol.* **328,** 575–592.

[27] Identification of Histidine Phosphorylations in Proteins Using Mass Spectrometry and Affinity-Based Techniques

By ANDREW R. S. ROSS

Abstract

Histidine phosphorylation plays a key role in prokaryotic signaling and accounts for approximately 6% of the protein phosphorylation events in eukaryotics. Phosphohistidines generally act as intermediates in the transfer of phosphate groups from donor to acceptor molecules. Examples include the bacterial phosphoenolpyruvate:sugar phosphotransferase system (PTS) and the histidine kinases found in two-component signal transduction pathways. The latter are utilized by bacteria and plants to sense and adapt to changing environmental conditions. Despite the importance of histidine phosphorylation in two-component signaling systems, relatively few proteins have so far been identified as containing phosphorylated histidine residues. This is largely due to the instability of phosphohistidines, which, unlike the phosphoesters formed by serine, threonine, and tyrosine, are labile and susceptible to acid hydrolysis. Nevertheless, it is possible to preserve and identify phosphorylated histidine residues in target proteins using appropriate sample preparation, affinity purification, and mass spectrometric techniques. This chapter provides a brief overview of such techniques, describes their use in confirming histidine phosphorylation of a known PTS protein (HPr), and suggests how this approach might be adapted for large-scale identification of histidine-phosphorylated proteins in two-component systems.

Introduction

Reversible phosphorylation is one of the most common and most important mechanisms by which the structure and function of a protein can become post-translationally modified. It is estimated that up to 30% of all proteins may be phosphorylated at any given time (Cohen, 2000). Site-specific phosphorylation of proteins affects localization, turnover, and enzymatic activity, as well as interactions with other proteins and DNA (Patel and Gelfand, 1996; Wolschin *et al.*, 2005). Phosphorylation occurs on the side chains of certain amino acid residues. The resulting phosphoamino acids fall into three categories: O-phosphates, which are formed by serine (Ser), threonine (Thr), and tyrosine (Tyr) and contain phosphoester linkages; N-phosphates formed by

METHODS IN ENZYMOLOGY, VOL. 423   0076-6879/07 $35.00
© 2007, Her Majesty the Queen in right of Canada.   DOI: 10.1016/S0076-6879(07)23027-7

histidine (His), lysine (Lys), and arginine (Arg), which contain phosphoamidate bonds; and the acyl-phosphate formed by aspartic acid (Asp). The addition and removal of phosphate groups are usually mediated by specific classes of enzymes, such as histidine kinases (phosphorylation of His) and serine/threonine phosphatases (dephosphorylation of Ser and Thr). Phosphohistidines, which are the least stable of the phosphoamino acids, may or may not require histidine phosphatases, depending on the protein and, in particular, the residues adjacent to phosphohistidine (Klumpp and Krieglstein, 2002).

In terms of physiological function, there are two main classes of protein phosphorylation. The first of these encompasses phosphorylation for the purpose of regulating enzymatic activity, and usually involves modification of Ser, Thr, and Tyr residues. The resulting phosphoesters are stable entities, and generally serve to regulate enzymatic catalysis without direct involvement in the catalytic mechanism. Reversible, multisite phosphorylation of Ser, Thr, and Tyr mediates numerous signal transduction pathways in eukaryotic cells (Cohen, 2000). The second class encompasses phosphorylation for the purpose of phosphate group transfer, and is generally restricted to phosphorylation of His residues. Phosphohistidines act as high-energy intermediates in the transfer of phosphate from phosphodonor to phosphoacceptor molecules (Stock *et al.*, 1989), a role for which these labile modifications are well suited. Examples include the bacterial phosphoenolpyruvate:sugar phosphotransferase system (Meadow *et al.*, 1990) and the histidine kinase enzymes found in two-component signal transduction pathways (Parkinson and Kofoid, 1992). The latter are utilized by bacteria, plants, and lower eukaryotes to sense and adapt to changing environmental conditions (Klumpp and Krieglstein, 2002). In bacteria, such adaptations may include changes in motility, cell morphology, and gene expression, as well as the establishment of virulence and antibiotic resistance.

Due to the importance of protein phosphorylation in regulating key biological processes, considerable effort has been put into developing procedures for identifying and mapping sites of phosphorylation, both for individual proteins and on a proteome-wide scale (Beausoleil *et al.*, 2004; de la Fuente van Bentem *et al.*, 2006; Nühse *et al.*, 2003). However, the substoichiometric nature of this modification continues to present major challenges; indeed, the phosphorylated form of a particular protein may represent only a small fraction of its total abundance. Furthermore, many proteins can undergo phosphorylation at different sites (Cohen, 2000), resulting in a number of potential phosphorylated isoforms of which several may be present at any given time. As a result, the full complement of phosphorylated proteins in a cell, tissue, or organism (the phosphoproteome) can be extremely complex.

For His-phosphorylated proteins, the situation is further complicated by the fact that phosphohistidines are susceptible to hydrolysis under the acidic conditions normally used in phosphoprotein and phosphopeptide analysis (Hess *et al.*, 1988; Matthews, 1995). In contrast, the phosphoesters formed by Ser, Thr, and Tyr residues resist acid treatment and so remain intact during extraction, purification, and analytical procedures. Consequently, phosphohistidines generally go undetected in conventional studies of protein phosphorylation (Klumpp and Krieglstein, 2002). Before deciding upon a strategy for large-scale identification of histidine phosphorylations, it is therefore necessary to undertake a critical review of existing techniques for targeting and analyzing phosphoproteins and sites of protein phosphorylation. Such techniques include sample fractionation, affinity purification, gel separation, and mass spectrometry (MS)-based analytical procedures.

Sample Fractionation

As a first step in addressing the low relative abundance of phosphorylated proteins, it is advisable to use some form of fractionation to reduce sample complexity during subsequent analytical steps (Gruhler *et al.*, 2005). Differential centrifugation is a popular technique for fractionation of biological samples and has been applied successfully to phosphoproteomic studies (Nühse *et al.*, 2003). Proteins and organelles differ in size, shape, and density, and can therefore be resolved from cell lysates or tissue homogenates by varying the speed of centrifugation. The forces created at low speeds are small (e.g., $600 \times g$) and only very large or dense particles will precipitate, forming a pellet from which proteins can be extracted and/or further purified. At high speed, most particles will precipitate and only soluble proteins will remain in solution. Figure 1 shows a stepwise fractionation protocol for eukaryotes based upon differential centrifugation. This fractionation technique causes minimal disruption to the sample and is, therefore, likely to preserve labile modifications, such as phosphohistidines. However, it should be borne in mind that the denaturing buffers and phosphatase inhibitors normally used to suppress enzyme activity during protein extraction will not guard against hydrolysis of phosphohistidines.

Subfractionation methods (e.g., for resolving mitochondrial and microsomal proteins) have also been developed to further simplify protein and proteome analysis (Hanson *et al.*, 2001; Nühse *et al.*, 2003). Linear and discontinuous sucrose density gradient fractionation procedures have traditionally been used to purify plasma membranes and mitochondrial complexes (Hodges *et al.*, 1972; VanPutte and Patterson, 2003). However, these methods are generally time-consuming, result in low yields, and are prone

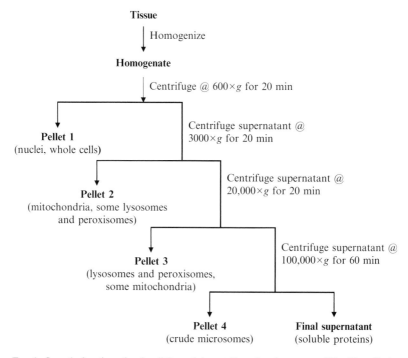

FIG. 1. Sample fractionation by differential centrifugation (courtesy of Dr. Uma K. Aryal, NRC Plant Biotechnology Institute). Plasma membrane proteins in crude microsomal fractions may be further resolved by aqueous two-phase partitioning.

to contamination. Affinity purification of plasma membranes from crude microsomal pellets has been achieved using an aqueous two-phase system comprising polyethylene glycol (PEG) and dextran (Nühse *et al.*, 2003; Schindler *et al.*, 2006). When mixed, these two polymers separate into upper (PEG) and lower (dextran) phases. Subsequent partitioning of anionic species into the lower phase results in the latter's becoming negatively charged. This, in turn, causes plasma membranes to migrate into the upper or interphase regions, effecting separation of subcellular membranes on the basis of charge rather than density (Larsson *et al.*, 1987; Persson and Jergil, 1992). As well as enhancing detection of phosphorylated and other low abundance proteins, fractionation by differential centrifugation and two-phase partitioning are very useful for subcellular localization of proteins, complementing informatics tools such as TargetP (Emanuelsson *et al.*, 2000) and SignalP (Bannai *et al.*, 2002) that are also available for this purpose. Two-phase partitioning may also assist in the detection of His-containing proteins, which migrate preferentially into the PEG phase (Wuenschell et al., 1990).

Phosphoprotein Enrichment

Regardless of whether or not sample fractionation procedures are employed, the purification and analysis of phosphoproteins can be greatly enhanced using affinity-based techniques. Monoclonal antibodies raised against specific protein modifications (e.g., phosphorylated Ser, Thr, or Tyr residues) can be used to isolate proteins carrying these modifications, either by immunoaffinity chromatography or immunoprecipitation (Liu et al., 2004). Such antibodies can also be used to visualize modified proteins separated on one- or two-dimensional electrophoresis gels (see later). Antibody purification of proteins containing phosphorylated Ser and Thr residues has met with limited success (Grønborg et al., 2002), due, in part, to the relatively small size of these phosphoamino acids. However, enrichment of Tyr-phosphorylated proteins using immunoprecipitation can be very efficient, enabling the identification of hundreds of phosphorylated proteins in biological tissues (Rush et al., 2005). A practical limitation with this approach is that nontarget proteins may be co-purified by interaction with the target proteins or with the antibodies themselves, which, in practice, are rarely (if ever) 100% specific for the target modification. Fortunately, many of the commercially available anti-phosphotyrosine antibodies are reactive toward phosphohistidine (Klumpp and Krieglstein, 2002), which also contains a relatively large aromatic structure. Such antibodies may therefore be effective in recovering both Tyr- and His-phosphorylated proteins, although this hypothesis has yet to be tested.

Immobilized metal-ion affinity chromatography (IMAC) has long been used for targeted enrichment of phosphorylated proteins (Andersson and Porath, 1986). This application exploits the particular affinity of phosphate groups for metal ions such as Fe^{3+} and Ga^{3+}, which arises from the predominantly electrostatic interaction between oxygen atom and trivalent metal ion (neither of which is easily polarized). Figure 2 shows the steps involved in a typical IMAC extraction procedure. Commercial kits based on Fe(III)- and Ga(III)-IMAC are available for selective enrichment of phosphoproteins from protein extracts (Wolschin et al., 2005). Immunoaffinity-based kits for the removal of high abundance proteins (HAP) are also available, and can enhance detection and analysis of phosphorylated and other low abundance proteins (LAP). Most of these commercial kits have been developed and optimized for use with mammalian (e.g., plasma) proteins. Consequently, the protocols developed for these products are not applicable to, or may require extensive modification for use with,

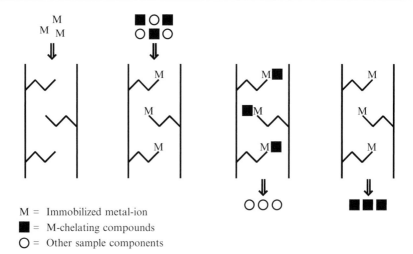

M = Immobilized metal-ion
■ = M-chelating compounds
O = Other sample components

FIG. 2. Immobilized metal-ion affinity chromatography (IMAC). Selected metal ions are loaded onto a chelating column. The sample is then passed through the column, and compounds with an affinity for the immobilized metal ions are selectively retained. Once the column has been washed, the retained compounds can be recovered by displacing them from the immobilized metal ions. This can be achieved by elution with a competing ligand or by raising or lowering pH, depending on the nature of the retained compounds.

plant and/or bacterial systems. Moreover, recovery of intact phosphoproteins using affinity-based methods relies upon the stability, and accessibility, of the target modification. Such methods may not, therefore, be suitable for the recovery of His-phosphorylated proteins unless experimental conditions can be adjusted to prevent hydrolysis of the phosphoamidate bond during extraction.

Metal oxide affinity chromatography (MOAC) has been used successfully to enrich intact proteins containing phosphorylated Ser, Thr, and Tyr residues (Laugesen *et al.*, 2006; Wolschin *et al.*, 2005). This technique compares favorably with commercial products for phosphoprotein enrichment. Furthermore, the incubation and elution buffers required for MOAC are nonacidic and may, therefore, preserve phosphohistidines, although this technique has yet to be evaluated for the recovery of His-phosphorylated proteins.

Gel Separation

Following extraction, fractionation, and/or phosphoprotein enrichment, individual proteins can be further resolved using one- or two-dimensional gel electrophoresis (1- or 2-DE), depending on the complexity of the sample or fraction. Gel electrophoresis allows the expression of individual proteins to be compared among samples and differentially expressed proteins to be targeted for identification and further analysis. This requires visualization of the gel-separated proteins using autoradiography, visible, or fluorescent staining methods. Autoradiography of ^{32}P-labeled proteins, following separation by gel electrophoresis and blotting onto a poly(vinylidine) difluoride (PVDF) membrane, is a well-established method for detecting newly synthesized phosphoproteins in systems that are amenable to metabolic radiolabeling, such as (bacterial) cell cultures. Treatment of the PVDF membrane with base, prior to autoradiography, hydrolyzes phosphoserine and phosphothreonine but leaves phosphotyrosine, phosphohistidine, and phospholysine residues intact (Klumpp and Krieglstein, 2002). Additional treatment with acid cleaves the phosphoamidate bonds, leaving only phosphotyrosine residues unhydrolyzed. Hence, one can discriminate among phosphoproteins containing different types of phosphoamino acids, although His-phosphorylated proteins that also contain phosphotyrosine residues may go undetected using this approach. However, proteins targeted using autoradiography may need to be excised from replicate, non-radiolabeled electrophoresis gels for further (MS) analysis in order to address such issues as sensitivity, recovery, and handling of radioactive samples (Conrads et al., 2002).

Monoclonal antibodies can also be used to detect gel-separated phosphoproteins in PVDF membranes. Again, this Western blotting approach provides information about which types of residues (e.g., Ser, Thr, or Tyr) are phosphorylated in each protein, as exemplified in Fig. 3. Unfortunately, the production of antibodies specific for phosphohistidine is hindered by the instability of this modification (Klumpp and Krieglstein, 2002). However, the reactivity of many anti-phosphotyrosine antibodies toward

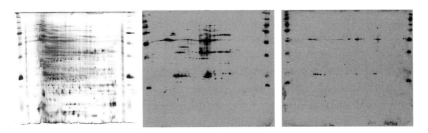

FIG. 3. A silver-stained 2-DE protein gel (left) and Western blots obtained from replicate gels using antibodies against phosphoserine (middle) and phosphothreonine residues (right, courtesy of Dr. Lianglu Wan, NRC Plant Biotechnology Institute). Horizontal spot trains are characteristic of multiple phosphorylation, which has a much greater effect on pI than on molecular weight. (See color insert.)

phosphohistidines, which could be 10 to 100 times more abundant that phosphotyrosine (Matthews, 1995), may provide a means of targeting these proteins for subsequent identification and mapping of phosphorylation sites by mass spectrometry (MS) (see later).

Western blotting and autoradiography are very sensitive techniques. Hence, proteins detected using these methods may not be of sufficient abundance for identification and structural analysis using MS, unless a number of replicate gels can be run and corresponding protein spots matched and combined. In contrast, proteins detected by direct in-gel methods such as Coomassie, silver, or fluorescent staining can usually be identified following excision of the individual bands or spots. Furthermore, the development of fluorescent probes for certain modifications (e.g., phosphorylation, glycosylation) means that gels can now be stained once for detection of the target modification, and again to visualize and compare expression levels for every protein in the gel (Vyetrogon et al., 2006). This approach does not provide site-specific information and is, therefore, of limited utility in targeting His-phosphorylated proteins, unless such proteins have been enriched prior to gel electrophoresis. Such information may, however, be obtained by comparing replicate gels run before and after treatment with acid, base, or site-specific phosphatases and observing changes in protein migration and/or visualization that are consistent with dephosphorylation.

Mass Spectrometry

Once the proteins of interest have been isolated using fractionation, affinity, and/or gel-based approaches, one can use mass spectrometry (MS) to identify target proteins and sites of modification (Schweppe et al., 2003). The conventional, "bottom-up" approach to protein MS involves the use of a site-specific protease, such as trypsin, to cleave the protein(s) into component peptides. Protein digestion, which normally takes several hours, can be performed in solution (using immobilized trypsin beads, for example) or in-gel, once the spots or bands of interest have been removed and the protein(s) destained, reduced, and alkylated to expose the proteolytic cleavage sites. Automation of this process significantly reduces contamination (especially from human keratin) and, together with automated gel spot excision, enables high-throughput preparation and digestion of protein samples (Ross et al., 2002). The resulting peptides are amenable to exact mass analysis using a time-of-flight (TOF) or Fourier transform-ion cyclotron resonance (FT-ICR) mass spectrometer equipped with a matrix-assisted laser desorption/ionization (MALDI) or electrospray ionization (ESI) source (Aebersold and Mann, 2003). MALDI and ESI are mild "desorption" ionization techniques that are capable of generating stable, intact molecular ions from proteins and peptides, from which mass and/or sequence information may be derived (see following text). Using a database search engine such as MS-Fit (http://prospector.ucsf.edu/) or Mascot (http://www.matrixscience.com/), one can compare and match the measured masses of these peptides with theoretical (tryptic) peptide masses generated *in silico* from a protein or gene sequence database such as SwissProt or NCBInr. This technique, known as peptide mass fingerprinting (PMF), is generally performed using a MALDI-TOF mass spectrometer, which provides rapid, simultaneous analysis of protonated $[M + H]^+$ molecular ions generated from the peptides in a protein digest (Yates, 1998). PMF usually works well if the sample contains one or two abundant proteins and the database search is well constrained, for example, by using data of high mass accuracy and/or specifying the organism(s) under investigation. Furthermore, by including phosphorylation of specific residues (e.g., Ser/Thr) as variable modifications in the search, one can identify possible sites of phosphorylation, if peptides carrying this modification appear as additional, significant matches.

However, to confirm sites of phosphorylation and identify phosphorylated and other proteins in complex mixtures (for example, in unresolved protein fractions or gel bands), one must resort to tandem mass spectrometry (MS/MS). This approach typically involves liquid chromatographic (LC) separation and on-line electrospray ionization (ESI), which generates multiply protonated $[M + nH]^{n+}$ molecular ions from the LC-separated peptides. Individual peptide (precursor) ions are then selected and fragmented using a tandem quadrupole, quadrupole-time of flight (Q-TOF), ion trapping, or FT-ICR mass spectrometer (Aebersold and Mann, 2003). The fragment (or product) ion spectra generated by these instruments encodes information about the amino acid sequences of the selected peptides, including modifications to any of the residues. These spectra can be decoded manually, or by using *de novo* sequencing software such as PEAKS (http://www.bioinformaticssolutions.com/), to determine the sequences of the peptides and any sites of phosphorylation, since this modification increases the mass of an amino acid residue by 80 Daltons (Da). They can also be used to identify the parent protein by matching the experimental MS/MS spectra with theoretical (tryptic) peptide fragment ions generated *in silico* from a protein or gene sequence database, using programs such as Mascot or Sequest (Yates, 1998).

The process of collision-induced dissociation (CID), commonly used to generated peptide fragments during MS/MS, may result in loss of phosphate, depending on the residue and the collision energy used. For example, loss of H_3PO_4 (98 Da), which results from β-elimination of the phosphate group, is diagnostic for peptides containing phosphorylated Ser and Thr residues (Schweppe *et al.*, 2003). In contrast, extensive CID results in the elimination of phosphotyrosine as a stable phosphorylated immonium ion with a mass-to-charge ratio (*m/z*) of 216, which is characteristic of peptides carrying this modification (Steen *et al.*, 2003). Figure 4 shows examples of product-ion MS/MS spectra obtained for standard peptides containing phosphorylated Ser, Thr, and Tyr residues using a Q-TOF mass spectrometer. As is apparent from these spectra, facile loss of phosphate from Ser and Thr tends to inhibit further fragmentation of the peptide, which, in turn, may prevent localization of residues carrying this modification. The replacement of phosphate groups with stable marker molecules has been used to enhance MS and MS/MS analysis, and to enable relative quantification of phosphorylated peptides and proteins (Conrads *et al.*, 2002; Wolschin *et al.*, 2005). Unfortunately, such techniques are currently applicable only to Ser and Thr phosphorylations. Nevertheless, loss of phosphate from Ser or Thr during MS/MS can be used to trigger selection and further fragmentation of the dephosphorylated peptide when performed on an ion trapping or FT-ICR instrument (Beausoleil *et al.*, 2004; Gruhler *et al.*, 2005). This technique, known as data-dependent MS/MS/MS (or MS^3), could

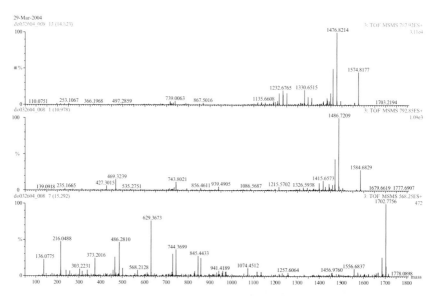

FIG. 4. Product-ion MS/MS spectra for Ser- and Thr-phosphorylated peptides (upper and middle panels), in which the most abundant fragment ion corresponds to neutral loss of phosphoric acid (98 Da) from the intact peptide, and for a Tyr-phosphorylated peptide (lower panel), which gives the diagnostic phosphotyrosine immonium ion (*m/z* 216). (See color insert.)

be used to detect and map sites of phosphorylation in His-phosphorylated peptides, provided that intact phosphopeptide precursor ions can be generated.

FT-ICR mass spectrometers are also capable of performing exact mass MS and MS/MS analysis on intact proteins (Cooper *et al.*, 2005), obviating the need for proteolysis. This "top-down" approach allows for a comparison of the actual mass of a modified protein with the theoretical mass of the matching, unmodified protein found in the database. When combined with the MS/MS data generated for the same protein, this approach provides a powerful method for identifying and locating known and novel modifications in target proteins. Furthermore, FT-ICR instruments are compatible with electron capture dissociation (ECD), a technique that preserves labile modifications while promoting fragmentation and sequencing of the peptide backbone (Emmett, 2003). Hence, FT-ICR MS and ECD appear well suited to the identification and structural analysis of proteins containing phosphorylated histidine residues, although the required instrumentation is very expensive and not yet widely available.

Phosphopeptide Enrichment

Despite the inherent sensitivity and specificity of mass spectrometry, the identification and structural analysis of phosphorylated peptides by MS and MS/MS can still be problematic, since phosphorylation is usually substoichiometric and tends to inhibit positive ionization by MALDI or ESI (a prerequisite for peptide fragmentation and sequencing). For this reason, researchers continue to investigate ways of enriching phosphorylated peptides from protein digests prior to MS analysis. Strong cation exchange (SCX) chromatography is often used as the first step in multidimensional liquid chromatography (MDLC), an alternative approach to gel-based protein analysis (Wolters *et al.*, 2001). Entire protein extracts or fractions are digested, usually with trypsin, and the resulting peptides loaded onto a SCX column. Solutions of increasing salt concentration are used to elute peptide fractions from the column, which are then separated by reversed-phase high performance liquid chromatography (RP-HPLC) and analyzed on-line by ESI-MS/MS, or off-line by MALDI-MS/MS. As with proteins, fractionation of peptides can assist in the detection of phosphorylation sites by reducing sample complexity during subsequent analytical steps (Gruhler *et al.*, 2005). Furthermore, whereas unmodified tryptic peptides (which have basic C-terminal residues) usually carry a net charge or 2+ or greater in solution, the presence of a negatively charged phosphate group generally limits this charge to 1+. Consequently, most phosphopeptides tend to elute from the SCX column before unphosphorylated peptides. This approach has been used successfully to enrich phosphorylated peptides for on-line MS analysis, greatly reducing background interference/ion suppression by unphosphorylated peptides (Beausoleil *et al.*, 2004). When combined with stable isotope labeling techniques such as SILAC (Gruhler *et al.*, 2005), this gel-free approach can also be used for relative quantification of phosphoproteins in different samples. Unfortunately, His-phosphorylated peptides are unlikely to survive the acidic conditions necessary for SCX or reversed-phase liquid chromatography.

Strong anion exchange (SAX) chromatography has also been evaluated for peptide fractionation (Nühse *et al.*, 2003) prior to enrichment of phosphopeptides using IMAC. The latter, which is based upon the affinity of oxygen-containing groups (e.g., phosphate, carboxylate) for trivalent metals ions, can be made more specific for phosphopeptides by converting carboxylic acid groups to methyl esters (Ficarro *et al.*, 2002). However, this process biases against IMAC recovery of singly phosphorylated peptides and may lead to unwanted peptide modifications (Wolschin *et al.*, 2005), making it unsuitable for phosphohistidines. SAX helps to compensate for this bias by permitting differential elution of singly and multiply phosphorylated peptides using different salt concentrations (Nühse *et al.*, 2003). Unfortunately, the recovery of His-phosphorylated peptides from SAX fractions is also compromised by the acidic conditions required for desalting by solid-phase extraction (SPE) prior to IMAC and/or MS analysis.

IMAC has been used extensively for selective recovery and enrichment of peptides containing phosphorylated Ser, Thr, and Tyr residues from protein digests. As with intact phosphoproteins, the metal ions most commonly used for this purpose are Fe^{3+} and Ga^{3+} (Nühse *et al.*, 2003; Posewitz and Tempst, 1999). IMAC media consist of acidic, metal-chelating functional groups bound to a solid support such as Sepharose, agarose, cellulose, polystyrene resin, or silica (Liu *et al.*, 2004). The most commonly used functional group is iminodiacetate (IDA), which is small, hydrophilic, and binds metal ions tightly while leaving coordination sites available for peptide or protein binding (Arnold, 1991). The most popular alternative, nitrilotriacetate (NTA), surrounds and binds metals more tightly that IDA, allowing strong metal chelators (e.g., His-tagged proteins) to be recovered by IMAC without stripping the metal from the column. Several products are commercially available for peptide purification by IMAC, including column packing materials, prepacked analytical or extraction columns, and preloaded extraction beads. Disposable pipette tips packed with IMAC media have also been developed, which allow rapid enrichment of phosphopeptides from small volumes of protein digests. These tips make it possible to extract His-phosphorylated peptides with sufficient speed to preserve the intact phosphohistidine residue for subsequent identification by MS (Napper *et al.*, 2003). Furthermore, it is possible to discriminate between peptides containing phosphohistidine and phosphoesters using IMAC (see following text), something that cannot be achieved using SCX, SAX, or MOAC, although the latter is effective in enriching phosphorylated peptides as well as intact phosphoproteins (Wolschin *et al.*, 2005).

Identification of Phosphohistidine in a Model Protein

Having reviewed the techniques currently available for targeting and analyzing phosphorylated proteins and peptides, it is apparent that only rapid and/or noninvasive methods are likely to be effective in recovering intact His-phosphorylated proteins, and identifying phosphohistidines within these proteins. The following describes a preliminary investigation into the use of two such methods, IMAC and MALDI-TOF MS, for selective recovery and identification of a His-phosphorylated peptide derived from the phosphocarrier protein HPr in *Escherichia coli*. HPr occupies a central role in the bacterial PTS, which mediates phosphorylation-dependent sugar uptake (Meadow *et al.*, 1990) as well as numerous regulatory roles in bacterial metabolism (Postma *et al.*, 1993; Titgemeyer, 1993). HPr undergoes phosphorylation at a conserved histidine residue (His-15) and functions as a phosphotransfer protein between Enzyme I, the initiating enzyme of the PTS, and a sugar-specific Enzyme IIA protein (Anderson *et al.*, 1971; Waygood *et al.*, 1985). The kinetics of HPr phosphorylation and the phosphohydrolysis properties of this protein are well established.

Phosphorylation and Digestion of HPr

Enzyme I, Enzyme IIA^{glc}, and HPr proteins from *E. coli* were purified to homogeneity using previously published protocols (Anderson *et al.*, 1991; Brokx *et al.*, 2000; Napper *et al.*, 2001), as was the CheY protein from *Salmonella typhimurium* (Stock *et al.*, 1985). Phosphorylation of HPr was achieved by combining 20 ng of Enzyme I with 10 μg (1.1 nmol) of HPr in 20 μl volumes of a reaction mixture containing 2 mM MgCl$_2$, 5 mM phosphoenolpyruvate, and 20 mM HEPES buffer (pH 7.0). Complete phosphorylation of HPr was achieved after incubation at 37° for 15 min. For unphosphorylated controls, the phosphodonor (phosphoenolpyruvate) was omitted from the mixture.

Rapid proteolysis was achieved by adding 2.5 μl of 0.4 mg/ml *Staphlococcus aureus* V8 protease (endoproteinase C) in 0.1 M bicine (pH 8.6) and 1 mM EDTA to a 20 μl volume of phosphorylation reaction mixture. Digests were incubated at 37° for just 30 min in an attempt to preserve the phosphohistidine residues, which undergo extensive hydrolysis during longer digestion periods (Anderson *et al.*, 1993) and are therefore incompatible with conventional trypsin digestion procedures. To estimate the final peptide concentration, aliquots of the V8 digest were separated by SDS-PAGE and the intensities of digested (peptide) and undigested (protein) bands quantified using a phosphoimager. The intensity ratio of digested to undigested bands was 10:1, corresponding to a maximum theoretical concentration of 44 pmol/μl for the phosphorylated HPr peptide VTITAPNGL(pH)TRPAAQFVKE (residues 6–25).

IMAC Conditions

Disposable metal-chelating pipette tips (ZipTip$_{MC}$), obtained from Millipore Corporation (Bedford, MA), were used to facilitate rapid extraction and enrichment of phosphorylated peptides. The tips were washed 10 times with 10 μl of 0.1% acetic acid in deionized water, then charged with metal ions by aspirating and dispensing 10 μl of a 100 mM metal salt solution 15 times. Several metal ions were evaluated for their ability to selectively extract and recover His-phosphorylated peptides by IMAC, including Ga^{3+}, Cu^{2+}, and Fe^{3+}. The charged tips were rinsed 5 times with 10 μl of deionized water, and 5 times with 10 μl of 0.1% acetic acid in 50% acetonitrile. Phosphopeptides were loaded onto the tips by aspirating and dispensing 10 μl of sample 10 times, then washing 10 times with 10 μl of 0.1% acetic acid in 50% acetonitrile. Phosphopeptides were eluted in 2 μl of 1% NH_4OH and immediately neutralized with 1 μl of 2% trifluoroacetic acid (TFA). Samples were then mixed with an equal volume of α-cyano-4-hydroxycinnamic acid (CHCA) matrix solution (5 mg/ml in 75% acetonitrile with 0.1 % TFA) and applied directly to the MALDI target plate. For comparison, the phosphorylated HPr digest was desalted by SPE using conventional C_{18} ZipTips (Millipore), according to the manufacturer's instructions, then combined with an equal volume of the same MALDI matrix solution, without performing IMAC.

MALDI-TOF MS Conditions

Phosphopeptides were analyzed by MALDI-TOF MS using a Voyager DE-STR instrument (Perseptive Biosystems, Framingham, MA) operating in the linear, reflectron, or post-source decay (PSD) modes with positive or negative ionization (Napper *et al.*, 2003). Reflectron TOF provides the high mass resolution necessary for exact mass measurements, whereas linear TOF provides greater sensitivity at the expense of resolution. PSD is a dissociation process that provides peptide fragment information (albeit at low mass resolution) from which sites of phosphorylation may be inferred.

Full scan spectra were obtained by combining and processing 100 scans, a process that takes less than a minute. For PSD analysis, the reflectron mirror ratio was adjusted incrementally, the resulting spectra combined, and fragment ion peaks assigned using the instrument software (a procedure can be performed automatically, and more rapidly, on newer MALDI-TOF instruments).

Enrichment of His-Phosphorylated Peptides

Positive-ion reflectron MALDI-TOF MS analysis of the V8 digest of phosphorylated HPr generated a spectrum containing most of the expected peptides, as shown in Fig. 5A. The His-containing peptide (6–25) was

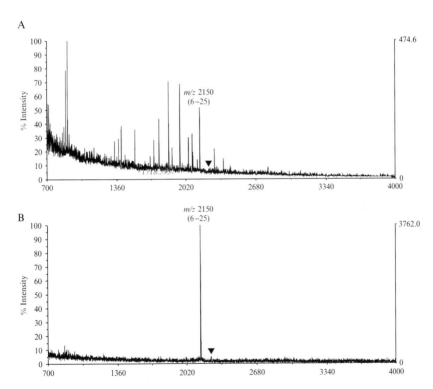

FIG. 5. Positive-ion reflectron MALDI-TOF MS spectra of (A) a complete V8 digest of the His-phosphorylated protein HPr, and (B) a Cu(II)-IMAC extract of the same digest. The His-containing peptide (6–25) is selectively recovered, but detected almost exclusively in nonphosphorylated form (*m/z* 2150). The expected position of the phosphorylated peptide (*m/z* 2230) is indicated by ▼. (See color insert.)

detected in unphosphorylated form (m/z 2150); however, the phosphory-lated form (m/z 2230) could not be observed, indicating that histidine phosphorylation is incomplete, labile, and/or inhibits ionization under experimental conditions. Analysis of the Cu(II)-IMAC extract of the phos-phorylated HPr digest contained a single intense peak at m/z 2150, as shown in Fig. 5B. Again, this corresponds to the predicted mass of the His-phosphorylated peptide without the phosphate group (HPO_3). When the same protocol was applied to a V8 digest of unphosphorylated HPr, no peptide ions were detected, even when operating in the more sensitive linear mode. This suggests that Cu(II)-IMAC extraction is selective for the phosphorylated form of the His-containing peptide, but that the phosphate group is lost before or during MALDI-TOF MS analysis.

Selectivity for Phosphorylated Histidine

Of the metal ions investigated, only Cu^{2+} proved effective in retain-ing and recovering peptides that contain phosphohistidine residues. To verify that peptides containing unphosphorylated histidine residues are not recovered using this procedure, a mixture containing four purified proteins (Enzyme I, Enzyme IIA^{glc}, and HPr from *E. coli*, and CheY from *Salmonella typhimurium*), each at a concentration of 1 mg/ml, were digested to completion with V8 protease and subjected to the Cu(II)-IMAC extrac-tion protocol. This mixture contains 11 unique peptides bearing His resi-dues; however, none of these was recovered using the Cu(II)-IMAC procedure.

Divalent transition metal ions are known to interact strongly with unmodified His residues; for example, Ni^{2+}–NTA is used routinely for purifying recombinant proteins in which a terminal poly(histidine) tag has been incorporated. Moreover, Cu^{2+}–IDA has a higher affinity for His-containing proteins than does Ni^{2+}–IDA (Arnold, 1991). Why, then, does Cu(II)-IMAC not recover peptides containing unmodified histidine using the protocol described above? To answer this question, we need to consider the acid/base properties of histidine and the pH of the sample during different steps in the IMAC process. The extraction of His-tagged proteins by IMAC is based on the interaction between divalent metal ions and the imidazole group of histidine. Under basic conditions, the uncharged imid-azole ring forms a complex with the immobilized metal ion. Elution of His-tagged proteins and peptides is then achieved by lowering pH and proto-nating the imidazole nitrogen (pK_a 6.0), which generates a positive charge that is repelled by the immobilized metal cation. During the aforemen-tioned Cu(II)-IMAC procedure, the binding of unphosphorylated histidines is prevented by loading the sample under the acidic conditions that result from mixing of the weakly buffered sample with residual acetic acid on the IMAC column. This reduces the sample pH from an initial value of 7.0 (20 mM HEPES) to around 3.5 (0.1 % acetic acid), at which point protonation of the imidazole ring negates interaction with the immobilized Cu^{2+} ions. Subsequent washing with 0.1% acetic in 50% acetonitrile serves to maintain the imidazole in protonated form, although this may also hydrolyze some of the intact phosphopeptides (see following text).

Detection of His-Phosphorylated Peptides

Previous studies have shown that switching from the positive to the negative ionization mode increases MS detection sensitivity for phosphorylated peptides relative to their unphosphorylated counterparts (Janek

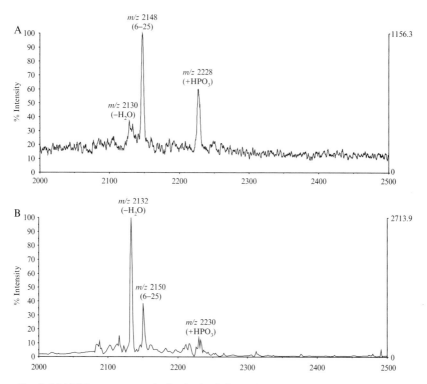

FIG. 6. MALDI mass spectra obtained using (A) negative-ion linear TOF MS analysis of the Cu(II)-IMAC extract from a V8 proteolytic digest of phosphorylated HPr, and (B) post-source decay (PSD) of the His-phosphorylated HPr peptide (m/z 2230) generated in positive-ion mode, both of which confirm recovery of the intact phosphopeptide. (See color insert.)

et al., 2001). As expected, negative-ion MALDI-TOF MS analysis of the Cu(II)-IMAC extract detected an ion of m/z 2228, corresponding to the deprotonated $[M - H]^-$ molecular ion of the intact HPr phosphopeptide, as shown in Fig. 6A. However, the unphosphorylated form of this peptide (m/z 2148) was also observed in the spectrum. This further suggests that the intact His-phosphorylated peptide is recovered using the Cu(II)-IMAC procedure but is susceptible to dephosphorylation during IMAC and/or MS analysis.

The same extract was subsequently analyzed in positive-ion mode using PSD with the ion gate set at m/z 2230, the expected value for the intact phosphopeptide. The resulting spectrum, shown in Fig. 6B, contains ions that correspond to sequential loss of HPO_3 (80 Da) and H_2O (18 Da) from the His-phosphorylated peptide, which is also observed in the spectrum. Similar results have been reported for collision-induced dissociation (CID) of His-phosphorylated peptide ions generated by MALDI (Janek *et al.*, 2001), confirming that the intact phosphopeptide was recovered by Cu(II)-IMAC. Figure 7 shows the different structures of phosphoester and phosphohistidine residues that give rise to neutral losses of 98 and 80 Da, respectively.

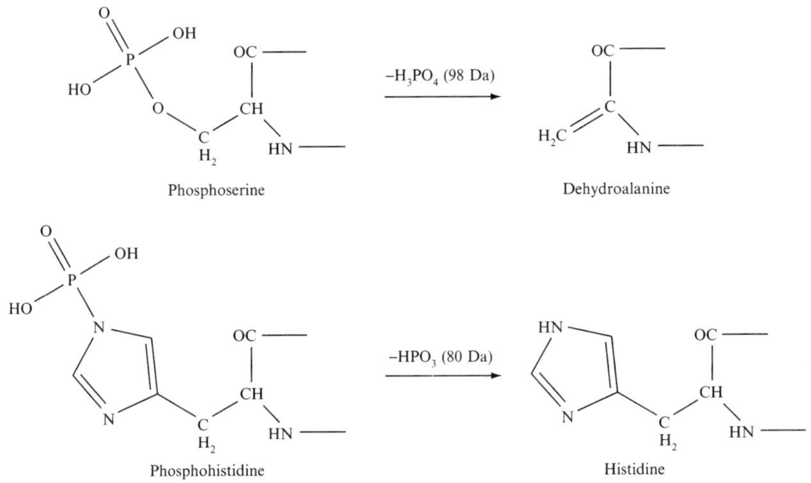

Fig. 7. Characteristic loss of H_3PO_4 and HPO_3 from, respectively, phosphoserine and phosphohistidine residues. The former (which also applies to phosphothreonine) can occur during tandem mass spectrometry, the latter during sample preparation and MS analysis. Neutral loss of HPO_3 can also occur during post-source decay of peptides containing phosphoserine and phosphothreonine residues.

Specificity for Phosphohistidines

To determine whether or not the protocol was specific for phosphopeptides containing phosphorylated His residues, the Cu(II)-IMAC procedure was applied to a tryptic digest of bovine β-casein. This protein yields two tryptic peptides that are phosphorylated at one and four Ser residues, respectively, and which produce ions of m/z 2062 and 3112 when analyzed by positive ion MALDI-TOF MS (Chalmers *et al.*, 2000; Posewitz and Tempst, 1999). Neither of these peptides was observed in Cu(II)-IMAC extracts, possibly because the strength of the interaction between phospho-serine and the immobilized Cu^{2+} ions prevented elution in 1% NH_4OH (Nühse *et al.*, 2003). Extraction of the β-casein digest using a Ga(III)-IMAC protocol optimized for the recovery of peptides containing phosphoester residues (Posewitz and Tempst, 1999) selectively recovered the singly phosphorylated peptide of m/z 2062, as confirmed by positive ion MALDI-TOF and PSD analysis. However, the Ga(III)-IMAC protocol did not recover the His-phosphorylated peptides from the phosphorylated HPr digest, which suggests that it may be possible to use Ga(III)-IMAC and Cu(II)-IMAC sequentially to extract peptides modified with phosphoester and phospho-histidine residues from mixed protein digests. Alternatively, stepwise elution of phosphohistidines and phosphoesters from Cu(II)-IMAC columns may be possible using basic solutions of increasing concentration, assuming that both types of phosphopeptides are indeed retained.

Differential Hydrolysis of Phosphohistidines

Histidine is unique among amino acids in that it can be phosphorylated at two different positions, namely, at atoms $N\Delta1$ and $N\varepsilon2$ of the imidazole ring. For free histidine, phosphorylation at the $N\Delta1$ position is much less stable than that at the $N\varepsilon2$ position due to the interaction of the $N\Delta1$ phosphate group with the positively charged amino group (Hultquist *et al.*, 1966). In peptides and proteins, the amino group of histidine is involved in peptide bonding, and the stability of the phosphate linkage may differ considerably from that of free histidine. Moreover, the local structural and electrostatic environment of the protein may have significant influence on the chemical properties of histidine and may vary in different protein contexts (Anderson *et al.*, 1993; Waygood *et al.*, 1985). For native HPr in *E. coli*, the rates of phosphohydrolysis for $N\Delta1$-phosphohistidine are three times greater than those for free $N\Delta1$-phosphohisitidine (Hultquist *et al.*, 1966), indicating that the protein serves to destabilize the linkage, perhaps to ensure efficient phosphotransfer. The pH-dependence of phosphohydrolysis for phosphorylated HPr is also quite distinctive, exhibiting a bell-shaped curve with greatest instability of the phosphoamidate bond between pH 5 and 8. The phosphopeptide generated by V8 digestion of phosphorylated HPr has phosphohydrolysis rates an order of magnitude lower than those of the intact protein, showing greatest stability at neutral to basic pH (Hultquist et al., 1966). This allows for the retention of phosphate during base elution of the chelating pipette tips.

Summary and Conclusions

Selective extraction of His-phosphorylated peptides from HPr protein digests can be achieved by immobilized Cu^{2+}-ion affinity chromatography using disposable metal-chelating pipette tips. On-tip acidification of the sample by residual acetic acid apparently inhibits binding of unphosphory-lated histidine residues during Cu(II)-IMAC, while recovery of the intact phosphopeptide via base elution is confirmed using negative-ion MALDI-TOF mass spectrometry. Subsequent PSD analysis of the protonated molecular ion shows characteristic loss of the HPO_3 moiety present in His-phosphorylated peptides. Possible refinements to this method include the use of "cooler" and/or less acidic matrix compounds, such as 2,4,6-trihydroxyacetophenone (THAP) combined with ammonium citrate (Wolschin et al., 2005) or 2,5-dihydroxybenzoic acid (DHB) combined with phosphoric acid (Kjellstrom and Jensen, 2004), to enhance detection and structural analysis of His-phosphorylated peptides by MALDI-TOF MS, or by MALDI-MS/MS using Q-TOF or TOF-TOF instrumentation.

Although disposable pipette tips are ideal for rapid IMAC, the characteristics of the column packing tend to vary from one tip to another. As a result, there is sometimes a lag in solvent uptake when aspirating solutions, which may result in air being accidentally drawn into the tip. If solutions are aspirated and dispensed with care, however, the packing material will remain properly conditioned, and consistent results should be obtained. An alternative approach is to use open tubular (OT) columns, in which the inner surface of a glass tube is uniformly modified by chemical attachment of a metal chelating group (Liu et al., 2004). OT columns can be used in the same way as disposable pipette tips, the high density and uniformity of the OT-IMAC medium providing enhanced specificity and reproducibility for phosphopeptide enrichment.

While preliminary results are encouraging, there is still much work to be done before routine, proteomewide identification of His-phosphorylated proteins can be achieved. Although fractionation of complex samples would certainly be advantageous, each additional step in the analytical process increases the chance of degrading phosphohistidine residues, unless mild and (preferably) basic conditions can be maintained throughout. This precludes techniques such as SCX, SPE, and derivatization procedures for enhancing recovery, detection, and/or relative quantification of phosphopeptides. However, differential centrifugation, two-phase partitioning, and/or MOAC may prove suitable for fractionation and co-enrichment of His-phosphorylated and other phosphorylated proteins. Gel electrophoresis combined with phosphate-specific fluorescent staining, or Western blotting for His- and Tyr-phosphorylated proteins, could then be used to resolve and detect phosphoproteins in the enriched fractions. Subsequent identification of His-phosphorylated proteins, and their component phosphohistidines, by mass spectrometry would be facilitated by Cu(II)-IMAC enrichment of His-phosphorylated peptides from in-gel digests. However, this approach requires the development of rapid in-gel digestion and peptide extraction procedures that minimize hydrolysis of phosphoamidate bonds. Alternatively, "top-down" analysis by ECD and FT-ICR MS may prove effective in identifying and mapping labile phosphohistidine residues in His-phosphorylated proteins, provided that individual proteins can be isolated or recovered intact from polyacrylamide gels (by electroelution, for example). The latter would also enable the use of rapid microcolumn digestion procedures (Slysz and Schriemer, 2005) for "bottom-up" analysis of His-phosphorylated proteins.

Until now, the identification of novel histidine kinases has been performed largely on the basis of sequence similarities among members of this enzyme class. Mild protein fractionation, gel separation, and affinity purification procedures, combined with high-resolution MS and data-dependent MS^n techniques, offer an alternative approach for identifying these and other His-phosphorylated proteins through direct analysis of histidine phosphorylations in proteolytic peptides and/or their parent proteins. Such techniques have the potential to increase significantly our understanding of the role played by histidine phosphorylation in two-component signaling systems. In conclusion, I should like to thank Dr. Scott Napper at the Vaccine and Infectious Disease Organization (VIDO) and Drs. Lianglu Wan and Uma K. Aryal at the NRC Plant Biotechnology Institute in Saskatoon for their assistance in preparing this article, which is contribution 48421 from the National Research Council of Canada.

References

Aebersold, R., and Mann, M. (2003). Mass spectrometry-based proteomics. *Nature* **422,** 198–207.

Anderson, B., Weigel, N., Kundig, W., and Roseman, S. (1971). Sugar transport. III Purification and properties of a phosphocarrier protein (HPr) of the phosphoenolpyruvate-dependent phosphotransferase system of *Escherichia coli. J. Biol. Chem.* **246,** 7023–7033.

Anderson, J. W., Bhanot, P., Georges, F., Klevit, R. E., and Waygood, E. B. (1991). Involvement of the carboxy-terminal residue in the active site of the histidine-containing protein, HPr, of the phosphoenolpyruvate:sugar phosphotransferase system of *Escherichia coli*. *Biochemistry* **30**, 9601–9607.

Anderson, J. W., Pullen, K., Georges, F., Klevit, R. E., and Waygood, E. B. (1993). The involvement of the arginine 17 residue in the active site of the histidine-containing protein, HPr, of the phosphoenolpyruvate:sugar phosphotransferase system of *Escherichia coli*. *J. Biol. Chem.* **268**, 12325–12333.

Andersson, L., and Porath, J. (1986). Isolation of phosphoproteins by immobilized metal (Fe^{3+}) affinity chromatography. *Anal. Biochem.* **154**, 250–254.

Arnold, F. H. (1991). Metal-affinity separations: A new dimension in protein processing. *Biotechnology* **9**, 151–156.

Bannai, H., Tamada, Y., Maruyama, O., Nakai, K., and Miyano, S. (2002). Extensive feature detection of N-terminal protein sorting signals. *Bioinformatics* **18**, 298–305.

Beausoleil, S. A., Jedrychowski, M., Schwartz, D., Elias, J. E., Villén, J., Li, J., Cohn, M. A., Cantley, L. C., and Gygi, S. P. (2004). Large-scale characterization of HeLa cell nuclear phosphoproteins. *Proc. Natl. Acad. Sci. USA* **101**, 12130–12135.

Brokx, S. J., Talbot, J., Goerges, F., and Waygood, E. B. (2000). Enzyme I of the phosphoenolpyruvate:sugar phosphotransferase system. *In vitro* intragenic complementation: The roles of Arg126 in phosphoryl transfer and the C-terminal domain in dimerization. *Biochemistry* **39**, 3624–3625.

Chalmers, M., Ross, A. R. S., Olson, D., and Gaskell, S. J. (2000). Comparing Ga(III)-IMAC methods for selective extraction and MALDI-TOF MS characterization of phosphopeptides. *Proceedings of the 48th ASMS Conference*, Long Beach, CA, June 12–15.

Cohen, P. (2000). The regulation of protein function by multisite phosphorylation—A 25-year update. *Trends Biochem. Sci.* **25**, 596–601.

Conrads, T. P., Issaq, H. J., and Veenstra, T. D. (2002). New tools for quantitative phosphoproteome analysis. *Biochem. Biophys. Res. Commun.* **290**, 885–890.

Cooper, H. J., Akbarzadeh, S., Health, J. K., and Zeller, M. (2005). Data-dependent electron capture dissociation FT-ICR mass spectrometry for proteomic analyses. *J. Proteome Res.* **4**, 1538–1544.

de la Fuente van Bentem, S., Roitinger, E., Anrather, D., Csaszar, E., and Hirt, H. (2006). Phosphoproteomics as a tool to unravel plant regulatory mechanisms. *Physiol. Plantarum* **126**, 110–119.

Emanuelsson, O., Nielsen, H., Brunak, S., and von Heijne, G. (2000). Predicting subcellular localization of proteins based on their N-terminal amino acid sequences. *J. Mol. Biol.* **300**, 1005–1016.

Emmett, M. R. (2003). Determination of post-translational modifications of proteins by high-sensitivity, high-resolution Fourier transform ion cyclotron resonance mass spectrometry. *J. Chromatogr. A.* **1013**, 203–213.

Ficarro, S. B., McCleland, M. L., Stukenberg, P. T., Burke, D. J., Ross, M. M., Shabanowitz, J., Hunt, D. F., and White, F. M. (2002). Phosphoproteome analysis by mass spectrometry and its application to *Saccharomyces cerevisiae*. *Nat. Biotechnol.* **20**, 301–305.

Grønborg, M., Kristianson, T. Z., Stensballe, A., Andersen, J. S., Ohara, O., Mann, M., Jensen, O. N., and Pandey, A. (2002). A mass spectrometry-based proteomic approach for identification of serine/threonine-phosphorylated proteins by enrichment with phospho-specific antibodies: Identification of a novel protein, Frigg, as a protein kinase A substrate. *Mol. Cell. Proteomics* **1**, 517–527.

Gruhler, A., Olsen, J. V., Mohammed, S., Mortensen, P., Faergeman, N. J., Mann, M., and Jensen, O. N. (2005). Quantitative phosphoproteomics applied to the yeast pheromone signaling pathway. *Mol. Cell. Proteomics* **4,** 310–327.

Hanson, B. J., Schulenberg, B., Patton, W. F., and Capaldi, R. A. (2001). A novel subfractionation approach for mitochondrial proteins: A three-dimensional mitochondrial proteome map. *Electrophoresis* **22,** 950–959.

Hess, J. F., Bourret, R. B., and Simon, M. I. (1988). Histidine phosphorylation and phosphoryl group transfer in bacterial chemotaxis. *Nature* **336,** 139–143.

Hodges, T. K., Leonard, R. T., Bracker, C. E., and Kennan, T. W. (1972). Purification of an ion-stimulated adenosine triphosphatase from plant roots: Association with plasma membranes. *Proc. Natl. Acad. Sci. USA* **69,** 3307–3311.

Hultquist, D., Moyer, R. W., and Boyer, P. D. (1966). The preparation and characterization of 1-phosphohistidine and 3-phosphohistidine. *Biochemistry* **5,** 322–331.

Janek, K., Wenschuh, H., Bienert, M., and Krause, E. (2001). Phosphopeptide analysis by positive and negative ion matrix-assisted laser desorption/ionization mass spectrometry. *Rapid Commun. Mass Spectrom.* **15,** 1593–1599.

Kjellstrom, S., and Jensen, O. N. (2004). Phosphoric acid as a matrix additive for MALDI MS analysis of phosphopeptides and phosphoproteins. *Anal. Chem.* **76,** 5109–5117.

Klumpp, S., and Krieglstein, J. (2002). Phosphorylation and dephosphorylation of histidine residues in proteins. *Eur. J. Biochem.* **269,** 1067–1071.

Larsson, C., Widell, S., and Kjellbom, P. (1987). Preparation of high purity plasma membranes. *Meth. Enzymol.* **148,** 558–568.

Laugesen, S., Messinese, E., Hem, S., Pichereaux, C., Grat, S., Ranjeva, R., Rossignol, M., and Bono, J. J. (2006). Phosphoprotein analysis in plants: A proteomic approach. *Phytochemistry* **67,** 2208–2214.

Liu, H., Stupak, J., Zheng, J., Keller, B. O., Brix, B. J., Fliegel, L., and Li, L. (2004). Open tubular immobilized metal ion affinity chromatography combined with MALDI MS and MS/MS for identification of protein phosphorylation sites. *Anal. Chem.* **76,** 4223–4232.

Matthews, H. R. (1995). Protein kinases and phosphatases that act on histidine, lysine, or arginine residues in eukaryotic proteins: A possible regulator of the mitogen-activated protein kinase cascade. *Pharmac. Ther.* **67,** 323–350.

Meadow, N. D., Fox, D. K., and Roseman, S. (1990). The bacterial phosphoenolpyruvate: glycose phosphotransferase system. *Annu. Rev. Biochem.* **59,** 497–542.

Napper, S., Brokx, S. J., Pally, E., Kindrachuk, J., Delbaere, L. T. J., and Waygood, E. B. (2001). Substitution of aspartate and glutamate for active center histidines in the *Escherichia coli* phosphoenolpyruvate:sugar phosphotransferase system maintain phosphotransfer potential. *J. Biol. Chem.* **276,** 41588–41593.

Napper, S., Kindrachuk, J., Olson, D. J. H., Ambrose, S. J., Dereniwsky, C., and Ross, A. R. S. (2003). Selective extraction and characterization of a histidine-phosphorylated peptide using immobilized copper(II)-ion affinity chromatograph and matrix-assisted laser desorption/ionization-time of flight mass spectrometry. *Anal. Chem.* **75,** 1741–1747.

Nühse, T. S., Stensballe, A., Jensen, O. N., and Peck, S. C. (2003). Large-scale analysis of *in vivo* phosphorylated membrane proteins by immobilized metal ion affinity chromatography and mass spectrometry. *Mol. Cell. Proteomics* **2,** 1234–1243.

Parkinson, J. S., and Kofoid, E. C. (1992). Communication modules in bacterial signaling proteins. *Annu. Rev. Genet.* **26,** 71–112.

Patel, H. R., and Gelfand, E. W. (1996). DNA-binding phosphoproteins induced after T cell activation: Effects of cyclosporin A. *Cell. Signalling* **8,** 253–261.

Persson, A., and Jergil, B. (1992). Purification of plasma membranes by aqueous two-phase affinity partitioning. *Anal. Biochem.* **204,** 131–136.

Posewitz, M. C., and Tempst, P. (1999). Immobilized gallium(III) affinity chromatography of phosphopeptides. *Anal. Chem.* **71,** 2883–2892.

Postma, P. W., Lengeler, J. W., and Jacobson, G. R. (1993). Phosphoenolpyruvate:carbohydrate phosphotransferase systems of bacteria. *Microbiol. Rev.* **57,** 543–594.

Ross, A. R. S., Lee, P. J., Smith, D. L., Langridge, J. I., Whetton, A. D., and Gaskell, S. J. (2002). Identification of proteins from two-dimensional polyacrylamide gels using a novel acid-labile surfactant. *Proteomics* **2,** 928–936.

Rush, J., Moritz, A., Lee, K. A., Guo, A., Goss, V. L., Spek, E. J., Zhang, H., Zha, X. M., Polakiewicz, R. D., and Comb, M. J. (2005). Immunoaffinity profiling of tyrosine phosphorylation in cancer cells. *Nat. Biotechnol.* **23,** 94–101.

Schindler, J., Lewandrowski, U., Sickmann, A., Friauf, E., and Nothwang, H. G. (2006). Proteomic analysis of brain plasma membranes isolated by affinity two-phase partitioning. *Mol. Cell. Proteomics* **5,** 390–400.

Schweppe, R. E., Haydon, C. E., Lewis, T. S., Resing, K. A., and Ahn, N. G. (2003). The characterization of protein post-translational modifications by mass spectrometry. *Acc. Chem. Res.* **36,** 453–461.

Slysz, G. W., and Schriemer, D. C. (2005). Blending protein separation and peptide analysis through real-time proteolytic digestion. *Anal. Chem.* **77,** 1572–1579.

Steen, H., Fernandez, M., Ghaffari, S., Pandey, A., and Mann, M. (2003). Phosphotyrosine mapping in Bcr/Abl oncoprotein using phosphotyrosine-specific immonium ion scanning. *Mol. Cell. Proteomics* **2,** 138–145.

Stock, A., Koshland, D. E., Jr., and Stock, J. (1985). Homologies between the *Salmonella typhimurium* Che Y protein and proteins involved in the regulation of chemotaxis, membrane protein synthesis, and sporulation. *Proc. Natl. Acad. Sci. USA* **82,** 7989–7993.

Stock, J., Ninfa, A., and Stock, A. M. (1989). Protein phosphorylation and regulation of adaptive responses in bacteria. *Microbiol. Rev.* **53,** 450–490.

Titgemeyer, F. (1993). Signal transduction in chemotaxis mediated by the bacterial phosphotransferase system. *J. Cell. Biochem.* **51,** 69–74.

VanPutte, R. D., and Patterson, C. O. (2003). Microalgal plasma membranes purified by aqueous two-phase partitioning. *Transactions of the Illinois State Academy of Sciences* **96,** 71–86.

Vyetrogon, K., Tebbji, F., Olson, D. J. H., Ross, A. R. S., and Matton, D.P (2006). A comparative proteome and phosphoproteome analysis of differentially regulated proteins during fertilization in the self incompatible species *Solanum chacoense* Bitt. *Proteomics* **7,** 232–247.

Waygood, E. B., Erikson, E. E., El-Kabbani, O. A. L., and Delbaere, L. T. J. (1985). Characterization of phosphorylated histidine-containing protein (HPr) of the bacterial phosphoenolpyruvate/sugarphosphotransferase system. *Biochemistry* **24,** 6938–6945.

Wolschin, F., Wienkoop, S., and Weckwerth, W. (2005). Enrichment of phosphorylated proteins and peptides from complex mixtures using metal oxide/hydroxide affinity chromatography (MOAC). *Proteomics* **5,** 4389–4397.

Wolters, D. A., Washburn, M. P., and Yates, J. R., III (2001). An automated multidimensional protein identification technology for shotgun proteomics. *Anal. Chem.* **73,** 5683–5690.

Wuenschell, G. E., Naranjo, E., and Arnold, F. H. (1990). Aqueous two-phase metal affinity extraction of heme proteins. *Bioprocess Engineering* **5,** 199–202.

Yates, J. R., III (1998). Mass spectrometry and the age of the proteome. *J. Mass Spectrom.* **33,** 1–19.

Author Index

A

Abbott, R. J. M., 53, 59, 60, 91
Abe, H., 216
Abelnour, A., 503
Abian, J., 206, 210, 216
Abragam, A., 56
Achtman, M., 516, 523
Adams, P. D., 102, 118, 119, 121
Adams, S. R., 292
Adamson, J., 136
Adler, G. K., 504
Adler, J., 416
Adriano, J. M., 208
Aebersold, R., 556, 557
Aguera y Arcas, B., 389
Aguilar, P. S., 223
Ahmer, B. M. M., 136
Ahn, N. G., 556, 558
Aiba, H., 167, 185, 195
Aizawa, S., 357
Aizawa, S.-I., 393
Akabas, M. H., 26, 27
Akbarzadeh, S., 559
Akiyama, Y., 238
Akke, M., 150, 154
Aksenov, S. V., 376, 389
Alarcon-Chaidez, F., 224, 227, 241
Albanesi, D., 223, 243
Albert, R., 389
Alex, L. A., 93, 94, 99, 104
Alexeyev, M. F., 238
Allard, J. D., 136
Alm, R. A., 514, 523
Alon, U., 389
Alonso, J. M., 205
Altenbach, C. A., 53, 54, 59, 60, 78, 126
Altendorf, K., 540
Altschul, S. F., 484
Ambrose, S. J., 561
Amemura, M., 496
Ames, P., 6, 27, 169, 259, 281, 366, 414, 417, 418, 419, 424, 425, 432, 436, 439, 440, 451
Ames, S. K., 161, 162

Amin, D. N., 300
Amy, N. K., 149, 156
Anand, G. S., 393
Anantharamaiah, G. M., 319
Andersen, J. B., 462, 553
Anderson, B., 561
Anderson, D. J., 59, 60, 487
Anderson, J. W., 561, 562, 567
Anderson, P. W., 70, 77
Anderson, T. R., 224
Andersson, L., 553
Angert, E. R., 458
Annan, R. S., 503
Anrather, D., 550
Antommattei, F. M., 268, 271, 291
Antonic, J., 59, 61, 91
Apostol, M. I., 487
Appleby, J. L., 341, 403, 516
Appleman, J. A., 168, 486
Aravind, L., 168, 169, 486
Arif, M., 534, 544, 545, 546
Arkin, A. P., 389
Armitage, J. P., 366, 388, 392, 393, 395, 397, 399, 401, 403, 410, 436
Arnold, F. H., 553, 560, 564
Aronoff-Spencer, E., 53
Arsyad, D. M., 224
Arvidson, S., 503
Asensio, J. G., 486
Ashby, M. K., 394
Asinas, A. E., 267, 271, 273, 276, 278, 281, 282, 283, 286, 290, 293
Aslund, F., 301
Astashkin, A. V., 59
Atkins, B., 318
Atkuri, R. V., 503
Aufhammer, S., 223
Austin, R. H., 150

B

Baar, C., 516
Bachhawat, P., 149
Backer, J. M., 53

Subject Index

A

AI-2, *see* LuxPQ

Alanine mutagenesis, *see* TonB

Antibiotic resistance, *Staphylococcus aureus*, 502–503

ArlRS
  gene regulation response in mutants
    DNA microarray studies
      arlR allelic replacement mutant construction, 504–505
      complementary DNA synthesis, fragmentation, and labeling, 506–508
      hybridization and scanning, 509
      overview, 504
      RNA purification, 505–506
      quantitative real-time polymerase chain reaction analysis, 510–512
    Staphylococcus aureus virulence factor regulation, 503

B

Bioluminescence resonance energy transfer, CheY–CheZ interactions in swimming bacteria
  advantages and limitations, 387–389
  apparatus and data acquisition, 385–386
  cell preparation, 386
  dose–response curves, 387
  overview, 382–385
  stability of luminescence intensity and spectrum, 386–387

BRET, *see* Bioluminescence resonance energy transfer

Bucket of light, *see* Soft agar assays

C

Carr-Purcell-Meiboom-Gill experiments, protein dynamics

relaxation dispersion for millisecond dynamics, 156–157

relaxation dispersion with exchange-free relaxation for microsecond dynamics, 157–161

CheA
  dephosphorylation, 340–341
  function, 366, 392–393, 414, 437, 439
  Helicobacter pylori chemotaxis, 526–527
  protein-interactions-by-cysteine modification, 6–7, 21–22
  pulsed dipolar electron spin resonance of CheA–CheW distances
    data analysis, 97, 99–100
    heterodimer studies, 104–105
    interdomain and interprotein distances, 102–104
    metric matrix distance geometry and rigid-body refinement, 100–102
    objectives of study, 90–91
    rigid body refinement with Crystallography and NMR System software
      error allocation scheme, 122–123
      overview, 117–120
      restraint addition, 125–126
      restraint type and number, 123–125
      spin label sites and effects, 121, 126–128
      weighting scheme for contact parameters, 123
    sample preparation, 96–97
    sensitivity, 83–84
    signaling complex in bacteria chemotaxis, 87–89
    spin-labeling, 95–96
    technical aspects, 97
    triangulation, 91–96
  receptor cytoplasmic domain fragment/ CheW/CheA assembly in liposomes
    assembly conditions, 289
    chemotaxis protein purification, 287
    cytoplasmic domain fragment variables binary mixtures, 281–283

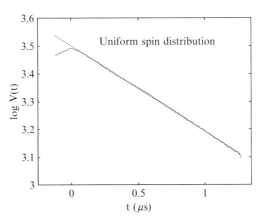

BORBAT AND FREED, CHAPTER 3, FIG. 8. An example of 4-pulse DEER (solid line) from a uniform distribution of spins in an isotropic sample illustrates the intermolecular signal given by Eq. (12). Dashed line is a fit to the straight line in the logarithmic plot. To achieve the uniform spin distribution, 0.01 mole percent of spin-labeled alamethicin was magnetically diluted with WT by a factor of 20 to avoid effects of its aggregation (unpublished, this lab).

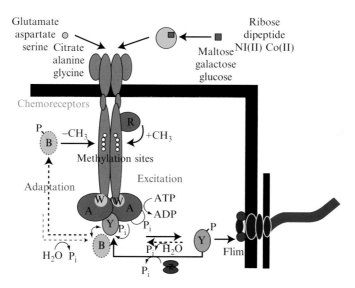

Borbat and Freed, Chapter 3, Fig. 17. The signaling bacterial receptor responds to chemical stimulants by activating histidine kinase CheA and, consequently, invoking a phosphorylation cascade that involves several other proteins. This ultimately affects the sense of rotation of the flagella motors and thus the swimming behavior. Receptors are coupled to the CheA dimers via the coupling protein CheW. A small conformational change, caused by stimulant binding at the periplasmic side of a $\sim$300 Å long transmembrane receptor, is relayed over $\sim$250 Å distance to the kinase at its distal cytoplasmic tip. The regulation is provided by enzymes, which (de)methylate or deamidate glutamate residues on the cytoplasmic part of the receptors. This may affect kinase activity via conformational changes of the signaling complex, cooperativity between receptors, or both, thereby changing the catalytic activity of the entire receptor array by orders of magnitude. (Adapted from Bilwes et al., 2003.)

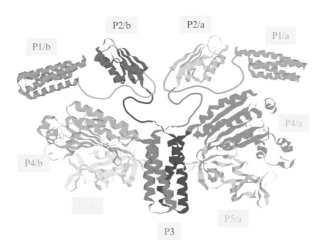

BORBAT AND FREED, CHAPTER 3, FIG. 18. A collage of the histidine kinase CheA. The whole kinase assembles as a dimer with the monomer composed of 5 domains P1–P5, connected by flexible links of various lengths. CheAΔ289 (lacking P1 and P2 domains) from *Thermotoga maritima* in the center was crystallized; P1 and P2 crystallized separately (PDB codes: 1TQG, 1U0S) were added to the figure to provide a complete view of the protein. Two P3 dimerization domains assemble into a 4-helix bundle; P4 is the catalytic domain, which phosphorylates conserved histidine of the P1 domain; P2 binds CheY or CheB, which, in turn, are phosphorylated by P1. Phosphorylated CheY(B) dissociates and controls flagella and receptor, respectively. P5 is a regulatory domain, which binds CheW and receptor and thus mediates regulation of kinase phosphotransfer activity. (Protein structure was rendered using Chine.)

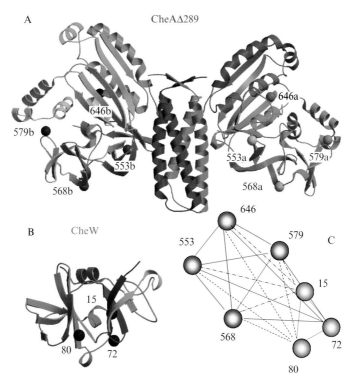

BORBAT AND FREED, CHAPTER 3, FIG. 20. Mutation sites selected for (A) CheAΔ289 (E646, S568, N553, D579) and (B) CheW (S15, S72, S80). Most of the sites are separated by more than 60Å, thus minimizing problems associated with these multi-spin cases. Additional sites were mutated at a later stage of the study in order to assess the global protein structure after two different conformations of CheAΔ354/CheW were provided by crystallography to confirm ESR-derived structure. (C) A suggested network of restraints for triangulation. Note that the putative binding site on the distal part of P5 domain detected by X-ray crystallography (Bilwes *et al.*, 1999) was under consideration here. (The structures were rendered using Mol Script.)

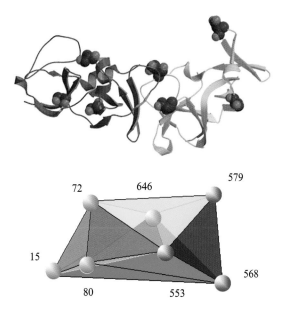

BORBAT AND FREED, CHAPTER 3, FIG. 22. (top) X-ray structure of P4/P5/CheW complex (P4 is not shown). Residues mutated to nitroxides for PDS are shown in a space-fill representation. (bottom) - Positions of nitroxide moieties found from PDS constraints correspond to that in the top figure. CheW appears to be slightly tilted and rotated about its long axis compared to the X-ray structure. The difference between X-ray and PDS is not large but cannot be discounted. Several reasons can be given: (1) the mutated residues should be replaced by nitroxide side-chains in their site-specific conformations to arrive at a better correspondence of PDS and X-ray derived structures; (2) Couplings between symmetrically positioned residues biased the long-distance constraints; (3) P4/P5/CheW was crystallized in the absence of P3, with which both CheW and P5 are expected to interact (Park *et al.*, 2006); (4) mutations at N553, and especially E646, might have had a small effect on the binding interface. Mutated sites were selected to optimize constraints in the originally presumed case of binding to the distal binding site of P5 (cf. Fig. 18), which turned out not to be the case. Thus, some chosen distances were outside the optimal range. (The ribbon structure was rendered with Mol Script; Delauney triangulation generated and rendered by MATLAB.)

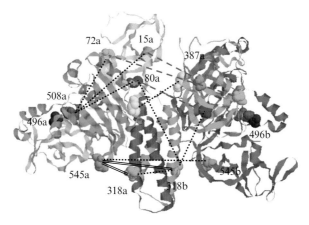

BORBAT AND FREED, CHAPTER 3, FIG. 23. Distances between several CheW mutants were measured to establish its position on the "top" of CheAΔ289. Heterodimers of P3 (S318C) and P5 (Q545C) mutants were used to establish the fixed position of the P5 domain, stabilized by interaction with P3. (Rendered with Chime.)

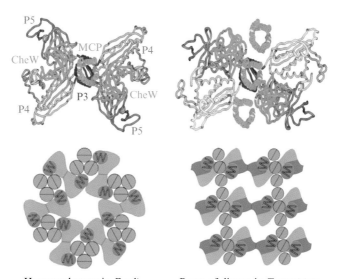

Hexagonal array in *E.coli*   Rows of dimers in *T. maritima*

BORBAT AND FREED, CHAPTER 3, FIG. 24. Receptor–kinase interactions. In *T. maritima*, the methyl-accepting domain of the receptor forms dimers which are 225 Å long, with a diameter of ∼20Å (Park *et al.*, 2006). The canyon formed by two CheW molecules sitting on top of a CheAΔ289 dimer is wide enough to accommodate one receptor dimer. Two additional receptor dimers can be positioned on either side of the center receptor giving a stoichiometry of three receptor dimers to one CheA dimer and two CheWs. The bottom figures compare two possible type of receptor arrays. (Adapted from Weis, 2006.)

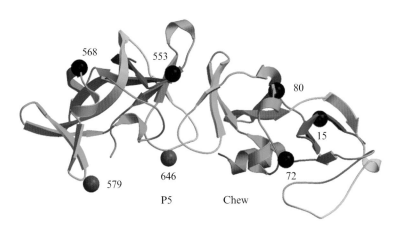

BHATNAGAR ET AL., CHAPTER 4, FIG. 1. Crystal structure of CheW-P5 complex showing positions of spin label sites (balls) along the polypeptide. Both proteins shown as ribbon representations colored blue to red from N to C terminus. Sites producing the most aberrant ESR restraints compared to the crystal structure shown in red.

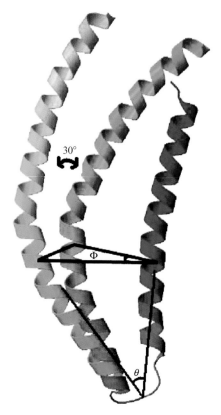

BHATNAGAR *ET AL.*, CHAPTER 4, FIG. 6. A comparison of helix orientations in α-synuclein from ESR-refinement and NMR. Orientation of two anti-parallel α-synuclein helixes (residues 3–34 and 44–94) as derived from NMR is shown in blue. Superposition of N terminal of helix from the rigid body refined structure, places the second helix rotated by angle of $30°$ with respect to the NMR structure.

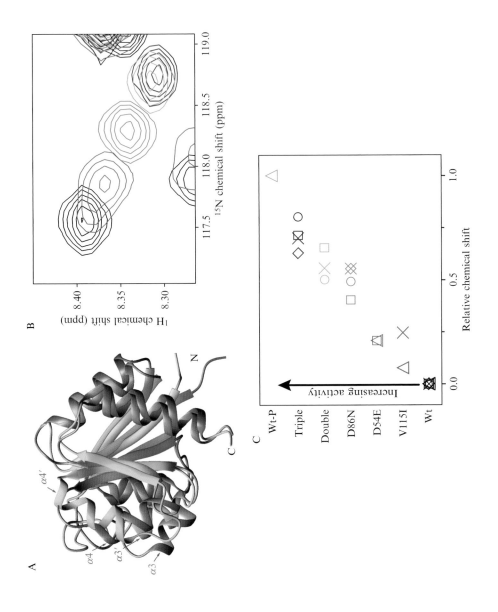

A

α4'
α4
α3'
α3
N
C

B

^1H chemical shift (ppm)

8.40
8.35
8.30

117.5   118.0   118.5   119.0

^{15}N chemical shift (ppm)

C

Wt-P
Triple
Double
D86N
D54E
V115I
Wt

Increasing activity

0.0   0.5   1.0

Relative chemical shift

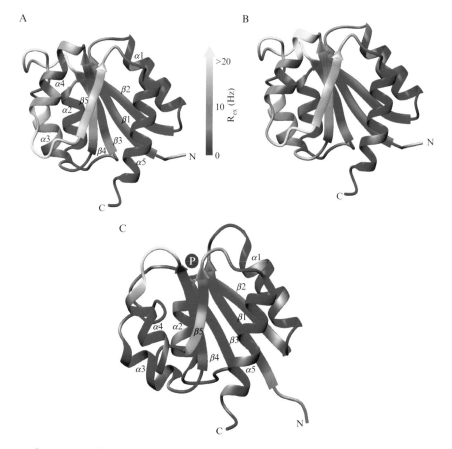

GARDINO AND KERN, CHAPTER 6, FIG. 2. Conformational exchange detected by standard transverse relaxation measurements. R_{ex} contributions are shown for (A) NtrC^r, (B) NtrC^r (D86N/A89T), and (C) P-NtrC^r as a continuous color scale. Reproduced with permission from Volkman *et al.*, 2001.

GARDINO AND KERN, CHAPTER 6, FIG. 1. Two-state model implied by NMR structure and chemical shift analysis. (A) The NMR structures of NtrC^r (blue) and P-NtrC^r (orange) are superimposed. The switch region is highlighted in lighter colors. (B, C) Chemical shift changes for NtrC^r forms with increasing activities: V115I (grey), D54E (red), D86N (green), D86N/ A89T (gold), D86N/AS89T/V115I (blue), P-NtrC^r (cyan) with respect to wild-type (black). Changes in peak position are shown for the backbone amide of D88 (B). The same pattern is observed for a number of residues as seen from the normalized chemical shifts (C), therefore reflecting the equilibrium between the two states. Reproduced with permission from Volkman *et al.*, 2001.

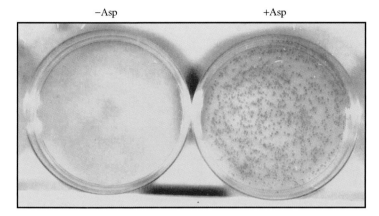

Yoshida *et al.*, Chapter 7, Fig. 3. Asp-dependent induction of *ompC-lacZ* is detected on X-gal plate. RU1012 cells harboring pIN-III-Taz construct are plated on X-gal agar plates with or without 5 m*M* Asp. Blue-colored colonies represent the cells expressing β-galactosidase.

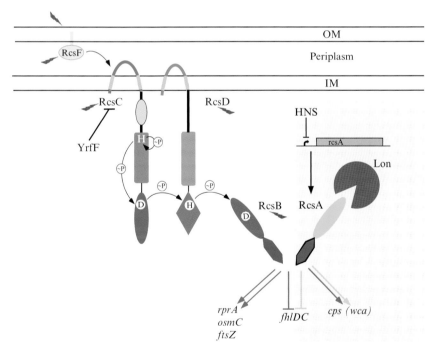

Majdalani and Gottesman, Chapter 16, Fig. 1. The Rcs phosphorelay. The core protein components of the phosphorelay include the RcsC sensor kinase, the RcsD histidine phospho-transfer protein, and the RcsB response regulator. Some targets of RcsB also require the accessory factor, RcsA. RcsA is subject to Lon-dependent proteolysis and its synthesis is regulated, in part by Hns. Upstream signaling to RcsC is sometimes, but not always, via the outer membrane lipoprotein RcsF. The jagged line indicates possible signaling inputs. See text and Majdalani and Gottesman, 2005, for further details.

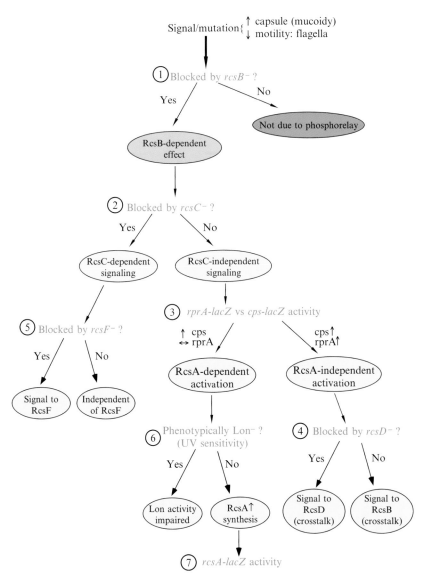

MAJDALANI AND GOTTESMAN, CHAPTER 16, FIG. 2. Decision tree for analyzing Rcs phosphorelay. The flow chart is keyed to the Steps for testing various models, as outlined in the text. Conclusions based on results are shown in blue ovals. The questions answered at each step are: Step 1: Is it dependent upon the Rcs phosphorelay? Step 2: Is it dependent upon RcsC activation? Step 3: Is there increased RcsA? Step 4: Is signaling directly to RcsD? Step 5: Is signaling RcsF dependent? Step 6: Is Lon activity impaired? Step 7: Is RcsA synthesis increased?

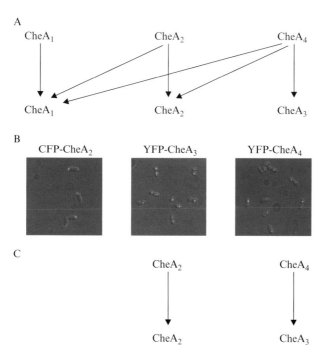

PORTER *ET AL.*, CHAPTER 18, FIG. 1. Reinterpretation of *in vitro* CheA phosphorylation reactions in *R. sphaeroides* after *in vivo* imaging. (A) Phosphorylation reactions that have been detected *in vitro* using purified proteins. (B) The localization of CheA$_2$, CheA$_3$, and CheA$_4$. CheA$_1$ does not appear to be expressed, so its localization could not be determined. (C) The auto-/heterophosphorylation reactions that are likely to occur *in vivo*. These reactions are the same as those in panel (A) except that all reactions involving CheA$_1$ have been removed because CheA$_1$ is not expressed and is not required for chemotaxis; also, the phosphorylation of CheA$_2$ by CheA$_4$ has been removed because these proteins localize to different regions of the cell.

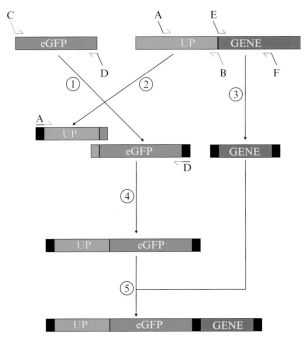

PORTER *ET AL.*, CHAPTER 18, FIG. 4. Diagram showing the production of a construct for introducing the gene encoding an N-terminal eGFP fusion into the *R. sphaeroides* genome. (1) Primers C and D are used to amplify the e*gfp* gene. (2) Primers A and B are used to amplify the ~500 bp region immediately upstream of the gene to be tagged. (3) Primers E and F are used to amplify the first ~500 bp of the gene to be tagged. (4) Primers B and C are exactly complementary to one another, facilitating an overlap extension reaction between the products of reactions 1 and 2. The overlap extension product is amplified using primers A and D. (5) The products of PCRs 3 and 4 are ligated into the suicide vector pK18*mobsacB* to generate the tagging construct. Restriction sites within primers are shown in black; the restriction sites used in primers D and E must be compatible.

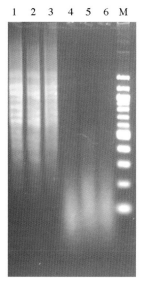

Ji *et al.*, Chapter 24, Fig. 1. Two percent agarose gel analysis of pre- and post-fragmented cDNA. Lane 1–3, total cDNA; Lane 4–6, fragmented cDNA; M:100 bp DNA marker.

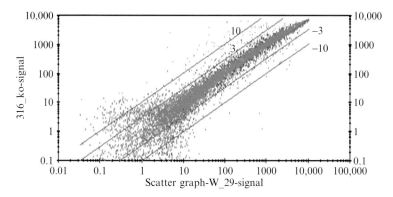

Ji *et al.*, Chapter 24, Fig. 2. Scatter plot of signal intensities from two GeneChips for *arlR* mutant and wild type strains. X axis, wild type strain; Y axis, *arlR* mutant. −3 and −10, fold of decrease after the mutation of arlR; 3 and 10, fold of increase after the mutation of *arlR*.

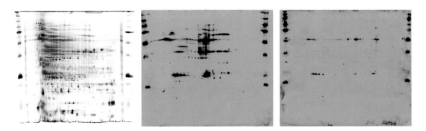

Ross, Chapter 27, Fig. 3. A silver-stained 2-DE protein gel (left) and Western blots obtained from replicate gels using antibodies against phosphoserine (middle) and phospho-threonine residues (right, courtesy of Dr. Lianglu Wan, NRC Plant Biotechnology Institute). Horizontal spot trains are characteristic of multiple phosphorylation, which has a much greater effect on pI than on molecular weight.

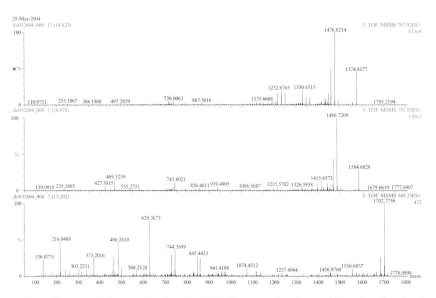

Ross, Chapter 27, Fig. 4. Product-ion MS/MS spectra for Ser- and Thr-phosphorylated peptides (upper and middle panels), in which the most abundant fragment ion corresponds to neutral loss of phosphate (98 Da) from the intact peptide, and for a Tyr-phosphorylated peptide (lower panel), which gives the diagnostic phosphotyrosine immonium ion (m/z 216).

A

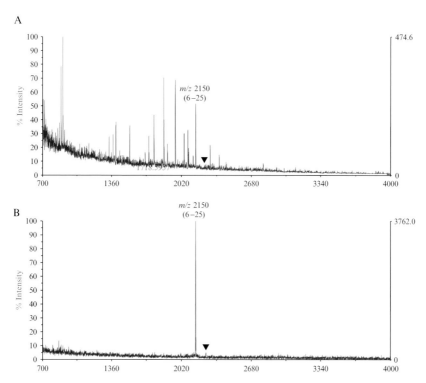

B

Ross, Chapter 27, Fig. 5. Positive-ion reflectron MALDI-TOF MS spectra of (A) a complete V8 digest of the His-phosphorylated protein HPr, and (B) a Cu(II)-IMAC extract of the same digest. The His-containing peptide (6–25) is selectively recovered, but detected almost exclusively in nonphosphorylated form (m/z 2150). The expected position of the phosphorylated peptide (m/z 2230) is indicated by ▼.

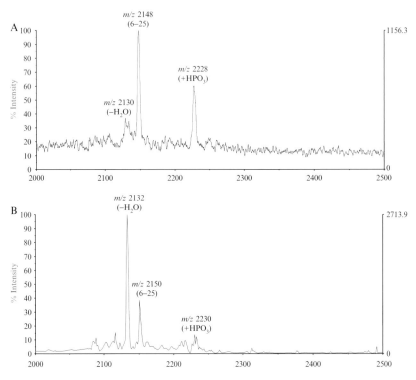

ROSS, CHAPTER 27, FIG. 6. MALDI mass spectra obtained using (A) negative-ion linear TOF MS analysis of the Cu(II)-IMAC extract from a V8 proteolytic digest of phosphorylated HPr, and (B) post-source decay (PSD) of the His-phosphorylated HPr peptide (*m/z* 2230) generated in positive-ion mode, both of which confirm recovery of the intact phosphopeptide.